Principles of
Engineering Economics with Applications

Engineering economics is an essential subject for engineers. A sound understanding of this subject is required for analyzing complex economic decision-making problems in several core engineering disciplines. Adapted to meet the syllabi requirements of most universities, the text introduces the fundamental concepts of engineering economics. It shows ways to calculate time value of money using cash-flow diagrams and it explains the procedure for making economy studies to select the best alternative. It also elaborates various methods to make replacement and retention decisions, calculate depreciation costs, evaluate public sector projects, perform economy studies considering inflation, arrive at make or buy decisions etc. It further explains project planning and scheduling through CPM and PERT. The concepts and applications of value engineering are also introduced. Various methods for making forecasts, cost estimation and analysis, and decision making under different environments are also discussed. The book is strong in its ability to relate abstract engineering and managerial concepts to real life situations.

Zahid A. Khan is a professor at the Department of Mechanical Engineering, Jamia Millia Islamia, New Delhi. He has published more than 100 articles in national and international journals. His research interests include optimization of design and manufacturing processes parameters, artificial neural network (ANN), fuzzy modelling, and environmental ergonomics.

Arshad N. Siddiquee is a professor at the Department of Mechanical Engineering, Jamia Millia Islamia, New Delhi. He served as Assistant Director at the All India Council of Technical Education (AICTE) from 2005 to 2007. He has published more than 100 articles in national and international journals. His research interests include materials structure property correlation, welding engineering, machining, optimization of design and process parameters using fuzzy modelling.

Brajesh Kumar is an associate professor at the National Institute of Financial Management (NIFM), Faridabad. His areas of interest include mathematical economics, business economics, managerial economics, computer applications in economics, research methodology, and econometrics.

Mustufa H. Abidi is a researcher at the Raytheon Chair for Systems Engineering (RCSE), Advanced Manufacturing Institute, King Saud University, Riyadh. His areas of interest include application of virtual reality techniques for sustainable product development, flexible manufacturing systems, micro-manufacturing, human-computer interaction, additive manufacturing, and reverse engineering.

Principles of
Engineering Economics
with Applications
2nd edition

Zahid A. Khan

Arshad Noor Siddiquee

Brajesh Kumar

Mustufa H. Abidi

CAMBRIDGE
UNIVERSITY PRESS

CAMBRIDGE
UNIVERSITY PRESS

University Printing House, Cambridge CB2 8BS, United Kingdom

One Liberty Plaza, 20th Floor, New York, NY 10006, USA

477 Williamstown Road, Port Melbourne, vic 3207, Australia

314 to 321, 3rd Floor, Plot No.3, Splendor Forum, Jasola District Centre, New Delhi 110025, India

79 Anson Road, #06–04/06, Singapore 079906

Cambridge University Press is part of the University of Cambridge.

It furthers the University's mission by disseminating knowledge in the pursuit of education, learning and research at the highest international levels of excellence.

www.cambridge.org
Information on this title: www.cambridge.org/9781108458856

First edition 2012

Second edition 2018

Printed in India by Rajkamal Electric Press

A catalogue record for this publication is available from the British Library

ISBN 978-1-108-45885-6 Paperback

Additional resources for this publication at www.cambridge.org/9781108458856

To Our Families

Contents

Foreword   *xv*
Preface   *xix*
Acknowledgments   *xxi*

1   Engineering Economics: A Prologue   1
  1.1   Introduction   1
  1.2   Introduction to Economics   1
  1.3   Need to Study Economics   1
  1.4   Circular Flow of Economic Activities   2
  1.5   Circular Flow of Income in Different Sectors   6
  1.6   Demand Theory   7
        1.6.1   Law of Demand   8
        1.6.2   Assumptions Used in Defining Demand   8
        1.6.3   Demand Schedule   8
        1.6.4   Demand Curve   9
        1.6.5   Determinants of Demand   9
  1.7   Elasticity of Demand   10
        1.7.1   Price Elasticity of Demand   10
        1.7.2   Income Elasticity of Demand   15
        1.7.3   Cross Elasticity   17
  1.8   Supply   20
        1.8.1   Factors Affecting Supply: The Determinants of Supply   20
        1.8.2   Law of Supply   21
        1.8.3   Supply Schedule   21
        1.8.4   Supply Curve   22
  1.9   Definition and Scope of Engineering Economics   22
        1.9.1   Meaning of Engineering Economics   22
        1.9.2   Definition of Engineering Economics   22
        1.9.3   Concepts of Engineering Economics   23
        1.9.4   The Scope of Engineering Economics   23
        1.9.5   Engineering Economics Environment   23
        1.9.6   Types of Efficiency   24
  1.10 Consumer and Producer Goods and Services   25
  1.11 Necessities, Luxuries and Relation between Price and Demand   25

1.12 Relation between Total Revenue and Demand   27
1.13 Cost Concepts   28
1.14 Relation between Cost and Volume   29
1.15 The Law of Supply and Demand   33
1.16 The Law of Diminishing Marginal Returns   34
1.17 Break-Even Analysis   34
1.18 Profit-Volume (P/V) Chart and P/V Ratio   38
1.19 Competition or Market Structure   41

2. **Fundamentals of Mathematics and Engineering Economics**   **45**
    2.1 Introduction   45
    2.2 Theory of Consumer Behavior   46
    2.3 Meaning of Utility   47
         2.3.1 Nature of the Utility Function   47
         2.3.2 Existence of Utility Function   48
         2.3.3 The Cardinal Marginal Utility Theory   49
         2.3.4 Equilibrium of the Consumer   50
    2.4 Meaning of Demand   51
         2.4.1 Demand Function   51
         2.4.2 Quantity Demanded   51
         2.4.3 Change in Demand   52
         2.4.4 Law of Demand   52
         2.4.5 Ordinary Demand Function   54
         2.4.6 Compensated Demand Function   55
         2.4.7 Reasons for Downward Slope of Demand Curve   55
    2.5 Concept of Elasticity   56
         2.5.1 Own Price Elasticity   56
         2.5.2 Determinants of Price Elasticity   58
         2.5.3 Income Elasticity of Demand   59
         2.5.4 Cross-Price Elasticity of Demand   60
         2.5.5 Engel Curve and Income Elasticity   61
         2.5.6 Relationship between Price Elasticity and Marginal Revenue 63
    2.6 Law of Diminishing Marginal Utility   63
    2.7 Principle of Equi-marginal Utility   65
    2.8 Indifference Curves Theory and Ordinal Utility Theory   67
         2.8.1 Indifference Curves   67
         2.8.2 Nature of Consumer Preferences   68
         2.8.3 Indifference Map   69
         2.8.4 Rate of Commodity Substitution   69
         2.8.5 Properties of ICs   70
         2.8.6 Budget Line   72
         2.8.7 Consumer's Equilibrium or Maximization of Utility   72
         2.8.8 Alternative Method of Utility Maximization   74

2.9	Application and Uses of Indifference Curves	76
	2.9.1	Income and Leisure Choice	76
	2.9.2	Revealed Preference Hypothesis	77
	2.9.3	Consumer's Surplus	80

3. Elementary Economic Analysis   **83**

3.1	Introduction: Theory of the Firm	83
3.2	Law of Supply	83
3.3	Concept of Elasticity of Supply	86
3.4	Meaning of Production	88
3.5	Production Function and its Types	90
	3.5.1	General Production Function	90
	3.5.2	Cobb-Douglas Production Function	91
	3.5.3	Properties of Cobb-Douglas Production Function	93
	3.5.4	CES Production Function	94
3.6	Producer's Equilibrium	94
3.7	Concept of Isoquants	96
3.8	Marginal Rate of Technical Substitution	99
3.9	The Elasticity of Substitution	100
3.10	Iso-cost Line	101
3.11	Producer's Surplus	102
3.12	Cost Minimization	102
3.13	Returns to Scale and Returns to Factor	107
3.14	Cost Theory and Estimation	108
3.15	Concept of Costs and their Types	108
3.16	Profits	110
	3.16.1	Normal Profits	111
	3.16.2	Economic Profits	111
3.17	Profit maximization	112
3.18	Market Structure and Degree of Competition	112
	3.18.1	Perfect Competition	113
	3.18.2	Monopoly	115
	3.18.3	Monopolistic Competition	117
	3.18.4	Oligopoly Models	118
	3.18.5	Monopsony	120

4. Interest Formulae and their Applications   **123**

4.1	Introduction	123
4.2	Why Return to Capital is Considered?	123
4.3	Interest, Interest Rate and Rate of Return	123
4.4	Simple Interest	125
4.5	Compound Interest	126
4.6	The Concept of Equivalence	126

4.7   Cash Flow Diagrams   127
4.8   Terminology and Notations/Symbols   129
4.9   Interest Formula for Discrete Cash Flow and Discrete Compounding 132
       4.9.1   Interest Formulae Relating Present and Future Equivalent Values
               of Single Cash Flows   132
       4.9.2   Interest Formulae Relating a Uniform Series (Annuity) to
               its Present and Future Worth   135
4.10   Interest Formulae Relating an Arithmetic Gradient Series to its Present
       and Annual Worth   142
       4.10.1 Finding P when given G   143
       4.10.2 Finding A when given G   144
4.11   Interest Formulae Relating a Geometric Gradient Series to its Present
       and Annual Worth   148
4.12   Uniform Series with Beginning-of-Period Cash Flows   154
4.13   Deferred Annuities or Shifted Uniform Series   156
4.14   Calculations Involving Uniform Series and Randomly Placed Single Amounts 159
4.15   Calculations of Equivalent Present Worth and Equivalent Annual
       Worth for Shifted Gradients   161
4.16   Calculations of Equivalent Present Worth and Equivalent Annual
       Worth for Shifted Decreasing Arithmetic Gradients   165
4.17   Nominal and Effective Interest Rates   168
4.18   Interest Problems with Compounding More-Often-Than-Once Per Year 173
       4.18.1 Single Amounts   173
       4.18.2 Uniform Series and Gradient Series   175
       4.18.3 Interest Problems with Uniform Cash Flows Less-Often-Than
               Compounding Periods   176
       4.18.4 Interest Problems with Uniform Cash Flows More-Often-Than
               Compounding Periods   178

5. Methods for Making Economy Studies   185
    5.1   Introduction   185
    5.2   Basic Methods   185
    5.3   Present Worth (P.W.) Method   186
    5.4   Future Worth (F.W.) Method   188
    5.5   Annual Worth (A.W.) Method   190
    5.6   Internal Rate of Return (I.R.R.) Method   196
    5.7   External Rate of Return (E.R.R.) Method   199
    5.8   Explicit Reinvestment Rate of Return (E.R.R.R.) Method   203
    5.9   Capitalized Cost Calculation and Analysis   204
    5.10   Payback (Payout) Method   207

6. Selection among Alternatives   212
    6.1   Introduction   212
    6.2   Alternatives having Identical Disbursements and Lives   212

6.3   Alternatives having Identical Revenues and Different Lives   221
        6.3.1   Comparisons using the Repeatability Assumption   221
        6.3.2   Comparisons using the Coterminated Assumption   225
6.4   Alternatives Having Different Revenues and Identical Lives   228
6.5   Alternatives Having Different Revenues and Different Lives   231
6.6   Comparison of Alternatives by the Capitalized Worth Method   234
6.7   Selection among Independent Alternatives   235

7.   Replacement and Retention Decisions   241
7.1   Introduction   241
7.2   Reasons for Replacement   241
7.3   Terminologies used in Replacement Study   242
7.4   Economic Service Life   242
7.5   Procedure for Performing Replacement Study   245
7.6   Replacement Study over a Specified Study Period   247

8.   Depreciation   254
8.1   Introduction   254
8.2   Depreciation Terminology   254
8.3   Methods of Depreciation   255
        8.3.1   Straight Line (SL) Method   255
        8.3.2   The Declining Balance (DB) Method   257
        8.3.3   Sum-of-the-Years'-Digits (SYD) Method   260
        8.3.4   The Sinking Fund Method   262
        8.3.5   The Service Output Method   264

9.   Economic Evaluation of Public Sector Projects   266
9.1   Introduction   266
9.2   Benefit/Cost Analysis of a Single Project   267
9.3   Selection between Two Mutually Exclusive Alternatives using
        Incremental B/C Analysis   269
9.4   Selection Among Multiple Mutually Exclusive Alternatives using
        Incremental B/C Analysis   271

10. Economics Study Considering Inflation   276
10.1   Introduction   276
10.2   Effects of Inflation   276
10.3   Present Worth Calculations Adjusted for Inflation   278
10.4   Future Worth Calculations Adjusted for Inflation   281
10.5   Capital Recovery Calculations Adjusted for Inflation   284

11. Make or Buy Decision   286
11.1   Introduction   286
11.2   Feasible Alternatives for Launching New Products   286
11.3   Decisive Factors for Make or Buy Decision   287

	11.3.1	Criteria for Make Decision	287
	11.3.2	Criteria for Buy Decision	288
11.4	Techniques used to Arrive at Make or Buy Decision		288
	11.4.1	Simple Cost Analysis	288
	11.4.2	Economic Analysis	290
	11.4.3	Break-Even Analysis	292

12. Project Management — **297**

12.1	Introduction		297
12.2	Phases of Project Management		297
	12.2.1	Planning	297
	12.2.2	Scheduling	298
	12.2.3	Monitoring and Control	298
12.3	Bar or Gantt Charts		298
12.4	Network Analysis Technique		300
12.5	Critical Path Method (CPM)		301
	12.5.1	Arrow Diagrams	301
	12.5.2	Activity Description	304
	12.5.3	Understanding Logic of Arrow Diagrams	305
	12.5.4	Dummy Activities	307
12.6	Guidelines for Drawing Network Diagrams or Arrow Diagrams		308
12.7	CPM Calculations		311
	12.7.1	Critical Path	312
	12.7.2	Critical Activities	312
	12.7.3	Non-critical Activities	312
	12.7.4	Earliest Event Time	313
	12.7.5	Latest Event Time	313
12.8	Calculation of the Earliest Occurrence Time of Events		313
12.9	Calculation of the Latest Occurrence Time of Events		318
12.10	Activity Times		324
	12.10.1	Earliest Start Time	325
	12.10.2	Earliest Finish Time	325
	12.10.3	Latest Finish Time	325
	12.10.4	Latest Start Time	326
12.11	Float		330
	12.11.1	Types of Float	331
	12.11.2	Negative Float	340
12.12	Identification of Critical Path		341
12.13	Program Evaluation and Review Technique (PERT)		343
	12.13.1	PERT Activity Time Estimates	343
	12.13.2	PERT Computations	344
	12.13.3	Computation of Probabilities of Completion by a Specified Date	352
12.14	Project Crashing		358

12.14.1 Cost Slope 359
12.14.2 Cost of Crashing 359

13. Value Engineering 375

13.1 Introduction 375
13.2 Concept of Value Engineering 375
13.3 Nature and Measurement of Value 378
13.3.1 The VE process 378
13.4 Origination Phase 379
13.4.1 Organization 379
13.4.2 Project Selection 379
13.4.3 The VE Team 380
13.5 Project or Study Mission 380
13.5.1 Product Definition and Documentation 380
13.6 Information Phase 380
13.6.1 Qualitative Analysis of Value: Function Analysis 380
13.6.2 Function Analysis Systems Technique (FAST) 381
13.6.3 Constraints Analysis 384
13.7 Quantitative Analysis of Value – State 1 Value Measurement 384
13.7.1 Cost Derivation 384
13.7.2 Worth or Importance Derivation 384
13.7.3 The Value Index 385
13.7.4 Value Measurement Techniques 385
13.8 Innovation Phase 386
13.8.1 Improvement of Value 386
13.9 Evaluation Phase 386
13.9.1 Pre-screening: Qualitative Analysis of Value 386
13.9.2 Quantitative Analysis of Value 386
13.10 Implementation Phase 387

14. Forecasting 388

14.1 Introduction 388
14.2 Basic Categories of Forecasting Methods 388
14.3 Extrapolative Methods 389
14.3.1 Components of Demand 389
14.3.2 Moving Average Method 390
14.3.3 Weighted Moving Average Method 391
14.3.4 Exponential Smoothing Methods 392
14.3.5 Adaptive Methods 403
14.4 Causal or Explanatory Methods 403
14.4.1 Regression Analysis 403
14.4.2 Simple Regression Analysis 404
14.4.3 Multiple Regression Analysis 413

14.5	Qualitative or Judgmental Methods	413
	14.5.1 Build-up Method	414
	14.5.2 Survey Method	414
	14.5.3 Test Markets	414
	14.5.4 Panel of Experts	414
14.6	Forecast Errors	415

15. Cost Estimation   422

15.1	Introduction	422
15.2	How Does an Organization Estimate Cost?	422
	15.2.1 Cost Estimates	423
	15.2.2 Cost Estimation Approach	423
	15.2.3 Accuracy of Estimates	424
	15.2.4 Cost Estimation Methods	424
15.3	Unit Method	424
15.4	Cost Indexes	426
15.5	Cost Estimation Relationships	427
	15.5.1 Cost-Capacity Equation	427
	15.5.2 Factor Method	428
	15.5.3 Learning Curve	429
15.6	Estimation and Allocation of Indirect Cost	432

16. Decision Making   439

16.1	Introduction	439
16.2	Terminologies used in Decision Making	439
16.3	Steps in Decision Making	439
16.4	Decision Making Environment	442
16.5	Decision Making under Uncertainty	442
	16.5.1 The Maximax Criterion	442
	16.5.2 The Maximin Criterion	443
	16.5.3 The Minimax Regret Criterion	443
	16.5.4 The Realism Criterion (Hurwicz's Rule)	444
	16.5.5 Criterion of Insufficient Reason (Laplace's Rule)	444
16.6	Decision Making under Risk	445
	16.6.1 Expected Monetary Value (EMV)	445
	16.6.2 Expected Opportunity Loss (EOL)	447
	16.6.3 Expected Value of Perfect Information (EVPI)	449
16.7	Marginal Analysis	449
16.8	Decision Trees	451

Appendix A	457
Appendix B	505
Appendix C	508
Bibliography	511
Index	514

Foreword

In the face of cut-throat competition of the present day, businesses the world over have become more and more technical. Alongside other professionals, engineers play a key role in running businesses successfully across the globe. They play an important role in decision-making, both in the manufacturing and service industries. Most of these decisions are made primarily on the basis of economic factors and their assessment. It is often seen that decision-makers do not possess the required knowledge and skills related to engineering, and thus, they frequently call upon engineers to make technical-economic analyses and suggest recommendations. Engineering Economy is an important subject for aspiring as well as practicing engineers today, as the techniques and models thus adopted assist engineers and managers in making well-thought-out decisions. They can use the knowledge of this subject to analyse and draw conclusions as they work on projects of all kinds of complexities.

The success of engineering and business projects is usually measured in terms of financial efficiency. A project would be able to achieve maximum financial efficiency if it is properly planned and operated with respect to its technical, social and financial requirements. Since it is the engineers who understand the technical requirements of a project, they are best placed to assimilate the technical details with their knowledge of engineering economy to do an effective economic analysis and arrive at a sound managerial decision.

The present volume, comprising 16 chapters, covers many such issues pertaining to economic analysis of projects. Chapter 1 summarizes the basic principles of engineering economy and its applications. Chapter 2 describes the fundamental concepts of mathematics and engineering economics, which will help readers learn the basic mathematical concepts required for economic analysis. The roles of factors involved in economic analysis have been discussed at length in Chapter 3. Chapter 4 describes the key concept of value of money, on which economic analyses are based. Topics such as simple and compound interests, cash flow diagrams, determination of equivalent cash flow at different points in time, nominal and effective interest rates have also been explained here. Chapter 5 describes the basic methods that can be used by engineers to perform economy-studies. The methods that can be used for selecting the best alternative out of many, have been presented in Chapter 6. Chapter 7 describes the procedure to be followed to decide whether an organization should continue to use existing physical assets (such as a machine) or whether if the asset should be replaced. The value of a physical asset depreciates, that is diminishes, with time; this concept of depreciation as well as the procedure for calculating depreciation costs have been described in Chapter 8. Chapter 9 describes different methods such as benefit-cost ratio for the economic evaluation of large public-sector projects. The concept of inflation and how it affects the worth of capital have been discussed in Chapter 10. Often organizations have to make decisions as to whether they should manufacture a component in-house or buy it from outside. The procedure of arriving at a make-or-buy-decision has been explained in Chapter 11. In Chapter 12, the focus is project management. Concepts such as CPM, PERT and project crashing have been described here to enable readers understand and apply these techniques for timely and economic completion of their projects. Chapter 13 presents a well-established technique, value engineering, adopted to reduce the cost of a product and increase its value. The success of an organization depends on how efficiently and effectively it can forecast the demand for its products.

Chapter 14, describes the underlying concepts, methods and models of forecasting. Chapter 15, explains the various types of costs and describes the different methods for cost estimation. The last chapter of the book, Chapter 16 discusses the various methods used for taking decisions under different decision-making environments. This book, highlights the principles and applications of economic analysis in a lucid manner, supported by a large number and wide range of engineering-oriented examples and end-of-chapter exercises. It covers the syllabi of undergraduate and postgraduate courses of major Indian and overseas universities. Special chapters such as Fundamentals of Mathematics and Engineering Economics, Elementary Economic Analysis, Project Management, Value Engineering, and Forecasting, covered in this book are rare in books of this kind, which makes it distinct from existing books.

Writing a book requires in-depth subject knowledge, dedication, sincere effort, sacrifices, and teaching and research experience. As head of the institution, I am aware that the first author of this volume, Dr Zahid Akhtar Khan, Professor in Mechanical Engineering, Jamia Millia Islamia, New Delhi, has more than 20 years of teaching and research experience. He has taught in overseas universities such as the University Sains Malaysia, Malaysia, and the King Abdulaziz University, Jeddah, Kingdom of Saudi Arabia. Apart from teaching, he is actively involved in research and development activities. He has published more than 45 research papers in reputed national and international journals and over 20 papers in the proceedings of conferences held in India and abroad. In addition, he has also contributed chapters in three books related to Mechanical Engineering. VDM Verlag, a German publishing company, has published one of his monographs. Dr Khan has supervised several MTech dissertations and BTech projects. Presently he is supervising five PhD and three MTech. students. He and his colleague Mr Arshad Noor Siddiquee have been instrumental in developing quite a few laboratories, including the Metrology Lab, in the department and in preparing proposals for financial grants. This year they have submitted a proposal for SAP (worth ~75 lakh) to the University Grants Commission. They, along with their team of students, have filed a patent with the Controller General of Patents, Design and Trademarks, Government of India, for the designing and development of a convertible wheel chair.

Dr Khan has been discharging additional duties as coordinator for the training and placement of postgraduate students; as member, sports committee of the faculty; in-charge of the faculty magazine 'Tech-Times'; in-charge of the Engineering workshop; member of the result analysis committee; member of the sub-purchase committee of the department; advisor of the students of Mechanical Engineering; and tabulator of the MTech and BTech results.

He has received international recognition: his biography has been published in 'Marqui's Who's Who in Science and Engineering, Tenth edition, 2008-09'. He has received the International Scientist of the Year 2008 award given by the International Biographical Centre, Cambridge, UK. He is a member of the Emerald Literati Network, UK, and is also on the panel of reviewers of international journals.

Arshad Noor Siddiquee, the second author of this book, has graduated from Government Engineering College, Jabalpur. He completed his MTech from the Indian Institute of Technology, Delhi, where he is currently pursuing this doctoral studies. He is presently working as an associate professor in the Department of Mechanical Engineering, Jamia Millia Islamia. He played a key role in the developmental phase of Glasgow University College in Oman during 1998-2001. He has had hands-on experience in the establishment and accreditation of technical institutions during his tenure at the All India Council for Technical Education (AICTE), New Delhi, in the capacity of an assistant director. Siddiquee has dexterously used his skills in making the Faculty of Engineering and Technology profile for ranking evaluation of institutions and also in making proposals for Petroleum Engineering and Aeronautical Engineering programmes. He has contributed chapters on engineering subjects to three books of reputed publishers and over 15 research papers to international journals. He is on the panel of reviewers for Elsevier and Springer journals.

Dr Brajesh Kumar, the third author, has worked in the Department of Expenditure, Ministry of Finance, Government of India, and is currently serving as an associate professor at the National Institute of Financial Management (NIFM), Faridabad. His areas of interest are managerial economics, financial econometrics and computer applications in economics. He has published several research works on managerial economics, and macro- and micro-economics. Dr Kumar is associated with various national and international organizations in different capacities; for instance, agro-expert, Federation of Indian Chambers of Commerce and Industry (FICCI); read group member, Centre for Trade and Development (CENTAD); and programme coordinator, civil servants from North-East Cadre.

I am extremely pleased to find that despite their most sincere involvement, commitment and dedication to teaching and research, the authors have put in so much effort in writing this extremely useful and timely book. This must have demanded of them time away from family, great sacrifices, pains and compromises. I have learnt that the range and content of the book has received excellent appreciation from its reviewers. It is an interesting fact that the market review of the publisher revealed that no single title in India is, so far, available to fulfill students' requirements in engineering economy. I have no doubt that this is a definitive text on the subject; that it would meet the genuine needs of students, teachers, and practising engineers and managers alike. I congratulate the authors for accomplishing this challenging task and wish them every success.

Najeeb Jung, IAS

Vice-Chancellor, Jamia Millia Islamia

Preface

ABOUT ENGINEERING ECONOMICS WITH APPLICATIONS

Engineers are required to provide economically feasible solutions to existing problems. To achieve this, engineers must possess knowledge of economy to evaluate the monetary consequences of the products, projects and processes that they design. Engineering design solutions do not exist in a vacuum but within the context of a business opportunity. Since almost every problem has multiple solutions, so the issue is: how does one rationally select a design with the most favorable economic result? The answer to this question is provided by engineering economy. Engineering economy, the analysis of the economic consequences of engineering decisions, is said to have originated in A. M. Wellington's *The Economic Theory of Railway Location*, published in 1887. Engineering economy is now considered a part of the education of every engineer, as it provides a systematic framework for evaluating the economic aspects of competing design solutions. Just as engineers model the effect of temperature on cutting tools or the thermodynamic response of an air compressor, they must also model the economic impact of their recommendations. What is 'engineering economy' and why is it so important? The initial reaction of many engineering students to this question is, 'money matters will be handled by someone else and I need not worry about these matters'. In reality, any engineering project must be, not only physically realizable but also economically affordable. Understanding and applying economic principles to engineering have never been more important. Engineering is more than a problem-solving activity focusing on the development of products, systems, and processes to satisfy a need or demand. Beyond function and performance, solutions must also be economically viable. Design decisions affect limited resources such as time, material, labor, capital and natural resources, not only initially i.e. during conceptual design but also through the remaining phases of the life cycle i.e. during detailed design, manufacture and distribution, service, retirement and disposal. Engineers should realize that the solution provided by them does not make sense and will not be acceptable, if it is not profitable.

EDUCATION LEVEL AND USE OF TEXT

The contents of this book have been designed in such a way that it serves two primary purposes: (i) to provide students with a sound understanding of the principles, basic concepts, and methodology of engineering economy; and (ii) to help students develop proficiency with these methods and with the processes for facilitating rational decisions they are likely to encounter in professional practice. Interestingly, an, engineering economics with applications, course may be a student's only college exposure to the systematic evaluation of alternative investment opportunities. In this respect, *Engineering Economics with Applications* is intended to serve as a basic text for classroom instruction and as well as a reference for use by practicing engineers in all areas (chemical, civil, computer, electrical, industrial, and mechanical engineering). The book is also useful for people engaged in the management of technical activities.

It is well suited for undergraduate as well as postgraduate courses in engineering economic analysis, project analysis, or engineering cost analysis. Additionally, because of its simple and easy to understand language, it is perfect for those who are studying the subject for the first time and on their own, and for those who simply want to review. The systematic approach used in the text design allows a practitioner unacquainted with economics and engineering principles to use the text to learn, understand, and correctly apply the principles and techniques for effective decision making.

SALIENT FEATURES OF THE BOOK

- Simple and easy to understand language.
- The concepts have been explained in a lucid manner.
- Numerous comprehensive real life examples appear throughout the book.
- Extended learning exercises, in the end-of-chapter problem sets.
- A large number of figures and diagrams enrich the text.
- Some special chapters such as 'Chapter 2: Fundamental Concepts of Mathematics and Engineering Economics', 'Chapter 3: Basic Mathematical Concepts for Economic Analysis', 'Chapter 12: Project Management', 'Chapter 14: Forecasting', 'Chapter 15: Cost Estimation' and 'Chapter 16: Decision Making', that generally do not appear in engineering economy texts are included in this book.

Acknowledgments

We are extremely grateful to the Almighty for thy blessings, which of course have been with us always, and for giving us the strength and dedication to complete this book to the best of our ability.

We are thankful to all people, including our colleagues and students, for extending their help and support in completing this book.

We are grateful to Raytheon Chair for Systems Engineering (RCSE), Advanced Manufacturing Institute for the funding. We would also like to thank our parent institutions for allowing us to complete this book.

We are extremely thankful to the Cambridge University Press, particularly Gauravjeet Singh Reen for his untiring efforts and continuous support, for timely publication of the book.

Last but not the least, we thank our beloved family members, who suffered a lot during completion of this book as we could not spend as much time with them we should have. We thank them for bearing with us.

1

Engineering Economics: A Prologue

1.1 INTRODUCTION

Engineering economics is concerned with the formulation, estimation and evaluation of the economic outcomes of alternatives that are available to accomplish a defined purpose. Engineering economics can be defined as a collection of mathematical techniques that simplify economic comparison. Engineers use the knowledge of engineering economics in analyzing, synthesizing and drawing conclusions as they work on projects of different sizes. In other words, the techniques and models of engineering economics assist engineers in making decisions. The success of engineering and business projects is normally measured in terms of financial efficiency. A project will be able to achieve maximum financial efficiency when it is properly planned and operated with respect to its technical, social, and financial requirements. Since engineers understand the technical requirements of a project, they can combine the technical details of the project and the knowledge of engineering economics to study and arrive at a sound managerial decision. The basic economic concepts that are essentially factors to be considered in making economy studies are briefly discussed in this chapter.

1.2 INTRODUCTION TO ECONOMICS

Economics is about choice and is at the heart of all decision-making. Individuals, businesses and governments are all faced with making choices in situations where resources are scarce. The principles of economics are applied to a wide range of fields, including business, finance, administration, law, local and national government and, indeed, in most of the aspects of everyday life. While studying economics, we examine topics of obvious importance to the well-being of all humans. Increasingly, policy debates in all areas are being cast in economic terms and understanding most of the current issues requires an understanding of economics.

1.3 NEED TO STUDY ECONOMICS

Studying economics gives insights into the general environment of resource allocation decisions, opportunity costs and project evaluation which are crucial in many areas. Often, these insights are not obvious, and can be counter-intuitive to those who don't apply economic reasoning.

Economics studies economic activities: Humans perform various types of activities in their daily lives like religious, social, political and economic activities. Economics does not deal with all of them. But humans spend the maximum time on economic activities. Therefore, economics is a subject which studies economic activities of humans.

Study of wants → efforts → wealth → satisfaction: Every human being is doing some business and every human being has some wants and these wants are unlimited. To fulfill these wants a person does efforts, by doing efforts he gets wealth and with this earned wealth he satisfies his wants.

Study of human behavior with relation to ends and scarce means: As long as a person is alive, his wants go on increasing. But the person cannot fulfill all the wants. The reason is that the resources required to fulfill these wants are limited. Besides the fact of scarcity of resources, we also find that resources have alternative uses. Hence economics is a subject which studies human behavior as a relationship between ends and scarce means which have alternative uses.

Economics studies problem of choice: Scarcity and choice go together. If things were available in abundance, then there would have been no problem of choice; the point is that "problems of choice" arise because of scarcity. We can summarize the basic economic problems by means of a chart shown in Fig. 1.1.

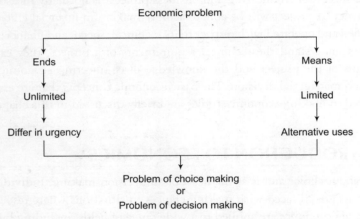

Figure 1.1   Summary chart of basic economic problems

1.4   CIRCULAR FLOW OF ECONOMIC ACTIVITIES

Circular Flow of economic activities is a flow which has neither a beginning nor an end.

Basic economic activities   An economic activity is a systematic endeavor to satisfy a material need. Material needs are the needs for goods and services. We know that human wants are unlimited. We are always busy satisfying one want or the other. "The vital processes or essentials functions like production, consumption, investment and distribution, as shown in Fig. 1.2, are those economic activities which are necessary for the working or survival of an economy."

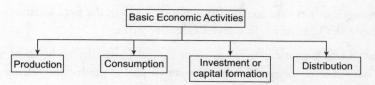

Figure 1.2 Vital economic activities

Production: Production as a process of creation of utility or value in goods or services (or both). Anatol Murad defined this as: "Production may be defined as the creation of utilities."

Factors of Production: Factors of production are the essential elements which cooperate with one another in the process of production. The various factors of production are shown in Fig. 1.3.

Land: It is that factor of production which is available to humankind as a free gift of nature.

Labor: It is the physical or mental effort of human beings in the process of production. Services of a doctor, lawyer, teacher, worker in the factory, all constitute labor.

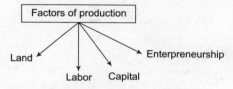

Figure 1.3 Factors of production

Capital: Capital is man-made material and is a source of production. It consists of the part of production which is used for further production.

Entrepreneurship: Entrepreneurship refers to the skills of the entrepreneur:
  (a) to organize business
  (b) to undertake risks of business

Consumption: In economics, consumption has a special meaning; it means the use of or utility of goods and services for the direct satisfaction of individual and collective wants.

For example: When you eat bread, you are using up the want-satisfying capacity of bread, that is, its utility. Different types of consumption are shown in Fig. 1.4.

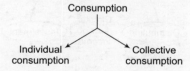

Figure 1.4 Types of consumption

Individual Consumption: It is that consumption which leads to the final satisfaction of the wants of an individual.

Collective consumption: It is that consumption which leads to the final satisfaction of collective wants. For example: Uses of roads, dams, bridges or parks.

Investment or Capital formation: Investment or capital formation is the third vital process or essential activity of an economy. "Investment is that part of production during a year which is not consumed but saved as capital formation for further production." The excess of production over consumption in an accounting year is called capital formation or investment.

$$I = Y - C$$

I = Investment, Y = Income, C = Consumption.

Some fundamental relationships:

1. Production = Consumption + Investment

$$Q = C + I$$

    Q = Production

2. Income = Consumption + Saving

$$Y = C + S$$

    S = Saving

3. Saving $\equiv$ Investment

$$S = I$$

$$Q = C + I$$

or $$Y = C + T \quad (\because Q = Y)$$

and $$Y = C + S$$

$$C + S = C + I$$

$\therefore$ $$S \equiv I$$

In other words, the circular flow of income can be explained using the flowchart shown in Fig. 1.5.

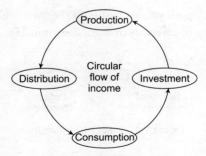

Figure 1.5   Circular flow of income

The flow of production, income and expenditure never stops. It is a circular flow without a beginning or an end. Production generates income, income generates demand for goods and services, and demand generates expenditure on the goods and services which leads to their production, so that the circle of production, consumption and expenditure always continues.

Performance of economic activities Economic activities are undertaken by the following sectors of the economy:

Household Sector: This sector includes households, who consumes goods and services, and provides factor services.

Firms or Business Sector: The firm produces goods and services by using factor services.

Government Sector: The government sector undertakes both consumption as well as production.

Flow of income There are two types of flow of income in an economy:
  ➢ Real flow of income
  ➢ Monetary flow of income

Real flow of income: It involves the flow of factor services from the household sector to the producing sector and the corresponding flow of goods and services from producing sector to household sector. This is explained in Fig. 1.6.

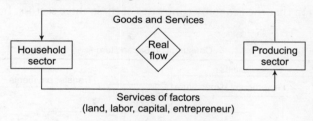

Figure 1.6   Real flow of income

Monetary flow of income: It refers to the flow of factor income e.g.: rent, interest, profit, wages and so on from the producing sector to the household sector as rewards for their factor services. The households spend their income on the goods and services produced by the producing sector. According to it, the money flows back to the producing sector. Fig. 1.7 explains this.

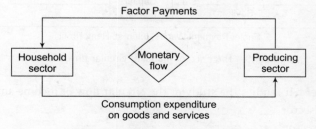

Figure 1.7   Monetary flow of income

1.5 CIRCULAR FLOW OF INCOME IN DIFFERENT SECTORS

Circular flow of Income is analyzed under three different situations based on certain simplifying assumptions:

Two sector model It studies the circular flow of Income between the household and producing sector on the assumption that there are only two sectors in the economy (Fig. 1.8).

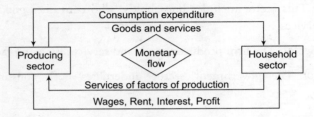

Figure 1.8 Two sector model

Three Sector Model It refers to the study of the circular flow of income among:
 (i) Household sector
 (ii) Producing sector
 (iii) Government sector

Here the assumption that the economy comprises of these three sectors. It is a closed economy (Fig. 1.9).

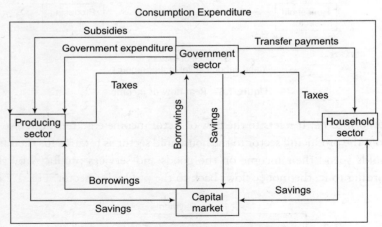

Figure 1.9 Three sector model of circular flow of income

Four Sector Model It studies the study of the circular flow of income among:
 (i) Household sector
 (ii) Producing sector

(iii) Government sector

(iv) Foreign sector or rest of the world.

In other words, it studies the flow of income in an open economy. The model studies all sectors of the economy, dropping all the simple assumptions made earlier (Fig. 1.10).

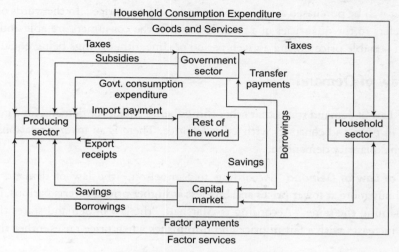

Figure 1.10   Four sector model

1.6   DEMAND THEORY

In economics, we are concerned with the demand for a commodity faced by a firm. Demand is regarded as the lifeline of a business enterprise. Demand analysis seeks to identify and measure the forces that determine sales.

Meaning of demand   In ordinary language, the terms need, desire, want and demand are used in the same sense. But in economics, all these terms have different meanings. For instance, a sick child needs medicine, a worker desires to have a car, but such needs and desire do not constitute demand.

Demand is the want of a person, which will become demand when he is ready to buy the goods at a given price and at a given point of time. So demand may, then, be defined as the quantity of a commodity which a consumer is willing and able to purchase at a given price, during some specific period of time.

There are seven essentials of demand:

1. Desire for a commodity.

2. Capacity to pay for it.

3. Willingness to pay for it.

4. Quantity bought and sold.

5. At a given price.
6. At a given time.
7. At a given place.

Definition of Demand   According to Veera Anstey, "the demand for a particular good is the amount that will be purchased at a given time and at a given price." In the words of Ferguson, "demand refers to the quantities of a commodity that the consumers are able and willing to buy at each possible price during a given period of time, other things being equal."

1.6.1   Law of Demand

Meaning   Law of demand states that other things being equal, the demand for a good extends with a fall in price and contracts with a rise in price. There is an inverse relationship with a price and the quantity demanded.

Definition of Law of Demand   According to Samuelson, "the law of demand states that people will buy more at lower prices and buy less at higher prices, ceteris paribus, or the other things remaining the same." According to Marshall, "the law of demand states that amount demanded increases with a fall in price and diminishes when price raises, other things being equal."

1.6.2   Assumptions Used in Defining Demand

(i)   There should be no change in the price of related goods.
(ii) There should be no change in the income of the consumer.
(iii) There should be no change in the taste and preference of the consumer.
(iv) The consumer does not expect any change in the price of the commodity in the near future.
(v)   There is no change in size or age-composition of the population.
(vi) There is no change in the range of goods available to the consumers.
(vii) There is no change in government policy.
(viii) There is no change in income of the consumer and the community.

1.6.3   Demand Schedule

A hypothetical individual demand schedule is given in Table 1.1.

Table 1.1   Demand schedule

Price per unit (₹)	Quantity demanded (units)
1	4
2	3
3	2
4	1

1.6.4 Demand Curve

Using demand schedule in Table 1.1, a demand curve is drawn in Fig. 1.11.

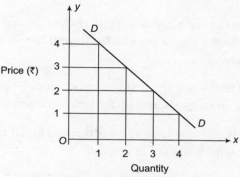

Figure 1.11   Individual demand curve

In Fig. 1.11, axis OX represents the quantity demanded and axis OY axis represents the price. DD is the demand curve. Each point on the demand curve expresses the relation between price and demand. At a price of ₹ 4 per unit, the demand is for 1 unit and at a price of ₹ 1 per unit, the demand is for 4 units. The demand curve slopes downwards from left to right, meaning that when price is high demand is low and when price is low demand is high.

1.6.5 Determinants of Demand

The demand for a commodity Q_x^d depends upon many factors as listed below:
1. Price of the commodity (Px)
2. Prices of the related goods (Pr)
3. Income of the consumer (Y)
4. Taste and preference of the consumer (T)
5. Expectation of price change of the commodity (E)
6. Size and composition of population (P)
7. Distribution of income (Y_d)

$$Q_x^d = (P_x, P_r, Y, T, E, P, Y_d)$$

Price of the commodity   Basically, demand for a commodity depends upon its price. If the price rises, the demand falls, and if the price falls, the demand rises.

Price of related goods   The demand for a commodity is also influenced by changes in the price of related goods like substitutes and complements.

Substitute goods: The demand of tea depends not only on its price, but also on the price of its substitute, coffee. If the price of coffee falls, while that of tea remains the same, the demand of tea falls.

Complementary goods: The demand for petrol depends not only on its own price, but also on the price of cars and scooters.

Income of the consumer   Income levels determine the demand to a great extent. Usually, there is a direct relationship between income and demand. In case of normal goods, if income rises, demand increases and if income falls demand decreases.

Price expectations   Demand for a commodity is also influenced by the expected changes in prices. If people anticipate a rise in price in the future, they buy more now and store the commodities, and vice-versa.

Taste and preference   These terms are used in a broad sense. They include fashions, habits, customs, etc. Demand for those goods goes up for which consumers develop a taste.

Population   Increase in population leads to more demand for all types of goods and services and decrease in population leads to a fall in demand.

Distribution of income   If income is equitably distributed, there will be more demand, if the income is not evenly distributed, then there will be less demand.

1.7   ELASTICITY OF DEMAND

Demand for a commodity is affected by several factors, including its own price, consumer's income, price of related goods, etc. Elasticity of demand is a measure of the sensitiveness of demand to changes in factors affecting demand. Elasticity of demand is classified into following broad categories:

   (i)   Price elasticity of demand
  (ii)   Income elasticity of demand
 (iii)   Cross elasticity of demand

1.7.1   Price Elasticity of Demand

Price elasticity can be defined as the responsiveness in the quantity demanded of a commodity to a change in its price. Total revenue (and hence the profits) of a firm can either increase or decrease due to change in price of the commodity which the firm produces. Thus, it is necessary to measure the probable effect of price changes on total revenue in order to minimize the uncertainty involved in the pricing decision made by the firm. The effect of price change on total revenue can be measured by the price elasticity of demand, which is mathematically defined as the ratio of the percentage changes in quantity demanded to the percentage change in price, as given in equation (1.1).

$$E_p = \frac{\%\Delta Q}{\%\Delta P} \qquad (1.1)$$

where E_p is price elasticity of demand
        Δ represents the change
        Q is quantity demanded
        P is price

1.7.1.1   Point and Arc Elasticity

Price elasticity can be computed by following two approaches, depending upon the available data and intended use. As a first approach, it can be computed to analyze the effect of discrete changes in price. This approach is referred to arc elasticity. For example, the effect of a price increase from ₹50 to ₹100 can be evaluated by computing the arc elasticity. The other approach is known as point elasticity. It can be used to evaluate the effect of very small price changes or to compute the price elasticity at a particular price.

The price elasticity of demand (E_p) is computed by using the Equation (1.2).

$$E_p = \frac{\frac{\Delta Q}{P}}{\frac{\Delta P}{P}} \qquad (1.2)$$

In this formula $\frac{\Delta Q}{Q}$ denotes percentage change in the quantity demanded and $\frac{\Delta P}{P}$ represents percentage change in the price.

By rearranging the terms, Equation (1.2) can be written as:

$$E_p = \frac{\Delta Q}{\Delta P} \cdot \frac{P}{Q} \qquad (1.3)$$

It may be noted that P and Q specify a particular point on the demand curve, while $\frac{\Delta Q}{\Delta P}$ is the reciprocal of the demand curve.

Point price elasticity of demand   Equation (1.3) gives the point price elasticity of demand, or the elasticity at a given point on the demand curve. Consider the demand curve for commodity X as shown in Fig. 1.12.

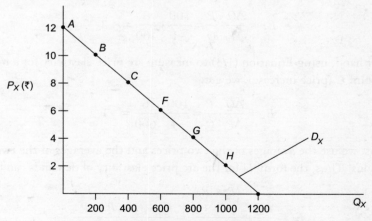

Figure 1.12   The point price elasticity of demand

In Fig. 1.12, $\dfrac{\Delta Q}{\Delta P} = \dfrac{-100}{1}$ at every point on D_x and the price elasticity at different points is given as

At point B,

$$E_p = \frac{-100}{1} \cdot \frac{10}{200} = -5$$

At point C,

$$E_p = \frac{-100}{1} \cdot \frac{8}{400} = -2$$

At point F,

$$E_p = \frac{-100}{1} \cdot \frac{6}{600} = -1$$

At point G,

$$E_p = \frac{-100}{1} \cdot \frac{4}{800} = -0.5$$

At point H,

$$E_p = \frac{-100}{1} \cdot \frac{2}{1000} = -0.2$$

It is evident from this calculation that the price elasticity of demand is different at different points on the demands curve. It is also clear from this calculation that for the linear demand curve as shown in Fig. 1.12, the absolute value of E_p, i.e., $|E_p|$ is greater than 1 above the geometric midpoint of the demand curve and therefore the demand curve is elastic. At the geometric midpoint, i.e. point F, $|E_p| = 1$, therefore the demand curve is *unitary elastic*. Below the geometric midpoint, the value of $|E_p|$ is less than 1 and hence the demand curve is inelastic.

Arc price elasticity of demand   On the demand curve, arc price elasticity of demand basically gives the measures of elasticity between two points of the curve. It is important to note that use of Equation (1.3) gives different values of E_p depending on whether the price rises or falls. For example, using Equation (1.3) to compute arc price elasticity for a movement from point C to point F (price decline) on the demand curve D_x shown in Fig. 1.12, we obtain:

$$E_p = \frac{\Delta Q}{\Delta P} \cdot \frac{P}{Q} = \frac{100}{-1} \cdot \frac{8}{400} = -2$$

On the other hand, using Equation (1.3) to measure arc price elasticity for a movement from point F to point C (price increase), we get:

$$E_p = \frac{\Delta Q}{\Delta P} \cdot \frac{P}{Q} = \frac{100}{-1} \cdot \frac{6}{600} = -1$$

To avoid this, we use the averages of the two prices and the averages of the two quantities in the calculations. Thus, the formula for the arc price elasticity of demands can be given as:

$$E_p = \frac{\Delta Q}{\Delta P} \cdot \frac{\dfrac{(P_1 + P_2)}{2}}{\dfrac{(Q_1 + Q_2)}{2}} = \frac{Q_2 - Q_1}{P_2 - P_1} \cdot \frac{P_2 + P_1}{Q_2 + Q_1} \tag{1.4}$$

In Equation (1.4), the subscripts 1 and 2 represent the original and new values respectively, of price and quantity or vice versa. For example, using Equation (1.4) to compute the arc price elasticity of D_x for a movement from point C to point F, we get:

$$E_p = \frac{Q_2 - Q_1}{P_2 - P_1} \cdot \frac{P_2 + P_1}{Q_2 + Q_1} = \frac{600 - 400}{6 - 8} \cdot \frac{6 + 8}{600 + 400} = -1.4$$

The same result is also obtained for the reverse movement from point F to point C.

$$E_p = \frac{Q_2 - Q_1}{P_2 - P_1} \cdot \frac{P_2 + P_1}{Q_2 + Q_1} = \frac{400 - 600}{8 - 6} \cdot \frac{8 + 6}{400 + 600} = -1.4$$

This means that between point C and F on D_x, a 1% change in price results, on an average, in a 1.4% opposite change in the quantity demanded of commodity X.

Point and arc elasticities can also be computed from a demand equation of a commodity. For example, consider the demand equation of a commodity X as:

$$Q_x = 1000 - 40P_x$$

(i) What is point price elasticity when $P = ₹15$ and (ii) what is the arc price elasticity when the price increases from ₹15 to ₹20?

Solution

$$Q_x = 1000 - 40P_x$$

(i)
$$E_p = \frac{\Delta Q_x}{\Delta P_x} \cdot \frac{P_x}{Q_x}$$

Or
$$E_p = \frac{dQ_x}{dP_x} \cdot \frac{P_x}{Q_x}$$

Now
$$\frac{dQ_x}{dP_x} = -40$$

And Q_x at $P_x = ₹15$ is

$$Q_x = 1000 - 40 \times 15 = 1000 - 600 = 400$$

Hence,
$$E_p = -40 \cdot \frac{15}{400} = -1.5$$

The interpretation is that a 1% increase in the price causes a 1.5% reduction in demand quantity.

(ii) $P_{x1} = ₹15; P_{x2} = ₹20$

$Q_{x1} = 400; Q_{x2} = 1000 - 40 \times 20 = 200$

$$E_p = \frac{Q_{x2} - Q_{x1}}{P_{x2} - P_{x1}} \cdot \frac{P_{x2} + P_{x1}}{Q_{x2} + Q_{x1}}$$

$$E_p = \frac{200 - 400}{20 - 15} \cdot \frac{20 + 15}{200 + 400}$$

$$E_p = \frac{-200}{5} \cdot \frac{35}{600} = -2.33\%$$

The interpretation is that a 1% increases in the price results, on an average, in a 2.33% reduction in the quantity demanded of commodity X.

1.7.1.2 Determinants of price elasticity

Generally, price elasticity of demand depends on three factors, which are

(i) Availability of substitutes (ii) Proportion of income spent on the goods or services and (iii) Length of time

Availability of substitutes   Price elasticity of demand is relatively higher for goods or services for which a large number of substitutes are available. This is because the customer has more choice to select from the existing substitutes in case price of the goods or services changes. For example, if the price of any goods or services is increased, then they will become more expensive and the customer would prefer to purchase their substitutes which are available at relatively lower price and therefore, the demand for the goods or services in question will reduce substantially. On the other, if sufficient number of substitutes are not available, then price increase or decrease will have almost negligible effect on the demand of the goods or services.

Proportion of income spent   Goods or services on which a small proportion of income is spent exhibit inelastic demand as compared to those which require expenditure of a high proportion of income. For example, if 1 kg salt fulfills one month's demand of a household and it costs only ₹ 30, and if the price of the salt is doubled, this change would not have a significant effect on the purchasing power of the household. Consequently, price changes have little effect on the household's demand for salt. On the other hand, demand will tend to be more elastic for goods and services that account for a substantial portion of total expenditure.

Length of time   Price elasticity of demand also depends on the length of time or time period. Usually, demand is more elastic in the long run than in the short run. This is because, in the long run, a customer will be able to find a large number of substitutes of the goods or services and a small increase in the price would cause replacement of this expensive goods or services with relatively less expensive ones, leading to reduction in the demand for the goods or services in question.

1.7.1.3 Price Elasticity and Decision Making

Price elasticity of demand provides useful information to the managers, on the basis of which they can take appropriate decision related to the price. For example, if at the existing price, the demand is elastic, then a small reduction in the price will result in substantial increase in the demand and, consequently, revenue will increase. In contrast, if the demand is inelastic at the existing price, then to increase revenue, price can be increased.

1.7.2 Income Elasticity of Demand

Income elasticity is defined as the responsiveness in the quantity demanded of a commodity to the change in income. Income elasticity gives a measure of change in quantity demanded due to change in income when other factors affecting demand are kept constant. Income elasticity is calculated from the formula given in Equation (1.5).

$$E_I = \frac{\Delta Q}{\Delta I} \cdot \frac{I}{Q} \tag{1.5}$$

where E_1 is income elasticity of demand

ΔQ is change in quantity demanded

ΔI is change in income

Q is quantity demanded

Similar to price elasticity of demand, income elasticity of demand can be expressed in terms of point or arc.

Equation (1.5) can be used to compute income elasticity of demand when there is a very small or no change in income. Equation (1.6) can be used to compute arc income elasticity of demand when there is a relatively large change in income.

$$E_1 = \frac{Q_2 - Q_1}{I_2 - I_1} \cdot \frac{(I_2 + I_1)}{(Q_2 + Q_1)} \tag{1.6}$$

where I_1 and Q_1 represent the initial level of income and demand, and I_2 and Q_2 are the changed values of income and demand, respectively.

Example 1.1

Suppose that the demand for two wheelers as a function of per capita income is given by the equation:

$$Q = 1,00,000 + 10(1)$$

(i) What is the income elasticity at a per capita income of ₹20,000?

(ii) What is the income elasticity when per capita income increases from ₹20,000 to ₹22,000?

Solution

(i)
$$E_I = \frac{\Delta Q}{\Delta I} \cdot \frac{I}{Q}$$
$$Q = 1,00,000 + 10(1)$$

For a per capita income of ₹ 20,000

$$Q = 1,00,000 + 10 \times 20,000 = 3,00,000$$

$$\frac{\Delta Q}{\Delta I} = \frac{dQ}{dI} = 10$$

$$E_1 = 10 \cdot \frac{20,000}{3,00,000} = 0.67$$

This result can be interpreted as 1% increase in the per capita income leads to 0.67% increase in the demand quantity.

(ii)

$$E_I = \frac{Q_2 - Q_1}{I_2 - I_1} \cdot \frac{I_2 + I_1}{Q_2 + Q_1}$$

$$Q = 1,00,000 + 10(1)$$

$$Q_1 = 1,00,000 + 10 \times 20,000 = 3,00,000$$

$$Q_2 = 1,00,000 + 10 \times 22,000 = 3,20,000$$

$$E_1 = \frac{3,20,000 - 3,00,000}{22,000 - 20,000} \cdot \frac{21,000 + 22,000}{3,10,000 + 3,20,000} = 0.682$$

The interpretation of this result is that over the income ₹ 20,000 to ₹ 22,000, each 1% increase in income causes 0.682 percent increase in quantity demanded.

1.7.2.1 *Inferior Goods, Necessities and Luxuries*

From the formula presented in Equation (1.5), it is evident that income elasticity of demand can be either negative or positive. A negative E_1 indicates that an increases in income leads to decrease in the quantity demanded and the goods and services for which it happens are categorized as *inferior* goods and services. For example, Korean bacon is a food which is very popular among low income group people as they can afford this cheap food due their tight budgets. However, as their income increases, perhaps they will stop consuming it and shift to chicken stew, which is relatively expensive. Thus, increase in income causes a decrease in demand for Korean bacon which becomes an inferior good.

Goods and services for which E_1 is positive are classified as *normal* goods and services. Depending upon the magnitude of E_1, normal goods and services can further be classified as *necessities* or *luxuries*. If $0 \le E_1 \le 1$, the percentage change in demand is positive, but less than or equal to the percentage change in income. Such goods and services are categorized as necessities and for these goods and services, demand is relatively unaffected by changes in income. Bread is a good example of a necessity good as it is a basic food which everybody eats. As the income of the people increases, they will consume more bread, but increase in consumption of bread is not proportionate to the increase in the income.

Goods and services for which $E_1 > 1$ are classified as luxury goods and services. For such goods and services, change in quantity demanded is proportionately greater than the change in income. For example, if $E_1 = 3$, a 1% increase in income would cause a 3% increase in demand. An expensive car, such as BMW, is an example of a luxury good. A rich person has more disposable income and only he or she can afford to buy luxury goods.

1.7.2.2   *Income Elasticity and Decision Making*

Based on the value of income elasticity of demand, the management of a business firm can take appropriate decisions at different stages of the business cycle. For example, when the economy is rising, the business of firms dealing in luxury goods and services will increase at a relatively faster rate than the rate of income growth and therefore they can increase their annual production yearly profit. However, during economic recession, the demand for luxury goods and services may decreases substantially. Firms that produce necessity goods and services may not gain as much benefit as those producing luxury goods and services during periods of economic prosperity. However, during economic recession, the demand for necessity goods and services will not be affected too much.

Using information about income elasticity of demand, firms can make their strategy to target customers to promote sales of their products. For example, suppose a firm is manufacturing expensive cars. Since expensive cars are luxury items, it is expected that people belonging to high income groups will buy the cars and they are the focus of attention. To promote sales of the cars, the firm can put advertisements those print and electronic media which are accessed by the high income groups. For example, the firm can advertise the car in English newspapers and TV channels that are seen these people.

1.7.3   Cross Elasticity

Demand for a product is also affected by the price of the related product. The responsiveness in the quantity demanded of a particular good and services to the change in the price of a related good and service is referred to as cross elasticity of demand. This can also be defined as the percentage change in the quantity demanded of a good due to 1% change in the price of some other related good. Equation (1.7) can be used to compute cross elasticity of demand when there is very small or no change in the price of some other related good.

$$E_c = \frac{\Delta Q_x}{\Delta P_y} \cdot \frac{P_y}{Q_x} \tag{1.7}$$

where x and y are two related goods.

Equation (1.8) can be used to compute cross elasticity of demand when there is a relatively large change in the price of the other related good.

$$E_c = \frac{Q_{x2} - Q_{x1}}{P_{y2} - P_{y1}} \cdot \frac{P_{y2} + P_{y1}}{Q_{x2} + Q_{x1}} \tag{1.8}$$

where the subscript 1 refers to the initial prices and quantities and 2 to the changed values.

Example 1.2

Suppose that the demand for a good x in terms of the price of some other related good y is given by $Q_x = 200 + 1.0P_y$

(i) Compute the cross elasticity of demand when the price of y is ₹100. Interpret the result.

(ii) Compute the cross elasticity of demand when the price of y is increased from ₹100 to ₹200.

Solution

(i)

$$E_c = \frac{\Delta Q_x}{\Delta P_y} \cdot \frac{P_y}{Q_x} = \frac{dQ_x}{dP_y} \cdot \frac{P_y}{Q_x}$$

$$Q_x = 200 + 1.0P_y$$

$$\frac{dQ_x}{dP_y} = 1.0$$

For $Py = ₹100$; $Q_x = 200 + 1.0 \times 100 = 300$

$$E_c = 1.0 \times \frac{100}{300} = 0.33$$

The interpretation is that a 1% increase in the price of y causes 0.33% percent increase in the quantity demanded of x.

(ii)

$$E_c = \frac{Q_{x2} - Q_{x1}}{P_{y2} - P_{y1}} \cdot \frac{P_{y2} + P_{y1}}{Q_{x2} + Q_{x1}}$$

$$P_{y1} = ₹100; P_{y2} = ₹200$$

$$Q_{x1} = 200 + 1.0 \times 100 = 300$$

$$Q_{x2} = 200 + 1.0 \times 200 = 400$$

$$E_c = \frac{400 - 300}{200 - 100} \cdot \frac{300}{700}$$

$$= \frac{100}{100} \cdot \frac{300}{700} = 0.43$$

The interpretation is that a 1% increase in the price of y, on an average, causes 0.43% increase in the quantity demanded of x.

1.7.3.1 Substitutes and Complements

From Equation (), it is clear that E_c can either be positive, i.e., $E_c > 0$ or negative, i.e., $E_c < 0$. A positive value of E_c indicates that an increase in the price of y causes an increase in the quantity demanded of x and the two goods are related to each other as substitutes. When the relationship is that of *substitutes*, one product can be used in place of the other. For example, Maruti Swift and Hyundai i20 cars can be considered as close substitutes. In case of increase in the price of Maruti Swift, it becomes relatively expensive and people will prefer to buy Hyundai i20, which will be relatively cheap and therefore, the demand for Hyundai i20 will increase.

A negative value, i.e., $E_c < 0$ is observed when an increase in the price of y leads to a decrease in the quantity demanded of x and this would happen when x and y are complements. Petrol and petrol-driven cars, tea and sugar, tennis balls and tennis rackets are pairs of goods that are complementary goods. Two goods x and y are classified as *complementary goods* when a person uses good y as he possesses good x. Thus, cross elasticity of demand defines relationship between two goods as either substitutes or complements.

1.7.3.2 *Cross Elasticity and Decision Making*

There are many firms that produce several products and they can compute cross elasticity of demand between them to establish the relationship between the products. Depending upon the relationship, the firm can take appropriate pricing decisions. When a company sells related products, the price of one product can influence the demand for the other products. For example, a company producing men's razors and razor blades will be able to sell more razor blades if it reduces the price of its razor.

Based on the value of cross elasticity of demand, boundaries can be established between industries. Cross elasticity of demand provides useful information for including firms in a particular industry. Firms whose products exhibit a high positive cross elasticity are included in an industry. But the goods and services with negative or small cross elasticity are considered to belong to different industries.

Example 1.3 illustrates the use of various elasticities in decision making.

Example 1.3

The demand for a product x is given by the following equation:

$$Q_x = 24000 - 10000P_x + 101 + 1000P_c$$

where P_x is the price of the product x, I is the per capita income and P_c is the price of the related products.
 (i) Determine what effect a price increase of product x would have on total revenue.
 (ii) Determine how the sale of the product x would change during period of rising incomes.
 (iii) What will be the impact on the sale of the product x if price of the related products is increased.
Assume that the initial values of P_x, I and P_y are ₹20, ₹20,000 and ₹12 respectively.

Solution

The demand function is given as:

$$Q_x = 24000 - 10000\,P_x + 101 + 1000\,P_c$$

Initial values of P_x, I and P_c are obtained as:

$$Q_x = 24000 - 10000 \times 10 + 10 \times 20000 \; 1000 \times 12 = 136000$$

(i) $E_p = \dfrac{\Delta Q_x}{\Delta P_x} \cdot \dfrac{P_x}{Q_x} = \dfrac{dQ_x}{dP_x} \cdot \dfrac{P_x}{Q_x} = -10000 \cdot \dfrac{10}{136000} = -0.735$

Since price elasticity of demand E_p is less than 1, the demand is inelastic and therefore raising the price of the product x would increase total revenue.

(ii) $E_I = \dfrac{\Delta Q_x}{\Delta I} \cdot \dfrac{I}{Q_x} = 10 \cdot \dfrac{20{,}000}{136000} = 1.47$

Since $E_1 > 1$, product x is a luxury good. Thus, as income increases, sales should increase more than the proportional increase of income.

(iii) $E_c = \dfrac{\Delta Q_x}{\Delta P_c} \cdot \dfrac{P_c}{Q_x} = 1{,}000 \cdot \dfrac{12}{1{,}36{,}000} = 0.088$

Since E_c is positive, i.e., $E_c > 0$, the related goods and product x are substitutes, therefore, if price of related products is increased, then demand for product x will increase.

1.8   SUPPLY

Meaning: The theory of supply is as much necessary as the theory of demand for the analysis of prices. Supply means the amount offered for sale at a given price during a certain period of time.

Definition: In the words of Thomas, "The supply of goods is the quantity offered for sale in a given market at a given time at various prices."

1.8.1   Factors Affecting Supply: The Determinants of Supply

$$Q_x^s = f(P_x, P_y, P_i, K, N, T, E \text{ and } G_p)$$

where

     P_x = Price of goods P_x
     P_y = Price of other goods P_y
     P_i = The price of factor inputs
     K = Technical knowhow
     N = Number of supplies
     T = Time periods
     E = Expectation of the consumer
     G_p = Government policy

The price of the commodity
The supply of a commodity depends upon the price of that commodity, other things being equal. Sellers of a commodity are normally willing to sell more if its price is higher, than if it is low.

Price of related goods   If the price of substitutes goes up, producers would be tempted to direct their customers to their products by increasing their supply.

Price of factor inputs   If there is an increase in the cost of production due to the increase in the cost of various factors of production like raw material and intermediate products, there will be a decrease in supply.

Improved technology   Improvement in techniques of production lowers the cost of production and, in turn, increases the supply over time. Technical knowhow changes, discoveries and innovations help to raise the productivity of the factors and this contributes to increasing the supply.

Number of sellers   The supply also depends upon the number of sellers. Entry of more sellers will increase and exits will decrease the supply.

Price expectations   If the sellers fear that the trend of a fall in price will continue in the future, they will adopt panic-selling and the supply will increase.

Government policy   The policies of the government are also important determinants of supply. A reduction in quotas and tariffs on foreign goods will open up the market to foreign producers and will tend to increase the supply.

1.8.2   Law of Supply

The law of supply is just the reverse of the law of demand. The law of supply expresses the relation between the price of a commodity and its supply. In other words, the law of supply states that, "other things being equal, when price rises, supply extends and when price falls, supply contracts."

Definition Law of Supply   According to Dooley, "the law of supply states that, other things being equal, the higher the price, the greater the quantity supplied or the lower the price, the smaller the quantity supplied." Thus, the relation between supply of a commodity and its price is a direct one. Symbolically, the law of supply can be shown as:

$$P \uparrow \rightarrow Q^s \uparrow$$
$$P \downarrow \rightarrow Q^s \downarrow$$

P = price, Q^s = Quantity supplied

1.8.3   Supply Schedule

A hypothetical firm's supply schedule is given in Table 1.2.

Table 1.2 Supply schedule

Price of Apple (₹)	Quantity Supplied (kg)
5	400
4	300
3	200
2	100
1	50

1.8.4 Supply Curve

Using the supply schedule in Table 1.2, the supply curve is plotted in Fig. 1.13.

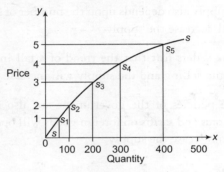

Figure 1.13 Supply curve

In Fig. 1.13, the quantity of apples offered for sale is represented on the x-axis and the price is represented on the y-axis. By plotting the quantity of apples supplied at various prices, we get the points S_1, S_2, S_3 and so on. By joining these points, the supply curve of the commodity is drawn. The supply curve S is upward sloping, which shows that more quantity is offered for sale at higher prices and less will be offered at lower prices.

1.9 DEFINITION AND SCOPE OF ENGINEERING ECONOMICS

1.9.1 Meaning of Engineering Economics

Engineering Economics uses the theories of economics and applies them to engineering decisions. For example, if an engineer at Maruti Udyog invents a new instrument that increases fuel mileage by 10%, how much can the company afford to spend to implement this invention? Engineering Economics can provide an answer to this question.

1.9.2 Definition

According to Eugene L. Grant, "Engineering Economics is the discipline concerned with economic aspects of engineering". It involves the systematic evaluation of the cost and benefits of the proposed technical projects.

1.9.3   Concepts of Engineering Economics

There are two concepts associated with Engineering Economics. These are:
  (i) Engineering and Science
  (ii) Engineering and Economics

Engineering and science   Engineering is not a science but is an application of science. It involves the use of the skill and ingenuity in adapting knowledge to the uses of the human race. According to Engineers Council for Professional Development, Engineering is the profession in which a knowledge of the mathematical and natural sciences gained by study, experience and practice, is applied with judgement to develop ways to utilize, economically, the materials and forces of nature for the benefit of mankind. In this definition, the applied nature of engineering activity has been emphasized.

Engineering and economics   Dr. Paul A. Samuelson, Nobel Laureate in economics, says that economists today agree on a general definition, something like the following: Economics is the study of how men and society end up choosing, with or without the use of money, to employ scarce productive resources that would have alternative uses, to produce various commodities and distribute them for consumption, now or in the future, among various people and groups in society. It analyzes the costs and benefits of improving patterns of resource allocation.

1.9.4   The Scope of Engineering Economics

The following is the scope or extent of Engineering Economics:
  ➤ Traditional economic theory is concerned with the optimum allocation of resources in our society.
  ➤ Engineering Economics deals with the optimum allocation of the enterprise's capital.
  ➤ Engineering Economics employs economic theory, mathematical programing, and statistical analysis to formulate and solve problems concerning the evaluation and selection of projects.
  ➤ A basic concept of Engineering Economics is the time value of money and the techniques associated with it like compounding, discounting and economic equivalence, which have applications in capital expenditure analysis, as well as financial analysis.
  ➤ Engineering Economics is also concerned with the impact of economic and institutional factors on capital investment decisions such as taxes, methods of depreciation, interest rates, and the availability of funds.
  ➤ From managerial point of view, Engineering Economics contributes to better analysis of cost of operations, capacity expansion and profitability.

1.9.5   Engineering Economics Environment

While dealing with decision problems, engineers encounter two different types of environments: physical and economic. The success of engineers depends upon their ability to manipulate goods and services by acquiring and applying physical laws. However, the worth of these goods and services lies in their utility, which is measured in economic terms.

Thus, knowledge of both physical laws and economics is required for making appropriate decisions.

1.9.6　Types of Efficiency

The efficiency of a system is generally defined as the ratio of its output to input. Efficiency can be classified as technical efficiency and economic efficiency as shown in Fig. 1.14.

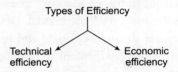

Figure 1.14　Types of efficiency

Technical efficiency　Technical efficiency is the ratio of the output to input of a physical system. The physical system may be a diesel engine, a machine working in a shop floor, a furnace, etc.

$$\text{Technical efficiency (\%)} = \frac{\text{Output}}{\text{Input}} \times 100$$

In practice, technical efficiency can never be more than 100%. This is mainly due to friction loss and incomplete combustion of fuel.

Economic efficiency　Economic efficiency is the ratio of output to input of a business system.

$$\text{Economic efficiency (\%)} = \frac{\text{Output}}{\text{Input}} \times 100 = \frac{\text{Worth}}{\text{Cost}} \times 100$$

Worth is the annual revenue generated by way of operating the business and Cost is the total annual expenses incurred in carrying out the business. For the survival and growth of any business, the economic efficiency should be more than 100%.

Economic efficiency, which is also called productivity, can be increased by the following ways:

1. Increased output for the same input.
2. Decreased input for the same output.
3. By a proportionate increase in the output which is more than the proportionate increase in the input.
4. By a proportionate decrease in the input which is more than the proportionate decrease in the output.
5. Through simultaneous increase in the output with decrease in the input.

1.10   CONSUMER AND PRODUCER GOODS AND SERVICES

The goods and services that are produced and utilized may be categorized into two groups. Consumer goods and services are those products or services that are directly used by consumers to satisfy their desire. Food, clothing, homes, air conditioners, fast moving consumer goods (FMCG), telephones, mobile phones and banking services are examples of consumer goods and services. The demand for such goods and services is directly related to the people and, therefore, it can be determined with considerable certainty.

Producer goods and services are those products or services that are used to produce consumer goods and services or other producer goods. Manufacturing infrastructure that includes building, machinery, instruments and control systems, locomotives and land transport used for the transportation of manufactured goods are some examples. The demand for producer goods and services is not as directly related to consumers like the demand for consumer goods and services. Due to this indirect relation, the demand for and production of producer goods may precede or lag behind the demand for the consumer goods that they will produce. For example, a company may set up a factory and install machines to produce products for which currently there is no demand, in the hope that it will be able to create such a demand through its advertising and marketing strategies. On the other hand, a company may wait until the demand for a product is well-established and then develop the required facilities to produce it.

1.11   NECESSITIES, LUXURIES AND RELATION BETWEEN PRICE AND DEMAND

Goods and services may be classified as necessities or luxuries. Necessities and luxuries are relative terms because a given good may be a necessity for a person but it may be considered to be a luxury for another. Income is an important factor which differentiates between necessity and luxury goods. In addition, there are other factors also which determine whether a good is a necessity or a luxury. For example, a car is a luxury good for a person who is living in a city which provides adequate public transport facilities for commuting. But the same car is a necessity good for a person living in a city where public transport is not available.

It has been observed that for most goods and services, there exists a relationship between the price that must be paid and the quantity that will be demanded or purchased. The general relationship is depicted in Fig. 1.15, which shows an inverse relationship between selling price, P and the quantity demanded, D. Fig. 1.15 is also known as law of demand, which states that as the selling price is increased, there will be less demand for the product, and as the selling price is decreased, the demand will increase. The rider to the law of demand is that all other factors, except price, that may affect the demand are constant.

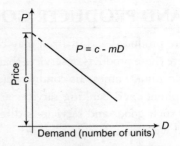

Figure 1.15   General relationship between price and demand

The relationship between price and demand can be expressed as a linear function:

$$P = c - mD \text{ for } 0 \le D \le \frac{c}{m} \tag{1.9}$$

where c is the intercept on the price axis and $-m$ is the slope of the line. Thus, m represents the amount by which demand increases for each unit decrease in P. Thus,

$$D = \frac{c - P}{m} \tag{1.10}$$

The relationship between price and demand shown in Fig. 1.15 is expected to be different for necessities and luxuries. It is easy to realize that the consumers would give up the consumption of luxuries if the price is increased, but it would be very difficult for them to reduce the consumption of necessities. Fig. 1.16 depicts the price-demand relationship for necessities and luxuries.

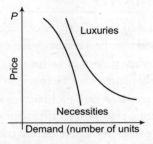

Figure 1.16   General relationship between price and demand for necessities and luxuries

The degree to which demand is affected due to change in price is referred to as the elasticity of the demand. The demand for products is said to be elastic when decrease in selling price results in considerable increase in their demand. On the other hand, if a change in selling price produces little or no effect on the demand, the demand is said to be inelastic. Thus, it is clear from Fig. 1.16 that luxuries exhibit elastic demand whereas, the demand for necessities is inelastic in nature.

1.12 RELATION BETWEEN TOTAL REVENUE AND DEMAND

The total revenue, R, that can be generated by selling a particular good during a given period is obtained by multiplying the selling price per unit, P, with the number of units sold, D. Thus,

$$R = \text{selling price per unit} \times \text{number of units sold} \qquad (1.11)$$

If the relationship between price and demand as given in Equation (1.9) is used,

$$R = P \times D$$

$$= (c - mD) \times D$$

$$= cD - mD^2 \text{ for } 0 \le D \le \frac{c}{m} \qquad (1.12)$$

The relationship between total revenue and demand for the condition expressed in Equation (1.12) may be depicted by the curve shown in Fig. 1.17.

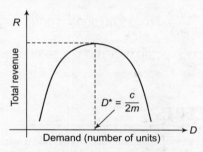

Figure 1.17   General relationship between total revenue and demand

In order to determine the optimum demand, D^* that would yield maximum total revenue, we differentiate Equation (1.12) with respect to D and equate it to zero as:

$$\frac{dR}{dD} = c - 2mD = 0 \qquad (1.13)$$

Thus,
$$D^* = \frac{c}{2m} \qquad (1.14)$$

The maximum total revenue can be obtained as:

$$R_{\text{max.}} = cD^* - mD^{*2} = \frac{c^2}{2m} - \frac{c^2}{4m} = \frac{c^2}{4m} \qquad (1.15)$$

The term $\dfrac{dR}{dD}$ is called the incremental or marginal revenue.

Example 1.4

A gear manufacturer estimates the following relationship between sales volume, D, and the unit selling price, P:

$$D = 50{,}000 - 0.02\,P$$

How many units of gear should the company produce and sell to maximize the total revenue?

Solution

We know that the total revenue is given as:

$$R = P \times D$$

$$= (25{,}00{,}000 - 50D) \times D$$

$$= 25{,}00{,}000D - 50D^2$$

For maximum total revenue $\dfrac{dR}{dD} = 0$

Thus,
$$\frac{dR}{dD} = 0 = 25{,}00{,}000 - 100\,D$$

$$D = D^* = 25{,}00{,}000/100 = 2{,}50{,}000 \text{ units}$$

1.13 COST CONCEPTS

The following are different types of cost that are involved in various kinds of engineering and managerial problems:

Investment cost   The investment cost, or first cost, refers to the capital required to start a project. It may be either a single lump sum, or a series of cash inputs that must be made at the beginning of the project. For example, if we want to purchase a new car, then the investment cost for acquiring it is the sum of the down payment, taxes and other charges involved in obtaining the ownership of the car. For a capital intensive project such as a construction project which takes several years to complete, the investment cost is usually spread over time in the form of progressive payments.

Fixed, variable and incremental costs   Fixed costs associated with a new or existing project are those costs that do not change over a wide range of activities of the project. Fixed costs are, at any time, the inevitable costs that must be paid regardless of the level of output and of the resources used. For example, interest on borrowed capital, rental cost of a warehouse, administrative salaries, license fees, property insurance and taxes are fixed costs.

The costs that vary proportionately to changes in the activity level of a new or existing project are referred to as variable costs. Examples of variable costs include raw material cost, direct labor cost, power cost, shipping charges, etc.

The additional cost that will result from increasing the output of a system by one or more unit(s) is called incremental cost. For example, if the cost of manufacturing 10 pieces is ₹2,000 and that of 11 pieces is ₹2,200, then the incremental cost is ₹200.

Sunk cost   Sunk cost refers to the cost that has occurred in the past and does not have any impact on the future course of action. For example, suppose you wanted to buy a second-hand car. You went to the showroom of a second-hand car dealer. You liked a car that costs ₹80,000. You made a down payment of ₹5,000 to the dealer, with a condition that, if in future, you change your decision and do not buy the car you liked, then ₹5,000 will not be given back to you. Next day your friend told you that he wants to sell his car. Your friend's car is almost of a similar condition as that of the car you saw in the dealer's showroom. Your friend offered you the car for a price of ₹70,000. Your friend's car is definitely cheaper than that of the dealer. In this situation you will not even think of ₹5,000 that you paid to the dealer. You will buy your friend's car. Thus, ₹5,000 in this case is a sunk cost, as it occurred in the past and it does not have any impact on your decision to buy car from your friend.

Opportunity cost   Opportunity cost or implicit cost refers to the value of the resources owned and used by a firm in its own production activity. The firm, instead of using the resources for its own purpose, could sell or rent them out and in turn can get monetary benefits. The amount that the firm is not getting by renting or selling the owned resources as it is using them for its own purpose represents the opportunity cost of the resources. Examples of opportunity costs include the salary that a manger could earn if he or she would have managed another firm, the return that the firm could receive from investing its capital in the most rewarding alternative use or renting its land and buildings to the highest bidder rather than using them itself.

1.14   RELATION BETWEEN COST AND VOLUME

In almost all firms, there are certain costs that remain constant irrespective of their level of output. For example, interest on borrowed capital, property taxes and many other overhead costs of firms remain constant regardless of the level of production. Such costs are referred to as fixed costs, FC. In addition, there are other costs that vary almost directly with the level of output, and these are called variable costs. For example, costs associated with direct labor, raw material and power consumption are variable costs, VC. The sum of the fixed costs and the variable costs is referred to as total cost, TC. The relationship of these costs with the number of units produced, i.e., demand, D, is depicted in Fig. 1.18.

Thus, at any demand D,

$$TC = FC + VC \tag{1.16}$$

Let, the variable cost per unit be v.

Thus,
$$VC = v \times D \tag{1.17}$$

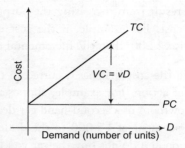

Figure 1.18   Relationship between fixed cost, variable cost and total cost with demand

When the total revenue-demand relationship shown in Fig. 1.17 and the total cost-demand relationship shown in Fig. 1.18 are combined together then it results in Fig. 1.19 for the case where $(c - v) > 0$. Fig. 1.19 clearly shows the volume for which profit and loss take place. Here, we wish to know the demand level that yields maximum profit.

We can define profit as:

$$\text{profit} = \text{total revenue} - \text{total cost}$$

$$= P \times D - (FC + VC)$$

$$= (c - mD) \times D - (FC + vD)$$

$$= -FC + (c - v) \times D - mD^2 \text{ for } 0 \le D \le \frac{c}{m}$$

We now take the first derivative with respect to D and equate it to zero:

$$\frac{d(\text{profit})}{dD} = c - v - 2mD = 0$$

The value of D that maximizes profit is

$$D* = \frac{c - v}{2m} \tag{1.18}$$

Example 1.5

An electrical fittings manufacturer has found that the unit selling price, P, and the sales volume, D, for one of its models of circuit-breakers has the following relationship:

$$P = 1,00,000 - 200D$$

It, therefore, wants to calculate the unit selling price to control the sales volume in accordance with the rate of production. The company's monthly fixed costs and the unit variable costs are ₹75,000 and ₹250 respectively. How many units should the company produce and sell to maximize the profit and what is the maximum profit?

Solution

$$\text{profit} = \text{total revenue} - \text{total cost}$$

$$= P \times D - (FC + VC)$$

$$= (1,00,000 - 200D) \times D - (₹75,000 + ₹250D)$$

$$= -₹75,000 + (1,00,000 - 250) \times D - 200D^2 \text{ for } 0 \le D \le \frac{c}{m}$$

We now take the first derivative with respect to D and equate it to zero:

$$\frac{d(\text{profit})}{dD} = (1,00,000 - 250) - 400D = 0$$

The value of D that maximizes profit is:

$$D* = \frac{(1,00,000 - 250)}{400} = 249.38 \approx 250 \text{ units}$$

$$(\text{profit})_{\text{max.}} = -75,000 + (1,00,000 - 250)(250) - (200)(250)^2$$

$$(\text{profit})_{\text{max.}} = ₹1,23,62,500$$

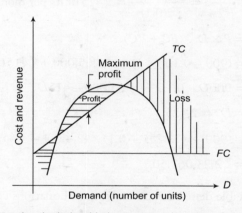

Figure 1.19   Functional relationship between total cost, total revenue and profit

Fig. 1.19 shows that the total revenue curve intersects the total cost line at two points. Thus, at these points total revenue is equal to the total cost and these points are called *breakeven points*. For a demand D, at the breakeven point:

total revenue = total cost

$$cD - mD^2 = FC + vD$$

$$-mD^2 + (c - v) \times D - FC = 0 \qquad (1.19)$$

Equation (1.19) is a quadratic equation with one unknown D; it can be solved for the breakeven points D_1 and D_2 which are the two roots of the equation.

$$D_1 = \frac{-(c - v) + \sqrt{(c - v)^2 - 4(-m)(-FC)}}{2(-m)} \qquad (1.20)$$

and

$$D_2 = \frac{-(c-v) - \sqrt{(c-v)^2 - 4(-m)(-FC)}}{2(-m)} \tag{1.21}$$

Example 1.6

The fixed cost, FC, of a company is ₹3,65,000 per month, and the variable cost, v is ₹415 per unit. The selling price per unit P is = ₹900 − 0.1D. Determine (a) optimal volume for this product and confirm that a profit occurs at this demand, and (b) the volumes at which breakeven occurs; what is the range of profitable demand?

Solution

Here, c = ₹900; m = 0.1; FC = ₹3,65,000; v = ₹415

(a) From Equation (1.10)

$$D^* = \frac{c-v}{2m} = \frac{900-415}{2(0.1)} = 2,425 \text{ units per month}$$

Now   profit $= P \times D - (FC + VC)$

$$= (900 - 0.1D) \times D - (₹3,65,000 + ₹415D)$$

$$= 900D - 0.1D^2 - 3,65,000 - 415D$$

Using $D = D^* = 2,425$

$$\text{profit} = (900) \times (2,425) - (0.1) \times (2,425)^2 - 3,65,000 - (415) \times (2,425)$$

$$= ₹2,23,062.50.$$

(b) The range of profitable demand is obtained from Equations (1.20) and (1.21)

$$D_1 = \frac{-(900-415) + \sqrt{(900-415)^2 - 4(-0.1)(-3,65,000)}}{2(-0.1)}$$

$$= 931.47 \approx 932 \text{ units per month}$$

$$D_2 = \frac{-(900-415) - \sqrt{(900-415)^2 - 4(-0.1)(-3,65,000)}}{2(-0.1)}$$

$$= 3,918.53 \approx 3,919 \text{ units per month}$$

Thus, the range of profitable demand is 932 to 3,919 units per month.

1.15 THE LAW OF SUPPLY AND DEMAND

The relationship between the price that customers pay for a product and the quantity of the product that they will buy is shown in Fig. 1.20. There also exists a similar relationship between the price at which a product can be sold and the quantity of the product that will be made available. If a supplier gets a high price for his products, then he will be willing to supply a large volume of them. On the other hand, if the price that a supplier gets for his products declines, then he will not be interested in the supply of a large quantity. It may be possible that the supplier, who is also a producer, will stop producing and may direct his efforts to other ventures. The relationship between price and the volume of product supplied or produced can be represented by the curve shown in Fig. 1.20.

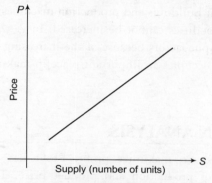

Figure 1.20   General relationship between price and Supply

Fig. 1.21 has been obtained by combining Fig. 1.15 and Fig. 1.20 and it explains the basic economic law of supply and demand which states that under conditions of perfect competition, the price at which a given product will be supplied and purchased is the price that will result in the supply and demand being equal, i.e., $D_1 = S_1$. Thus, it is evident from Fig. 1.21 that at price P_1 the supply and demand are equal. Price P_1 is known as *equilibrium price*. A change in either the level of supply or the level of demand would cause a change in the equilibrium price.

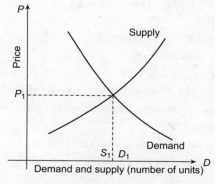

Figure 1.21   General relationship between price, supply and demand

1.16 THE LAW OF DIMINISHING MARGINAL RETURNS

This law refers to the amount of extra output that results from the successive additions of equal units of variable input to some fixed or limited amount of other input factor. The law states that when additional variable input factors are added to the fixed or limited input factors of production, then in the beginning the output increases in a relatively larger proportion, but beyond an output level, such addition of variable inputs will result in a less than proportionate increase in output.

We observe the result of this principle in our day-to day—life, too, but quite often we do not realize that what we have encountered is related to this law. Some examples where we can observe the result of the law diminishing return include internal combustion engine, electric motor, office and apartment buildings and production machines, where the capacity of these machines and other such facilities cannot be increased. In view of the fact that capital and markets are also limited input factors because of the increasing cost of procuring them, the law of diminishing return occupies an important place in making many investment related decisions.

1.17 BREAK-EVEN ANALYSIS

Every business firm strives to maximize profit. Profit has a strong relationship with production cost, volume of production and product price. Therefore, it is necessary to examine the relationship of profit with these factors. Breakeven analysis, which is also known as Cost Volume Profit (CVP) analysis, is an effective method to establish cost-volume-profit relationship. Break-even analysis involves the study of revenues and costs of a firm in relation to its volume of production or sales. It involves determination of a specific volume at which the firm's cost and revenues are equal. Break-even analysis shows the relation of fixed cost, variable cost, output volume, selling price, etc., with the firm's profit. This analysis helps the management to understand the effect of change in fixed cost, variable cost and selling price on the output or sales volume.

Break-even analysis assumptions The following assumptions are used in break-even analysis:
   (i) The relationship between cost/volume such as fixed cost, variable cost, sales, etc., and production volume or sales is linear.
  (ii) Only fixed costs and variable costs are considered in the analysis.
 (iii) Selling price of the product remains constant at all sales level.
  (iv) All information related to the cost/volume/revenue are deterministic.
   (v) The cost is affected by quantity only.
  (vi) The rate of increase in variable cost is constant.
 (vii) Production and sales quantities are equal.

Break-even point   In order to understand the meaning of break-even point (BEP), we need to draw a break-even chart. A break-even chart plots cost/revenue against production volume/sales volume. To make this chart, production volume/sales volume is taken on x-axis and cost/revenue is taken on y-axis. Fig. 1.22 shows a typical break-even chart.

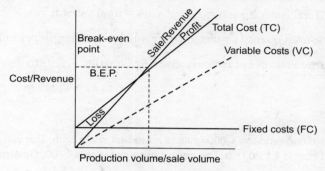

Figure 1.22   Break-even chart

It can be seen from Fig. 1.22 that fixed costs, variable costs, revenue are all linear as they are represented by straight lines, according to the assumption. Further, the sales/revenue line intersects the total cost line at a particular point, which is known as break-even point. At this point total cost is equal to sales/revenue, which means there is no profit. Thus, break-even point refers to the production volume/sales volume at which total cost is equal to sales/revenue and hence there is neither a profit nor a loss. The quantity produced and sold above break-even quantity yields profit and below break-even quantity results in loss. Mathematically, a formula for break-even quantity can be obtained as:

FC represents the fixed costs,

Q_B is breakeven quantity,

VC is the variable cost, and

TC is total cost which is sum of fixed costs and variable costs.

i.e.,
$$TC = FC + VC$$

R is the revenue.

P is the selling price per unit.

v is the variable cost per unit.

At break-even point,

$$R = TC$$

$$P.Q_B = FC + VC$$

$$P.Q_B = FC + v.Q_B$$

$$Q_B = FC/(P - v) \qquad\qquad (1.22)$$

Here $(P - v)$ represents the contribution margin per unit. Total contribution can be obtained by subtracting total variable costs from actual sales, i.e.,

Total contribution = Actual sales − total variable costs

Thus,   Break-even quantity = Fixed cost/contribution margin per unit

Example 1.7 illustrate the use of formula given in Equation (1.22) to determine break-even quantity.

Example 1.7

A manufacturing firm can produce 1,000 units of a product in a month. The variable cost per unit is ₹ 100 and the fixed cost is ₹ 1,20,000. The selling price per unit is ₹ 700. Determine the break-even quantity.

Solution

$v = ₹ 100; FC = ₹ 1,20,000; P = ₹ 700$

$$Q_B = FC/(P - v) = 1,20,000/(700 - 100) = ₹ 200 \text{ units.}$$

Break-even point in terms of sales value:

In terms of sales value, break-even point (S_B) is given as,

$$S_B = \text{Fixed Cost/Contribution margin ratio}$$

$$S_B = \left[\frac{FC/(P - v)}{P} \right]$$

$$S_B = \left[\frac{FC}{1 - (v/P)} \right]$$

We know that, $R = P \cdot Q_B$

$$P = R/Q_B$$

and $v = VC/Q_B$

Therefore, $S_B = \dfrac{FC}{1 - \dfrac{VC/QB}{R/QB}}$

$$S_B = \frac{FC}{1 - \dfrac{VC}{R}}$$

$$S_B = \frac{120000}{1 - 100/700} = ₹ 1,40,000$$

Break-even point as a percentage of capacity:

Let, Q_{max} be the full capacity of the firm

Q_B is the break-even point

$$\text{BEP as percentage of capacity} = \frac{Q_B}{Q_{max}} \times 100$$

$$= \frac{FC}{(P-v)Q_{max}} \times 100 \qquad (1.23)$$

Applying Equation (1.23) we can determine the break-even point as a percentage of capacity as,

$$\text{BEP as percentage of capacity} = \frac{1,20,000}{(700-100) \times 1000} \times 100$$

$$= \frac{1,20,000}{600 \times 1000} \times 100 = 20 \ \%$$

The formula given in Equation (1.22) can be modified to determine the quantity which is required to be produced and sold to achieve a set or target profit.

Let Q_T be the target quantity

B_T be the target profit

$$B_T = R - TC$$

$$= P \cdot Q_T - v \cdot Q_T - FC$$

$$Q_T = \frac{FC + BT}{(P-v)} \qquad (1.24)$$

Example 1.8 illustrates the use of the formula given in Equation (1.24) to determine the target quantity.

Example 1.8

Consider the previous example. How many units should be produced and sold to achieve a profit of ₹60,000?

Solution

$$Q_T = \frac{FC + BT}{(P-v)} = \frac{1,20,000 + 60,000}{700 - 100} = \frac{1,80,000}{100} = 300 \text{ units.}$$

Margin of safety   Margin of Safety is defined as the difference between the sales at existing level of output and break-even sales.

Margin of safety = sales at existing level of output – Break-even sales

In terms of ratio, margin of safety (M/S) ratio is given as,

$$\text{M/S ratio} = \frac{\text{margin of safety}}{\text{sales at existing production level}}$$

This ratio defines the economic soundness of a firm. Higher this ratio, more sound is the economic soundness of the firm. If the margin of safety of a firm is high, then it will keep earning profit even in case of a small reduction in production.

Margin of safety can also be obtained as,

$$\text{M/S} = \frac{\text{Profit}}{\text{Total Contribution}} \times \text{Actual sales}$$

For example in Example 1.7, the production level is 1,000 units and break-even production level is 200 units. Thus margin of safety,

$$\text{MS} = \text{Existing production level} - \text{Break-even production}$$

$$= 1000 - 200 = 800 \text{ units}$$

In monetary terms, i.e., in terms of rupees

$$\text{MS} = 800 \times ₹\, 700 = ₹\, 56,00,000$$

In ratio terms $$\text{M/S ratio} = \frac{\text{margin of safety}}{\text{sales at existing production level}}$$

$$= \frac{₹\, 56,00,000}{700 \times 1000} = 8$$

1.18 PROFIT-VOLUME (P/V) CHART AND P/V RATIO

Profit-volume (P/V) graph is a graph which is obtained by plotting quantity on x-axis and profit on the y-axis. This graph is useful in comparing different processes and systems. When there is no production, i.e., production level is zero, the y-coordinate is negative and it is basically equal to fixed cost. A typical P/V graph is shown in Fig. 1.23. The slope of the line AB depends on profit volume ratio.

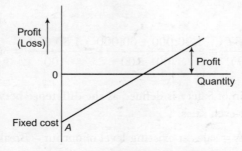

Figure 1.23   Profit-Volume (P/V) graph

Profit volume ratio is defined as,

$$\text{P/V ratio} = \frac{\text{Total Contribution}}{\text{Actual sales}} = \frac{\text{Actual sales} - \text{Total variable costs}}{\text{Actual sales}}$$

The relationship between break-even point (BEP) and P/V ratio is given as:

$$BEP = \frac{FC}{\dfrac{P}{V}\text{ratio}} \tag{1.25}$$

The relationship between margin of safety (MS) and *P/V* ratio is given as:

$$MS = \frac{\text{Profit}}{\dfrac{P}{V}\text{ratio}} \tag{1.26}$$

Example 1.9

The data for a manufacturing firm is given below:

$$\text{Fixed cost} = ₹40,00,000$$
$$\text{Variable cost per unit} = ₹200$$
$$\text{Selling price per unit} = ₹400$$

Determine:
(a) The break-even sales quantity
(b) The break-even sales amount
(c) If the actual production quantity is ₹1,20,000, find (i) total contribution, and (ii) margin of safety.

Solution

$FC = ₹40,00,000$; $v = ₹200$; $P = ₹400$

(a) Break-even quantity, $Q_B = \dfrac{FC}{P-v} = \dfrac{40,00,000}{400-200} = 20,000$ units

(b) Break even sales, $S_B = \dfrac{FC}{1-v/P} = \dfrac{40,00,000}{1-200/400} = ₹80,00,000$

(c) (i) Total contribution = Actual sales − Total variable cost
$$= P \times Q - v \times Q$$

Here $Q = 1,20,000$

Total contribution $= (P - v) \times Q = (400 - 200) \times 1,20,000$

(ii) *Margin of safety*

Method 1
$$M/S = \text{Actual sales} - \text{Break even sales}$$

$$= 400 \times 1,20,000 - 400 \times 20,000$$

$$= 4,80,00,000 - 80,00,000$$

$$= ₹4,00,00,000$$

Method 2

$$M/S = \frac{Profit}{Total\ Contribution} \times Actual\ sales$$

$$Profit = Actual\ sales - (FC + v \times Q)$$

$$= 1,20,000 \times 400 - (40,00,000 + 200 \times 1,20,000)$$

$$= 4,80,00,000 - (40,00,000 + 2,40,00,000)$$

$$= ₹2,00,00,000$$

$$M/S = \frac{2,00,00,000}{2,40,00,000} \times 4,80,00,000 = ₹4,00,00,000$$

In ratio terms, M/S ratio $= \dfrac{margin\ of\ safety}{sales\ at\ existing\ quantity} = \dfrac{4,00,00,000}{4,80,00,000} = 0.833$

In terms of percentage of sales = 83.33%

Example 1.10

Consider the following data of a company:
Sales = ₹2,40,000
Fixed cost = ₹50000
Variable cost = ₹90,000
Compute the following:
(a) Total contribution
(b) Profit
(c) Break-even point (BEP)
(d) Margin of safety (M/S)

Solution

(a) Total contribution = Sales − variable cost

$$= ₹2,40,000 - ₹90,000 = ₹1,50,000$$

(b) Profit = Sales − (variable cost + fixed cost)

$$= ₹2,40,000 - (₹90,000 + ₹50,000) = ₹1,00,000$$

(c) BEP $= \dfrac{Fixed\ cost}{\dfrac{P}{V}\ ratio}$

$$P/V \text{ ratio} = \frac{\text{Total Contribution}}{\text{Actual sales}} = \frac{₹1,50,000}{₹2,40,000} = 0.625 = 62.5\%$$

$$BEP = \frac{₹50,000}{0.625} = ₹80,000$$

(d) $\quad M.S = \dfrac{\text{Profit}}{\dfrac{P}{V}\text{ratio}} = \dfrac{₹1,00,000}{0.625} = ₹1,60,000$

1.19 COMPETITION OR MARKET STRUCTURE

Consumer goods are made available in the market which consists of potential buyers and sellers of these goods. Competition is basically the market structure which refers to the competitive environment in which buyers and sellers of the product operate. Following are the four types of market structures:

Perfect competition This refers to the market structure in which (i) there are many buyers and sellers of a product, each too small to affect the price of the product; (ii) the product is homogeneous; (iii) there is a perfect mobility of the resources; and (iv) Firms, suppliers of inputs, and consumers have perfect knowledge of market conditions.

Perfect competition is an ideal form of market structure because there are many factors that put some degree of restriction on the action of buyers or sellers, or both. It is easier to formulate general economic laws for perfect market conditions and that is why most of the economic laws have been stated for this ideal form of market structure.

Monopoly Perfect monopoly is at the opposite extreme of perfect competition. Monopoly refers to the market structure in which a single firm sells the product for which no close substitutes are available. In addition, the firm is able to prevent the entry of all other suppliers into the market. A monopolistic firm can manipulate the price and quantity of the goods the way it likes and the buyer is completely at the mercy of the supplier. When a firm is classified as a monopoly, then a governmental body regulates the rates that the firm can charge for its products or services.

Perfect monopoly seldom exists because there is hardly any unique product for which good substitutes are not available in the market and, also, there are many government regulations that stop the practice of monopoly if it is found to exploit the buyers.

Monopoly is considered to be advantageous as it avoids duplication of cost-intensive facilities which results in lower prices for product and services.

Oligopoly This refers to the market structure that exists when there are so few suppliers of either homogeneous or heterogeneous products that the action taken by one supplier results in similar action by other suppliers. For example, consider a small city in which there are a few petrol stations. If the owner of one station increases the price of petrol by ₹2 per liter, then the others will probably do the same and still retain the same level of competition.

Monopolistic competition   In this type of market structure, there are many firms selling a heterogeneous product. As evident from the name, monopolistic competition has elements of both competition and monopoly. The competitive element arises from the fact that there are many suppliers in the market. The monopoly element arises because the product of each supplier is a little different from the product of other suppliers. This form of market structure is commonly observed in the services sector. For example, consider the large number of barber shops in a given area; each one of them provides similar, but not identical, services to the customer.

PROBLEMS

1. Distinguish between consumer goods and producer goods. Why is it difficult to estimate demand for producer goods?

2. Explain the difference between elasticity of demand for necessities and the elasticity of demand for luxuries.

3. Describe and differentiate between different types of market structures.

4. With the help of suitable examples explain fixed costs and variable costs faced by a manufacturing firm.

5. State and explain the law of supply and demand.

6. With the help of a suitable example, explain the law of diminishing marginal returns.

7. A company manufacturing vacuum flasks has determined a relationship between unit selling price, P, of the flask (in rupees) and monthly sales volume, D as:

$$D = 10,000 - \frac{10P}{5}$$

   The company has a monthly fixed cost of ₹60,000 and variable cost per unit ₹165. What sales target should the company set in order to maximize the profit and what is the maximum profit?

8. An industry producing molded luggage has established that the unit selling price of one of its models and its annual demand bears an empirical relation as:

$$P = 50,000 - 0.25D$$

   The annual fixed and the unit variable cost associated with the model are ₹10,00,000 and ₹750 respectively. How many units should the company produce and sell to maximize profit and what is the maximum profit?

9. A consumer durables manufacturer has found that the sales volume, D, and the unit selling price, P, for one of its items has the following relationship:

$$D = 50,000 - 0.5P$$

   It wants to calculate the unit selling price to control the sales volume in accordance with the rate of production. The company's monthly fixed costs and the unit variable costs are ₹60,000 and ₹1,000 respectively. How many units should the company produce and sell to maximize profit and what is the maximum profit?

10. A pharmaceutical firm has estimated that the sales volume, D, of one of its tablets is highly sensitive to its unit selling price, P. It has also established that the total revenue, R, is related to the sales volume as:

$$R = cD - mD^2 \text{ for } 0 \le D \le (c/m)$$

The firm has a monthly fixed cost and unit variable cost of ₹75,000 and ₹18 respectively. If $c = 500$ and $m = 0.01$.

(i) What is the optimal number of units that should be produced and sold per month?

(ii) What is the maximum profit per month?

(iii) What is the range of profitable demand?

11. The relationship between price and quantity demanded for a product is given as:

$$P = 300 - \frac{Q}{2000}$$

(i) Determine the price elasticity of demand at $P = ₹100$ and at $P = ₹150$

(ii) How does the point price elasticity of demand vary with the price?

12. The demand function for a product is given as

$$P = 500 - \frac{Q}{1000}$$

(i) What is the arc elasticity of demand as price increases from ₹100 to ₹120?

(ii) What is the arc elasticity of demand as price decreases form ₹120 to ₹100?

13. For each of the following equations, Determine whether demand is elastic, inelastic or unitary elastic at the given price:

(i) $Q = 50 - 2P$ at $P = ₹20$

(ii) $Q = 30 - 5P$ at $P = ₹2$

14. The management of a manufacturing firm believes that the price elasticity of demand of its product is -2.0. Currently, the product is priced at ₹400 and the quantity demanded is 8000 per month. If the price is increased to ₹500, how many units will the firm be able to sell each month?

15. Demand for a product is given by $Q = 4000 - 200 P$

The product is initially priced at ₹70.

(i) Compute the point price elasticity of demand at $P = ₹70$

(ii) If the objective is to increase total revenue, should the price be increased or decreased? Explain.

(iii) Compute the arc price elasticity for a price decrease from ₹60 to ₹50

(iv) Compute the arc price elasticity for a price decrease from ₹60 to ₹50

16. The price quantity relationship for a product is given as $P = 100 - 10Q$

(i) At what output rate is the demand unitary elastic?

(ii) Over what range of output is the demand elastic?

(iii) At the current price, 9 units are demanded in each period.

If the objective is to increase total revenue, should the price be increased or decreased? Explain.

17. The price elasticity of demand of a product is -0.2 and the income elasticity of demand is 0.6. At a price of ₹50 per unit and a per capita income of ₹50,000, the demand for the product is 5,00,000 units per year.

    (i) Is the product in question an inferior good, a necessity, or a luxury? Explain.

    (ii) If the per capita income increases to ₹60,000, what will be the quantity demanded of the product?

    (iii) If the price of the product increases to ₹60 and per capita income remains at ₹50,000, what will be the quantity demanded?

18. SAIL is a major producer of steel. The demand for its steel is given by the following equation:

$$Q_s = 10,000 - 200\,P_s + 0.21 + 200\,P_a$$

where Q_s is steel demand in thousands of tons per year, P_s is the price of steel in rupees per kg, I is per capita income, and P_a is the price of aluminum in rupees per kg. Initially, the price of steel is ₹110 per kg, per capita income is ₹50,000 and the price of aluminum is ₹350 per kg.

    (i) How much steel will be demanded at the initial price and income?

    (ii) What is the point price elasticity at the initial values?

    (iii) What is the point income elasticity at the initial values?

    (iv) What is the point cross elasticity between steel and aluminum? Are they substitutes or complements?

19. The price of oil is ₹4,000 per barrel and the price elasticity is constant and equal to –0.5. An oil embargo reduces the quantity available by 30%. What is the percentage increase in the price of oil? Use the arc elasticity of demand formula.

20. Apex industries Ltd. produces a particular consumer durable good.

Fixed cost = ₹1,50,00,000

Variable cost per unit = ₹900

Selling price per unit = ₹1250

Find the following:

    (a) The break-even sales quantity

    (b) The break-even sales amount

    (c) If the actual production quantity is 2,40,000, compute

        (i) Total contribution (ii) Margin of safety (M/S)

21. Consider the following data for the year 2015 of Excel Ltd.:

Sales = ₹25,00,000

Fixed cost = ₹11,00,000

Variable cost = ₹10,00,000

Compute the following:

    (a) Total contribution

    (b) Profit

    (c) Break-even point

    (d) Margin of safety (M/S)

❑❑❑

2

Fundamentals of Mathematics and Engineering Economics

2.1 INTRODUCTION

The activities to generate income are termed as economic activities, which are responsible for the origin and development of Economics as a subject. Economics originated as a science of statecraft and the emergence of political economy. Adam Smith (Father of Economics) stated that economics is the science of wealth and Economy is concerned with the production, consumption, distribution and investment of goods and services. We might take our definition of economics from him. He titled his famous book, in 1776, as *An Inquiry into the Nature and Causes of the Wealth of Nations*. That is not a bad description of the subject matter of economics, but many economists have tried to find a more logical or scientific definition.

There are various stages and definitions of economics found in literature. The major classifications of the definition of economics are as follows:
- Wealth Definition (Adam Smith)
- Welfare Definition (Alfred Marshall)
- Scarcity Definition (L. Robbins)
- Growth Oriented Definition (Paul A. Samuelson)
- Need Oriented Definition (Jacob Viner)

Wealth Concept   Adam Smith, who is generally regarded as the father of economics, defined economics as "*a science which enquires into the nature and cause of wealth of a nation.*" He emphasized the production and growth of wealth as the subject matter of economics. This definition takes into account only material goods.

Welfare Concept   According to A. Marshall, "*Economics is a study of mankind in the ordinary business of life; it examines that part of individual and social action which is most closely connected with the attainment and with the use of material requisites of well-being*". Thus, it is on one side a study of wealth; and on other; and more important side, a part of the study of man.

Scarcity Concept   According to Lionel Robbins, "*Economics is the science which studies human behavior as a relationship between ends and scarce means which have alternate uses.*"

Growth and Development Concept   According to Paul A. Samuelson, *"Economics is the study of how men and society choose, with or without the use of money, to employ the scarce productive resources which have alternative uses, to produce various commodities over time and distribute them for consumption now and in future among various people and groups of society."*

Need Oriented Definitions   According to Jacob Viner, *"Economics is what economists do."*

The significance and advantages of studying economics constitute both theoretical as well as practical aspects. The theoretical advantages are: it increases the knowledge-base of day-to-day economic life and develops the analytical attitude to examine and understand the economic behavior and phenomena. The practical advantages of studying the economics are: it has a significance for the consumers, producers, workers, politicians, academicians, administrators and engineers in effective manpower planning, fixing prices of finished goods and commodities and solving the distribution problems of an economy.

In this chapter, an attempt has been made to introduce some of the basic terminology and key assumptions imposed in the theory of economics in general and engineering economics in particular. The standard analytical framework adopted in economics is discussed using mathematical terms, followed by the methodology for studying engineering economics, as well as some key points one should give attention to. Mathematics is a set of tools which facilitates the derivation and expositions of the economic theories. It is useful for translating verbal arguments into concise and consistent forms. It provides the economists with the set of tools often more powerful than ordinary speech. The calculus and the simple concepts of simultaneous equations are the major mathematical tools used in describing economic theories in this chapter.

2.2   THEORY OF CONSUMER BEHAVIOR

The goal of economic theory is to characterize the economic behavior based on the assumption that economic agents have stable preferences. Given those preferences, there are constraints placed on their resources and the changes in behavior are due to changes in these constraints. In this section, we use this approach to develop a theory of consumer behavior based on the simplest assumptions possible. Along the way, we develop the tool of demand analysis, which attempts to characterize how consumers, firms, governments, etc., react to changes in the constraints they face. The present section is designed to examine the role of the consumers and their behavior in the markets. The objective of a consumer is to attain maximum satisfaction derived from the commodities purchased or consumed. In other words, every rational consumer wishes to buy commodities which are available at a low price and can maximize his satisfaction.

Over the years, many theories have been developed to explain the consumer's demand for a good and thus derive a valid demand theorem. The cardinal utility analysis is the oldest theory of demand which provides an explanation of consumer's demand for a product and derives the law of demand which establishes an inverse relationship between quantity demanded of a commodity and its price, assuming other factors affecting demand remain unchanged. Contrary to the cardinal utility approach, alternative theories like the ordinal utility analysis, are also available. This chapter will discuss all the relevant approaches to explain the relationships between quantity demanded of a commodity and its price.

The traditional theory of demand starts with the examination of the behavior of the consumer, since the market demand is assumed to be the summation of the demands of individual consumers. So, let's examine the derivation of demand for an individual consumer. The consumer is assumed to be rational. Given the income and the prices of the various commodities, a consumer plans to spend his income to attain the maximum possible satisfaction or utility, which is known as the *axiom of utility maximization.* It is also assumed that the consumer has full level of knowledge of all the information relevant to his decision. To attain this objective, the consumer must be able to compare the utility, i.e., *satisfaction derived from consuming commodities*, derived from various "baskets of goods" which he or she can buy with his income.

There are two basic approaches to the problem of comparison of utilities, such as, the cardinalist and the ordinalist utility approach. The cardinalist school assumes that utility can be measured. Some economists have suggested that utility can be measured in monetary units, by the amount of money the consumer is willing to sacrifice for another unit of a commodity. Others have suggested the measurement of utility in subjective units is called *utils.* The ordinalist school assumes that utility is not measurable, but can be ranked or ordered into an ordinal magnitude. The consumer need not know in specific units the utility of various commodities to make his choice. The main ordinal theories are the *indifference curves approach* and the *revealed preference hypothesis*.

2.3   MEANING OF UTILITY

The assumption of rationality is the customary point of start in the theory of the consumer's behavior. The consumer is assumed to choose among the available alternatives in such a manner that the satisfaction derived from consuming commodities, in the broadest sense, is as large as possible. This implies that he or she is aware of the alternatives facing him or her and is capable of evaluating them. All information pertaining to the satisfaction that a consumer derives from various quantities of commodities is contained in his *utility function*.

2.3.1   Nature of the Utility Function

Assume that a consumer's purchases are limited to two commodities. Then his ordinal utility function is $U = f(q_1, q_2)$, where q_1 and q_2 are the quantities of the two commodities q_1 and q_2 which he or she consumes. It is assumed that $f(q_1, q_2)$ is continuous, has continuous first- and second-order partial derivatives, and is a regular, strictly quasi-concave function.

Furthermore, it is assumed that the partial derivatives of U are strictly positive. This means that the consumer will always desire more of both commodities *pareto sense*. Non-negative consumption levels normally constitute the domain of the utility function, though in some cases the domain is limited to positive levels.

Sometimes it is easier to work directly with the preference relation and its associated sets. Especially when one wants to use calculus methods, it is easier to work with preferences that can be represented by a utility function, *i.e.*, a function $U: Q_i \rightarrow \mathbb{R}$ such that $q_1 \geq q_2$ if and only if $U(q_1) \geq U(q_2)$. Following are the different forms of utility functions:

Cobb-Douglas Utility Function   A utility function that is used frequently for illustrative and empirical purposes is the Cobb-Douglas utility function,

$$U = f(q_1, q_2, \ldots, q_n) = q_1^{\alpha_1} \cdot q_2^{\alpha_2} \cdot q_3^{\alpha_3} \cdot \ldots \cdot q_n^{\alpha_n}$$

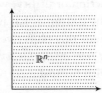

with $\alpha_n > 0$, $n = 1, \ldots, n$. This utility function (Fig. 2.1) represents a preference ordering that is continuous, strictly monotonic, and strictly convex in $\mathbb{R}_n$.

Figure 2.1   Consumption set $\mathbb{R}^n$

Linear Utility Function   A utility function that describes perfect substitution between commodities is the linear utility function,

$$U = f(q_1, q_2, \ldots, q_n) = \alpha_1 \cdot q_1 + \alpha_2 \cdot q_2 + \alpha_3 \cdot q_3 + \cdots + \alpha_n \cdot q_n$$

with $\alpha_n \geqq 0$ for all $n = 1, \ldots, n$ and $\alpha_n > 0$ for at least n. This utility function represents a preference ordering that is continuous, monotonic, and convex in $\mathbb{R}^n$.

Leontief Utility Function   A utility function that describes perfect complementary relations between commodities is the Leontief utility function,

$$U = (q_1, q_2, \ldots, q_n) = \min\{\alpha_1 \cdot q_1, \alpha_2 \cdot q_2, \alpha_3 \cdot q_3, \ldots, \alpha_n \cdot q_n\}$$

with $\alpha_n \geqq 0$ for all $n = 1, \ldots, n$ and $\alpha_n > 0$ for at least n. This represents a preference that all commodities should be used together in order to increase consumer utility. This utility function represents a preference ordering that is also continuous, monotonic, and convex in $\mathbb{R}^n$.

2.3.2   Existence of Utility Function

Each preference orderings cannot be represented by utility functions, but it can be shown that any upper semi-continuous preference ordering can be represented by an upper semi-continuous utility function. The following proposition depicts the existence of a utility function when a preference ordering is continuous and strictly monotonic.

Suppose preferences are complete, reflexive, transitive, continuous and strongly monotonic. Then there exists a continuous utility function $U: \mathbb{R}^n \rightarrow \mathbb{R}$ which represents those preferences.

Proof: Let e be the vector in $\mathbb{R}^n$ consisting of all ones. Then, given any vector x, let $U(x)$ be that number such that $x \sim U(x)e$. We have to show that such a number exists and is unique.

Let $B = \{t$ in $\mathbb{R}: te \geq x\}$ and $W = \{t$ in $\mathbb{R}: x \geq te\}$.

Then strong monotonicity implies B is non-empty; W is certainly non-empty since it contains 0. Continuity implies both sets are closed. Since the real line is connected, there is some t_x such that $t_x e \sim x$. We have to show that this utility function actually represents the underlying preferences.

Let

$$U(x) = t_x \text{ where } t_x \underline{e} \sim \text{x}$$

$$U(y) = t_y \text{ where } t_y e \sim \text{y}$$

Then if $t_x < t_y$, strong monotonicity shows that $t_x e > t_y e$, and transitivity shows that $x \sim t_x e < t_y e \sim y$.

Similarly, if $x > y$, then $t_x e > t_y e$ so that t_x must be greater than t_y.

Finally, we show that the function U defined above is continuous.

Suppose $\{x_k\}$ is a sequence with $x_k \geq x$. We want to show that $U(x_k) \geq U(x)$.

Suppose it is not. Then we can find $e > 0$ and an infinite number of k's such that $U(x_k) > U(x) + e$ or an infinite set of k's such that $U(x_k) < U(x) - e$.

Without loss of generality, let us assume the first of these.

This means that $x_k \sim U(x_k)e > (U(x) + e)e \sim x + ee$.

So by transitivity, $x_k > x + ee$.

But for a large k in our infinite set, $x + ee > x_k$, so $x + ee > x_k$, which is a contradiction.

Thus, U must be continuous.

2.3.3 The Cardinal Marginal Utility Theory

The concept of subjective measurement of utility is attributed to Gossen (1854), Jevons (1871) and Walras (1874). Marshall, in 1890, also assumed independent and additive utilities, but his position on utility is not clear in several aspects. The Cardinal Marginal Utility Theory is based on the following assumptions:

  (i) *Consumer's behavior is assumed rational*: The consumer aims at the maximization of his utility, subject to the constraint imposed by his given income.

 (ii) *Utility is assumed in Cardinal approach*: The utility of commodity is measurable and the most convenient measure is money.

(iii) *The assumption of constant marginal utility of money*: This is necessary if the monetary unit is used as the measure of utility. The essential feature of a standard unit of measurement is that it be constant.

(iv) *The diminishing marginal utility*: The utility gained from successive units of a commodity diminishes. This is the axiom of diminishing marginal utility.

(v) *Total utility*: The total utility of a basket of goods depends on the quantities of the individual commodities. If there are n commodities in the bundle with quantities $x_1, x_2, x_3, \ldots, x_n$, the total utility is $U = f(x_1, x_2, x_3, \ldots, x_n)$. In very early versions of the theory of consumer behavior, it was assumed that the total utility is additive, $U = U_1(x_1) + U_2(x_2) + \ldots + U_n(x_n)$. The additivity assumption was dropped in later versions of the cardinal utility theory. Additivity implies independent utilities of the various commodities in the bundle, an assumption clearly unrealistic, and unnecessary for the cardinal theory.

2.3.4   Equilibrium of the Consumer

Using the simple model of a single commodity x, the consumer can either buy x or retain his money income M. Under these conditions, the consumer is in equilibrium when the marginal utility of x is equated to its market price (P_x). Symbolically, we have,

$$MU_x = P_x.$$

If the marginal utility of x is greater than its price, the consumer can increase his welfare by purchasing more units of x. Similarly, if the marginal utility of x is less than its price, the consumer can increase his total satisfaction by cutting down the quantity of x and keeping more of his income unspent. Therefore, he attains the maximization of his utility when $MU_x = P_x$. If there are more commodities, the condition for the equilibrium of the consumer is the equality of the ratios of the marginal utilities of the individual commodities to their prices.

$$\frac{MUx_1}{Px_1} = \frac{MUx_2}{Px_2} = \frac{MUx_3}{Px_3} = \ldots = \frac{MUx_n}{Px_n}$$

Mathematical derivation of the equilibrium of the consumer   The utility function is,

$$U = f(Q_n^x)$$

where utility is measured in monetary units.

If the consumer buys Q_n^x, his expenditure is $Q_n^x \times P_n^x$.

Assumingly, the consumer seeks to maximize the difference between his utility and his expenditure U,

$$Q_n^x \times P_n^x$$

The necessary condition for a maximum is that the partial derivative of the function with respect to Q_n^d be equal to zero. Thus,

$$\frac{\partial U}{\partial Q_n^x} - \frac{\partial (Q_n^x \times P_n^x)}{\partial Q_n^x} = 0$$

Rearranging, we get,

$$\frac{\partial U}{\partial Q_n^x} = P_n^x$$

Or,

$$MU_n^x = P_n^x$$

The utility derived from spending an additional unit of money must be the same for all commodities. If the consumer derives greater utility from any one commodity, he can increase his welfare by spending more on that commodity and less on the others, until the above equilibrium condition is fulfilled.

2.4   MEANING OF DEMAND

Markets as allocative mechanism require non-attenuated property rights (exclusive, enforceable, transferable) and "voluntary" transactions. This includes all "potential buyers and sellers". The behavior of buyers is represented by "demand" (i.e., benefits side of the model) and the behavior of sellers is represented by "supply" (i.e., cost side of the model). The markets include all geographic boundaries of the market and they are defined by the nature of the product and the characteristics of the buyers, conditions of entry into the market and the competition and underlying substitutes.

The concept of demand is defined as "a schedule of the quantities of a good that buyers are willing and able to purchase at each possible price during a period of time, *ceteris paribus* (meaning all other things held constant)". It can also be perceived as a schedule of the maximum prices buyers are willing and able to pay for each unit of a good.

2.4.1   Demand Function

Demand function is the functional relationship between the price of the good and the quantity of that good purchased in a given time period, while income, other prices and preferences being held constant. A change in income, prices of other goods or preferences will alter (shift) the demand function.

2.4.2   Quantity Demanded

A change in the price of the good under consideration will change the quantity demanded.

$$Q_x^d = f(P_x, \text{ holding } M, P_r, \text{ preferences constant})$$

where:

M = income of the individual

P_r = prices of related goods

∂P_x causes a change in X (∂Q_x^d); this is the change in quantity demanded

2.4.3 Change in Demand

If M, P_r, or preferences change, the demand function (relationship between P_x and Q_x^d) will change. These are sometimes called *demand shifters*. There is a difference between a *change in demand* and a *change in quantity demanded*. A change in demand means *shift of the function* whereas a change in quantity demanded denotes a *move on the function*.

2.4.4 Law of Demand

The theory and empirical evidence suggest that the relationship between Price and Quantity is an inverse or negative relationship. At higher prices, quantity purchased is smaller and at lower prices the quantity purchased is greater. The demand relationship can be demonstrated as in Table 2.1.

Table 2.1 Demand relationship

Demand Schedule	
Cup of tea purchased each day between 07 to 09 AM	
Price per cup	Cup purchased
₹0.00	20
₹0.50	15
₹0.75	12.5
₹1.00	10
₹1.25	7.5
₹1.50	5
₹1.75	2.5
₹2.00	0

$\partial P_x > 0$ [+0.75]

$\partial Q_x^d < 0$ [−0.75]

Table 2.1 shows that the demand is a schedule of quantities that will be purchased at a schedule of prices during a given time period, *ceteris paribus*. As the price is increased, the quantity purchased decreases. This demand relationship can be expressed as an equation:

$$P_x = 2 - 0.1Q_x^d \quad \text{or} \quad Q = 20 - 10P: Q_x^d = f(P_x,...)$$

But we plot P_x on the Y-axis and Q_x^d on the X-axis. This is a convention due to Marshallian approach to define demand.

The demand relationship can be expressed as a table (Table 2.1) or an equation,

Either $P_x = 2 - 0.1Q_x^d$ or $Q_x^d = 20 - 10 P_x$

The data from the table or equation can be plotted on a graph as in Fig. 2.2.

The demand function can be represented as a table, an equation or a graph.

The demand equation $P_x = 2 - 0.1Q_x^d$ was graphed in Fig. 2.3. A change in quantity demanded is a movement on the demand function caused by a change in the independent variable (price).

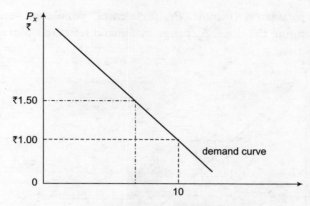

Figure 2.2 Demand curve

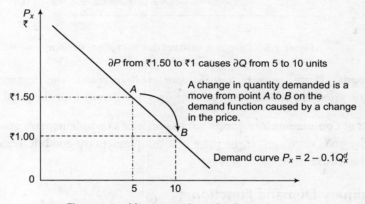

Figure 2.3 Movement along the demand curve

A change in any of the parameters (income, price of related goods, preferences, population of buyers, etc.) will cause a shift of the demand function. In this example in Fig. 2.4, the intercepts have changed, but the slope has remained constant.

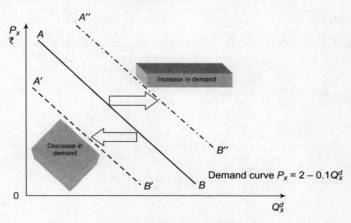

Figure 2.4 Changes in demand

A change in the parameters (income, Pr, preferences, population, etc.) might alter the relationship by changing the slope. A change in demand refers to a movement or shift of the entire demand function.

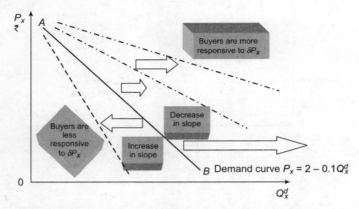

Figure 2.5   Change in demand due to changes in slope

Substitute goods   If the price of a substitute good increases, the demand for the good increases. If the price of a substitute good decreases, the demand for the good decreases.

Complements or Complementary goods   If the price of a complementary good increases, the demand for the good decreases. If the price of a complementary good decreases, the demand for the good increases.

2.4.5   Ordinary Demand Function

The ordinary demand function, also known as Marshallian demand function, gives the quantity of the commodity that a customer will buy as a function of commodity prices and income.

Assume that the utility function is $U = Q_1 \times Q_2$ and the budget constraint $M - P_1Q_1 - P_2Q_2 = 0$.

Given this expression:

$\phi = Q_1 \times Q_2 + \lambda(M - P_1Q_1 - P_2Q_2)$ and setting its partial derivatives equal to zero:

$$\frac{\partial \phi}{\partial Q_1} = Q_2 - P_1 = 0$$

$$\frac{\partial \phi}{\partial Q_2} = Q_1 - P_2 = 0$$

$$\frac{\partial \phi}{\partial \lambda} = M - P_1 \cdot Q_1 - P_2 \cdot Q_2 = 0$$

Solving for Q_1 and Q_2, we get demand functions,

$$Q_1 = \frac{M}{2P_1}$$

and

$$Q_2 = \frac{M}{2P_2}$$

2.4.6   Compensated Demand Function

Under a situation, when the government imposes taxes and gives subsidy on any commodity, the consumer behaves in such a way as to leave his utility unchanged after a price change. In other words, the consumer compensates his demand functions which drives compensated demand curve. The compensated demand curve is obtained by minimizing the consumer's expenditures, subject to the constraint that his utility is at the fixed level U^o.

Assume again that utility function is $U = Q_1 \times Q_2$.

Given this expression:

$\phi = P_1 Q_1 + P_2 Q_2 + \lambda(U^o - Q_1 \times Q_2)$ and setting its partial derivatives equal to zero:

$$\frac{\partial \phi}{\partial Q_1} = P_1 - Q_2 = 0$$

$$\frac{\partial \phi}{\partial Q_2} = P_2 - Q_1 = 0$$

$$\frac{\partial \phi}{\partial \lambda} = U^o - Q_1 \cdot Q_2 = 0$$

Solving for Q_1 and Q_2, we get demand functions,

$$Q_1 = \sqrt{\frac{U^o P_2}{P_1}}$$

and

$$Q_2 = \sqrt{\frac{U^o P_1}{P_2}}$$

These functions are homogeneous of degree zero in their corresponding prices.

2.4.7   Reasons for Downward Slope of Demand Curve

There is an inverse relationship between quantity demanded and its price, other things remaining constant. This is the reason which keeps the demand curve download sloping. Due to income and substitution effects, when the price of a commodity increases, quantity demanded of that commodity decreases. When the price of a commodity falls, the consumer can buy more quantity of the commodity with his given income. Hence, as a result of the decrease in price of a commodity, the consumer's real income increases. In other words, the consumer's purchasing power increases. The expansion in purchasing power induces the consumer to buy more of that commodity, which is called the *income effect*. Another reason is the *substitution effect* where, due to the decrease in price of the commodity, it becomes relatively cheaper than other commodities. This induces the consumer to substitute the commodity whose price has fallen for other commodities which have now become relatively dearer, thus purchasing more of the cheaper commodity. The substitution effect is more important than the income effect. Marshall defined the reason for downward sloping demand curve with the substitution effect only, ignoring the income effect. Hicks and Allen came out with an alternative theory of demand as *indifference curve*, which explains this downward sloping demand curve with the help of both income and substitution effects.

2.5 CONCEPT OF ELASTICITY

Elasticity is a concept borrowed from physics, i.e., *Hooke's law*. It is a measure of how responsive a dependent variable is to a small change in an independent variable(s). It is defined as a ratio of the percentage change in the dependent variable to the percentage change in the independent variable. It can be computed for any two related variables. It can be computed to show the effects of a change in price on the quantity demanded, i.e., a change in quantity demanded is a movement on a demand function; a change in income on the demand function for a good; a change in the price of a related good on the demand function for a good; a change in the price on the quantity supplied; and a change of any independent variable on a dependent variable.

2.5.1 Own Price Elasticity

This is sometimes also called as *price elasticity*. It can be computed at a point on a demand function or as an average (arc) between two points on a demand function. The e_p, η, ε are common symbols used to represent price elasticity. Price elasticity (e_p) is related to revenue, seeking answer to the question: "How will changes in price effect the total revenue?"

Elasticity as a measure of responsiveness: the law of demand says that as the price of a good increases, the quantity that will be bought decreases, but it does not tell us by how much. If you change price by 15%, by what percentage will the quantity purchased change? Elasticity provides this answer.

$$e_p \equiv \frac{\% \text{ change in quantity demanded}}{\% \text{ change in price}}$$

Or
$$e_p = \frac{\%\Delta Q_x^d}{\%\Delta P_x}$$

At a point in demand function it can be calculated as:

$$e_p = \frac{\%\Delta Q_x^d}{\%\Delta P_x} = \frac{\dfrac{Q_{x_2}^d - Q_{x_1}^d}{Q_{x_1}^d} \times 100}{\dfrac{P_{x_2} - P_{x_1}}{P_{x_1}} \times 100} = \frac{\dfrac{Q_{x_2}^d - Q_{x_1}^d}{Q_{x_1}^d}}{\dfrac{P_{x_2} - P_{x_1}}{P_{x_1}}} = \frac{\dfrac{\Delta Q_x^d}{Q_{x_1}^d}}{\dfrac{\Delta P_x}{P_{x_1}}} = \frac{\Delta Q_x^d}{Q_{x_1}^d} \times \frac{P_{x_1}}{\Delta P_x}$$

The price elasticity is always negative, but we take the modulus of the calculated value of price elasticity. $|e_p| > 1$ means the demand of a good is relatively price elastic; $|e_p| < 1$ means demand of a good is relatively price inelastic; $|e_p| = 1$ means demand of the good is unitary price elastic; $|e_p| = \infty$ means the demand of the good is perfectly price elastic; $|e_p| = 0$ means demand of a good is perfectly price inelastic.

It is important to notice that at higher prices, the absolute value of the price elasticity of demand, $|e_p|$, is greater. Total revenue (TR) is price times quantity; TR $= P_x \times Q_x^d$. Where the total revenue is a maximum, $|e_p|$ is equal to 1. In the range where $|e_p| < 1$, (less than 1 or inelastic), TR increases as price increases, TR decreases as P decreases. In the range where $|e_p| > 1$, (greater than 1 or elastic), TR decreases as price increases, TR increases as P decreases.

To solve the problem of a point elasticity that is different for every price quantity combination on a demand function, arc price elasticity can be used. This arc price elasticity is an average or midpoint elasticity between any two prices. Typically, the two points selected would be representative of the usual range of prices in the time frame under consideration. The formula to calculate the average or arc price elasticity is:

$$e_p = \frac{\Delta Q_x^d}{Q_{x_1}^d + Q_{x_2}^d} \times \frac{P_{x_1} + P_{x_2}}{\Delta P_x}$$

Relationship between elasticity of demand and total revenue can also be presented using a graph. For example, graphing $Q_x^d = 120 - 4\,P_x$, when e_p is -1, TR is a maximum. When $|e_p| > 1$ [elastic], TR and P_x move in opposite directions (P_x has a negative slope, TR a positive slope). When $|e_p| < 1$ [inelastic], TR and P_x move in the same direction (P_x and TR both have a negative slope). Arc or average e_p is the average elasticity between two point (and prices); point e_p is the elasticity at a point or price. Price elasticity of demand describes how responsive buyers are to change in the price of the good. The more elastic, the more responsive to ΔP_x.

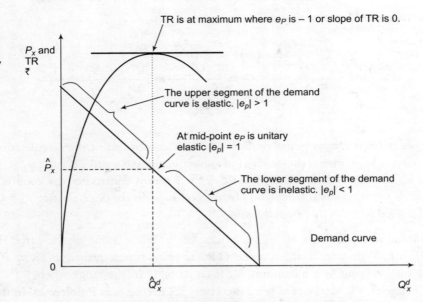

Figure 2.6 Price elasticity of demand and total revenue

2.5.2 Determinants of Price Elasticity

Availability of substitutes affects elasticity: greater availability of substitutes makes a good relatively more elastic. The proportion of the expenditures on the good to the total budget affects elasticity: lower proportion tends to increase relative elasticity. Time to adjust to the price changes affects elasticity: longer time period means there are more adjustment possible and increases relative elasticity. Price elasticity for "brands" tends to be more elastic than for other category of goods.

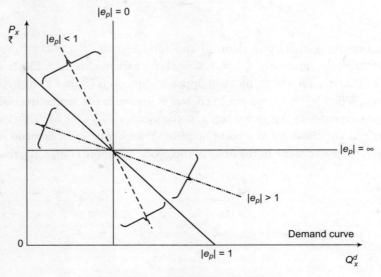

Figure 2.7 Price elasticity of demand

Fig. 2.7 shows various zones under which the demand curve may pass through. Given that, the corresponding elasticity of demand is marked.

2.5.3 Income Elasticity of Demand

Income elasticity is a measure of the change in demand (a *shift* of the demand function) that is caused by a change in income.

$$e_y \equiv \frac{\% \text{ change in quantity demanded}}{\% \text{ change in income}}$$

Or
$$e_y = \frac{\% \Delta Q_x^d}{\% \Delta Y}$$

At a point in the demand function, it can be calculated as:

$$e_y = \frac{\% \Delta Q_x^d}{\% \Delta Y} = \frac{\dfrac{Q_{x_2}^d - Q_{x_1}^d}{Q_1^d} \times 100}{\dfrac{Y_2 - Y_1}{Y_1} \times 100} = \frac{\dfrac{Q_{x_2}^d - Q_{x_1}^d}{Q_1^d}}{\dfrac{Y_2 - Y_1}{Y_1}} = \frac{\dfrac{\Delta Q_x^d}{Q_1^d}}{\dfrac{\Delta Y}{Y_1}} = \frac{\Delta Q_x^d}{Q_1^d} \times \frac{Y_1}{\Delta Y}$$

where y = income

The increase in income, ΔY, increases demand to $A''B''$. The increase in demand results in a larger quantity being purchased at the same Price (P_1). At a price of P_1, the quantity demanded, given the demand curve AB, is Q_1. AB is the demand function when the income is Y_1. For a *normal good* an increase in income to Y_2 will *shift* the demand to the right. This is an increase in demand to $A''B''$. If $\% \Delta Y > 0$ and $\% \Delta Q > 0$, then $e_y > 0$ (it is positive).

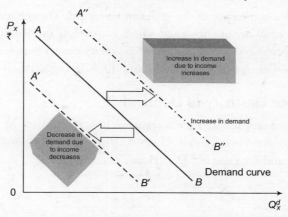

Figure 2.8 Changes in demand due to changes in other demand determinants

A decrease in income is associated with a decrease in the demand for a normal good. For a decrease in income $[-\Delta Y]$, the demand decreases, i.e., shifts to the left; at the price (P_1), a smaller Q_2 will be purchased. At income Y_1, the demand AB represents the relationship between P and Q. At a price (P_1) the quantity (Q_1) is demanded. Since $\% \Delta Y < 0$ (negative) and $\%\Delta Q < 0$ (negative), so, $e_y > 0$ is positive. For either an increase or decrease in income, e_y is positive. A positive relationship (positive correlation) between ΔY and ΔQ is evidence of a normal good.

When income elasticity is positive, the good is considered a *normal good*. An increase in income is correlated with an increase in the demand function. A decrease in income is associated with a decrease in the demand function. For both increases and decreases in income, e_y is positive.

$$+e_y \equiv \frac{-\%\Delta Q_x^d}{-\%\Delta Y}$$

The greater the value of e_y, the more responsive buyers are to a change in their incomes. When the value of e_y is greater than 1, it is called a *superior good*. The $|\%\Delta Q_x^d|$ is greater than the $|\%\Delta Y|$. Buyers are very responsive to changes in income. Superior goods are also called *luxury goods*.

There is another classification of goods where changes in income shift the demand function in the *opposite* direction. An increase in income $(+\Delta Y)$ reduces demand. An increase in income reduces the amount that individuals are willing to buy at each price of the good. Income elasticity is negative $(-e_y)$. The greater the absolute value of $-e_y$, the more responsive buyers are to changes in income.

Decreases in income increase the demand for inferior goods. A decrease in income $(-\Delta Y)$ increases the demand. A decrease in income $(-\Delta Y)$ results in an increase in demand, the income elasticity of demand is negative. For both increases and decreases in income, the income elasticity is negative for inferior goods. The greater the absolute value of e_y, the more responsive buyers are to changes in income.

Income elasticity (e_y) is a measure of the effect of an income change on demand. When $e_y > 0$ (positive), it is a normal or superior good where an increase in income increases demand, a decrease in income decreases demand. If income elasticity ranges between $0 > e_y > 1$. then it is a normal good; if $e_y > 1$ then it is a superior good; if $e_y < 0$, (negative) it is an inferior good.

2.5.4   Cross-Price Elasticity of Demand

Cross-price elasticity of demand (e_{xy}) is a measure of how responsive the demand for a good is to changes in the prices of related goods. Given a change in the price of good Y is P_y, what is the effect on the demand for good X, (Q_x^d)? Here e_{xy} is defined as:

$$e_{xy} \equiv \frac{\%\Delta Q_x^d}{\%\Delta P_y}$$

$$e_{xy} = \frac{\%\Delta Q_x^d}{\%\Delta P_y} = \frac{\dfrac{Q_{x2}^d - Q_{x1}^d}{Q_1^d} \times 100}{\dfrac{P_{y2} - P_{y1}}{P_{y1}} \times 100} = \frac{\dfrac{Q_{x2}^d - Q_{x1}^d}{Q_1^d}}{\dfrac{P_{y2} - P_{y1}}{P_{y1}}} = \frac{\dfrac{\Delta Q_x^d}{Q_1^d}}{\dfrac{\Delta P_y}{P_{y1}}} = \frac{\Delta Q_x^d}{Q_1^d} \times \frac{P_{y1}}{\Delta P_y}$$

If goods are substitutes, e_{xy} will be positive. The greater the coefficient, the more likely they are good substitutes. For complements, the cross elasticity is negative for price increase or decrease. If $e_{xy} > 0$ (positive), it suggests substitutes; the higher the coefficient the better the substitute. If $e_{xy} < 0$ (negative), it suggests the goods are complements, the greater the absolute value the more complementary the goods are. If $e_{xy} = 0$, it suggests the goods are not related. The e_{xy} can be used to define markets in legal proceedings.

2.5.5 Engel Curve and Income Elasticity

The own elasticity of demand for $Q_x^d (\varepsilon_p)$ is defined as the proportionate rate of change of Q_x^d divided by the proportionate rate of change of its own price with P_2 and y^0 constant:

$$\varepsilon_p = \frac{\partial (\ln Q_x^d)}{\partial (\ln P_1)} = \frac{P_1}{Q_{x_1}^d} \frac{\partial Q_x^d}{\partial P_1}$$

A numerically large value for elasticity implies that quantity is proportionately very responsive to price changes. Commodities which have numerically high elasticities $\varepsilon_p < -1$ are called luxuries, whereas those with numerically small elasticities ($\varepsilon_p > -1$) are called necessities. Price elasticities of demand are pure numbers independent of the units in which prices and outputs are measured. The elasticity ε_p is negative if the corresponding demand curve is downward sloping.

The consumer's expenditure on $Q_{x_1}^d$ is $P_1 Q_{x_1}^d$, and

$$\frac{\partial (P_1 Q_{x_1}^d)}{\partial P_1} = Q_{x_1}^d + P_1 \frac{\partial Q_{x_1}^d}{\partial P_1} = Q_{x_1}^d \left(1 + \frac{P_1}{q_1} \frac{\partial Q_{x_1}^d}{\partial p_1} \right) = Q_{x_1}^d (1 + \varepsilon_p)$$

The consumer's expenditures on $Q_{x_1}^d$ will increase with P_1, if $\varepsilon_p > -1$, remain unchanged if $\varepsilon_p = -1$, and decrease if $\varepsilon_{11} < -1$.

A cross-price elasticity of demand for the ordinary demand function relates the proportionate change in one quantity to the proportionate change in the other price. For example,

$$\varepsilon_{xy} = \frac{\partial (\ln Q_x^d)}{\partial (\ln P_y)} = \frac{P_y}{Q_x^d} \frac{\partial Q_x^d}{\partial Q_y^d}$$

Cross-price elasticities may be either positive or negative. Taking the total differential of the budget constraint ($M - P_x Q_x^d - P_y Q_y^d = 0$) and letting $d\gamma^0 = dp_y = 0$,

$$P_y dQ_y^d + Q_y^d dP_y + P_x dP_x = 0$$

Multiplying through by $P_y Q_y^d Q_x^d / \gamma^0 Q_y^d Q_x^d dP_y$, and rearranging terms, $\alpha_y \varepsilon_p + \alpha_y \varepsilon_{xy} = -\alpha_1$ where $\alpha_y = P_y Q_y^d / \gamma^0$ and $\alpha_x = P_x Q_x^d / \gamma^0$ are the proportions of total expenditures for the two goods.

The equation $(P_y dQ_y^d + Q_y^d dP_y + P_x dP_x = 0)$ is called the *Cournot aggregation condition*. If the own-price elasticity of demand for Q_y^d is known, the Cournot aggregation condition can be used to evaluate the cross-price elasticity of demand for Q_x^d.

If $\varepsilon_p = -1$, $\varepsilon_{xy} = 0$. If $\varepsilon_p = -1$, $\varepsilon_{xy} > 0$, and if $\varepsilon_p > -1$, $\varepsilon_{xy} < 0$.

Own, and cross-price elasticities of demand for compensated demand functions can be defined in an analogous manner by inserting compensated rather than ordinary demand functions in $(Q_1 - P_2 = 0)$ and $\left(\varepsilon_{xy} = \dfrac{\partial(\ln Q_x^d)}{\partial(\ln P_y)} = \dfrac{P_y}{Q_x^d} \dfrac{\partial Q_x^d}{\partial Q_y^d} \right)$.

Equation $(\alpha_y \varepsilon_p + \alpha_y \varepsilon_{xy} = -\alpha_1)$ does not hold for compensated demand functions. Taking the total differential of the utility function and letting $dU = 0$, $f_y dQ_y^d + f_x Q_y^d dP_x = 0$

Using the first-order condition $P_y/P_x = f_y/f_x$, multiplying through by $P_y Q_y^d Q_x^d / \gamma^0 Q_y^d Q_x^d dP_y$ and rearranging terms, $\alpha_y \xi_y + \alpha_x \xi_{xy} = 0$ where the compensated price elasticities are denoted by ξ_y and ξ_{xy}. Since $\xi_y = 0$, it follows that $\xi_{xy} > 0$. Returning to the example $U = Q_x^d \cdot Q_y^d$, the own- and cross-price elasticities for the ordinary demand function are:

$$\varepsilon_{11} = -\frac{p_1}{q_1} \frac{\gamma^0}{2p_1^2} = -\frac{p_1}{\gamma^0 / 2p_1} \frac{\gamma^0}{2p_1^2} = -1$$

$$\varepsilon_{21} \frac{p_1}{p_2} 0 = 0$$

This is a special case. Not all demand functions have unit own and zero cross elasticities or even constant elasticities. In general, elasticities are a function of p_1, p_2 and γ^0. The reader can verify that the compensated elasticities for this example are $\xi_y = -\frac{1}{2}$ and $\xi_{xy} = \frac{1}{2}$.

An income elasticity of demand for an ordinary demand function is defined as the proportionate change in the purchases of a commodity relative to the proportionate change in income with prices constant:

$$\eta_y = \frac{\partial(\ln Q_y^d)}{\partial(\ln Y)} = \frac{y}{q^1} \frac{\partial \varphi(P_y, P_x, Y)}{\delta Y}$$

where n_y denotes the income elasticity of demand for Q_y^d. Income elasticities can be positive, negative, or zero, but are normally assumed to be positive. Taking the total differential of the budget constraint,

$$P_y dQ_y^d + P_x dP_x = dY$$

Multiplying through by Y/Y, multiplying the first term on the left by Q_y^d/Q_y^d, the second by Q_x^d/Q_x^d, and dividing through by dY, $\alpha_y \eta_y + \alpha_x \eta_x = 1$ which is called the *Engel aggregation condition*. The sum of the income elasticities weighted by total expenditure proportions equals to unity. Income elasticities cannot be derived for compensated demand functions since income is not an argument of these functions.

2.5.6 Relationship between Price Elasticity and Marginal Revenue

The crucial relationship for the theory of pricing is that the marginal revenue is related to the price elasticity of demand. Mathematically, the relationship is as follows:

$$MR = P_x \times \left(1 - \frac{1}{\varepsilon_p}\right)$$

Let the demand function be $P_x = f(Q_x^d)$. The Total Revenue is $TR = P_x \times Q_x^d = [f(Q_x^d)] \times Q_x^d$.

Given the TR, $MR = \dfrac{d(P_x \times Q_x^d)}{dQ_x^d} = P_x \times \dfrac{dQ_x^d}{dQ_x^d} + Q_x^d \times \dfrac{dP_x}{dQ_x^d} = P_x + Q_x^d \times \dfrac{dP_x}{dQ_x^d}$

Whereas the price elasticity of demand is $\varepsilon_p = -\dfrac{dQ_x^d}{dP_x} \times \dfrac{P_x}{Q_x^d}$

Rearranging, we get

$$-\varepsilon_p \times \frac{Q_x^d}{P_x} = \frac{dQ_x^d}{dP_x}$$

$$\frac{1}{-\varepsilon_p} \times \frac{P_x}{Q_x^d} = \frac{dP_x}{dQ_x^d}$$

Substituting $\dfrac{dP_x}{dQ_x^d}$ in MR, we get

$$MR = P_x + Q_x^d \times \frac{dP_x}{dQ_x^d} = P_x + Q_x^d \times \left(\frac{1}{-\varepsilon_p} \times \frac{P_x}{Q_x^d}\right) = P_x + \left(\frac{P_x}{-\varepsilon_p}\right) = P_x \times \left(1 - \frac{1}{\varepsilon_p}\right)$$

2.6 LAW OF DIMINISHING MARGINAL UTILITY

In neoclassical economics, the goal of consumer behavior is utility maximization. This is consistent with maximization of net benefits. Consumer choice among various alternatives is subject to constraints: income or budget, prices of goods purchased, and preferences. Total utility (*TU*) is defined as the amount of utility an individual derives from consuming a given quantity of a good during a specific period of time.

$$TU = f(Q, \text{preferences, and others factors})$$

Table 2.2   Utility schedule

Utility Schedule		
Q_x	TU_x	MU_x
1	20	12
2	30	10
3	39	9
4	47	8
5	54	7
6	60	6
7	65	5
8	65	0
9	55	-10

Based on Table 2.2, the Total Utility curve is plotted in Fig. 2.9.

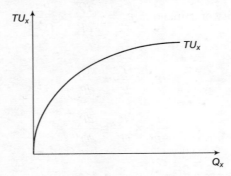

Figure 2.9   Total utility curve

Nature of Total Utility   When more and more units of a good are consumed in a specific time period, the utility derived tends to increase at a decreasing rate. Eventually, some maximum utility is derived and additional units cause total utility to diminish. As an example, think of eating free burgers. It is possible for total utility to initially increase at an increasing rate.

Marginal Utility   Marginal utility (MU_x) is the change in total utility associated with a 1 unit change in consumption. As total utility increases at a decreasing rate, MU_x declines. As total utility declines, MU_x is negative. When TU_x is a maximum, MU_x is 0. This is sometimes called the *satiation point* or the point of *absolute diminishing utility*.

Marginal Utility (MU_x) is the change in total utility (ΔTU_x) caused by a one unit change in quantity (ΔQ_x).

$$MU_x = \frac{\partial TU_x}{\partial Q_x}$$

Remember that the MU_x is associated with the midpoint between the units as each additional unit is added.

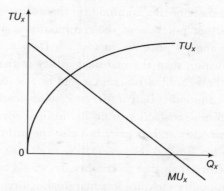

Figure 2.10   Total utility and marginal utility curve

The MU_x is the slope of TU_x or the rate of change in TU_x associated with a one unit change in quantity. Using calculus, we can say that MU_x is the change in TU_x as change in quantity approaches 0. Where $MU_x = 0$, TU_x is a maximum. Both MU_x and TU_x are determined by the *preferences* or utility function of the individual and the quantity consumed.

Utility cannot be measured directly, but individual choices reveal information about the individual's preferences. Surrogate variables like, age, gender, ethnic background, religion, etc., may be correlated with preferences. There is a tendency for TU_x to increase at a decreasing rate (MU_x declines) as more of a good is consumed in a given time period, i.e., diminishing marginal utility.

Initially, it may be possible for TU_x to increase at an increasing rate, in which case MU_x will increase; MU_x is the slope of TU_x which is increasing. Eventually, as more and more of a good is consumed in a given time period, TU_x continues to increase but at a decreasing rate, in which case MU_x decreases. This is called the *point of diminishing marginal utility*.

2.7   PRINCIPLE OF EQUI-MARGINAL UTILITY

The extension of the law of diminishing marginal utility applied to two or more commodities is the *law of equi-marginal utility*. The law of equi-marginal utility is also known with various other names such as *law of equilibrium utility*, *law of substitution*, *law of maximum satisfaction*, *law of indifference*, *proportionate rule* and *Gossen's second law*. In *cardinal utility analysis*, this law is stated by *Lipsey*. According to this, "the household maximizing the utility will so allocate the expenditure between commodities that the utility of the last penny spent on each item is equal." A consumer's wants are unlimited. However, the disposable income at any time is limited. Therefore, he faces a choice among many commodities that he would like to pay. He consciously (or unconsciously) compresses the satisfaction which he obtains from the purchase of the commodity and the price which he pays for it. If he thinks the utility of the commodity is greater or at least equal to the loss of utility of the money paid, he buys that commodity.

As he consumes more and more of that commodity, the utility of the successive units starts diminishing. He stops further purchase of the commodity at a point where the marginal utility of the commodity and its price are just equal. If he continues to purchase further from this point of equilibrium, then the marginal utility of the commodity would be less than that of price and the household satisfaction would be at a loss. A consumer would be in equilibrium with a single commodity only if $MU_x = P_x$ mathematically. A prudent consumer, in order to get the maximum satisfaction from his limited means, compares not only the utility of a particular commodity and its price but also the utility of the other commodities which he could buy with his limited resources. If he finds that a particular expenditure in one use is yielding less utility than that in another, he would attempt to transfer a unit of expenditure from the commodity yielding less marginal utility. The consumer would reach his equilibrium position when it would not be possible for him to increase the total utility by using different goods. The position of equilibrium would be reached when the marginal utility of each good is in proportion to its price and the ratio of the prices of all goods is equal to the ratio of their respective marginal utilities. The consumer would then maximize the total utility from his income when the utility from the last rupee spent on each good is the same. Symbolically,

$$\frac{MUx_1}{Px_1} = \frac{MUx_2}{Px_2} = \frac{MUx_3}{Px_3} = \ldots = \frac{MUx_n}{Px_n}$$

where $x_1, x_2, x_3, \ldots, x_n$ are various commodities consumed.

The law of equi-marginal utility is based on following assumptions:

(i) The marginal utilities of different commodities are independent of one another and diminish with more and more purchases.

(ii) The marginal utility of money remains constant to the consumer as he spends more and more of it on the purchase of goods.

(iii) The utility derived from consuming commodities is measured cardinally.

(iv) The consumer's behavior is rational.

This law is also known as the *law of maximum satisfaction* because a consumer tries to get the maximum satisfaction from his limited resources by planning his expenditure so that the marginal utility of a rupee spent in one use is the same as the marginal utility of a rupee spent in another use. This law is also known as the *law of substitution* because the consumer continues substituting one good for another till he gets the maximum satisfaction. This law is also known as the *law of indifference* because the maximum satisfaction has been achieved by equating the marginal utility in all the uses. Then the consumer becomes indifferent to readjust his expenditure unless some change takes place in his income or the prices of the commodities, or any other determinants.

The law of equi-marginal utility is of great practical importance. The application of the principle of substitution extends over almost every field of economic enquiry. Every consumer who is consciously trying to get the maximum satisfaction from his limited resources acts upon this principle of substitution. The same is the case with the producer. In the field of exchange and in the theory of distribution, too, this law plays a vital role. In short, despite its limitations, the law of maximum satisfaction is a meaningful general statement of how consumers behave.

In addition to its application to consumption, the law of equi-marginal utility applies equally well to the theory of production and the theory of distribution. In the theory of production, it is applied to the substitution of various factors of production to the point where marginal return from all the factors are equal. The government can also use this analysis for evaluation of its different economic policies. The equal marginal rule guides an individual in the spending of his savings on different types of assets. The law of equi-marginal utility also guides an individual in the allocation of his time between work and leisure.

2.8 INDIFFERENCE CURVES THEORY AND ORDINAL UTILITY THEORY

Hicks and Allen (1934) developed a theory of utility which was the juxtaposition of ordinal utility analysis. The basic tool of ordinal utility analysis is the indifference curve (IC) approach where the consumer is only capable of comparing the different levels of satisfaction. In such a case, he may not be able to quantify the exact amount of satisfaction, but utility can be defined on relative sense: whether the new basket is superior or better than the earlier basket of commodities in consumption. To define the equilibrium state of the consumer given a particular basket of commodities, it is important to introduce the concept of indifference curves and of their slope, and the budget line concept.

2.8.1 Indifference Curves

A consumer's preference among consumption bundles may be illustrated with indifference curves. An indifference curve shows bundles of goods that make the consumer equally happy. In other words, the locus of combination of basket of goods consumed by an individual giving same level of satisfaction is known as indifference curve. The consumer is indifferent, or equally happy, with the combinations shown at points A, B, and C because they are all on the same curve. The slope at any point on an indifference curve is the marginal rate of substitution. This is the rate at which a consumer is willing to substitute one good for another. It is the amount of one good that a consumer requires as a compensation to give up one unit of the other good.

The indifference curves theory is based on the following assumptions:

(i) The consumer is assumed to be rational as he aims at the maximization of his utility at his given income and market prices.

(ii) There is perfect knowledge of the relevant information on the commodities and their prices to the consumer.

(iii) The utility is defined as ordinal where a consumer can rank his preferences according to the satisfaction of each combination of goods.

(iv) The marginal rate of substitution of one good for another is diminishing. Preferences are ranked in terms of the indifference curves and assumed to be convex to the origin. The slope of the indifference curve is called the marginal rate of substitution of the commodities (MRS or RCS).

(v) The total utility of the consumer depends on the quantities of the commodities consumed.

$$U = f(Q_1^x, Q_2^x, Q_3^x, ..., Q_n^x, Q_1^y, Q_2^y, Q_3^y, ..., Q_n^y)$$

(vi) The choice of the consumer is consistent and transitive.
Symbolically,
if $A > B$, then $B > A$ (consistency assumption)
if $A > B$, and $B > C$, then $A > C$ (transitivity assumption)

2.8.2 Nature of Consumer Preferences

The consumer is assumed to have preferences on the consumption bundles in X so that he could compare and rank various commodities available in the economy. When we write $A \geq B$, we mean "the consumer thinks that the bundle A is at least as good as the bundle B". We want the preferences to order the set of bundles. Therefore, we need to assume that they satisfy the following standard properties:

(i) *Complete*: For all A and B in X, either $A \geq B$ or $B \geq A$ or both.

(ii) *Reflexive*: For all A in X, $A \geq A$.

(iii) *Transitive*: For all A, B and C in X, if $A \geq B$ and $B \geq C$, then $A \geq C$.

The *first* assumption just says that any two bundles can be compared, the *second* is trivial and says that every consumption bundle is as good as itself, and the *third* requires the consumer's choice be consistent. A preference relation that satisfies these three properties is called a *preference ordering*.

Given an ordering B describing "weak preference", we can define the *strict preference* by $A > B$ to mean not $B \geq A$. We read $A > B$ as "A is strictly preferred to B". Similarly, we define a notion of indifference by $A \sim B$ if and only if $A \geq B$ and $B \geq A$. The set of all consumption bundles that are indifferent to each other is called an indifference curve. For a two-goods case, the slope of an indifference curve at a point measures marginal rate of substitution between goods Q_1^x and Q_2^x. For a L-dimensional case, the marginal rate of substitution between two goods is the slope of an indifference surface, measured in a particular direction. The other assumptions on consumers' preferences are:

(iv) *Continuity*: For all B in X, the upper and lower segments are closed. It follows that the strictly upper and lower segments are open sets.

An interesting preference ordering is the so-called *lexicographic ordering* defined on $\mathbb{R}$. Essentially the lexicographic ordering compares the components one at a time, beginning with the first, and determines the ordering based on: the first time a different component is found, the vector with the greater component is ranked highest.

There are two more assumptions, namely, monotonicity and convexity, which are often used to guarantee sound behavior of consumer demand functions. Following are the various types of monotonicity properties used in the consumer theory:

(v) *Weak Monotonicity*: If $Q_1^x \geqq Q_2^x$ then $Q_1^x \geq Q_2^x$.

(vi) *Monotonicity*: If $Q_1^x > Q_2^x$, then $Q_1^x > Q_2^x$ always.

(vii) *Strong Monotonicity*: If $Q_1^x \geqq Q_2^x$ and $Q_1^x \neq Q_2^x$ then $Q_1^x > Q_2^x$.

Following are the assumption weaker than either kind of monotonicity or strong monotonicity:

(viii) *Local Non-Satiation*: Given any Q_1^x in A and any $\varepsilon_p > 0$, then there is some bundle Q_2^x in A with $|Q_1^x - Q_2^x| < \varepsilon$ such that $Q_2^x > Q_1^x$.

(ix) *Non-Satiation*: Given any Q_1^x in A, then there is some bundle Q_2^x in A such that $Q_2^x > Q_1^x$.

The convexity properties used in the consumer theory are as follows:

(x) *Strict Convexity*: Given Q_1^x, Q_2^x in A such that $Q_2^x \geq Q_1^x$, then it follows that $\tau Q_2^x + (1 - \tau) Q_2^x > y$ for all $0 < \tau < 1$.

(xi) *Convexity*: Given Q_1^x, Q_2^x in A such that $Q_2^x > Q_1^x$, then it follows that $\tau Q_1^x + (1 - \tau) Q_2^x > Q_1^x$ for all $0 < \tau < 1$.

(xii) *Weak Convexity*: Given Q_1^x, Q_2^x in A such that $Q_2^x > Q_1^x$, then it follows that $\tau Q_1^x + (1 - \tau) Q_2^x \geqq Q_1^x$ for all $0 \leqq \tau < 1$.

2.8.3 Indifference Map

The indifference map shows all the indifference curves on two-dimensional coordinate space, which rank the preferences of the consumer. Any point on the same indifference curve gives an equal level of satisfaction. An indifference curve which is near origin of the two-dimensional coordinate space will always have lower level of satisfaction compared to an indifference curve which is farther or farthest from the origin. Fig. 2.11 shows the indifference map.

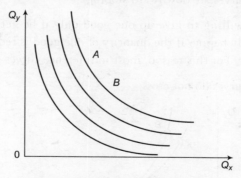

Figure 2.11 The indifference map

2.8.4 Rate of Commodity Substitution

Both commodities on x- and y-axis can be substituted one for another. This becomes the basis of indifference curve being rectangular hyperbolic and downward sloping. The slope of the indifference curve at any point on it is negative, which is known as the marginal rate of substitution or rate of commodity substitution. For any line passing tangent to the indifference curve, at the point of tangency, the slope of the indifference curve is equal to $-\dfrac{dQ_y}{dQ_x} = MRS_{xy} = RCS_{xy}$. The concept of marginal utility is implicit in the MRS_{xy} or RCS_{xy} which is equal to the ratio of the marginal utilities derived from the commodity x and y. Symbolically,

$$MRS_{xy} = RCS_{xy} = \frac{MU_x}{MU_y}$$

or

$$MRS_{yx} = RCS_{yx} = \frac{MU_y}{MU_x}$$

This implies that the diminishing marginal rate of substitution and indifference curve is convex to the origin.

2.8.5   Properties of ICs

Properties of Indifference Curves   The properties of usual indifference curve are: higher indifference curves are preferred to lower ones; indifference curves are downward sloping; indifference curves do not cross; and indifference curves are bowed inward (that is, convex to the origin)).

Property 1: Higher indifference curves are preferred to lower ones.

Remark: Consumers usually prefer more of something rather than less of it. The higher indifference curves represent larger quantities of goods than the lower indifference curves.

Property 2: Indifference curves are downward sloping.

Remark: A consumer is willing to give up one good only if he gets more of the other good in order to remain equally happy. If the quantity of one good is reduced, the quantity of the other good must increase. For this reason, most indifference curves slope downward.

Property 3: Indifference curves do not cross.

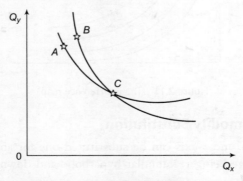

Figure 2.12   Intersecting indifference curve

Property 4: Indifference curves are bowed inward.

People are more willing to trade away goods that they have in abundance and less willing to trade away goods of which they have little.

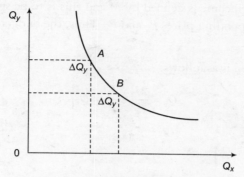

Figure 2.13 Indifference curve

Perfect Substitutes: If two commodities are perfect substitutes to each other, then the indifference curve becomes a straight line with a negative slope. Fig. 2.14 shows a perfect substitute case.

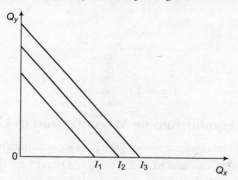

Figure 2.14 Perfect substitute

Perfect Complements: If two commodities are perfect complements to each other, then the indifference curve would be the shape of a right angle. Fig. 2.15 shows a perfect complement case.

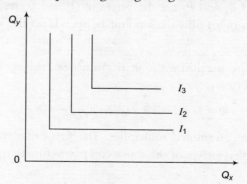

Figure 2.15 Perfect complement case

2.8.6   Budget Line

The consumer has a given level of income which restricts him to attain the maximum level of satisfaction. The income of the individual acts as a constraint on the maximization of utility. Suppose an individual's income is defined by M and this is to be spent on two commodities, say, x and y, with corresponding prices P_x and P_y. Thus, the budget constraint would be $M = P_x Q_x - P_y Q_y$.

The budget line equation is as follows:

$$Q_y = \frac{M}{P_y} - \frac{P_x}{P_y} \times Q_x$$

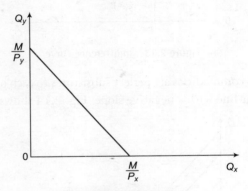

Figure 2.16   Budget line

2.8.7   Consumer's Equilibrium or Maximization of Utility

The rational behavior of the consumer is to purchase a combination of Q_x and Q_y from which he derives the highest level of satisfaction. He wants to maximize the utility. With his limited income, he is not able to purchase unlimited amounts of the commodities. The consumer's budget constraint can be written as:

$$M = p_x q_x + p_y q_y$$

where M is income and P_x and P_y are the prices of Q_x and Q_y respectively. The amount he spends on x commodity $(p_x q_x)$ plus the amount he spends on the y commodity $(p_y q_y)$ equals his income M.

The consumer's desires to maximize the utility function, subject to the budget constraint, using the Lagrange function is:

$$\phi = f(q_x, q_y) + \lambda(M - p_x q_x - p_y q_y)$$

where λ is as yet an undetermined multiplier. The first-order conditions are obtained by setting the first partial derivatives of the above composite function with respect to q_x, q_y and λ equals to zero:

$$\frac{\partial \phi}{\partial q_x} = f_1 - \lambda p_x = 0$$

$$\frac{\partial \phi}{\partial q_y} = f_2 - \lambda p_y = 0$$

$$\frac{\partial \phi}{\partial \lambda} = M - p_x q_x - p_y q_y = 0$$

Transposing the second terms in the first two equations to the right and dividing the first by the second yields:

$$\frac{f_1}{f_2} = \frac{p_x}{p_y}$$

The ratio of the marginal utilities must be equal to the ratio of prices of both the commodities for a maximum. Since f_1/f_2 is the *MRS*, the first-order condition for a maximum is expressed by the equality of the *MRS* and the price ratio. The first two equations of the partial derivatives may also be written as:

$$\frac{f_1}{p_x} = \frac{f_2}{p_y} = \lambda$$

Here, f_1 and f_2 are the marginal utility derived from commodities x and y, which when divided by price must be the same for all commodities. This ratio gives the rate at which satisfaction would increase if an additional rupee was spent on a particular commodity. If more satisfaction could be gained by spending an additional rupee on Q_x rather than Q_y, the consumer would not be maximizing utility. The Lagrange multiplier λ can be interpreted as the marginal utility of income. Since the marginal utilities of commodities are assumed to be positive, the marginal utility of income is positive.

The second-order condition as well as the first-order condition must be satisfied to ensure that a maximum is actually reached. The second direct partial derivatives of the utility function is denoted by f_{11} and f_{22} and the second cross partial derivatives are denoted by f_{12} and f_{21}. The second-order condition for a constrained maximum requires that the relevant bordered-Hessian determinant should be positive.

$$\begin{vmatrix} f_{11} & f_{12} & -p_x \\ f_{21} & f_{22} & -p_y \\ -p_x & -p_y & -0 \end{vmatrix} > 0$$

On expanding the above matrix, we get:

$$2f_{12}p_x p_y - f_{11}p_y^2 - f_{22}p_x^2 > 0$$

Substituting $p_x = f_1/\lambda$ and $p_y = f_2/\lambda$ into above equation and multiplying through by $\lambda^2 > 0$:

$$2f_{12}f_1f_2 - f_{11}f_2^2 - f_{22}f_1^2 > 0$$

The above equation satisfies the assumption of regular strict quasi-concavity. This assumption ensures that the second-order condition is satisfied at any point at which the first-order condition is satisfied.

2.8.8   Alternative Method of Utility Maximization

Given the market prices and his income, the consumer aims at the maximization of his utility. Assume that there are n commodities available to the consumer, with given market prices $P_1, P_2,..., P_n$. The consumer has a money income (M), which he spends on the available commodities.

Formally, the problem may be states as follows:

Maximize $U = (q_1, q_2,..., q_n)$

subject to $\sum_{i=1}^{n} q_i P_i = q_1 P_1 + q_2 P_2 + \cdots + q_n P_n = M$

We use the Lagrangian multiplier method for the solution of this constrained maximum. The steps involved in this method are outlined as follows:

Rewriting the constraint in the form:

$$(q_1 P_1 + q_2 P_2 + \cdots + q_n P_n - M) = 0$$

Multiplying the constraint by a constant λ, which is the Lagrangian multiplier

$$\lambda(q_1 P_1 + q_2 P_2 + \cdots + q_n P_n - M) = 0$$

Subtracting the above constraint from the utility function and obtaining the *composite* function:

$$\phi = U - \lambda(q_1 P_1 + q_2 P_2 + \cdots + q_n P_n - M)$$

It can be shown that the maximization of the *composite* function implies maximization of the utility function. The first condition for the maximization of a function is that its partial derivatives be equal to zero. Differentiating ϕ with respect to $(q_1,..., q_n)$ and λ, and equating to zero we get:

$$\frac{\partial \phi}{\partial q_1} = \frac{\partial U}{\partial q_1} - \lambda(P_1) = 0$$

$$\frac{\partial \phi}{\partial q_2} = \frac{\partial U}{\partial q_2} - \lambda(P_2) = 0$$

$$\vdots \qquad \vdots \qquad \vdots \qquad \vdots$$

$$\frac{\partial \phi}{\partial q_n} = \frac{\partial U}{\partial q_n} - \lambda(P_n) = 0$$

$$\frac{\partial \phi}{\partial \lambda} = -(q_1 P_1 + q_2 P_2 + \ldots + q_n P_n - M) = 0$$

From the above equations we get:

$$\frac{\partial U}{\partial q_1} = \lambda P_1$$

$$\frac{\partial U}{\partial q_2} = \lambda P_2$$

$$\vdots \qquad \vdots \qquad \vdots$$

$$\frac{\partial U}{\partial q_n} = \lambda P_n$$

But:

$$\frac{\partial U}{\partial q_1} = MU_1, \quad \frac{\partial U}{\partial q_2} = MU_2, \ldots, \frac{\partial U}{\partial q_n} = MU_n$$

Substituting and solving for λ we find:

$$\lambda = \frac{MU_1}{P_1} = \frac{MU_2}{P_2} = \ldots = \frac{MU_n}{P_n}$$

Alternatively, we may divide the preceding equation corresponding to commodity x by the equation which refers to commodity y, and obtain:

$$\frac{MU_x}{MU_y} = \frac{P_x}{P_y} = MRS_{xy}$$

We observe that the equilibrium conditions are identical in the cardinalist approach and in the indifference curves approach. In both theories we have

$$\frac{MU_1}{P_1} = \frac{MU_2}{P_2} = \ldots = \frac{MU_n}{P_x} = \frac{MU_y}{P_y} = \ldots = \frac{MU_n}{P_n}$$

Thus, although in the indifference curves approach cardinality of utility is not required, the *MRS* requires knowledge of the ratio of the marginal utilities, given that the first-order condition for any two commodities may be written as:

$$\frac{MU_x}{MU_y} = \frac{P_x}{P_y} = MRS_{x,y}$$

The above condition is said to be the consumer's equilibrium condition.

2.9   APPLICATION AND USES OF INDIFFERENCE CURVES

Indifference curve analysis may be used to explain various other economic relations of consumer behavior. It has been used to explain the concept of consumer's surplus, substitutability and complementarity, supply curve of labor of an individual, several principles of welfare economics, burden of different forms of taxation, gains from trade, welfare implications of subsidy granted by the government, index number issues, mutual advantage of exchange of goods and services between two individuals and several other similar economic phenomena. The present section deals with only income and leisure choice model, revealed preference theory, and consumer surplus.

2.9.1   Income and Leisure Choice

On the basis of the work performed by the individual, he is paid remuneration/wages/salary/honorarium. The optimum amount of work that he performs could be derived from the analysis of utility maximization. Individual's demand curve for income from this analysis can also be derived. Assume that the individual's satisfaction depends on the income and leisure. Then his utility function is:

$$U = f(L, Y)$$

where L denotes leisure and Y denotes income. For the sustenance of any individual, both income and leisure are required. It is assumed that the individual derives utility from the commodities he purchases with his income. For the derivation of the utility function, it is assumed that he buys the n number of commodities at constant prices, and hence income is treated as generalized purchasing power of that individual. The marginal rate of substitution of income for leisure is:

$$MRTS_{YL} = -\frac{dY}{dL} = \frac{f_1}{f_2}$$

The amount of work performed by the individual is denoted by w and its wage rate by r. According to definition, $L = t - w$, where t is the total amount of time available for work. Thus, the budget constraint is:

$$Y = r \times w$$

Substituting the definition of income and leisure into utility function, we get:

$$U = f(t - w, r \times w)$$

In order to maximize the utility, taking first-order derivative of this derived utility function with respect to w and setting it equals to zero, we get:

$$\frac{dU}{dw} = -f_1 + f_2 r = 0$$

and therefore:

$$-\frac{dy}{dL} = \frac{f_1}{f_2} = r$$

which states that the rate of substitution of income for leisure equals the wage rate. The second-order condition states:

$$\frac{d^2U}{dw^2} = f_{11} - 2f_{12}r + f_{22}r^2 < 0$$

The above equation is a relation in terms of w and r and is based on the individual consumer's optimizing behavior. It is therefore the consumer's supply curve for work and states how much he will perform at various wage rates. Since the supply of work is equivalent to the demand for income, the second-order derivative is indirectly providing the consumer's demand curve for income. For example, let the utility function defined for a time period of one day is given by $U = 48L + LY - L^2$. Then:

$$U = 48(t - w) + (t - w)wr - (t - w)^2$$

and setting the derivative equals to zero:

$$\frac{dU}{dw} = -48 - wr + r(t - w) + 2(t - w) = 0$$

$$w = \frac{t(r + 2) - 48}{2(r + 1)}$$

and Y may be obtained by substituting in utility equation. The second-order condition is fulfilled, since

$$\frac{d^2U}{dw^2} = -2(r + 1) < 0$$

for any positive wage. At this derived wage rate, the individual's utility would be at the maximum with a balance between income and leisure.

2.9.2 Revealed Preference Hypothesis

Paul Samuelson, in his remarkable work in 1947, introduced the term *revealed preference*. After that various literature came out discussing the concept, use and application of the revealed preference hypothesis. This hypothesis is considered as a major turning point in the theory of demand, while establishing the law of demand without the use of indifference curves and the assumptions associated with that. To discuss the revealed preference hypothesis, let's first look into the assumptions associated with it, as given below:

(i) As usual, the consumer is assumed to behave normally and rationally, meaning thereby that he would exercise his preference for a basket of goods rationally.

(ii) The consumer's behavior would be consistent and transitive. Symbolically, if $A > B$, then $B \not> A$ it indicates consistency and if at any given situation $A > B$ and $B > C$ then $A > C$ shows transitive characteristics.

(iii) The consumer, while choosing a set of goods in a given budget situation, would reveal his preferences for a particular selected collection. The selected basket is revealed to be preferred among all other alternative baskets available in a budget constraint. The selected basket maximizes the utility of the consumer.

The theory of revealed preference allows prediction of the consumer's behavior without specification of an explicit utility function, provided that he conforms to some simple axioms. The existence and nature of his utility function can be deduced from his observed choices among commodity baskets.

Assume that there are n number of commodities and their respective particular set of prices are $p_1, p_2, p_3, \ldots, p_n$ which is denoted by P_n and the corresponding quantities bought by the consumer is denoted by Q_n. The consumer's total expenditure is defined by the product of price per unit and quantity bought of the commodity as $P_n Q_n$, symbolically:

$$\sum_{i=1}^{n} P_n Q_n$$

Consider an alternative basket of commodities Q_m that could have been purchased by the consumer but was not. The total cost of Q_m at prices P_m must be no greater than the total cost of Q_n, such as, $P_n Q_m \leqq P_n Q_n$. Since Q_n is at least as expensive a combination of commodities as Q_m, and since the consumer refused to choose combination Q_m, Q_n is revealed to be preferred to Q_m.

2.9.2.1 *Weak Ordering versus Strong Ordering*

If Q_n is revealed to be preferred to Q_m, the latter must never be revealed to be preferred to Q_n. The only way in which Q_m can be revealed to be preferred to Q_n is to have the consumer purchase the combination Q_m in some price situation in which he could also afford to buy Q_n.

Q_m is revealed to be preferred if $P_m Q_n \leqq P_m Q_m$

The $P_m Q_n \leqq P_m Q_m$ axiom states that it can never hold if $P_n Q_m \leqq P_n Q_n$ does.

It implies the opposite of $P_m Q_n \leqq P_m Q_m$ or $P_n Q_m \leqq P_n Q_n$ implies that $P_m Q_n > P_m Q_m$.

If Q_n is revealed to be preferred to Q_m which is revealed to be preferred to $Q_o, Q_p, \ldots$, which is revealed to be preferred to Q_k, which must never be revealed to be preferred to Q_n. This axiom ensures the transitivity of revealed preferences, but is stronger than the usual transitivity condition.

2.9.2.2 *Characterization of Revealed Preference*

Since the revealed preference conditions are a complete set of the restrictions imposed by utility-maximizing behavior, they must contain all of the information available about the underlying preferences. It is obvious now to use the revealed preference relations to determine the preferences among the observed choices, x^n, for $n = 1, \ldots, N$. However, it is less obvious to use the revealed preference relations to tell us about preference relations between choices that have never been observed.

Let's try to use revealed preference to "bound" the indifference curve through x. First, we observe that y is revealed preferred to x. Assume that preferences are convex and monotonic. Then all the bundles on the line segment connecting x and y must be at least as good as x, and all the bundles that lie to the northeast of this bundle are at least as good as x. Call this set of bundles $R^P(x)$, for "revealed preferred" to x. It is not difficult to show that this is the best "inner bound" to the upper contour set through the point x.

To derive the best outer bound, we must consider all possible budget lines passing through x. Let R^W be the set of all bundles that are revealed worse than x for all these budget lines. The bundles in R^W are certain to be worse than x no matter what budget line is used.

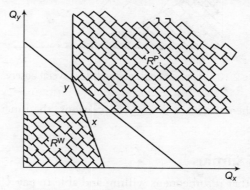

Figure 2.17   Inner and outer bounds

In Fig. 2.17, R^P is the inner bound to the indifference curve through x0; the consumption of R^W is the outer bound.

The outer bound to the upper contour set at x is then defined to be the complement of this set: NRW = all bundles not in RW. This is the best outer bound in the sense that any bundle not in this set cannot ever be revealed preferred to x by a consistent utility-maximizing consumer. Why? Because by construction, a bundle that is not in $NR^W(x)$ must be in $R^W(x)$, in which case it would be revealed *worse* than x.

In the case of a single observed choice, the bounds are not very tight. But with many choices, the bounds can become quite close together, effectively trapping the true indifference curve between them. It is worth tracing through the construction of these bounds to make sure that you understand where they come from. Once we have constructed the inner and outer bounds for the upper contour sets, we have recovered, essentially, all the information about preferences that is not aimed in the observed demand behavior. Hence, the construction of R^P and R^W is analogous to solving the integrability equations.

Our construction of R^P and R^W until this point has been graphical. However, it is possible to generalize this analysis to multiple goods. It turns out that determining whether one bundle is revealed preferred or revealed worse than another involves checking to see whether a solution exists to a particular set of linear inequalities.

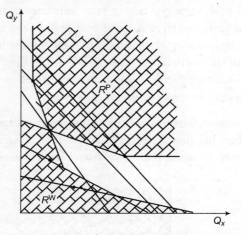

Figure 2.18   Inner and outer bounds for several observations

Fig. 2.18 shows that when there are several observations, the inner bound and outer bound can be quite tight.

2.9.3 Consumer's Surplus

It is important to note that if someone is willing and able to pay ₹5.00 for the first unit and if the market price established by Supply and Demand was ₹2.00, the buyer would purchase at ₹2.00, even though he was willing to pay ₹5.00 for the first unit. He receives utility that he did not have to pay for (₹5.00 – ₹2.00). This is called *consumer surplus*. At market equilibrium, consumer surplus will be the area above the market price and below the demand function. A demand function $P_x = f(Q_x^d)$ represents the different prices consumers are willing to pay for different quantities of a good. If equilibrium in the market is at (Q_e, P_e), then the consumers who would be willing to pay more than P_e benefit. Total benefit to consumers is represented by the consumer's surplus. This is shown in Fig. 2.19.

Mathematically,

$$\text{Consumer's Surplus} = \int_0^{Q^e} f(Q)dQ - Q_e P_e$$

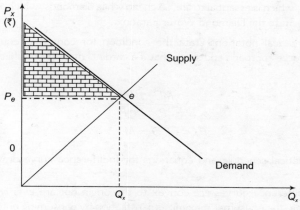

Figure 2.19   Consumer's surplus

PROBLEMS

1. Derive the market demand schedule for this commodity and draw the market demand curve assuming everything else constant and the demand function of a commodity is commodity per time period and P_x for the price of the commodity.

2. Express the law of demand in simple mathematical language. How do we arrive at the expression $Q^d_x = f(P_x)$ *ceteris paribus*?

3. Given the demand function $Q^d_x = 12 - 2P_x$ (P_x is given in rupees), derive the individual's demand schedule and the individual's demand curve. Find out the maximum quantity this individual will ever demand of commodity X per time period.

4. Define elasticity of demand. What does it measure in general? What do the price elasticity of demand, the income elasticity of demand and the cross elasticity of demand measure in general?

5. Why don't use the slope of the demand curve (i.e., $\Delta P/\Delta Q$) or its reciprocal (i.e., $\Delta P/\Delta Q$) to measure the responsiveness in the quantity of a commodity demanded to a change in its price?

6. Show that when $Q^d_y = 600/P_y$ (a rectangular hyperbola), the total expenditures on commodity Y remain unchanged as P_y falls and derive the value of e_p along the hyperbola.

7. Does income elasticity of demand (e_y) measure movements along the same demand curve or shifts in demand? Find the income elasticity of demand for the entire market and give some examples of luxury goods.

8. Find the cross elasticity of demand between burger (X) and pizza (Y) and between pizza (X) and momos (Z) using hypothetical data. State the *ceteris paribus* conditions in finding e_{xy} and e_{xz}.

9. Draw the MU_x curve geometrically from the TU_x curve. Explain the shape of the MU curve in terms of the shape of the TU_x curve. What is the relevant portion of the TU_x curve?

10. State the constraints or limitations the consumer face in maximizing the total utility from personal expenditures and express it mathematically the condition for consumer equilibrium.

11. Why is water, which is essential to life, so cheap while diamonds, which are not essential to life, so expensive? State the Diamond Water Paradox.

12. Using mathematical notations state the condition for consumer equilibrium given by the indifference curve approach and show that if a cardinal measure of utility exists, the condition reduces to

$$\frac{MU_x}{P_x} = \frac{MU_y}{P_y}$$

$$P_xQ_x + P_yQ_y = M$$

13 Using hypothetical schedule and equations for Indifference curve draw a diagram showing that

(a) if the indifference curves are convex to the origin but are everywhere flatter than the budget line, the consumer maximizes satisfaction by consuming only commodity Y,

(b) if the indifference curves are convex to the origin but are everywhere steeper than the budget line, the consumer maximizes satisfaction by consuming only commodity X, and

(c) if the indifference curves are concave to the origin, the consumer maximizes satisfaction by consuming either only commodity X or only commodity Y.

14. Derive the equilibrium condition using calculus given the utility function $U = u(X,Y)$ and budget constraint $P_xX + P_yY = M$.

❑❑❑

3

Elementary Economic Analysis

3.1 INTRODUCTION: THEORY OF THE FIRM

Consumers and producers are the two sides of markets. The present chapter will discuss producers' side of the market. The firm is the basic unit of activity on the production side of the market. The task of the firm is to take inputs and turns them into goods or commodities. In the neoclassical model, the objective of the firm is to maximize profits and the firm is assumed to be one firm among many others. The firm chooses the production plan from all feasible plans that maximize the profit earned. Because of this, as in the consumer model, prices are exogenous in the neoclassical production model. Firms are unable to affect the prices of either their inputs or their outputs.

Our study of production is divided into three parts: *First*, we will consider production from a purely technological point of view, characterizing the firm's set of feasible production plans in terms of its production set Y. *Second*, we will assume that the firm produces a single output using multiple inputs, and we will study its profit maximization and cost minimization problems using a production function to characterize its production possibilities. *Finally*, we will consider a special class of production models, where the firm's production function exhibits constant returns to scale.

3.2 LAW OF SUPPLY

The supply function is defined as a schedule of quantities of a good that will be produced and offered for sale at a schedule of prices during a given time, *ceteris paribus*. Generally, producers are willing to offer greater quantities of a good for sale if the prices are higher. There is a positive (direct) relationship between price of the commodity and quantity supplied of that commodity, *ceteris paribus*. This is known as the *law of supply*. The information can be represented on a graph by plotting each price-quantity combination.

$$Q_x^s = f(P_x, P_{inputs}, \text{technology, other factors})$$

Supply schedule Table 3.1 represents supply of a good and its price relationship, assuming other things remain unchanged.

Table 3.1   Supply-price relationship

Supply Schedule	
Cups of tea supplied each day between 07 to 09 am	
Price per cup	**Cup supplied**
₹0.00	0
₹0.50	2.5
₹0.75	5
₹1.00	7.5
₹1.25	10
₹1.50	12.5
₹1.75	15
₹2.00	20

$\partial P_x > 0$

$[+0.75]$

$\partial Q_x^s > 0$

$[+ 7.5]$

Supply curve   Table 3.1 is graphically presented in Fig. 3.1 to explain the relationship between price and quantity supplied.

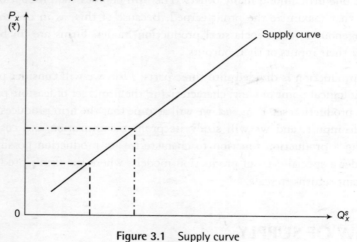

Figure 3.1   Supply curve

Change in Quantity Supplied   A change in the price of the good causes a change in the quantity supplied. The change in the price of the good causes a *movement on the supply function*, not a change or *shift of the supply function*.

Change in Supply   A change in supply, like a change in demand, refers to a change in the relationship between the price and quantity supplied. A change in supply is caused by a change in any variable, other than price, that influences supply. A change in supply can be represented by a shift of the supply function on a graph. There are many factors that influence the willingness of producers to supply a good, such as, technology, prices of inputs, returns in alternative choices, taxes, expectations, weather, number of sellers, and so on. A change in the price (P_x) causes a *change in quantity supplied*; and a change in any other variable causes a *change in supply*.

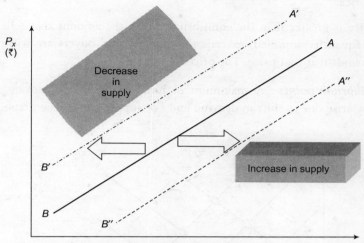

Figure 3.2   Changes in supply curve

Equilibrium   The term equilibrium is defined by *Webster's Encyclopedic Unabridged Dictionary of the English Language* as "a state of rest or balance due to the equal action of opposing forces". In other words, it is an equal balance between any powers, influences, or forces. In a market, equilibrium is said to exist when the forces of supply (sellers) and demand (buyers) are in balance, and the actions of sellers and buyers are coordinated. The quantity supplied equals the quantity demanded.

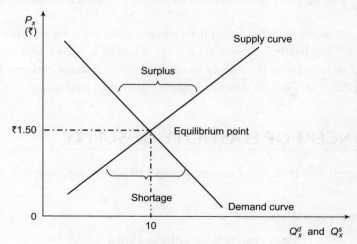

Figure 3.3   Equilibrium point

Given a demand function which represents the behavior or choices of buyers and a supply function that represents the behavior of sellers, where the quantity that people want to buy is equal to the quantity that the producers want to sell, there is an equilibrium quantity. The price that represents the equilibrium of the preferences of the buyers and sellers is the equilibrium price.

When the price is greater than the equilibrium price, the amount that sellers want to sell at that price (quantity supplied) exceeds the amount that buyers are willing to purchase (quantity demanded) at that price. The price is "too high."

Feasible equilibrium points   A maximum of nine feasible combinations of equilibrium situations may arise due to shift in demand and / or supply curve. The graph is presented in Fig. 3.4.

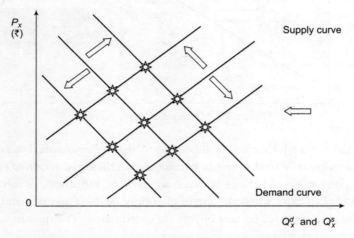

Figure 3.4   Equilibrium points on changes in demand and supply

The supply and demand model is a simplistic model which provides insights into the effects of events which are related to a specific market. Whether an event will tend to cause the price of a good to increase or decrease is of importance to decision makers. To estimate the magnitude of price and quantity changes, more sophisticated models are needed.

3.3   CONCEPT OF ELASTICITY OF SUPPLY

Elasticity of supply is a measure of how responsive sellers are to changes in the price of the good.

Elasticity of supply e_p is defined as:

$$e_p = \frac{\%\text{age change in quantity supplied}}{\%\text{age change in price}}$$

or,
$$e_p = \frac{\%\Delta Q_x^s}{\%\Delta P_x}$$

$$e_p = \frac{\%\Delta Q_x^s}{\%\Delta P_x} = \frac{\dfrac{Q_{x_2}^s - Q_{x_1}^s}{Q_{x_1}^s} \times 100}{\dfrac{P_{x_2} - P_{x_1}}{P_{x_1}} \times 100} = \frac{\dfrac{Q_{x_2}^s - Q_{x_1}^s}{Q_{x_1}^s}}{\dfrac{P_{x_2} - P_{x_1}}{P_{x_1}}} = \frac{\dfrac{\Delta Q_x^s}{Q_{x_1}^s}}{\dfrac{\Delta P_x}{P_{x_1}}} = \frac{\Delta Q_x^s}{Q_{x_1}^s} \times \frac{P_{x_1}}{\Delta P_x}$$

Given a supply function, at a price P_{x_1}, $Q_{x_1}^s$ is produced and offered for sale. At a higher price (P_{x_2}), a larger quantity, $Q_{x_2}^s$, will be produced and offered for sale. The increase in price (ΔP_x), induces a larger quantity goods (ΔQ_x^s) for sale. This is shown in Figure 3.5. The more responsive sellers are to ΔP_x, the greater the absolute value of e_p and the supply function is flatter or more elastic.

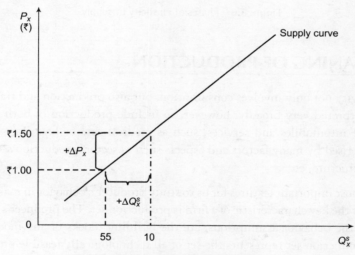

Figure 3.5 Elasticity of supply

The supply function is a model of sellers' behavior, which is influenced by:
  (i) technology;
 (ii) prices of inputs;
(iii) time for adjustment in the market period: whether it is short, long, or very long;
 (iv) expectations; and
  (v) anything that influences costs of production, such as, taxes and regulations, etc.

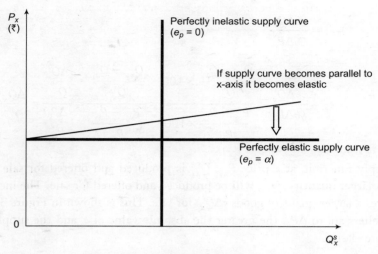

Figure 3.6   Phases of elasticity of supply

3.4   MEANING OF PRODUCTION

Economic activity not only involves consumption, but also production and trade. Production should be interpreted very broadly, however, to include production of both physical goods such as rice or automobiles and services, such as medical care or financial services. A firm can be characterised by many factors and aspects such as sector, production scale, ownership, organization structure, etc.

But what are most important features for us to study producer's behavior in making choices? It is assumed that the key characteristic of a firm is production set. The producer's characteristics, together with the behavior assumptions, are the building blocks of any model of producer theory. The production set represents the set of all technologically feasible production plans. The behavior assumption expresses the guiding principle the producer uses to make choices. It is generally assumed that the producer seeks to identify and select a production plan that is the most profitable.

Production is the process of transforming inputs to outputs. Typically, inputs consist of labor, capital equipment, raw materials, and intermediate goods purchased from other firms. An output consists of finished products or services, or intermediate goods to be sold to other firms. Often, alternative methods are available for producing the same output, using different combinations of inputs. A firm produces outputs from various combinations of inputs. The most general way is to think of the firm as having a production possibility set. Suppose the firm has N possible goods to serve as inputs and/or outputs. If a firm uses Y_j^i units of a good j as an input and produces y_j^o of the good as an output, then the net output of good j is given by $y_j = y_j^o - y_j^i$.

A production plan is simply a list of net outputs of various goods. We can represent a production plan by a vector y in $\mathbb{R}^N$ where y_j is negative if the jth good serves as a net input and positive if the jth good serves as a net output. The set of all technologically feasible production plans is called the firm's *production possibilities* set and will be denoted by Y, a subset of $\mathbb{R}^N$. The set Y is supposed to describe all patterns of inputs and outputs that are technologically feasible. It gives us a complete description of the technological possibilities facing the firm. When we study the behavior of a firm in certain economic environments, we may want to distinguish between production plans that are "immediately feasible" and those that are "eventually" feasible. We will generally assume that such restrictions can be described by some vector z in $\mathbb{R}^L$. The restricted or short-run production possibilities set will be denoted by $Y(z)$; this consists of all feasible net output bundles consistent with the constraint level z.

Relationships among Total Product (TP_L) , Marginal Product (MP_L) and Average Product (AP_L)

In many production processes Q initially increases at an increasing rate. This is due to division of labor and a better mix of the variable input with the fixed inputs. As Q $[TP_L]$ increases at an increasing rate, MP increases. As Q $[TP_L]$ increases at a decreasing rate, MP_L decreases. Where $0Z$ is tangent to TP_L , AP_L is a maximum; $AP_L = MP_L$. When TP_L is a maximum, MP_L is zero. When TP_L is decreasing, MP_L is negative. {$MP > AP$, AP rises}. Fig. 3.7 shows the relationships amongt TP_L , MP_L and AP_L.

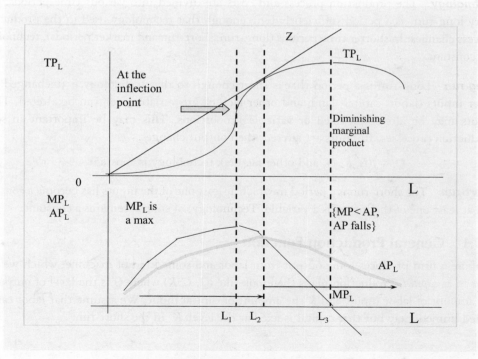

Figure 3.7   Relationships among TP_L , MP_L and AP_L

3.5   PRODUCTION FUNCTION AND ITS TYPES

A production function expresses the relationship between a set of inputs and the output of a good or service. The relationship is determined by the nature of the good and technology. A production function is like a recipe for cookies; it tells you the quantities of each ingredient, how to combine and cook, and how many cookies you will produce.

$$Q_X = f(L, K, N, \text{technology, and other factors})$$

Q_X = quantity of output
L = labor input
K = capital input
N = natural resources (e.g., land)

Decisions about alternative ways to produce a good X require that we have information about how each variable influences Q_X. One method used to identify the effects of each variable on output is to vary one input at a time. The use of the *ceteris paribus* convention allows this analysis. The time period used for analysis also provides a way to determine the effects of various changes of inputs on the output.

Technology   The production process and costs are divided into various time periods. The "very long-run" is a period sufficiently long enough that technology used in the production process changes. In shorter time periods (long-run, short-run and market periods), technology is a constant.

Long-run   Long-run is a period that is short enough so that technology is unchanged. All other inputs (labor, capital, land, and other factors) are variable, i.e., can be altered. These inputs may be altered in fixed or variable proportions. This may be important in some production processes. If inputs are altered, then output changes.

$$Q_x = f(L, K, N, \text{and other factors}), \text{technology is constant}$$

Short-run   The short-run is a period in which at least one of the inputs has become a constant and at least one of the inputs is a variable. Technology, of course, remains a constant.

3.5.1   General Production Function

Suppose a firm produces some output from labor and some kind of machine, which we will refer to as *capital*. Production plans then look like (Q, L, K) where Q is the level of output, L the amount of labor input, and K the amount of capital input. We assume that labor can be varied immediately but that capital is fixed at the level $\overline{K}$ in the short-run.

Then,

$$Q(\overline{K}) = f(A, L, K) \text{ where } Q : K = \overline{K}$$

is a *short-run production possibilities set*.

Let α be a parameter such that $0 < \alpha < 1$. Then the Cobb-Douglas technology is defined in the following manner:

$$Q = f(A, x_1, x_2) \text{ where } P^3\colon A \, x^{\alpha}_1 \, x^{1-\alpha}_2$$

$$V(A) = f(x_1, x_2) \text{ where } P^2\colon A \, x^{\alpha}_1 \, x^{1-\alpha}_2$$

$$Q(A) = f(x_1, x_2) \text{ where } P^2\colon A \, x^{\alpha}_1 \, x^{1-\alpha}_2$$

$$Q(z) = f(A, x_1, x_2) \text{ where } P^3\colon A \, x^{\alpha}_1 \, x^{1-\alpha}_2, \, x_2 = z$$

$$T(A, x_1 x_2) = A$$

$$f(x_1, x_2) = x^{\alpha}_1 \, x^{1-\alpha}_2.$$

Let $\alpha > 0$ and $\alpha > 0$ be parameters. Then the Leontief technology is defined in the following manner:

$$Q = f(A, x_1, x_2) \text{ where } \mathbb{R}^2\colon A \min (\alpha x_1, \beta x_2)$$

$$V(A) = f(x_1, x_2) \text{ where } \mathbb{R}^2\colon A \min (\alpha x_1, \beta x_2)$$

$$Q(A) = f(x_1, x_2) \text{ where } \mathbb{R}^2\colon A \min (\alpha x_1, \beta x_2)$$

$$T(A, x_1, x_2) = A - \min (\alpha x_1, \beta x_2)$$

$$f(x_1, x_2) = \min (\alpha x_1, \beta x_2).$$

In the case of long-run production function, all factors determining output are variable, which is defined as:

$$Q = f(A, L, K)$$

is a *long-run production possibilities set*.

3.5.2 Cobb-Douglas Production Function

The Cobb-Douglas production function is the most popular in applied research, because it is easiest to handle mathematically. The usual Cobb-Douglas function is of the form:

$$Q = A \cdot L^{\alpha} \cdot K^{\beta}$$

The marginal product of factors is:

$$MP_L = \frac{\partial Q}{\partial L} = \alpha \cdot A \cdot L^{\alpha-1} \cdot K^{\beta}$$

$$= \alpha (A \cdot L^{\alpha} \cdot K^{\beta}) L^{-1}$$

$$\alpha \cdot \frac{Q}{L} = \alpha(AP_L)$$

where AP_L = the average product of labor.

Similarly,

$$MP_K = \beta \cdot \frac{Q}{K} = \beta(AP_K)$$

The marginal rate of substitution is:

$$MRS_{L,K} = \frac{\partial Q / \partial L}{\partial Q / \partial K} = \frac{\alpha\left(\dfrac{Q}{L}\right)}{\beta\left(\dfrac{Q}{K}\right)} = \frac{\alpha}{\beta} \cdot \frac{K}{L}$$

The elasticity of substitution is equal to 1.

$$\sigma = \frac{d(\,K/L)/(K/L)}{d(MRS)/(MRS)} = 1$$

Proof: Substitute the MRS and obtain:

$$\sigma = \frac{d(\,K/L)/(K/L)}{\dfrac{d\left(\dfrac{\alpha}{\beta} \cdot \dfrac{K}{L}\right)}{\left(\dfrac{\alpha}{\beta}\right)\left(\dfrac{K}{L}\right)}}$$

$$= \frac{d\left(\dfrac{K}{L}\right)\left(\dfrac{\alpha}{\beta}\right)}{\left(\dfrac{\alpha}{\beta}\right)d\left(\dfrac{K}{L}\right)} = 1$$

given that b_1/b_2 is constant and does not affect the derivative.

In a Cobb-Douglas function, factor intensity is measured by the ratio α/β. The higher the ratio, the more labor intensive the technique. Similarly, the lower the ratio b_1/b_2 the more capital intensive the technique.

The efficiency in the organization of the factors of production is measured by the coefficient A. It is clear that if two firms have the same K, L, α and β and still produce different quantities of output, the difference can be due to the different organization and entrepreneurship of the firms, which results in different efficiencies. The more efficient firm will have a larger A than the less efficient one.

3.5.3 Properties of Cobb-Douglas Production Function

Though the production possibility of different processes can differ widely in structure, many technologies share certain general properties. If it can be assumed that these properties are satisfied, special theoretical results can be derived. The important properties are defined as follows:

(i)   *Possibility of Inaction:* $0 \in Y$. it means that no action on production is a possible production plan.

(ii) *Closeness:* Y is closed. The possibility set Q is closed means that, whenever a sequence of production plans A_i, $i = 1, 2,...$, are in Q and $A_i \rightarrow A$, then the limit production plan A is also in Q. It guarantees that points on the boundary of Q are feasible. Where Q is closed, it implies that the input requirement set $V(A)$ is a closed set for all $A \geq 0$.

(iii) *Free disposal or monotonicity:* If $A \in Q$ implies that $A' \in Q$ for all $Q' \geq A$, then the set Q is said to satisfy the free disposal or monotonicity property. Free disposal implies that commodities, either inputs or outputs, can be thrown away. This property means that if $A \in Q$, then Q includes all vectors in the negative orthant translated to A, i.e., there are only inputs, but no outputs. A weaker requirement is that we only assume that the input requirement is monotonic: If L is in $V(A)$ and $L' \geq L$, then L' is in $V(A)$. Monotonicity of $V(A)$ means that, if L is a feasible way to produce A units of output and L' is an input vector with at least as much of each input, then L' should be a feasible way to produce A.

(iv) *Irreversibility:* $Q \cap \{-Q\} = \{0\}$. It means a production plan is not reversible unless it is a non-action plan.

(v)   *Convexity:* Q is convex if whenever A and A' are in Q, the weighted average $nA + (1 - n)$ A is also in Q for any n with $0 \leq n \geq 1$. Convexity of Q means that, if all goods are divisible, it is often reasonable to assume that two production plans A and A' can be scaled downward and combined. However, it should be noted that the convexity of the production set is a strong hypothesis.

(vi) *Strict Convexity:* A is strictly convex if $A \in Q$ and $A' \in Q$, then $nA + (1 - n)A' \in intQ$ for all $0 < n < 1$, where $intQ$ denotes the interior points of Q. As the strict convexity of Q can guarantee the profit maximising production plan is unique provided it exists. A weak and more reasonable requirement is to assume that $V(A)$ is a convex set for all outputs.

(vii) *Convexity of Input Requirement Set:* If L and L' are in $V(A)$, then $nL + (1 - n)L'$ is in $V(A)$ for all $0 \leq n \geq 1$. That is, $V(A)$ is a convex set. Convexity of $V(A)$ means that, if L and L' both can produce A units of output, then any weighted average $nL + (1 - n)L'$ can also produce A units of output.

3.5.4 CES Production Function

The constant elasticity of substitution (*CES*) production function has the form:

$$A = [\alpha x_1^p + \beta x_2^p]^{\frac{1}{p}}$$

The *CES* function exhibits constant returns to scale. It has a constant elasticity of substitution.

Proof: The marginal rate of technical substitution is given by:

$$MRTS = -\left(\frac{x_1}{x_2}\right)^{p-1}$$

so that,

$$\frac{x_2}{x_1} = |MRTS|^{\frac{1}{1-P}}$$

Taking logs, we see that:

$$\ln\frac{x_2}{x_1} = \frac{1}{1-p}\ln|MRTS|$$

Applying the definition of σ using the logarithmic derivative,

$$\sigma = \frac{d\ln(x_2/x_1)}{d\ln|MRTS|} = \frac{1}{1-p}$$

3.6 PRODUCER'S EQUILIBRIUM

The most commonly used forms of production function is the Cobb-Douglas form:

$$Q = A \cdot L^\alpha \cdot K^\beta$$

Given this production function and the cost equation:

$$C = wL + rK$$

deriving the cost function, that is, the total cost as a function of output:

$$C = f(X)$$

solving the constrained output maximization problem:

Maximize $Q = A \cdot L^\alpha \cdot K^\beta$

Subject to $\overline{C} = wL + rK$

where cost is the constraint forming the "composite" function

$$\varphi = Q + \lambda(\bar{C} - wL - rK)$$

where λ = Lagrangian multiplier

The first condition for maximization is that the first derivatives of the function with respect to L, K and λ be equal to zero:

$$\frac{\partial \varphi}{\partial L} = \alpha \frac{Q}{L} - \lambda w = 0$$

$$\frac{\partial \varphi}{\partial K} = \beta \frac{Q}{K} - \lambda r = 0$$

$$\frac{\partial \varphi}{\partial \lambda} = (\bar{C} - wL - rK) = 0$$

From above equation we obtain:

$$\alpha \frac{Q}{L} = \lambda w \quad \text{and} \quad \beta \frac{Q}{K} = \lambda r$$

Dividing these expressions we obtain:

$$\frac{\alpha}{\beta} \cdot \frac{K}{L} = \frac{w}{r}$$

Solving for K:

$$K = \frac{w}{r} \cdot \frac{\alpha}{\beta} L$$

Substituting K into the production function we obtain:

$$Q = A \cdot L^{\alpha} \cdot K^{\beta}$$

$$Q = AL^{\alpha} \left[\frac{w}{r} \frac{\beta}{\alpha} L \right]^{\beta}$$

$$Q = A \left[\left(\frac{w}{r} \right) \left(\frac{\beta}{\alpha} \right) \right]^{\beta} L^{(\alpha + \beta)}$$

The term in brackets is the constant term of the function which includes the three coefficients of the production function, A, α, β, and the prices of the factors of production.

Solving the above form of the production function for L, we find:

$$\frac{1}{A \left(\frac{w}{r} \frac{\beta}{\alpha} \right)^{\beta}} Q = L^{(\alpha + \beta)}$$

or

$$\left[\frac{1}{A\left(\dfrac{w}{r} \dfrac{\beta}{\alpha} \right)^{\beta}} \cdot X \right]^{1/(\alpha+\beta)} = L$$

or

$$L = \left(\frac{r\alpha}{w\beta} \right)^{\beta/(\alpha+\beta)} \left(\frac{X}{A} \right)^{1/(\alpha+\beta)}$$

Substituting the value of L from above expressions for capital we obtain

Substituting expression of above equations into the cost equation $\overline{C} = wL + rK$ we find

$$\overline{C} = \left(\frac{1}{A} \right)^{1/(\alpha+\beta)} \left[w \left(\frac{r\alpha}{w\beta} \right)^{\beta/(\alpha+\beta)} + r \left(\frac{w\beta}{r\alpha} \right)^{\alpha/(\alpha+\beta)} \right] \cdot Q^{1/(\alpha+\beta)}$$

Rearranging the expression above we obtain:

$$\overline{C} = \left\{ \left(\frac{1}{A} \right) \left[\left(\frac{\alpha}{\beta} \right)^{\beta} + \left(\frac{\beta}{\alpha} \right)^{\alpha} \right] \right\}^{1/(\alpha+\beta)} \cdot \left\{ \frac{w^{\alpha/(\alpha+\beta)} \cdot r^{\beta}}{(\alpha+\beta)} \right\} \cdot Q^{1/(\alpha+\beta)}$$

This is the cost function, that is, the cost expressed as a function of output is Q; the production function coefficients are A, α, β where the sum $\alpha + \beta$ is a measure of the returns to scale; and the prices of factors are w, r.

If prices of factors are given the usual assumption in the theory of the firm, cost depends only on output Q, which expresses graphically the cost function:

$$C = f(Q) \text{ ceteris paribus}$$

Here *ceteris paribus* implies that all other determinants of costs, that is, the production technology and the prices of factors, remain unchanged. If these factors change, the cost curve will shift upwards or downwards.

3.7 CONCEPT OF ISOQUANTS

An isoquant is the locus of combination of all the technically efficient methods or all the combinations of factors of production for producing a given level of output. On the basis of degree of substitutability of factor inputs, an isoquant can have various shapes. The following are the different shapes of an isoquant:

(i) When the factors of production are perfectly substitutable, then the isoquant is linear. This is also known as linear isoquant.

(ii) When the factors of production are perfect complements, then the isoquant takes the shape of a right angle. This type of isoquant is known as Leontief isoquant or input-output isoquant.

(iii) When the factors of production have limited substitutability, then the isoquant will have kinks. This is also known as kinked isoquant or activity analysis isoquant or linear programming isoquant.

(iv) When the factors of production have continuous substitutability, then the isoquant will have a smooth convex curve. This type isoquant is called a convex isoquant.

The isoquant gives all input bundles that produce exactly y units of output.

(i) There is perfect knowledge of the relevant information on the factor inputs and its prices to the producer.

(ii) The output or production set are defined as ordinal, where producer can rank the production possibility sets of each combination of inputs.

(iii) The marginal rate of technical substitution of one input for another is diminishing. Production possibilities are ranked in terms of the isoquant and assumed to be convex to the origin. The slope of the isoquant curve is called the marginal rate of technical substitution of the inputs ($MRTS$ or RTS).

(iv) The total output depends on the quantities of the inputs employed.

$$Q = f(Q_1^L, Q_2^L, Q_3^L, \ldots, Q_n^L, Q_1^K, Q_2^K, Q_3^K, \ldots, Q_n^K)$$

(v) The possibility of the production is consistent and transitive. Symbolically,

if $A > B$, then $B > A$ [consistency assumption]

if $A > B$, and $B > C$, then $A > C$ [transitivity assumption]

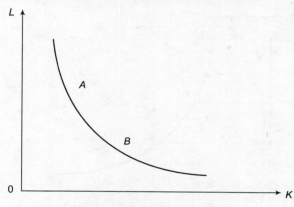

Figure 3.8   Convex isoquant

Properties of Isoquant   The properties of Isoquant are: higher Isoquant are preferred to lower ones; Isoquant are downward sloping; Isoquant do not cross; and Isoquant are bowed inward.

Property 1: Higher Isoquant are preferred to lower ones.

Remark: Producers usually prefer more of something to less of it and a higher Isoquant represents larger production possibility set compared to a lower Isoquant.

Property 2: Isoquant is downward sloping.

Remark: A producer is willing to give up one factor input only if he gets more of the other factor inputs in order to get the level of production or output. If the quantity of one factor input is reduced, the quantity of the other factor input must increase. For this reason, Isoquant slope downward.

Property 3: Isoquant does not cross each other.

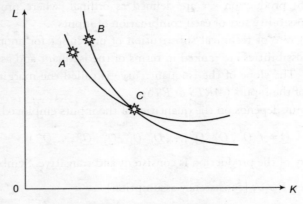

Figure 3.9   Intersecting Isoquant

Property 4: Isoquant is bowed inward.

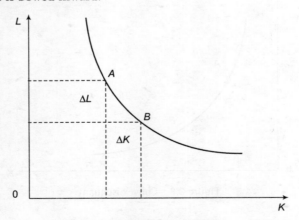

Figure 3.10   Slope of the Isoquant

Perfect Substitutes   If two factor inputs are perfect substitutes to each other, then the Isoquant becomes a straight line with negative slope. Figure 3.11 shows a perfect substitute case.

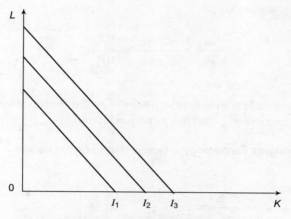

Figure 3.11 Perfect substitute case

Perfect Complements If two factor inputs are perfect complements to each other, then the Isoquant would be right angle shape. Figure 3.12 shows perfect complement case.

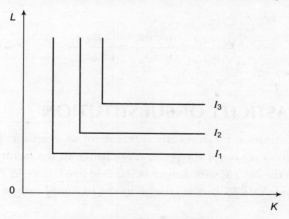

Figure 3.12 Perfect complement case

3.8 MARGINAL RATE OF TECHNICAL SUBSTITUTION

Suppose that technology is summarized by a smooth production function and that we are producing at a particular point $Q = f(x_1, x_2)$. Suppose that we want to increase a small amount of input 1 and decrease some amount of input 2 so as to maintain a constant level of output. Deriving the marginal rate of technical substitution of an isoquant is similar to marginal rate of substitution on the indifference curve. Differentiating the production function when output keeps constant, we have:

$$0 = \frac{\partial f}{\partial x_1} dx_1 + \frac{\partial f}{\partial x_2} dx_2$$

which can be solved for:

$$\frac{dx_2}{dx_1} = \frac{\partial f / \partial x_1}{\partial f / \partial x_2} = \frac{MP_{x_1}}{MP_{x_2}}$$

This gives us an explicit expression for the marginal rate technical substitution, which is the rate of marginal production of x_1 and marginal production of x_2.

MRTS for a Cobb-Douglas Technology   Given that $f(x_1, x_2) = x_1^{\alpha} x_2^{1-\alpha}$, we can take the derivatives to find:

$$\frac{\partial f(x)}{\partial x_1} = ax_1^{\alpha-1}x_2^{1-a} = a\left[\frac{x_2}{x_1}\right]^{1-a}$$

$$\frac{\partial f(x)}{\partial x_2} = (1-a)x_1^{\alpha}x_2^{-a} = 1 - a\left[\frac{x_2}{x_1}\right]^{a}$$

It follows that:

$$\frac{\partial x_2(x_1)}{\partial x_1} = \frac{\partial f / \partial x_1}{\partial f / \partial x_2} = \frac{a}{1-a}\frac{x_1}{x_2}$$

3.9   THE ELASTICITY OF SUBSTITUTION

The elasticity of substitution measures the curvature of an isoquant. More specifically, the elasticity of substitution measures the percentage change in the factor ratio divided by the percentage change in the $MRTS$, with output being held fixed. If we let $\Delta(x_2/x_1)$ be the change in the factor ratio and $\Delta MRTS$ be the change in the technical rate of substitution, we can express this as:

$$\sigma = \frac{\dfrac{\Delta(x_2 / x_1)}{x_2 / x_1}}{\dfrac{\Delta MRTS}{MRTS}}$$

This is a relatively natural measure of curvature. It expresses how the ratio of factor inputs changes as the slope of the isoquant changes. If a small change in slope gives us a large change in the factor input ratio, the isoquant is relatively flat, which means that the elasticity of substitution is large.

In practice, we think of the percent change as being very small and take the limit of this expression as Δ goes to zero. Hence, the expression for σ becomes:

$$\sigma = \frac{MRTSd(x_2 / x_1)}{(x_2 / x_1)dMRTS} = \frac{dIn(x_2 / x_1)}{dIn|MRTS|}$$

The absolute value sign in the denominator is to convert the $MRTS$ to a positive number so that the logarithm makes sense.

For the Cobb-Douglas Production Function, We have seen above that:

$$MRTS = \frac{a}{1-a}\frac{x_2}{x_1}$$

$$MRT\frac{x_2}{x_1} = \frac{1-a}{a}MRTS$$

It follows that:

$$In\left(\frac{x_2}{x_1}\right) = \frac{1-a}{a} + In|MRTS|$$

This in turn implies:

$$\sigma = \frac{dIn(x_2/x_1)}{dIn|MRTS|} = 1$$

3.10 ISO-COST LINE

The producer has a given level of resources or budget which restricts him to attain maximum level of output or production. The cost of the production acts as a constraint on the maximization of output. Suppose the cost is defined by C and this is to be borne on two inputs L and K, with corresponding prices w and r. Thus, the iso-cost line equation would be $C = wL + rK$.

The budget line equation is as follows:

$$L = \frac{C}{w} - \frac{r}{w} \times K$$

The iso-cost line is represented in Fig. 3.13.

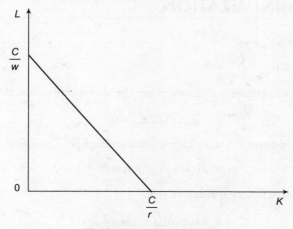

Figure 3.13   Iso-cost line

3.11 PRODUCER'S SURPLUS

Suppose someone is willing and able to pay ₹5.00 for a unit of a good. If the market price established by Supply and Demand was ₹2.00, the producer would receive ₹5.00 even though they were willing to receive ₹2.00 for the unit. The profit that they thus receive is [₹5.00 − ₹2.00]. This is called *producer's surplus*. At market equilibrium, producer surplus will be the area beyond the equilibrium level of quantity and below the demand function. The total gain to the producer is represented by the producer's surplus.

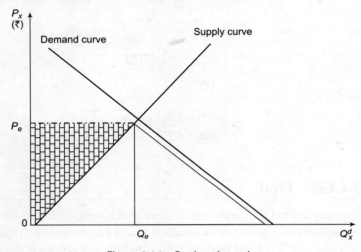

Figure 3.14   Producer's surplus

Mathematically,

$$\text{Producer's Surplus} = Q_e P_e - \int_0^{Q^e} f(Q)dQ$$

3.12 COST MINIMIZATION

Minimize $C = f(Q) = wL + rK$

Subject to $\overline{Q} = f(L, K)$

Using constrained optimization rules, rewriting the above expression as:

$$\overline{Q} - f(L, K) = 0$$

Multiplying the above equation with Lagrangian multiplier, λ, we get:

$$\lambda(\overline{Q} - f(L, K)) = 0$$

Applying the composite function rule, we get:

$$\phi = C - \lambda(\overline{Q} - f(L, K))$$

$$\phi = (wL + rK) - \lambda(\overline{Q} - f(L, K))$$

Taking first order derivatives of ϕ with respect to L, K and λ and setting it equals to zero, we get:

$$\frac{\partial \phi}{\partial L} = w - \lambda \frac{\partial f(L, K)}{\partial L} = 0$$

$$\frac{\partial \phi}{\partial L} = w - \lambda \frac{\partial Q}{\partial L} = 0$$

$$\frac{\partial \phi}{\partial K} = r - \lambda \frac{\partial f(L, K)}{\partial K} = 0$$

$$\frac{\partial \phi}{\partial K} = r - \lambda \frac{\partial Q}{\partial K} = 0$$

$$\frac{\partial \phi}{\partial \lambda} = -(\overline{Q} - f(L, K)) = 0$$

From the first two derivatives, we get:

$$w = \lambda \frac{\partial Q}{\partial L} \Rightarrow \lambda = w \cdot \frac{\partial L}{\partial Q}$$

$$r = \lambda \frac{\partial Q}{\partial K} \Rightarrow \lambda = r \cdot \frac{\partial K}{\partial Q}$$

Equating both λ, we get:

$$\lambda = w \cdot \frac{\partial L}{\partial Q} = r \cdot \frac{\partial K}{\partial Q} \Rightarrow \frac{w}{r} = \left(\frac{\partial L}{\partial Q} \middle/ \frac{\partial K}{\partial Q} \right) = MRTS_{L,K}$$

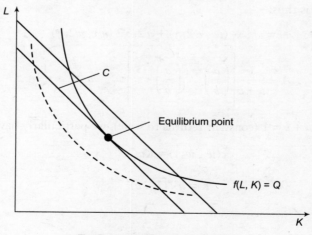

Figure 3.15 Cost minimization

Second condition for minimization would be, then as:

$$\frac{\partial^2 Q}{\partial L^2} < 0, \frac{\partial^2 Q}{\partial K^2} < 0 \text{ and } \left(\frac{\partial^2 Q}{\partial L^2}\right)\left(\frac{\partial^2 Q}{\partial K^2}\right) > \left(\frac{\partial^2 Q}{\partial L \partial K}\right)^2$$

Cost Function for the Cobb-Douglas Technology   Consider the cost minimization problem:

$$c(w, y) = \min_{x_1, x_2}^{w_1, x_1} + w_2 x_2$$

such that $Ax_1^a x_2^b = y$

Solving the constraint for x_2, we see that this problem is equivalent to:

$$\min_{x_1}^{w_1, x_1} + w_2 A - \frac{1}{b_y} \frac{1}{b_{x_1}} - \frac{a}{b}$$

The first-order condition is:

$$w_1 - \frac{a}{b} w_2 A - \frac{1}{b_y} \frac{1}{b_{x_1}} \frac{a+b}{b} = 0$$

which gives us the conditional input demand function for factor 1:

$$x_1(w_1, w_2, y) = A - \frac{1}{a+b}\left[\frac{aw_1}{bw_1}\right]^{\frac{b}{a+b}} \frac{1}{y^{a+b}}$$

The other conditional input demand function is:

$$x_2(w_1, w_2, y) = A - \frac{1}{a+b}\left[\frac{aw_2}{bw_1}\right]^{\frac{b}{a+b}} \frac{1}{y^{a+b}}$$

The cost function is thus:

$$c(w_1, w_2, y) = w_1, x_1\,(w_1, w2, y) + w2, x2\,(w1, w2, y)$$

$$A - \frac{1}{a+b}\left[\left(\frac{a}{b}\right)^{\frac{b}{a+b}}\left(\frac{a}{b}\right)^{\frac{-a}{a+b}}\right] w_1^{\frac{a}{a+b}} w_2^{\frac{b}{a+b}} \frac{1}{y^{a+b}}$$

When $A = 1$ and $a + b = 1$ (constant returns to scale), we particularly have

$$c(w_1, w_2 y) = K w_1^a w_{1-a}^2 y$$

where $K = a^{-a}(1 - a)^{a-1}$

The Cost Function for the CES Technology

Suppose that $f(x_1, x_2) = \left[w_1^{\frac{p}{p-1}} w_2^{\frac{p}{p-1}} \right] \left[w_1^{\frac{p}{p-1}} w_2^{\frac{p}{p-1}} \right]^{-\frac{1}{p}}$. The cost minimization problem is:

$$\min w_1 x_1 + w_2 x_2$$

such that $x_1^p + x_2^p = y^p$

The first-order conditions are:

$$w_1 - \lambda p x_1^{p-1} = 0$$

$$w_2 - \lambda p x_2^{p-1} = 0$$

$$w_2 - \lambda p x_2^{p-1} = 0$$

$$x_1^p + x_2^p = y$$

Solving the first two equations for x_1^p and x_2^p, we have:

$$x_1^p = w_1^{\frac{p}{p-1}} (\lambda p)^{\frac{-p}{p-1}}$$

$$x_2^p = w_2^{\frac{p}{p-1}} (\lambda p)^{\frac{-p}{p-1}}$$

Substitute this into the production function to find:

$$(\lambda p)^{\frac{-p}{p-1}} \left(w_1^{\frac{p}{p-1}} + w_2^{\frac{p}{p-1}} \right) = y^p$$

Solve this for $(\lambda p)^{\frac{-p}{p-1}}$ and substitute into above equations. This gives us the conditional input demand functions:

$$x_1(w_1, w_2, y) = w_1^{\frac{p}{p-1}} \left[w_1^{\frac{p}{p-1}} + w_2^{\frac{-p}{p-1}} \right]^{-\frac{1}{p}} y$$

$$x_2(w_1, w_2, y) = w_2^{\frac{p}{p-1}} \left[w_1^{\frac{p}{p-1}} + w_2^{\frac{-p}{p-1}} \right]^{-\frac{1}{p}} y$$

Substituting these functions into the definition of the cost function yields:

$$c(w_1, w_2, y) = w_1 x_1(w_1, w_2, y) + w_2 \, x_2(w_1, w_2, y)$$

$$= \left[w_1^{\frac{p}{p-1}} + w_2^{\frac{-p}{p-1}} \right] \left[w_1^{\frac{p}{p-1}} + w_2^{\frac{p}{p-1}} \right]^{-\frac{1}{p}}$$

$$= y \left[w_1^{\frac{p}{p-1}} + w_2^{\frac{p}{p-1}} \right]^{\frac{p-1}{p}}$$

This expression looks a better if we set $r = \rho/(\rho - 1)$ and write:

$$c(w_1, w_2, y) = y[w_1^y + w_2^y]^{\frac{1}{y}}$$

Note that this cost function has the same form as the original *CES* production function with r replacing ρ. In the general case where:

$$f(x_1, x_2) = \left[(a_1 x_1)^p + (a_2 x_2)^p \right]^{\frac{1}{p}}$$

Similar computations can be done to show that:

$$c(w_1, w_2, y) = \left[\left(\frac{w_1}{a_1} \right)^y \left(\frac{w_2}{a_2} \right)^y \right]^{\frac{1}{y}} y$$

Cost function for the Leontief Technology   Suppose $f(x_1, x_2) = \min\{ax_1, bx_2\}$. Since we know that the firm will not waste any input with a positive price, the firm must operate at a point where $y = ax_1 = bx_2$. Hence, if the firm wants to produce y units of output, it must use y/a units of good 1 and y/b units of good 2 no matter what the input prices are. Hence, the cost function is given by:

$$c(w_1, w_2, y) = \frac{w_1 y}{a} + \frac{w_2 y}{b} = y \left(\frac{w_1}{a} + \frac{w_2}{b} \right)$$

Cost function for the Linear Technology   Suppose that $f(x_1, x_2) = ax_1 + bx_2$, so that factors 1 and 2 are perfect substitutes. Since the two goods are perfect substitutes, the firm will use whichever is cheaper. Hence, the cost function will have the form:

$$c(w_1, w_2, y) = \min\{w_1/a, w_2/b\}y$$

In this case the cost-minimization problem typically involves a boundary.
One of the two factors will be used in a zero amount. It is easy to see the answer to this particular problem by comparing the relative steepness of the iso-cost line and isoquant curve.

If $\dfrac{a_1}{a_2} < \dfrac{w_1}{w_2}$, the firm only uses x_2 and the cost function is given by $c(w_1, w_2, y) = w_2 x_2 = w_2 \dfrac{y}{a^2}$.

If $\dfrac{a_1}{a_2} > \dfrac{w_1}{w_2}$, the firm only uses x_1 and the cost function is given by $c(w_1, w_2, y) = w_1 x_1 = w_1 \dfrac{y}{a_1}$.

3.13   RETURNS TO SCALE AND RETURNS TO FACTOR

Suppose that we are using some vector of inputs L to produce some output Q and we decide to scale all inputs up or down by some amount $n \geq 0$. Returns to scale refer to how output responds when all inputs are varied in the same proportion so that they consider long-run production processes. There are three possibilities technology exhibits: (1) constant returns to scale; (2) decreasing returns to scale, and (3) increasing returns to scale. Formally, we have:

A production function $f(L)$ is said to exhibit :
  i. constant returns to scale if $f(nL) = nf(L)$ for all $n \geq 0$
  ii. decreasing returns to scale if $f(nL) < nf(L)$ for all $n > 1$
  iii. increasing returns to scale if $f(nL) > nf(L)$ for all $n > 1$

Constant returns to scale means that doubling inputs exactly doubles outputs, which is often a reasonable assumption to make about technologies. Decreasing returns to scale means that doubling inputs results in less than doube outputs. Increasing returns to scale means that doubling inputs results in more than double outputs.

It is important to note that a technology has constant returns to scale if and only if its production function is homogeneous of degree 1. Constant returns to scale is also equivalent to the statement q in Q implies nq is in Q for all $n \geq 0$; or equivalent to the statement x in $V(q)$ implies nL is in $V(nq)$ for all $n > 1$.

The various kinds of returns to scale defined above are global in nature, which may well happen that a technology exhibits increasing returns to scale for some values of L and decreasing returns to scale for other values. Thus, in many circumstances a local measure of returns to scale is useful. To define locally returns to scale, we first define elasticity of scale. The elasticity of scale measures the percent increase in output due to a one percent increase in all inputs, that is, due to an increase in the scale of operations. Let $q = f(L)$ be the production function. Let n be a positive scalar, and consider the function $q(n) = f(nL)$. If $n = 1$, we have the current scale of operation; if $n > 1$, we are scaling all inputs up by n; and if $n < 1$, we are scaling all inputs down by n. The elasticity of scale is given by

$$e(L) = \dfrac{\dfrac{dq(n)}{q(n)}}{\dfrac{dn}{n}}$$

n evaluated at $n = 1$.

Rearranging this expression, we have

$$e(L) = \frac{dq(n)}{dn}\frac{n}{q}\bigg|_{n=1} = \frac{df(nL)}{dn}\frac{n}{f(nL)}\bigg|_{n=1}$$

Evaluating the expression at $n = 1$ to calculate the elasticity of scale at the point L, we have the following local returns to scale: A production function $f(L)$ is said to exhibits locally increasing, constant, or decreasing returns to scale as $e(L)$ is greater, equal, or less than 1.

3.14　COST THEORY AND ESTIMATION

Cost functions are the pecuniary relationships between outputs and the costs of production, defined as:

Cost = $f(Q_X$ (inputs, technology), prices of inputs, and other factors)

Cost functions are determined by input prices and production relationships. It is necessary to understand production functions if you are to interpret cost data. Costs are incurred as a result of production. The important concept of cost is opportunity cost (marginal cost). These are the costs associated with an activity. When inputs or resources are used to produce one good, the other goods they could have been used to produce are sacrificed. Costs may be in real or monetary terms, expressed into implicit costs and/or explicit costs. The next best foregone alternative is known as opportunity cost or transfer earnings.

$$\text{Total Cost} = \text{Total Fixed cost} + \text{Total Variable Cost}$$

$$TC = TFC + TVC$$

$$TC = f(Q)$$

$$AC = TC/Q$$

$$MC = dTC/dQ$$

3.15　CONCEPT OF COSTS AND THEIR TYPES

Let us consider the structure of the cost function. The cost function can always be expressed simply as the value of the conditional factor demands.

$$c(w, y) = w_x x(w, y)$$

In the short-run, some of the factors of production are fixed at predetermined levels. Let x_f be the vector of fixed factors, let x_v be the vector of variable factors, and break up w into $w = (w_v, w_f)$, the vectors of prices of the variable and the fixed factors. The short-run conditional factor demand functions will generally depend on x_f, so we write them as $xv(w, y, x_f)$. Then the short-run cost function can be written as:

$$c(w, y, x_f) = w_v x_v(w, y, x_f) + w_f x_f.$$

The term $w_v x_v(w, y, x_f)$ is called short-run variable cost (SVC), and the term $w_f x_f$ is the fixed cost (FC). We can define various derived cost concepts from these basic units:

short-run total cost $= STC = w_v x_v(w, y, x_f) + w_f x_f$

short-run average cost $= SAC = \dfrac{c(w, y, x_f)}{y}$

short-run average variable cost $= SAVC = \dfrac{w_v x_x(w, y, x_f)}{y}$

short-run average fixed cost $= SAFC = \dfrac{w_f x_f}{y}$

short-run marginal cost $= SMC = \dfrac{\partial c(w, y, x_f)}{\partial y}$

When all factors are variable, the firm will optimize in the choice of x_f. Hence, the long-run cost function only depends on the factor prices and the level of output as indicated earlier. We can express this long-run function in terms of the short-run cost function in the following way:

Let $x_f(w, y)$ be the optimal choice of the fixed factors, and let $x_v(w, y) = x_v(w, y, x_f(w, y))$ be the long-run optimal choice of the variable factors. Then the long-run cost function can be written as:

$$c(w, y) = w_v x_v(w, y) + w_f x_f(w, y) = c(w, y, x_f(w, y)).$$

Similarly, we can define the long-run average and marginal cost functions:

long-run average cost $= LAC = \dfrac{c(x, y)}{y}$

long-run marginal cost $= LMC = \dfrac{\partial c(w, y)}{\partial y}$

The long-run average cost equals long-run average variable cost since all costs are variable in the long-run; long-run fixed costs are zero for the same reason. The average variable cost (AVC) and marginal cost (MC) are mirror images of the AP and MP functions.

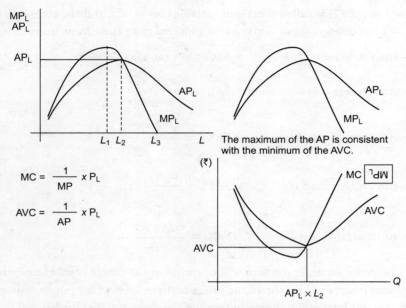

Figure 3.16   Relations between MP and MC, and AP and AC

Implicit Costs   Opportunity costs or *MC* should include all costs associated with an activity. Many of the costs are implicit and difficult to measure. A production activity may adversely affect a person's health. This is an implicit cost that is difficult to measure. Another activity may reduce the time for other activities. It may be possible to make a monetary estimate of the value of all the implicit costs.

Explicit Costs   Explicit costs are those costs where there is an actual expenditure involved. The costs of labor or interest payments are examples. Some implicit costs are estimated and used in the decision process. Depreciation is an example.

3.16   PROFITS

The difference between total revenue and total cost is total profit.

Symbolically,

$$\pi = TR - TC$$

where

     π = Total Profit
     $TR = f(Q) = P.Q$ = Total Revenue
     $TC = f(Q)$ = Total Cost

Thus,

$$\pi = f(Q)$$

The objective of an entrepreneur is to attain maximum profit. Mathematically,

Max. π Subject to: a given level of Cost.

$$\pi = f(Q) = TR - TC$$

If we divide both sides by Q, we get:

$$\frac{\pi}{Q} = \frac{f(Q)}{Q} = \frac{TR - TC}{Q} = \frac{TR}{Q} - \frac{TC}{Q}$$

That is,   Average Profit = Average Revenue − Average Cost

If we measure the total change in profit, then we get:

$$\Delta\pi = \Delta TR - \Delta TC$$

And dividing both sides by ΔQ, we get

$$\frac{\Delta\pi}{\Delta Q} = \frac{\Delta TR}{\Delta Q} - \frac{\Delta TC}{\Delta Q}$$

Marginal Profit = Marginal Revenue − Marginal Cost

Properties of the Profit Function   The Profit Function has the following properties:
  (i)   Non-decreasing in output prices, non-increasing in input prices. If $P_i' \geqq P_i$ for all outputs and $P_j' \leqq P_j$, for all inputs, then $\pi(P') \geq \pi(P)$.
  (ii) Homogeneous of degree 1 in P: $\pi(nP) = n\pi(P)$ for all $n \geq 0$.
  (iii) Convex in P: Let $P'' = nP + (1 - n)P'$ for $0 \leq n \leq 1$. Then $\pi(P'') \leq n\pi(P) + (1 - n)P$ (P').
  (iv) Continuous in P: The function $\pi(P)$ is continuous, at least when $\pi(P)$ is well-defined and $P_i > 0$ for $i = 1,..., n$.

3.16.1   Normal Profits

In neoclassical economics, all costs should be included: wages represent the cost of labor, interest represents the cost of capital, rent represents the cost of land, normal profit (π) represents the cost of entrepreneurial activity which includes risk.

3.16.2   Economic Profits

An economic profit arises when revenue exceeds the opportunity cost of inputs, which includes the cost of capital that is met by normal profits. If a firm is making an economic loss means its economic profit is negative. It follows that all costs are not being met in full and it would be better for the firm to leave the industry in the long-run.

3.17 PROFIT MAXIMIZATION

An important implication of the firm choosing a profit-maximizing production plan is that there is no way to produce the same amounts of outputs at a lower total input cost. Thus, cost minimization is a necessary condition for profit maximization.

$$\text{Max. } \pi = f(Q) = TR - TC$$

Taking first order derivative and setting it equals to zero, we get:

$$\frac{d\pi}{dQ} = \frac{dTR}{dQ} - \frac{dTC}{dQ} = 0$$

$$\frac{dTR}{dQ} = \frac{dTC}{dQ}$$

$$MR = MC$$

This is the necessary condition for profit maximization under any forms of market.

The sufficient condition depends on the second order derivative:

$$\frac{d^2TR}{dQ^2} < \frac{d^2TC}{dQ^2}$$

This shows that the slope of the marginal revenue must be greater than slope of marginal cost. In other words, MC curve must be rising.

3.18 MARKET STRUCTURE AND DEGREE OF COMPETITION

The equilibrium analysis of price and quantity determination in a single market or group of closed related markets by the actions of the individual agents for different markets structures is called a partial equilibrium analysis. Attention is drawn on modelling the market behavior of the firm in the present section. How is the price at which a firm sells its output or the price at which it is willing to purchase inputs determined? In certain situations, the price-taking behavior might be a reasonable approximation to optimal behavior, but in other situations, the price-setting process may not be optimal.

The definition of market is that it is a meeting place where buyer and seller meet each other for the exchange of goods and/or services. On the basis of degree of competition, it is classified into various forms, such as, pure competition, perfect competition, monopoly, monopolistic competition, oligopoly, duopoly, monopsony and bilateral monopoly.

Table 3.2 Forms of market

Different forms of market					
Market Forms	Number of Firms	Products Differentiated or Homogenous	Price a decision variable	Free Entry	Distinguished by
Perfect Competition	Large	Homogenous	No	Yes	Price Competition only
Monopoly	One	A single, unique product	Yes	No	Still constrained by market demand
Monopolistic Competition	Many	Differentiated	Yes, but limited	Yes	Price and quality competition
Oligopoly	Few	Either	Yes	Limited	Strategic behavior

3.18.1   Perfect Competition

Under pure competition there are a large number of independent sellers of some uniform product. In this situation, when each firm sets the price in which it sells its output, it will have to take into account not only the behavior of the consumers, but also the behavior of the other producers. In this market, no player has any control over the market in the determination of prices.

The perfectly competitive markets are based on the following assumptions:

(i)   there are a large number of buyers and sellers, which shows price-taking behavior.

(ii)   there is free entry and exit from market. There is unrestricted mobility of resources among industries.

(iii)   the products sold in such market are homogeneous. All the firms in the industry produce an identical product in the consumers' eyes.

(iv)   there is perfect knowledge of prices among buyers and sellers in the market. In other words, the details of all relevant information are common knowledge for the players in the market.

3.18.1.1   Equilibrium Condition

A competitive firm is one that takes the market price of output as being given. Let $\overline{P}$ be the market price. Then the demand curve facing an ideal competitive firm takes the form:

$$\partial(P) = \begin{bmatrix} 0 & \text{if } P > \overline{P} \\ \text{any amount} & \text{if } P = \overline{P} \\ \infty & \text{if } P < \overline{P} \end{bmatrix}$$

A competitive firm is free to set whatever price it wants and produce whatever quantity it wants to produce. However, if a firm is in a competitive market, it is clear that each firm that sells the product must sell it for the same price: for if any firm attempted to set its price at a level greater than the market price, it would immediately lose all its customers. If any firm set its price at a level below the market price, all the consumers would immediately come to it. Thus, each firm must take the market price as a given, exogenous variable when it determines its supply decision.

Since the competitive firm must take the market price as given, its profit maximization problem is simple. The firm only needs to choose output level y so as to solve:

$$\text{Max. } \pi = f(Q) = TR - TC$$

Taking first order derivative and setting it equals to zero, we get:

$$\frac{d\pi}{dQ} = \frac{dTR}{dQ} - \frac{dTC}{dQ} = 0$$

$$\frac{dTR}{dQ} = \frac{dTC}{dQ}$$

where $TR = P.Q$ and P is given information, so treated as a constant.

Substituting this into above equation, we get

$$\frac{d(\bar{P} \cdot Q)}{dQ} = \frac{dTC}{dQ}$$

$$\bar{P} \cdot \frac{dQ}{dQ} = \frac{dTC}{dQ}$$

$$\bar{P} = \frac{dTC}{dQ}$$

Since
$$AR = \frac{TR}{Q} = \frac{(\bar{P} \cdot Q)}{Q} = \bar{P}$$

So for equilibrium in perfectly competitive market:

$$MR = MC = AR = \boldsymbol{\bar{P}}$$

This is the necessary condition for profit maximization under any form of market.

The sufficient condition depends on the second order derivative.

$$\frac{d^2 TR}{dQ^2} < \frac{d^2 TC}{dQ^2}$$

This shows that the slope of the marginal revenue must be greater than slope of marginal cost. In other words, the MC curve must be rising as shown in Fig. 3.17

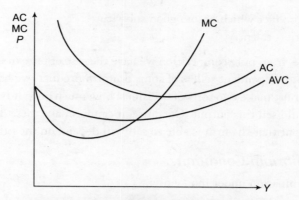

Figure 3.17 Equilibrium in perfect competition

The market price is determined by the requirement that the total amount of output that the firms wish to supply will be equal to the total amount of output that the consumers wish to demand. Formally, we can say: *A partial equilibrium price P is a price where the aggregate quantity demanded equals the aggregate quantity supplied.* That is, it is the solution of the following equation:

$$\sum_{i=1}^{n} x_i(P) = \sum_{j=1}^{j} y_i(P)$$

Once this equilibrium price is determined, we can go back to look at the individual supply schedules of each firm and determine the firm's level of output, its revenue, and its profits as shown in Fig. 3.18. The first figure has positive profits, the second has zero profits and the third has negative profits. Even though the third firm has negative profits, it may make sense for it to continue to produce as long as its revenues cover its variable costs.

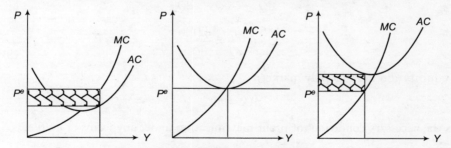

Figure 3.18 Profits in perfect completion

3.18.2 Monopoly

A monopolist is the supply-side of the market and has complete control over the amount offered for sale. Profits will be maximized at the level of output where marginal revenue equals marginal cost. The following are the characteristics of monopoly:

i. There is only one seller but many buyers.

ii. There is one product which has no good substitutes.

iii. There are barriers to entry.

At the opposite pole, from pure competition we have the case of pure monopoly. Here instead of a large number of independent sellers of some uniform product, we have only one seller. A monopolistic firm must make two sorts of decisions: how much output it should produce, and at what price it should sell this output. Of course, it cannot make these decisions unilaterally. The amount of output that the firm is able to sell will depend on the price that it sets.

3.18.2.1 *Equilibrium Condition*

For price out determination under monopoly:

$$\text{Max. } \pi = f(Q) = TR - TC$$

Taking first order derivative and setting it equals to zero, we get:

$$\frac{d\pi}{dQ} = \frac{dTR}{dQ} - \frac{dTC}{dQ} = 0$$

$$\frac{dTR}{dQ} = \frac{dTC}{dQ}$$

where $TR = P.Q$ and P is the given information, so it is treated as constant.

Substituting this into above equation, we get:

$$\frac{d(P \cdot Q)}{dQ} = \frac{dTC}{dQ}$$

$$P \cdot \frac{dQ}{dQ} + Q\frac{dP}{dQ} = \frac{dTC}{dQ}$$

$$P + Q\frac{dP}{dQ} = \frac{dTC}{dQ}$$

So for equilibrium in monopoly market:

$$MR = MC$$

This is the necessary condition for profit maximization under any forms of market.

The sufficient condition depends on the second order derivative.

$$\frac{d^2TR}{dQ^2} < \frac{d^2TC}{dQ^2}$$

This shows that the slope of the marginal revenue must be greater than slope of marginal cost. In other words, *MC* curve must be rising.

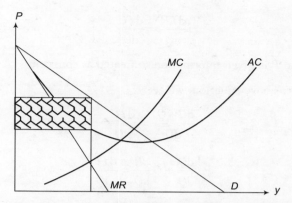

Figure 3.19 Profit-maximizing monopolist's price and output

3.18.3 Monopolistic Competition

Imperfect competition refers to those market structures that fall between perfect competition and pure monopoly. There are types of *Imperfectly Competitive Markets*: Monopolistic Competition, where many firms are selling products that are similar but not identical and Oligopoly, where there are only a few sellers, each offering a similar or identical product as the others. Markets that have some features of competition and some features of monopoly are called monopolistic competition. Following are the characteristics of monopolistic competition:

i. *Many sellers:* There are many firms competing for the same group of customers.

ii. *Product differentiation:* Each firm produces a product that is at least *slightly different* from those of other firms. Rather than being a price-taker, each firm faces a *downward-sloping demand curve.*

iii. *Free entry and exit:* Firms can enter or exit the market without restriction.

iv. The number of firms in the market adjusts until economic profits are zero.

The monopolistically competitive firms which are making economic profits in the short-run encourage new firms to enter the market. This increases the number of products offered and reduces demand faced by the firms already in the market.

Incumbent firms' demand curves shift to the left. Demand for the incumbent firms' products fall, and their profits decline.

3.18.3.1 Equilibrium Condition

For price out determination under monopoly:

$$\text{Max. } \pi = f(Q) = TR - TC$$

Taking first order derivative and setting it equals to zero, we get:

$$\frac{d\pi}{dQ} = \frac{dTR}{dQ} - \frac{dTC}{dQ} = 0$$

$$\frac{\mathrm{d}TR}{\mathrm{d}Q} = \frac{\mathrm{d}TC}{\mathrm{d}Q}$$

where $TR = P \cdot Q$ and P is given information so treated as constant.

Substituting this into above equation, we get:

$$\frac{\mathrm{d}(P \cdot Q)}{\mathrm{d}Q} = \frac{\mathrm{d}TC}{\mathrm{d}Q}$$

$$P \cdot \frac{\mathrm{d}Q}{\mathrm{d}Q} + Q\frac{dP}{dQ} = \frac{\mathrm{d}TC}{\mathrm{d}Q}$$

$$P + Q\frac{dP}{dQ} = \frac{\mathrm{d}TC}{\mathrm{d}Q}$$

So for equilibrium in monopoly market:

$$MR = MC$$

This is the necessary condition for profit maximization under any forms of market.

The sufficient condition depends on the second order derivative.

$$\frac{d^2TR}{dQ^2} < \frac{\mathrm{d}^2TC}{dQ^2}$$

This shows that the slope of the marginal revenue must be greater than slope of marginal cost. In other words, MC curve must be rising. This is shown in Fig. 3.20.

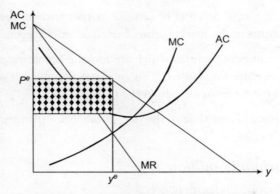

Figure 3.20   Monopolistic competition equilibrium

3.18.4   Oligopoly Models

Oligopoly is the study of market interactions with a few number of firms. Such an industry usually does not exhibit the characteristics of perfect competition, since individual firms' actions can in fact influence market price and the actions of other firms. The modern study of this subject is grounded almost entirely in the theory of games. It is of two types:

i. Non-Collusive Oligopoly Model

ii. Collusive Oligopoly Model

Some of the important models are discussed in this section.

3.18.4.1 Cournot's Duopoly Model

A fundamental model for the analysis of oligopoly was the oligopoly model that was proposed by Cournot, a French economist, in 1838. A Cournot equilibrium, is a special set of production levels that have the property that no individual firm has an incentive to change its own production level if other firms do not change theirs. To formally present this equilibrium concept, suppose there are J firms producing a single homogeneous product. If firm j produces output level q_j, the firm's cost is $c_j(q_j)$. There is a single market inverse demand function. The total supply is $= \sum_{j=1}^{j} q_j$. The profit to firm j is $p(\hat{q})q_j - c_j(q_j)$. A set of output levels $q_1, q_2,...,$ q_J constitutes a Cournot equilibrium if for each $j = 1, 2,..., J$ the profit to firm j cannot be increased by changing q_j alone. Accordingly, the Cournot model can be regarded as a one-shot game: the profit of firm j is its payoff, and the strategy space of firm j is the set of outputs, and thus a Cournot equilibrium is just a pure strategy Nash equilibrium. Then the first-order conditions for the interior optimum are:

$$p'(\hat{q})q_j + p(\hat{q}) - c'_j(q_j) = 0, j = 1, 2,..., j.$$

The first-order condition for firm determines firm j optimal choice of output as a function of its beliefs about the sum of the other firms' outputs, denoted by $\hat{q} - j$, i.e., the FOC condition can be written as:

$$p'(q_j + \hat{q}_{-j})q_j + p(\hat{q}) - c'_j (q_j) = 0, j = 1, 2,..., j.$$

The solution to the above equation is called the reaction function to the total outputs produced by the other firms. Reaction functions give a direct characterization of Cournot equilibrium. A set of output levels $q_1, q_2,..., q_J$ constitutes a Cournot equilibrium if for each reaction function given, $q_j = Q_j(\hat{q} - 1) j = 1, 2,..., j.$

An important special case is that of duopoly which is an industry with just two firms. In this case, the reaction function of each firm is a function of just the other firm's output. Thus, the two reaction functions have the form $Q_1(q_2)$ and $Q_2(q_1)$.

3.18.4.2 Bertrand Duopoly Model

Another model of oligopoly of some interest is the Bertrand model. The Cournot model and Stackelberg model take the firms' strategy spaces as being quantities, but it seems equally natural to consider what happens if price is chosen as the relevant strategic variable. Almost 50 years after Cournot, another French economist, Joseph Bertrand (1883), offered a different view of a firm under imperfect competition and is known as the Bertrand model of oligopoly. Bertrand argued that it is much more natural to think of firms competing in their choice of price, rather than quantity. This small difference completely changes the character of market equilibrium. This model is striking, and it contrasts starkly with what occurs in the Cournot model: with just two firms in a market, we obtain a perfectly competitive outcome in the Bertrand model! The Cournot game and the Bertrand game have a radically different structure. In the Cournot game, the payoff to each firm is a continuous function of its strategic choices; in the Bertrand game, the payoffs are discontinuous functions of the strategies.

3.18.4.3   Stackelberg Duopoly Model

There are alternative methods for characterising the outcome of an oligopoly. One of the most popular of these is that of quantity leadership, also known as the Stackelberg model. Consider the special case of a duopoly. In the Stackelberg formulation one firm, say firm 1, is considered to be the leader and the other, firm 2, is the follower. The leader may, for example, be the larger firm or may have better information. If there is a well-defined order for firms committing to an output decision, the leader commits first. Given the committed production level q_1 of firm 1, firm 2, the follower, will select q_2 using the same reaction function as in the Cournot theory. That is, firm 2 finds q_2 to maximize $\pi_2 = p(q_1 + q_2)q_2 - c_2(q_2)$, where $p(q_1 + q_2)$ is the industrywide inverse demand function. This yields the reaction function $Q_2(q_1)$. Firm 1, the leader, accounts for the reaction of firm 2 when originally selecting q_1. In particular, firm 1 selects q_1 to maximize $\pi_2 = p(q_1 + Q_2(q_1))q_1 - c_1(q_1)$. That is, firm 1 substitutes $Q_2(q_1)$ for q_2 in the profit expression.

Stackelberg equilibrium does not yield a system of equations that must be solved simultaneously. Once the reaction function of firm 2 is found, firm 1's problem can be solved directly. Usually, the leader will do better in a Stackelberg equilibrium than in a Cournot equilibrium.

3.18.4.4   Cartels

In the above modes, each firm maximizes its own profits and makes its decisions independently of the other firms. These are called Non-Collusive Oligopoly Models. However, if they coordinate their actions, the industry structure where the firms start colluding to some degree in setting their prices and outputs is called a cartel. This is also called Collusive Oligopoly Model.

3.18.5   Monopsony

Output markets can be classified as *competitive* or *monopolistic*" depending on whether firms take the market price as given or whether firms take the demand behavior of consumers as given. There is similar classification for inputs markets. If firms take the factor prices as given, then we have competitive factor markets. If, instead, there is only one firm which demands some factor of production and it takes the supply behavior of its suppliers into account, then we say we have a monopsonistic factor market. The behavior of a monopsonist is analogous to that of a monopolist.

PROBLEMS

1. Given the supply function $Q^d_x = 30P_x$ (P_x is given in rupees), derive
    (a) the producer's supply schedule and
    (b) the producer's supply curve.
    (c) State the variables which is kept as constant in the given supply function?
    (d) Find out the minimum price that this producer must be offered in order to induce him to start supplying commodity X to the market?

2. Suppose that from the equilibrium condition, the government decides to collect a tax of ₹5 per unit sold, from each of the 4,000 identical sellers of commodity X.
    (a) State its effect on the equilibrium price and quantity of commodity X.
    (b) Find out the total amount of taxes collected by the government.
    (c) Draw the TP, AP_L, and MP_L curves as smooth curves on the same coordinate space.
    (d) Explain the shape of the AP_L and MP_L curves in terms of the shape of the TP curve.
    (e) What does the law of diminishing returns state in terms of "labor" and "land,"?

3. Explain briefly how each of followings might arise:
    (a) constant returns to scale,
    (b) increasing returns to scale, and
    (c) decreasing returns to scale.

4. Starting with the general production function $Q = f(L, K)$, which states that output Q is a function of or depends on the quantity of labor (L) and capital (K) used in production, derive using calculus the expression for the slope of the isoquant.

5. Let labor be only variable input in the short-run and its price is constant, explain (a) the AVC curve and (b) the MC curve in terms of the shape of the AP_L and MP_L curves, respectively.
    (a) State the relationship between the long-run and the short-run.
    (b) Derive LAC curve.

6. Differentiate between followings:
    (a) decreasing returns to scale and increasing cost industries,
    (b) increasing returns to scale and decreasing cost industries, and
    (c) constant returns to scale and constant cost industries.

7. (a) State the difference between the pure monopolist and the perfectly competitive firm.
    (b) State the assumptions applied in determining the short-run equilibrium output of the pure monopolists.

8. Compare the long-run equilibrium point of a pure monopolist with that of a perfectly competitive firm and industry. Should the government break up a monopoly into a great number of perfectly competitive firms?

9.   (a) Define monopolistic competition and state examples.

     (b) Identify the competitive and monopolistic elements in monopolistic competition.

     (c) How to define the market demand curve, the market supply curve, and the equilibrium price under monopolistic competition?

10. Define oligopoly. State the important characteristic in oligopolistic markets. Explain the issues involved in price output determination under Oligopoly.

❏❏❏

4

Interest Formulae and their Applications

4.1 INTRODUCTION

It has been recognized that the value of money is different at different time periods and this concept is referred to as *time value of money*. In this regard, it is recognized that a rupee today is worth more than a rupee one or more years from now because of the interest or profit it can earn. Thus, it can be said that interest is the manifestation of time value of money.

4.2 WHY RETURN TO CAPITAL IS CONSIDERED?

Capital is required in engineering, other business projects and ventures for their operations. The capital thus required may be classified into two broad categories.

Equity Capital   It refers to the capital owned by individuals who have invested their money or property in business project or venture in the hope of receiving a profit.

Debt Capital   It is often called borrowed capital and is obtained from lenders for investment. In return, the lenders receive interest from the borrowers.

Return to capital in the form of interest and profit is an essential factor of engineering economy studies owing to the following reasons:
  ➢ Interest and profit pay the providers of capital for forgoing its use during the time the capital is being used.
  ➢ Interest and profit are payment for the risk the investor takes in permitting another person, or an organization, to use his or her capital.

4.3 INTEREST, INTEREST RATE AND RATE OF RETURN

Interest, in general, is computed by taking the difference between an ending amount of money and the beginning amount. In case of zero or negative difference, interest does not accrue. Interest is either paid or earned. In situations where a person or organization borrows money, interest is paid on the borrowed capital. On the other hand, interest is earned when a person or organization saved, invested or lent money and obtains the return of a larger amount. Interest paid on borrowed capital is determined by the following relation:

$$\text{Interest} = \text{Amount owed now} - \text{Principal amount} \qquad (4.1)$$

Interest rate is defined as the interest paid over a specific time unit as a percentage of original principal amount.

Thus,

$$\text{Interest rate (\%)} = \frac{\text{Interest accrued per time unit}}{\text{Original amount}} \times 100 \qquad (4.2)$$

The time unit of the rate is called the interest period. The most commonly used interest period to state the interest rate is 1 year. However, shorter time period such as 2% per month can also be used. Interest period of the interest rate should always be stated. If only the rate is stated e.g., 10%, a 1 year interest period is assumed.

Example 4.1

You borrow ₹ 11,000 today and must repay a total of ₹ 11,550 exactly 1 year later. What is the interest amount and the interest rate paid?

Solution

Interest amount is obtained by using Equation (4.1) as:

$$\text{Interest} = ₹11,550 - ₹11,000 = ₹550$$

Interest rate is computed by using Equation (4.2) as:

$$\text{Interest rate (\%)} = \frac{₹550}{₹11,000} \times 100 = 5\% \text{ per year}$$

Example 4.2

You plan to borrow ₹ 25,000 from a bank for 1 year at 8% interest for your personal work. Calculate the interest and the total amount due after 1 year.

Solution

Total interest accrued is computed by using Equation (4.2) as:

$$\text{Interest} = ₹25,000 \times 0.08 = ₹2,000$$

Total amount due is the sum of principal and interest.

Thus, the total amount due after 1 year = ₹ 25,000 + ₹ 2,000 = ₹ 27,000.

From the point of view of a lender or an investor, interest earned is the difference between final amount and the initial amount.

Thus,

$$\text{Interest earned} = \text{total amount now} - \text{original amount} \qquad (4.3)$$

Interest earned over a specified period of time is expressed a percentage of the original amount and is called rate of return (ROR).

Thus,

$$\text{ROR (\%)} = \frac{\text{Interest accrued per time unit}}{\text{Original amount}} \times 100 \qquad (4.4)$$

4.4   SIMPLE INTEREST

When the total interest paid or earned is linearly proportional to the principal, the interest rate and the number of interest periods for which the principal is committed, the interest and the interest rate are said to be simple. Simple interest is not used frequently in commercial practice.

When simple interest is used, the total interest, I paid or earned is computed from the following relation:

$$I = (P)(n)(i) \qquad (4.5)$$

where,

P is the amount borrowed or lent.
n is the number of interest periods e.g. years
i is the interest rate per interest period

Total amount paid or earned at the end of n interest periods is $P + I$.

Example 4.3

You borrow ₹1,500 from your friend for three years at a simple interest rate of 8% per year. How much interest you will pay after three years and what is the value of total amount that will be paid by you?

Solution

The interest I paid is computed by using Equations (4.5). Here, $P = ₹1,500$; $n = 3$ years; $i = 8\%$ per year.

Thus, $I = (₹1,500)\,(3)\,(0.08) = ₹360.$

Total amount paid by you at the end of three years would be ₹1,500 + ₹360 = ₹1,860.

4.5   COMPOUND INTEREST

Whenever the interest charge for any interest period e.g. a year, is based on the remaining principal amount plus any accumulated interest charges up to the beginning of that period, the interest is said to be compound interest.

Example 4.4

Show the effect of compounding interest for Example 4.3, if the interest is compounded at the rate of 8% per year.

Solution

The effect of compounding is shown in Table 4.1.

Table 4.1   Compounding of interest

Period	(1) Amount owed at the beginning of period	(2) = (1) × 8% Interest amount for period	(3) = (1) + (2) Amount owed at the end-of-period
1	₹1,500	₹120	₹1,620
2	₹1,620	₹129.60	₹1,749.60
3	₹1,749.60	₹139.97	₹1,889.57

It can be seen from the above calculations that a total of ₹1,889.57 would be due for repayment at the end of third year. This amount can be compared with ₹1,860 for the same problem given in Example 4.3 with simple interest. There is a difference of ₹29.57 and this difference is due to the effect of compounding which is essentially the calculation of interest on previously earned interest.

The general formula for calculating total amount due after a particular period is given below:

Total amount due after a number of years = principal × (1 + interest rate)$^{\text{number of years}}$ (4.6)

Compound interest is much more common in practice than simple interest and is used in engineering economy studies.

4.6   THE CONCEPT OF EQUIVALENCE

The time value of money and the interest rate considered together helps in developing the concept of economic equivalence, which means that different sums of money at different times would be equal in economic value. For example, if the interest rate is 5% per year, ₹1,000 today is equivalent to ₹1,050 one year from today. Thus, if someone offers you ₹1,000 today or ₹1,050 one year from today, it would make no difference which offer you accept from an economic point of view. It should be noted that ₹1,000 today or ₹1,050 one year from today are equivalent to each other only when the interest rate is 5% per year. At a higher or lower interest rate, ₹ 1,000 today is not equivalent to ₹ 1,050 one year from today. Similar to future equivalence, this concept can also be applied with the same logic to determine equivalence for previous years. A total of ₹ 1,000 now is equivalent to ₹ 1,000/1.05 = ₹ 952.38 one year ago at an interest rate of 5% per year. From these illustrations, we can state the following: ₹ 952.38 last year, ₹ 1,000 now and ₹ 1,050 one year from now are equivalent at an interest rate of 5% per year.

4.7 CASH FLOW DIAGRAMS

Cash flow diagrams are important tools in engineering economy because they form the basis for evaluating alternatives. They are used in economic analysis of alternatives that involve complex cash flow series. A cash flow diagram is a graphical representation of cash flows drawn on a time scale. The diagram includes what is known, what is estimated and what is needed i.e. once the cash flow diagram is complete, one should be able to work out the problem by looking at the diagram. The cash flow diagram employs the following conventions:

1. The horizontal line is a time scale, with progression of time moving from left to right. The period (e.g., year, quarter and month) can be applied to intervals of time. Consider Fig. 4.1, which shows a typical cash flow time scale for 5 periods. Cash flow diagram time $t = 0$ is the present and the end of interval 1 is the end of time period 1.

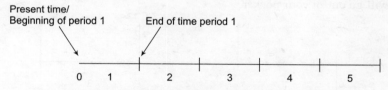

Figure 4.1   A typical cash flow time scale for 5 periods

2. The arrows signify cash flows and are placed at the end of the period when the end-of-period convention is used. The end-of-period convention means that all cash flows are assumed to occur at the end of an interest period. In order to make distinction, downward arrows represent expenses (negative cash flows or cash outflows). A few typical examples of negative cash flows are first cost of assets, engineering design cost, annual operating costs, periodic maintenance and rebuild costs, loan interest and principal payments, major expected/unexpected upgrade costs, income taxes etc. Upward arrows represent receipts (positive cash flows or cash inflows). A few typical examples of positive cash flows are revenues, operating cost reductions, asset salvage value, receipt of loan principal, income tax savings, receipts from stock and bond sales, construction and facility costs savings, saving or return of corporate capital funds etc. Fig. 4.2 depicts a receipt (cash inflow) at the end of year 1 and equal disbursements (cash outflows) at the end of years 2, 3 and 4.

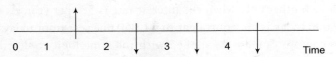

Figure 4.2   Example of positive and negative cash flows

3. The cash flow diagram is dependent on the point of view. For example, if you borrow ₹1,000 from your friend now and repay him in equal yearly installments in 3 years, then the cash flow diagram from your point of view would be as shown in Fig. 4.3 and from your friend's point of view it would be as shown in Fig. 4.4.

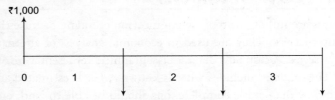

Figure 4.3   Cash flow diagram from your point of view

Fig. 4.3 shows that there is a positive cash flow of ₹1,000 at the present time as it is a cash inflow for you. Fig. 4.3 also shows that there are 3 equal negative cash flows at the end of year 1, 2, and 3. These cash flows are negative for you as you will repay the borrowed money and the money will go out of your pocket.

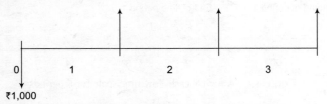

Figure 4.4   Cash flow diagram from your friend's point of view

Fig. 4.4 is obtained by reversing the direction of arrows of Fig. 4.3 owing to the fact that a positive cash flow for you is obviously a negative cash flow for your friend and vice-versa.

When cash inflows and cash outflows occur at the end of a given interest period, the net cash flow can be determined from the following relationship:

$$\text{Net cash flow} = \text{Receipts} - \text{Disbursements}$$
$$= \text{Cash inflows} - \text{Cash outflows}$$

For example, consider the cash flow diagram shown in Fig. 4.5.

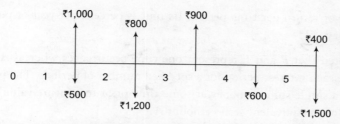

Figure 4.5 Cash flow diagram with cash inflows and cash outflows

The net cash flow at the end of interest period 1 = ₹ 1,000 – ₹ 500 = ₹ 500.

The net cash flow at the end of interest period 2 = ₹ 800 – ₹ 1,200 = – ₹ 400.

The net cash flow at the end of interest period 3 = ₹ 900 – ₹ 0 = ₹ 900.

The net cash flow at the end of interest period 4 = ₹ 0 – ₹ 600 = – ₹ 600.

The net cash flow at the end of interest period 5 = ₹ 400 – ₹ 1,500 = – ₹ 1,100.

The cash flow diagram in terms of net cash flows is shown in Fig. 4.6.

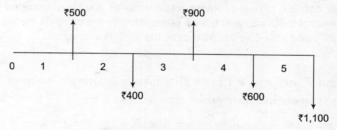

Figure 4.6 Net cash flows

4.8 TERMINOLOGY AND NOTATIONS/SYMBOLS

The equations and procedures that are used in engineering economy employs the following terms and symbols.

P = Equivalent value of one or more cash flows at a reference point in time called the present or time 0. P is also referred to as present worth ($P.W.$), present value ($P.V.$), net present value ($N.P.V.$), discounted cash flow ($D.C.F$) and capitalized cost ($C.C.$). Its unit is rupees.

F = Equivalent value of one or more cash flows at a reference point in time called the future. F is also referred to as future worth ($F.W.$), future value ($F.V.$). Its unit is also rupees.

A = Series of consecutive, equal end-of-period amounts of money. A is also referred to as annual worth ($A.W.$), annuity and equivalent uniform annual worth ($E.U.A.W$). Its unit is rupees per year, rupees per month.

n = Number of interest periods. Its unit is years or month or days.

i = interest rate or return per time period. Its unit is percent per year or percent per month or percent per day.

It should be noted that P and F represent one time occurrences whereas A occurs with the same value once each interest period for a specified number of periods. The present value P in fact, represents a single sum of money at some time prior to a future value F or prior to the first occurrence of an equivalent series amount A.

It should further be noted that symbol A always represents a uniform amount that extends through consecutive interest periods. Both conditions must exist before a series can be represented by A.

The interest rate i is assumed to be a compound rate, unless specifically stated as simple interest. i is expressed in percent per interest period, for example 10% per year. Unless otherwise stated, assume that the rate applies throughout the entire n years or interest periods. The following examples illustrate the meaning of the above discussed terms and symbols:

Example 4.5

You plan to borrow ₹ 15,000 to help in buying a two wheeler. You have arranged to repay the entire principal plus interest of 8.5% per year after 5 years. Identify the symbols involved and their values for the total amount owed after 5 years. Also draw the cash flow diagram.

Solution

In this case P and F are involved since all amounts are single payment. n and i are also involved. So, the symbols and their values are:

$P = ₹ 15,000 \qquad i = 8.5\%$ per year $\qquad n = 5$ years $\qquad F = ?$

The cash flow diagram for this case is shown in Fig. 4.7(a) and (b).

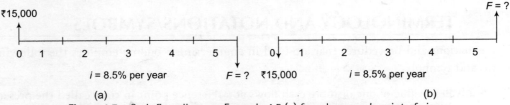

Figure 4.7 Cash flow diagram, Example 4.5 (a) from borrower's point of view
(b) from lender's point of view

Example 4.6

Assume you borrow ₹ 50,000 now at 8% per year for 10 years and must repay the borrowed money in equal yearly payments. Determine the symbols involved and their values. Draw the cash flow diagram.

Solution

$P = ₹ 50,000 \qquad n = 10$ years $\qquad i = 8\%$ per year $\qquad A = ?$

The cash flow diagram is shown in Fig. 4.8 (lender's perspective).

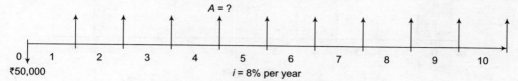

Figure 4.8   Cash flow diagram, Example 4.6

Example 4.7

You deposit ₹50,000 into your saving account that pays 8% per year. You plan to withdraw an equal end-of-year amount of ₹10,000 for 5 years, starting next year. At the end of sixth year, you plan to close your account by withdrawing the remaining money. Define the symbols involved and draw the cash flow diagram.

Solution

$P = ₹50,000$, $A = ₹10,000$, $F = ?$ at the end of year 6, $n = 5$ years for A series and 6 for the F value, $i = 8\%$ per year

The cash flow diagram from depositor's point of view is shown in Fig. 4.9.

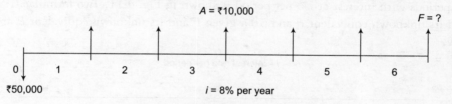

Figure 4.9   Cash flow diagram, Example 4.7

Example 4.8

You want to deposit a particular amount now in your saving accounts so that you can withdraw an equal annual amount of ₹15,000 per year for the first 5 years starting 1 year after the deposit, and a different equal annual withdrawl of ₹25,000 per year for the following 5 years. How much you should deposit? The rate of interest is 8% per year. Define the symbols involved and draw the cash flow diagram.

Solution

$P = ?$ $n = 10$ years $A_1 = ₹15,000$ per year for first 5 years

$A_2 = ₹25,000$ per year for the next five years i.e. from year 6 to year 10

$i = 8\%$ per year

The cash flow diagram from your perspective is shown in Fig. 4.10.

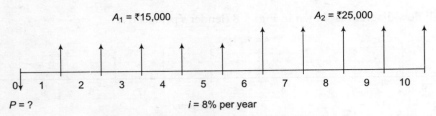

$A_1 = ₹15,000$ $A_2 = ₹25,000$

$P = ?$ $i = 8\%$ per year

Figure 4.10   Cash flow diagram, Example 4.8

4.9   INTEREST FORMULA FOR DISCRETE CASH FLOW AND DISCRETE COMPOUNDING

The word discrete cash flow emphasizes that the cash flow follows the end-of-period convention and occurs at the end of a period. Similarly, discrete compounding refers to the compounding of interest once each interest period. The interest formulae for these cases are given below.

4.9.1   Interest Formulae Relating Present and Future Equivalent Values of Single Cash Flows

A cash flow diagram that involves a present single sum P, and future single sum F, separated by n periods with interest at i% per period is shown in Fig. 4.11. Two formulae (i) a given P and its unknown equivalent F and (ii) a given F and its unknown equivalent P are given below:

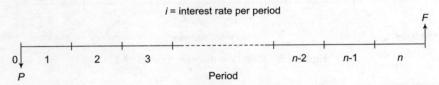

i = interest rate per period

Period

Figure 4.11   Cash flow diagram relating present equivalent and future equivalent of single cash flow

4.9.1.1   *Finding F when Given P*

If an amount P is invested at time $t = 0$ as shown in Fig. 4.11 and i% is the interest rate per period, the amount will grow to a future amount of $P + Pi = P(1 + i)$ by the end of one period. Thus, if F_1 is the amount accumulated by the end of one period, then,

$$F_1 = P(1 + i)$$

At the end of the second period, the amount accumulated F_2 is the amount after 1 period plus the interest from the end-of-period 1 to the end-of-period 2 on the entire F_1.

Thus,

$$F_2 = F_1 + F_1 i$$
$$= P(1 + i) + P(1 + i)i$$

$$= P(1 + i)^2$$

Similarly, the amount of money accumulated at the end-of-period 3 i.e.

$$F_3 = F_2 + F_2 i = P(1 + i)^2 + p(1 + i)^2 i = P(1 + i)^3$$

From the preceding values, it is evident that the amount accumulated at the end-of-period n i.e. F would be $P(1 + i)^n$.

Thus,

$$F = P(1 + i)^n \tag{4.7}$$

The factor $(1 + i)^n$ in Equation (4.7) is commonly called the *single payment compound amount factor* (SPCAF). A standard notation has been adopted for this factor as well as for other factors that will appear in the subsequent sections. The notation includes two cash flow symbols, the interest rate, and the number of periods. The general form of the notation is $(X/Y, i\%, n)$. The letter X represents what is to be determined, the letter Y represents what is given. Thus, the notation for the single payment compound amount factor which is $(1 + i)^n$ is $(F/P, i\%, n)$. F/P means find F when P is given. Using this factor F can be written as:

$$F = P(F/P, i\%, n) \tag{4.8}$$

Thus, $(F/P, 5\%, 10)$ represents the factor that is used to calculate future amount F accumulated in 10 periods (may be year), if the interest rate is 5% per period.

To simplify engineering economy calculations, tables of factor values have been prepared for interest rates from 0.25% to 50% and time periods from 1 to large n values, depending on the i value. These tables are given at the end of this book in Appendix A. In these tables, several factors are arranged in different columns and the interest period is arranged in the first column of the tables. For a given factor, interest rate and interest period, the correct value of a factor is found at the intersection of the factor name and n. For example, the value of the factor $(F/P, 5\%, 10)$ is found in the F/P column of table 10 at period 10 as 1.6289.

4.9.1.2 Finding P when Given F

From Equation (4.7), $F = P(1 + i)^n$. Solving this for P gives the following relationship:

$$P = F(1 + i)^{-n} \tag{4.9}$$

The factor $(1 + i)^{-n}$ is called the *single payment present worth factor* (SPPWF). The notation for this factor is $(P/F, i\%, n)$ and thus, P can be calculated using this factor as:

$$P = F(P/F, i\%, n) \tag{4.10}$$

The numerical values of the factor $(P/F, i\%, n)$ are also given in tables for a wide range of values of i and n.

Example 4.9

A company manufactures transducers. It is investigating whether it should update certain equipment now or wait to do it later. If the cost now is ₹10,00,000, what will be the equivalent amount 5 years from now at an interest rate of 8% per year?

Solution

The cash flow diagram is shown in Fig. 4.12.

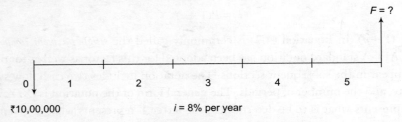

Figure 4.12   Cash flow diagram, Example 4.9

$$F = P(F/P, i\%, n)$$

$$= ₹10,00,000(F/P, 8\%, 5)$$

$$= ₹10,00,000(1.4693)$$

$$= ₹14,69,300$$

Example 4.10

A transport company is considering installing temperature loggers in all of its refrigerated trucks for monitoring temperature during transit. If the system will reduce insurance claims by ₹5,00,000 two years from now, how much should the company be willing to spend now if it uses an interest rate of 10% per year?

Solution

The cash flow diagram is shown in Fig. 4.13.

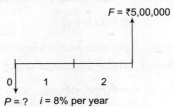

Figure 4.13   Cash flow diagram, Example 4.10

$$P = F(P/F, i\%, n)$$

$$= ₹5,00,000(P/F, 10\%, 2)$$

$$= ₹5,00,000(0.8246)$$

$$= ₹4,13,200$$

Example 4.11

A company is examining its cash flow requirements for the next 5 years. It expects to replace a few machines and office computers at various times over the 5-year planning period. The company expects to spend ₹45,000 two years from now, ₹40,000 three years from now and ₹25,000 five years from now. What is the present worth of the planned expenditure at an interest rate of 6% per year?

Solution

The cash flow diagram is shown in Fig. 4.14.

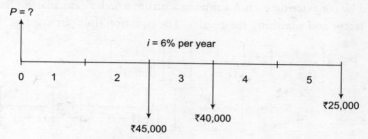

Figure 4.14 Cash flow diagram, Example 4.11

$$P = F(P/F, i\%, 2) + F(P/F, i\%, 3) + F(P/F, i\%, 5)$$

$$= ₹45,000(P/F, 6\%, 2) + ₹40,000(P/F, 6\%, 3) + ₹25,000(P/F, 6\%, 5)$$

$$= ₹45,000(0.8900) + ₹40,000(0.8396) + ₹25,000(0.7473)$$

$$= ₹92,316.5$$

4.9.2 Interest Formulae Relating a Uniform Series (Annuity) to its Present and Future Worth

The cash flow diagram involving a series of uniform disbursement, each of amount A, occurring at the end of each period for n periods with interest at $i\%$ per period is shown in Fig. 4.15. Such a uniform series is often called an annuity. The following points pertaining to the occurrence of uniform series, A should be noted:

➤ Present worth P occurs one interest period before the first A.
➤ Future worth F occurs at the same time as the last A i.e. n periods after P.

The occurrence of P, F and A can be observed in Fig. 4.15. The following four formulas relating A to P and F are developed.

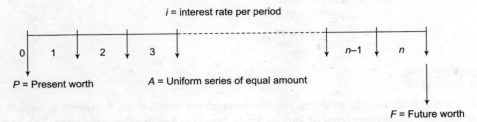

Figure 4.15   General cash flow diagram for uniform series (annuity)

4.9.2.1   *Finding P when given A*

Fig. 4.15 shows that a uniform series (annuity) each of equal amount A exists at the end of each period for n periods with interest $i\%$ per period. The formula for the present worth P can be determined by considering each A values as a future worth F, calculating its present worth with the P/F factor and summing the results. The equation thus obtained is:

$$P = A\left[\frac{1}{(1+i)^1}\right] + A\left[\frac{1}{(1+i)^2}\right] + A\left[\frac{1}{(1+i)^3}\right] + \cdots$$

$$+ A\left[\frac{1}{(1+i)^{n-2}}\right] + A\left[\frac{1}{(1+i)^{n-1}}\right] + A\left[\frac{1}{(1+i)^n}\right]$$

The terms in brackets are the P/F factors for periods 1 through n, respectively; A being common, factor it out. Thus,

$$P = A\left[\frac{1}{(1+i)^1} + \frac{1}{(1+i)^2} + \frac{1}{(1+i)^3} + \cdots + \frac{1}{(1+i)^{n-2}} + \frac{1}{(1+i)^{n-1}} + \frac{1}{(1+i)^n}\right] \quad (4.11)$$

In order to simplify Equation (4.11) so as to get the P/A factor, multiply each term of Equation (4.11) by $\dfrac{1}{(1+i)}$. This results in Equation (4.12) given below. Then subtract Equation (4.11) from Equation (4.12) and simplify to obtain expression for P.

$$\frac{P}{(1+i)} = A\left[\frac{1}{(1+i)^2} + \frac{1}{(1+i)^3} + \frac{1}{(1+i)^4} + \cdots + \frac{1}{(1+i)^{n-1}} + \frac{1}{(1+i)^n} + \frac{1}{(1+i)^{n+1}}\right] \quad (4.12)$$

Equation (4.11) is subtracted from Equation (4.12) to get Equation (4.13)

$$\frac{P}{(1+i)} - P = \frac{A}{(1+i)^2} + \frac{A}{(1+i)^3} + \frac{A}{(1+i)^4} + \cdots + \frac{A}{(1+i)^{n-1}} + \frac{A}{(1+i)^n} + \frac{A}{(1+i)^{n+1}}$$

$$- \frac{A}{(1+i)^1} - \frac{A}{(1+i)^2} - \frac{A}{(1+i)^3} \cdots \frac{A}{(1+i)^{n-2}} - \frac{A}{(1+i)^{n-1}} - \frac{A}{(1+i)^n}$$

$$P\left[\frac{-i}{(1+i)}\right] = A\left[\frac{1}{(1+i)^{n+1}} - \frac{1}{(1+i)^1}\right]$$

$$P = \frac{A}{-i}\left[\frac{1}{(1+i)^n} - 1\right]$$

$$P = A\left[\frac{(1+i)^n - 1}{i(1+i)^n}\right] \quad i \neq 0 \tag{4.13}$$

The term in brackets in Equation (4.13) is called the *uniform series present worth factor* (USPWF) and its notation is *P/A*.

Thus,

$$P = A(P/A, i\%, n) \tag{4.14}$$

4.9.2.2 Finding A when given P

The formula to obtain *A* when *P* is given is obtained by reversing the Equation (4.13). Thus,

$$A = P\left[\frac{i(1+i)^n}{(1+i)^n - 1}\right] \quad i \neq 0 \tag{4.15}$$

The term in brackets of Equation (4.15) is called the *capital recovery factor* (CRF) and its notation is *A/P*.

Thus,

$$A = P(A/P, i\%, n) \tag{4.16}$$

The formulas given by Equation (4.13) and Equation (4.15) are derived with the present worth *P* and the first uniform annual amount *A*, one year (period) apart. That is, the present worth *P* must always be located one period prior to the first *A*.

Example 4.12

How much money you should deposit in your savings account now to guarantee withdrawal of ₹3,000 per year for 10 years starting next year? The bank pays an interest rate of 9% per year.

Solution

The cash flow diagram is shown in Fig. 4.16.

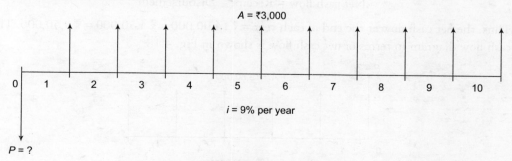

Figure 4.16 Cash flow diagram, Example 4.12

Here, $P = ?$

$$A = ₹3,000, n = 10 \text{ years}, i = 9\% \text{ per year}$$
$$P = A(P/A, i\%, n)$$
$$= ₹3,000(P/A, 9\%, 10)$$
$$= ₹3,000(6.4177)$$
$$= ₹19,253.10$$

Example 4.13

How much money can a company borrow to finance an ongoing project if it expects revenues of ₹14,00,000 per year over a 5 year period? Expenses associated with the project are expected to be ₹4,50,000 per year and interest rate is 10% per year.

Solution

The cash flow diagram is shown in Fig. 4.17.

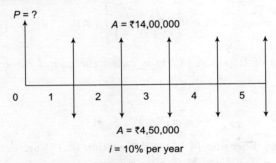

Figure 4.17   Cash flow diagram, Example 4.13

$$P = ₹14,00,000(P/A, 10\%, 5) - ₹4,50,000(P/A, 10\%, 5)$$
$$= ₹14,00,000(3.7908) - ₹4,50,000(3.7908)$$
$$= ₹36,01,260$$

This problem can also be solved by first finding the net cash flow at the end of each year. The net cash flow is determined as:

$$\text{Net cash flow} = \text{Receipt} - \text{Disbursement}$$

Thus, the net cash flow at the end of each year = ₹14,00,000 − ₹4,50,000 = ₹9,50,000. The cash flow diagram in terms of net cash flow is shown in Fig. 4.18.

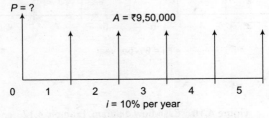

Figure 4.18   Net cash flow diagram, Example 4.13

$$P = ₹9,50,000(P/A, 10\%, 5)$$
$$= ₹9,50,000(3.7908)$$
$$= ₹36,01,260$$

Example 4.14

A company wants to have enough money to purchase a new machine in 4 years. If the machine will cost ₹12,50,000 how much should the company set aside each year if the account earns 8% per year?

Solution

The cash flow diagram is shown in Fig. 4.19.

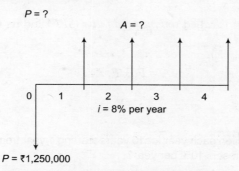

Figure 4.19   Cash flow diagram, Example 4.14

$$A = P(A/P, i\%, n)$$
$$= ₹12,50,000(A/P, 8\%, 4)$$
$$= ₹12,50,000(0.30192)$$
$$= ₹3,77,400$$

4.9.2.3   Finding F when given A

From Equation (4.9), $P = F(1 + i)^{-n}$, substituting for P in Equation (4.13) we can determine the expression for F when A is given as:

$$F(1 + i)^{-n} = A\left[\frac{(1+i)^n - 1}{i(1+i)^n}\right]$$

$$F = A\frac{1}{(1+i)^{-n}}\left[\frac{(1+i)^n - 1}{i(1+i)^n}\right]$$

$$F = A\left[\frac{(1+i)^n - 1}{i}\right] \tag{4.17}$$

The term in the bracket is called *uniform series compound amount factor* (*USCAF*) and its notation is *F/A*.

Thus,

$$F = A(F/A, i\%, n) \tag{4.18}$$

It is important to note here that *F* occurs at the same time as last *A* and n periods after *P* as shown in Fig. 4.15.

4.9.2.4 Finding A when given F

Taking Equation (4.17) and solving for *A*, it is found that:

$$A = F\left[\frac{i}{(1+i)^n - 1}\right] \tag{4.19}$$

The term in the brackets is called *sinking fund factor* (*SFF*) and its notation is *A/F*.

Thus,

$$A = F(A/F, i\%, n) \tag{4.20}$$

Example 4.15

A company invests ₹5 million each year for 10 years starting 1 year from now. What is the equivalent future worth if rate of interest is 10% per year?

Solution

The cash flow diagram is shown in Fig. 4.20.

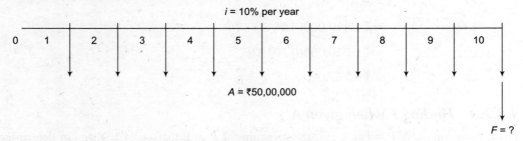

Figure 4.20   Cash flow diagram, Example 4.15

$$F = A(F/A, i\%, n)$$

$$= ₹50,00,000(F/A, 10\%, 10)$$

$$= ₹50,00,000(15.9374)$$

$$= ₹7,96,87,000$$

Example 4.16

How much money must you deposit every year starting 1 year from now at 6% per year in order to accumulate ₹3,00,000 five years from now?

Solution

The cash flow diagram is shown in Fig. 4.21.

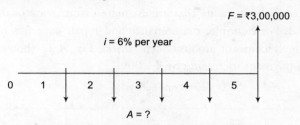

Figure 4.21   Cash flow diagram, Example 4.16

$$A = F(A/F, i\%, n)$$
$$= ₹3,00,000(A/F, 6\%, 5)$$
$$= ₹3,00,000(0.17740)$$
$$= ₹53,220.$$

Table 4.2 summarizes the standard notations and equations for the six factors i.e. *F/P, P/F, P/A, A/P, F/A* and *A/F* to provide information at one place to the readers.

Table 4.2   *F/P, P/F, P/A, A/P, F/A* and *A/F* factors: notations and equations

S. No.	Factor		Find/Given	Standard Notation Equation	Equation with Factor Formula
	Notation	Name			
1.	$(F/P, i\%, n)$	Single payment compound amount factor	F/P	$F = P(F/P, i\%, n)$	$F = P(1 + i)^n$
2.	$(P/F, i\%, n$	Single payment present worth factor	P/F	$P = F(P/F, i\%, n)$	$P = F(1 + i)^{-n}$
3.	$(P/A, i\%, n)$	Uniform series present worth factor	P/A	$P = A(P/A, i\%, n)$	$P = A\left[\dfrac{(1 + i)^n - 1}{i(1 + i)^n}\right]$
4.	$(A/P, i\%, n)$	Capital recovery factor	A/P	$A = P(A/P, i\%, n)$	$A = P\left[\dfrac{i(1 + i)^n}{(1 + i)^n - 1}\right]$
5.	$(F/A, i\%, n)$	Uniform series compound amount factor	F/A	$F = A(F/A, i\%, n)$	$F = A\left[\dfrac{(1 + i)^n - 1}{i}\right]$
6.	$(A/F, i\%, n)$	Sinking fund factor	A/F	$A = F(A/F, i\%, n)$	$A = F\left[\dfrac{i}{(1 + i)^n - 1}\right]$

4.10　INTEREST FORMULAE RELATING AN ARITHMETIC GRADIENT SERIES TO ITS PRESENT AND ANNUAL WORTH

A cash flow series that either increases or decreases by a constant amount forms an arithmetic gradient series. In engineering economy analysis, such situations are observed in problems involving receipts or disbursements that are estimated to increase or decrease by a constant amount each period. For example, maintenance and repair expenses of an equipment may increase by a relatively constant amount each period. Fig. 4.22 shows a cash flow diagram with end-of-period payment increasing by ₹ 1,000.

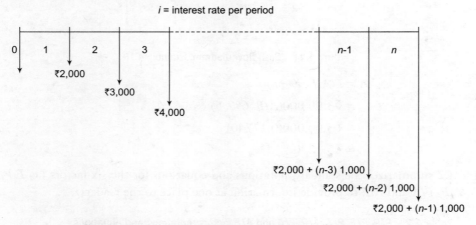

Figure 4.22　Cash flow diagram of an arithmetic gradient series

It can be seen from Fig. 4.22 that the payment at the end of period 1 is ₹ 2,000 and then this amount increases by ₹ 1,000 in each of the subsequent periods. The cash flow at the end-of-period 1 i.e. ₹ 2,000 is called the *base amount* and the *constant amount* i.e. ₹ 1,000 by which subsequent amount increases is called *arithmetic gradient G*. The cash flow at the end-of-period n (CF_n) may be calculated as:

$$CF_n = \text{base amount} + (n-1)G$$

For example, for Fig. 4.22, the cash flow at the end-of-period 4 is:

$$CF_4 = ₹ 2,000 + (4-1).1,000 = ₹ 5,000$$

It is important to note that the base amount appears at the end-of-period 1 and the gradient begins between period 1 and 2. The timing for the flow of gradient is shown in Table 4.3.

Table 4.3 Timing for the flow of gradient

End-of-period	Gradient
1	0
2	G
3	$2G$
⋮	⋮
⋮	⋮
⋮	⋮
⋮	⋮
$(n-2)$	$(n-3)G$
$(n-1)$	$(n-2)G$
n	$(n-1)G$

A generalized increasing arithmetic gradient cash flow diagram is shown in Fig. 4.23. This is called conventional gradient.

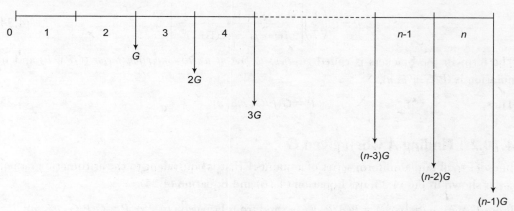

Figure 4.23 Cash flow diagram of an arithmetic gradient series

In relation to conventional arithmetic gradient series, the following formulas are derived.

4.10.1 Finding *P* when given *G*

In Fig. 4.23, the present worth at period 0 of only the gradient is equal to the sum of the present worths of the individual values, where each value is considered a future amount.

$$P = G(P/F, i\%, 2) + 2G(P/F, i\%, 3) + 3G(P/F, i\%, 4) + \ldots + [(n-3)G](P/F, i\%, n-2)$$
$$+ [(n-2)G](P/F, i\%, n-1) + [(n-1)G](P/F, i\%, n)$$

Using the *P/F* formula and factoring out G:

$$P = G\left[\frac{1}{(1+i)^2} + \frac{2}{(1+i)^3} + \frac{3}{(1+i)^4} + \cdots + \frac{n-3}{(1+i)^{n-2}} + \frac{n-2}{(1+i)^{n-1}} + \frac{n-1}{(1+i)^n}\right] \quad (4.21)$$

Multiplying both sides of Equation (4.21) by $(1+i)^1$ yields:

$$P(1+i)^1 = G\left[\frac{1}{(1+i)^1} + \frac{2}{(1+i)^2} + \frac{3}{(1+i)^3} + \cdots + \frac{n-3}{(1+i)^{n-3}} + \frac{n-2}{(1+i)^{n-2}} + \frac{n-1}{(1+i)^{n-1}}\right] \quad (4.22)$$

Subtract Equation (4.21) from Equation (4.22) and simplify:

$$iP = G\left[\frac{1}{(1+i)^1} + \frac{1}{(1+i)^2} + \frac{1}{(1+i)^3} + \cdots + \frac{1}{(1+i)^{n-2}} + \frac{1}{(1+i)^{n-1}} + \frac{1}{(1+i)^n}\right] - G\left[\frac{n}{(1+i)^n}\right]$$

$$\quad (4.23)$$

After simplification,

$$P = G\left(\frac{1}{i}\right)\left[\frac{(1+i)^n - 1}{i(1+i)^n} - \frac{n}{(1+i)^n}\right] \quad (4.24)$$

The term in the brackets is called *gradient to present worth conversion factor* (GPWF) and its notation is (P/G, i%, n).

Thus, $P = G(P/G, i\%, n)$ (4.25)

4.10.2   Finding A when given G

In order to obtain a uniform series of amount A that is equivalent to the arithmetic gradient series shown in Fig. 4.23, use Equation (4.16) and Equation (4.25).

From Equation (4.16): $A = P(A/P, i\%, n)$ and from Equation (4.25): $P = G(P/G, i\%, n)$.

Therefore, $A = G(P/G, i\%, n)(A/P, i\%, n)$.

$$A = G\left(\frac{1}{i}\right)\left[\frac{(1+i)^n - 1}{i(1+i)^n} - \frac{n}{(1+i)^n}\right]\left[\frac{i(1+i)^n}{(1+i)^n - 1}\right]$$

$$A = G\left[\frac{1}{i} - \frac{n}{(1+i)^n - 1}\right] \quad (4.26)$$

The term in brackets is called the *gradient to uniform series conversion factor* and its notation is (A/G, i%, n).

Thus, $A = G(A/G, i\%, n)$ (4.27)

It is important to note that the use of the above two gradient conversion factors requires that there is no payment at the end of first period.

The total present worth P_T at period 0 for an arithmetic gradient series is computed by considering the base amount and the gradient separately as explained below:

➤ The base amount is the uniform series amount A that begins in period 1 and extends through period n. Compute its present worth P_A by using the factor $(P/A, i\%, n)$.

➤ For an increasing gradient, the gradient begins in period 2 and extends through period n. Compute its present worth P_G by using the factor $(P/G, i\%, n)$. Add P_G to P_A to get P_T.

➤ For a decreasing gradient, the gradient also begins in period 2 and extends through period n. Compute its present worth P_G by using the factor $(P/G, i\%, n)$. Subtract P_G from P_A to get P_T.

The general equations for calculating total present worth P_T of conventional arithmetic gradient are:

$$P_T = P_A + P_G \text{ and } P_T = P_A - P_G \tag{4.28}$$

Similarly, the equivalent total annual series are:

$$A_T = A_A + A_G \text{ and } A_T = A_A - A_G \tag{4.29}$$

where A_A is the annual base amount and A_G is the equivalent annual amount of the gradient series.

Example 4.17

For the cash flow diagram shown in Fig. 4.24, compute (i) present worth and (ii) annual series amounts.

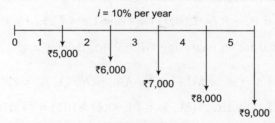

Figure 4.24   Cash flow diagram with increasing arithmetic gradient series, Example 4.17

Solution

(i) Fig. 4.24 is divided into two cash flow diagrams as shown in Fig. 4.25(a) and (b).

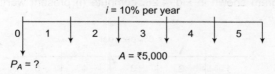

Figure 4.25(a)   Cash flow diagram in terms of base amount A

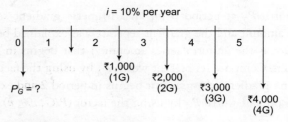

Figure 4.25(b)   Cash flow diagram in terms of Gradient G

Now, present worth P_A of the cash flow diagram shown in Fig. 4.25(a) is obtained as:

$$P_A = A(P/A, i\%, n)$$

$$= ₹5,000(P/A, 10\%, 5)$$

$$= ₹5,000(3.7908)$$

$$= ₹18,954$$

Present worth P_G of the cash flow diagram shown in Fig. 4.25(b) is obtained as:

$$P_G = G(P/G, i\%, n)$$

$$= ₹1,000(P/G, 10\%, 5)$$

$$= ₹1,000(6.8618)$$

$$= ₹6,861.8$$

Since this problem involves an increasing arithmetic gradient series,

$$P_T = P_A + P_G = ₹18,954 + ₹6,861.8 = ₹25,815.8$$

(ii) We also know that for an increasing arithmetic gradient series, $A_T = A_A + A_G$. In this problem,

$A_A = ₹5,000$ and A_G can be found using A/G factor i.e. $A_G = G(A/G, i\%, n)$.

$$A_G = ₹1,000(A/G, 10\%, 5) = ₹1,000(1.8101) = ₹1,810.1.$$

Thus, $A_T = A_A + A_G = ₹5,000 + ₹1,810.1 = ₹6,810.1$

Example 4.18

For the cash flow diagram shown in Fig. 4.26, compute (i) present worth and (ii) annual series amounts.

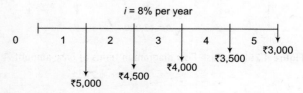

Figure 4.26   Cash flow diagram with decreasing arithmetic gradient series, Example 4.18

Solution

(i) Cash flow diagram shown in Fig. 4.26 is partitioned in two parts as shown in Fig. 4.27(a) and Fig. 4.27(b).

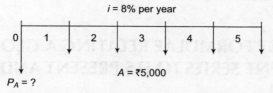

Figure 4.27(a) Cash flow diagram in terms of base amount A

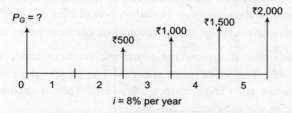

Figure 4.27(b) Cash flow diagram in terms of gradient G

Present worth P_A of cash flow diagram shown in Fig. 4.27(a) is obtained as:

$$P_A = A(P/A, i\%, n)$$

$$= ₹5,000(P/A, 8\%, 5)$$

$$= ₹5,000(3.9927)$$

$$= ₹19,963.5$$

Present worth P_G of the cash flow diagram shown in Fig. 4.27(b) is obtained as:

$$P_G = G(P/G, i\%, n)$$

$$= ₹500(P/G, 8\%, 5)$$

$$= ₹500(7.3724)$$

$$= ₹3,686.2$$

Since this problem involves a decreasing arithmetic gradient series,

$$P_T = P_A - P_G = ₹19,963.5 - ₹3,686.2 = ₹16,277.3$$

(ii) We also know that for a decreasing arithmetic gradient series, $A_T = A_A - A_G$. In this problem, $A_A = ₹5,000$ and A_G can be found using A/G factor i.e. $A_G = G(A/G, i\%, n)$.

$$A_G = ₹500(A/G, 10\%, 5) = ₹500(1.8465) = ₹923.25$$

Thus, $A_T = A_A - A_G = ₹5,000 - ₹923.25 = ₹4,076.75$

A_T can also be found for P_T using the factor A/P as:

$$A_T = P_T(A/P, i\%, n) = ₹16,277.3(A/P, 8\%, 5) = ₹16,277.3(0.25046) = ₹4,076.81$$

Round off accounts for a difference of ₹0.06.

4.11 INTEREST FORMULAE RELATING A GEOMETRIC GRADIENT SERIES TO ITS PRESENT AND ANNUAL WORTH

There are problems in which cash flow series such as operating costs, construction costs and revenues either increase or decrease from period to period by a constant percentage, say 5% per year. This uniform rate of change defines geometric gradient series cash flows.

Fig. 4.28 depicts cash flow diagrams for geometric gradient series with increasing and decreasing uniform rates. The series starts in period 1 at an initial amount A_1, which is not considered a base amount as in the arithmetic gradient. Then it increases or decreases by a constant percentage g in the subsequent periods.

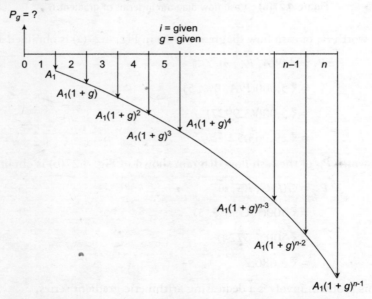

Figure 4.28(a)   Cash flow diagram of increasing geometric gradient series

The present worth P_g can be found by considering each individual cash flow as future cash flow F and finding P for given F. Thus, for the cash flow diagram shown in Fig. 4.28(a)

$$P_g = \frac{A_1}{(1+i)^1} + \frac{A_1(1+g)}{(1+i)^2} + \frac{A_1(1+g)^2}{(1+i)^3} + \frac{A_1(1+g)^3}{(1+i)^4}$$

$$+ \frac{A_1(1+g)^4}{(1+i)^5} + \cdots + \frac{A_1(1+g)^{n-2}}{(1+i)^{n-1}} + \frac{A_1(1+g)^{n-1}}{(1+i)^n} \qquad (4.30)$$

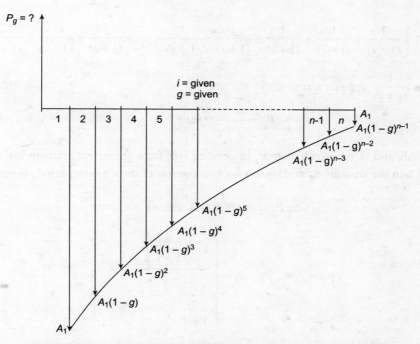

Figure 4.28(b) Cash flow diagram of decreasing geometric gradient series

$$P_g = A_1 \left[\frac{1}{(1+i)} + \frac{(1+g)}{(1+i)^2} + \frac{(1+g)^2}{(1+i)^3} + \frac{(1+g)^3}{(1+i)^4} + \frac{(1+g)^4}{(1+i)^5} + \cdots + \frac{(1+g)^{n-2}}{(1+i)^{n-1}} + \frac{(1+g)^{n-1}}{(1+i)^n} \right]$$

(4.31)

Multiply both sides of Equation (4.31) by $\dfrac{(1+g)}{(1+i)}$, subtract Equation (4.31) from the result, factor out P_g and obtain

$$P_g \left(\frac{1+g}{1+i} - 1 \right) = A_1 \left[\frac{(1+g)^n - 1}{i(1+i)^{n+1}} - \frac{1}{(1+i)} \right]$$

Solve for P_g and simplify,

$$P_g = A_1 \left[\frac{1 - \left(\dfrac{1+g}{1+i} \right)^n}{i - g} \right] \quad g \neq i$$

(4.32)

The term in brackets in Equation (4.32) is the geometric gradient present worth factor for values of g not equal to the interest rate i. The standard notation used is $(P/A, g\%, i\%, n)$ when $g = i$, substitute for g in Equation (4.31) to obtain

$$P_g = A_1\left(\frac{1}{(1+i)} + \frac{1}{(1+i)} + \frac{1}{(1+i)} + \frac{1}{(1+i)} + \frac{1}{(1+i)} + \cdots + \frac{1}{(1+i)} + \frac{1}{(1+i)} + \frac{1}{(1+i)}\right)$$

The term $\dfrac{1}{(1+i)}$ appears n times, so

$$P_g = \frac{n A_1}{(1+i)} \quad \text{for } g = i \tag{4.33}$$

The formula and factor to calculate P_g in period $t = 0$, for a geometric gradient series starting in period 1 in the amount A_1 and increasing by a constant rate of g each period are summarized below:

$$P_g = A_1(P/A, g\%, i\%, n) \tag{4.34}$$

$$(P/A,\ g\%,\ i\%,\ n) = \begin{cases} \dfrac{1 - \left(\dfrac{1+g}{1+i}\right)^n}{i - g}; & g \neq i \\[4mm] \dfrac{n}{1+i} & ; & g = i \end{cases}$$

It should be noted that for a decreasing geometric gradient series as shown in Fig. 4.28(b), P_g can be calculated by the same formulas as given above but $-g$ should be used in place of g.

It is possible to derive formulas and factor for the equivalent A and F. However, it is easier to determine the P_g amount and then multiply by the A/P or F/P factors as given below:

$$A = P_g(A/P, i\%, n) \tag{4.35}$$

$$F = P_g(F/P, i\%, n) \tag{4.36}$$

Example 4.19

Consider the end-of-year geometric gradient series shown in Fig. 4.29. Determine its present worth P, future worth F and annual worth A. The rate of increase is 15% per year after the first year and the interest rate is 20% per year.

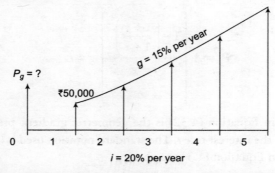

Figure 4.29   Cash flow diagram for geometric gradient series, Example 4.19

Solution

In this problem $A_1 = ₹50,000$; $g = 15\%$ per year; $i = 20\%$ per year; $n = 5$ years

$$P_g = A_1 \left[\frac{1 - \left(\frac{1+g}{1+i}\right)^n}{i - g} \right]$$

$$= ₹50,000 \left[\frac{1 - \left(\frac{1+0.15}{1+0.20}\right)^5}{0.20 - 0.15} \right]$$

$$= ₹1,91,680.657$$

$$F = P_g (F/P, i\%, n)$$

$$= ₹1,91,680.657(F/P, 20\%, 5)$$

$$= ₹1,91,680.657(2.4883)$$

$$= ₹4,76,958.98$$

$$A = P_g (A/P, i\%, n)$$

$$= ₹1,91,680.657(A/P, 20\%, 5)$$

$$= ₹1,91,680.657(0.33438)$$

$$= ₹64,094.18$$

Example 4.20

A person planning his retirement will deposit 10% of his salary each year into a stock fund for 15 years. If his salary this year is ₹3,00,000 (i.e. at end of year 1) and he expects his salary to increase by 10% each year, what will be the present worth of the fund if it earns 8% per year?

Solution

The cash flow diagram for this problem is shown in Fig. 4.30.

The person deposits 10% of his first year's salary (i.e. $0.1 \times ₹3,00,000 = ₹30,000$ into the fund and then this amount decreases by 10% in each of the subsequent years. Thus, in this problem $A_1 = ₹30,000$; $g = 10\%$ per year; $i = 8\%$ per year; $n = 15$ years; $P_g = ?$

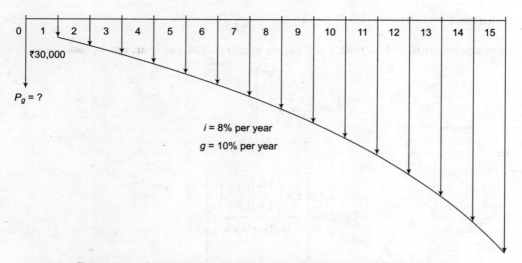

Figure 4.30   Cash flow diagram for geometric gradient series, Example 4.20

$$P_g = A_1 \left[\frac{1 - \left(\dfrac{1+g}{1+i} \right)^n}{i-g} \right]$$

$$= ₹\, 30,000 \left[\frac{1 - \left(\dfrac{1+0.10)}{1+0.08} \right)^{15}}{0.08 - 0.10} \right]$$

$$P_g = ₹\, 4,75,264.245$$

Example 4.21

The future worth in year 10 of a geometric gradient series of cash flows was found to be ₹ 4,00,000. If the interest rate was 10% per year and the annual rate of increase was 8% per year, what was the cash flow amount in year 1.

Solution

The cash flow diagram for this problem is shown in Fig. 4.31.

In this problem $F = ₹\, 4,00,000$; $g = 8\%$ per year; $i = 10\%$ per year; $n = 10$ years; $A_1 = ?$

In order to find cash flow amount in year 1 i.e. A_1, we will compute present worth at $t = 0$, for the given F of ₹ 4,00,000 using the factor P/F.

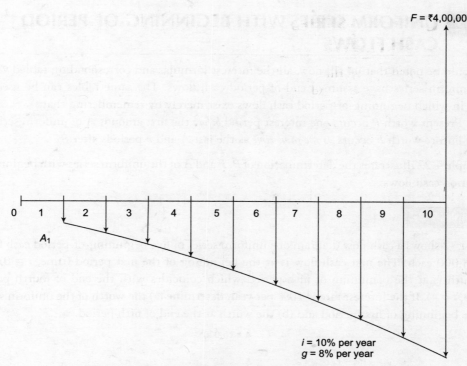

Figure 4.31   Cash flow diagram for geometric gradient series, Example 4.21

Thus,

$$P = F(P/F, i\%, n)$$

$$= ₹4,00,000(P/F, 10\%, 10)$$

$$= ₹4,00,000(0.3855)$$

$$= ₹1,54,200$$

This P is in fact Pg and for this A_1 is found as:

$$P_g = A_1 \left[\frac{1 - \left(\dfrac{1+g}{1+i}\right)^n}{i - g} \right]$$

$$1,54,200 = A_1 \left[\frac{1 - \left(\dfrac{1+0.08}{1+0.1}\right)^{10}}{0.10 - 0.08} \right]$$

$$A_1 = ₹18,396.56$$

4.12 UNIFORM SERIES WITH BEGINNING-OF-PERIOD CASH FLOWS

It should be noted that up till now, all the interest formulas and corresponding tabled values for uniform series have assumed end-of-period cash flows. The same tables can be used for cases in which beginning-of-period cash flows exist merely by remembering that:

➤ Present worth P occurs one interest period *before* the first amount A of uniform series
➤ Future worth F occurs *at the same time* as the last A and n periods after P.

Example 4.22 illustrates the determination of P, F and A of the uniform series with beginning-of-period cash flows:

Example 4.22

Fig. 4.32 shows a cash flow diagram of a uniform series of five beginning-of-period cash flows of ₹5,000 each. The first cash flow is at the beginning of the first period (time, $t = 0$) and the fifth is at the beginning of fifth period which coincides with the end of fourth period (time, $t = 4$). If the interest rate is 10% per year, determine (a) the worth of the uniform series at the beginning of first period and (b) the worth at the end of fifth period.

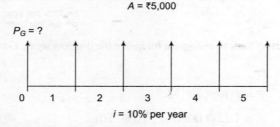

Figure 4.32   Cash flow diagram of uniform series with beginning-of-period cash flows, Example 4.22

Solution

To find worth of the uniform series at $t = 0$ i.e. P_0, we can take one imaginary time period $t = -1$, as shown in Fig. 4.33. By doing so, the first amount of the uniform series appears one period after P_{-1}:

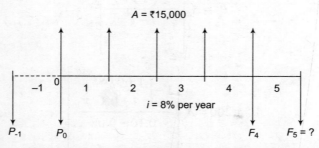

Figure 4.33   Cash flow diagram of uniform series with beginning-of-period cash flows, Example 4.22

P_{-1} can be calculated by using the factor P/A as:

$$P_{-1} = A(P/A, i\%, n)$$

$$= ₹5,000(P/A, 10\%, 5)$$

$$= ₹5,000(3.7908)$$

$$= ₹18,954$$

But, we have to find worth at $t = 0$ i.e. P_0. To find we P_0 consider P_{-1} as present worth and P_0 obviously as future worth and use the factor F/P i.e.

$$P_0 = P_{-1}(F/P, 10\%, 1)$$

$$= ₹18,954(1.1000)$$

$$= ₹20,849.4$$

To find F_5 consider either P_{-1} or P_0 as present worth and use the factor F/P.

If we consider P_{-1}, then:

$$F_5 = P_{-1}(F/P, 10\%, 6)$$

$$= ₹18,954(1.7716)$$

$$= ₹33,578.91$$

If we consider P_0, then:

$$F_5 = P_0(F/P, 10\%, 5)$$

$$= ₹20,849.4(1.6105)$$

$$= ₹33,577.96$$

Small difference in the value of F_5 is observed due to rounding-off the values given in interest table.

P_0 can also be found as:

$$P_0 = ₹5,000 + ₹5,000(P/A, 10\%, 4)$$

$$= ₹5,000 + ₹5,000(3.1699)$$

$$= ₹20,849.5$$

In order to obtain the value of F_5, we first find the value of F_4 as:

$$F_4 = A(F/A, 10\%, 5)$$

$$= ₹5,000(6.1051)$$

$$= ₹30,525.5$$

For F_5 this F_4 becomes P and therefore F_5 can be found out using factor F/P as:

$$F_5 = F_4(F/P, 10\%, 1)$$

$$= ₹30,525.5(1.1000)$$

$$= ₹33,578.05$$

4.13 DEFERRED ANNUITIES OR SHIFTED UNIFORM SERIES

If the occurrence of annuity does not happen until some later date, the annuity is known as deferred annuity. In other words, when a uniform series begins at a time other than the end-of-period 1, it is called a shifted uniform series. Several methods can be used to determine the equivalent present worth P for the case of deferred annuity. For example, P of the uniform series shown in Fig. 4.34 can be determined by any one of the following methods:

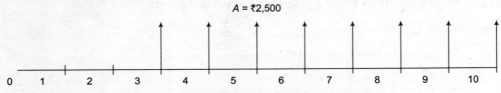

Figure 4.34   Cash flow diagram of shifted uniform series

> Consider each amount of the uniform series as future worth F and use the P/F factor to find the present worth of each receipt at $t = 0$, and add them.
> Consider each amount of the uniform series as present worth P and use the F/P factor to find the future worth of each receipt at $t = 10$ add them and then find the present worth of the total amount at $t = 0$ using $P = F_{10}(P/F, i\%, 10)$.
> Use the F/A factor to find the future amount, F_{10} at $t = 10$ i.e. $F_{10} = A(F/A, i\%, 8)$ and then compute P at $t = 0$ using $P = F_{10}(P/F, i\%, 10)$.
> Use the P/A factor to compute the present worth, P_2 which will be located one period before the first amount of A i.e. at $t = 2$, not at $t = 0$ and then find the present worth at $t = 0$ using the factor $(P/F, i\%, 2)$.

The last method is generally used for calculating the present worth of a shifted uniform series.

Example 4.23

A CAD center just purchased a CAD software for ₹50,000 now and annual payments of ₹5,000 per year for 8 years starting 4 years from now for annual upgrades. What is the present worth of the payments if the interest rate is 8% per year?

Solution

The cash flow diagram is shown in Fig. 4.35.

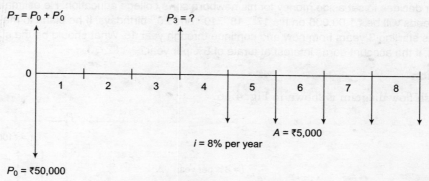

Figure 4.35 Cash flow diagram, Example 4.23

Let us first calculate the present worth of the uniform series at $t = 3$ i.e. P_3 using the factor $(P/A, i\%, 5)$.

Thus,

$$P_3 = A(P/A, 8\%, 5)$$

$$= ₹5,000(P/A, 8\%, 5)$$

$$= ₹5,000(3.9927)$$

$$= ₹19,963.5$$

Consider P_3 as future worth F, compute its present worth at $t = 0$ i.e. P_0' using the factor $(P/F, i\%, 3)$.

Thus,

$$P_0' = P_3(P/F, 8\%, 3)$$

$$= ₹19,963.5(P/F, 8\%, 3)$$

$$= ₹19,963.5(0.7983)$$

$$= ₹15,847.03$$

At time $t = 0$, there is already a cash flow P_0 of ₹50,000 and therefore the total present worth P_T at $t = 0$ is

$$P_T = P_0 + P_0'$$

$$= ₹50,000 + ₹15,847.03$$

$$= ₹65,847.03$$

Example 4.24

A father decides to set aside money for his newborn son's college education. He estimates that his son's needs will be ₹1,00,000 on his 17th, 18th, 19th and 20th birthdays. If he plans to make uniform deposits starting 3 years from now and continue through year 16. What should be the size of each deposit, if the account earns interest at a rate of 8% per year?

Solution

The cash flow diagram is shown in Fig. 4.36.

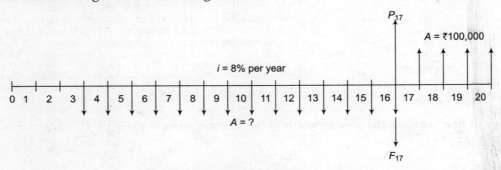

Figure 4.36 Cash flow diagram, Example 4.24

For the uniform series, each of amount ₹1,00,000, compute the present worth at $t = 17$ i.e. P_{17} using the factor $(P/A, i\%, 4)$

Thus,

$$P_{17} = A(P/A, i\%, 4)$$

$$= ₹1,00,000(P/A, 8\%, 4)$$

$$= ₹1,00,000(3.3121)$$

$$= ₹3,31,210$$

For the unknown uniform series, each of amount A, determine the future worth F at $t = 17$ i.e. F_{17} using the factor $(F/A, i\%, 14)$

Thus,

$$F_{17} = A(F/A, i\%, 14)$$

$$= A(F/A, 8\%, 14)$$

$$= A(24.2149)$$

Now,

$$P_{17} = F_{17}$$

$$₹3,31,210 = ₹(24.2149)\,A$$

$$A = (₹3,31,210/24.2149)$$

$$= ₹13,677.94$$

4.14 CALCULATIONS INVOLVING UNIFORM SERIES AND RANDOMLY PLACED SINGLE AMOUNTS

In situations, where a cash flow includes both uniform series and randomly placed single amount, the equivalent present worth, equivalent future worth and equivalent annual worth are calculated by using suitable factors discussed in the previous sections. Example 4.25 illustrates this approach.

Example 4.25

Determine the present equivalent value at time $t = 0$ and the equivalent annual value in the cash flow diagram shown in Fig. 4.37 when $i = 10\%$ per year. Try to minimize the number of interest factors.

Solution

In order to minimize the number of interest factors, we will take positive cash flow of ₹5,000 each at the end of year 4 and year 8 to make a uniform series of amount ₹5,000 right from the end of year 1 to the end of year 10. But we also have to take a single negative cash flow of amount ₹10,000 at the end of year 4 as well as at the end of year 8. The new cash flow diagram is shown in Fig. 4.38.

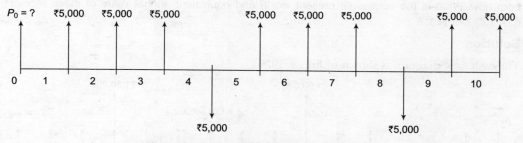

Figure 4.37   Cash flow diagram, Example 4.25

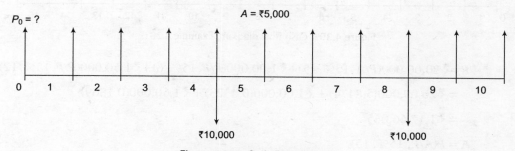

Figure 4.38   Cash flow diagram

Equivalent present worth at time $t = 0$, is obtained as:

$$P_0 = A(P/A, i\%, 10) - ₹10,000(P/F, i\%, 4) - ₹10,000(P/F, i\%, 8)$$

$$= ₹5,000(P/A, 10\%, 10) - ₹10,000(P/F, 10\%, 4) - ₹10,000(P/F, 10\%, 8)$$

$$= ₹5,000(6.1446) - ₹10,000(0.6830) - ₹10,000(0.4665)$$

$$= ₹19,228$$

The equivalent uniform series is obtained using P_0 and converting it to A using factor A/P.

Thus,

$$A = P(A/P, i\%, n)$$

$$= ₹19,228(A/P, 10\%, 10)$$

$$= ₹19,228(0.16275)$$

$$= ₹3,129.36$$

Example 4.26

A small company receives an annual profit of ₹20,00,000 per year for 15 years beginning 1 year from now. In addition, it also receives ₹1,00,000 six years from now and ₹1,50,000 twelve years from now. What is the equivalent present worth and equivalent annual value of these receipts, if i = 15% per year?

Solution

The cash flow diagram is shown in Fig. 4.39.

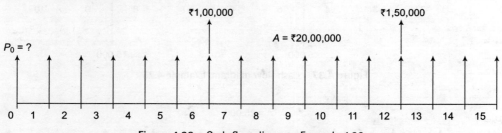

Figure 4.39   Cash flow diagram, Example 4.26

$$P = ₹20,00,000(P/A, 15\%, 15) + ₹1,00,000(P/F, 15\%, 6) + ₹1,50,000(P/F, 15\%, 12)$$

$$= ₹20,00,000(5.8474) + ₹1,00,000(0.4323) + ₹1,50,000(0.1869)$$

$$= ₹1,17,66,065$$

$$A = P(A/P, 15\%, 15)$$

$$= ₹1,17,66,065(0.17102)$$

$$= ₹20,12,232.44$$

Example 4.27

Determine the present equivalent value, future equivalent value and the annual equivalent value in the cash flow diagram shown in Fig. 4.40.

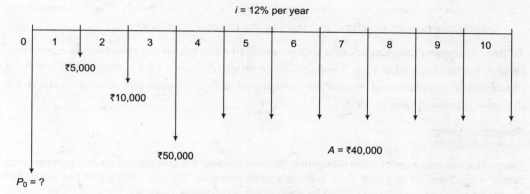

Figure 4.40   Cash flow diagram, Example 4.27

Solution

$$P_0 = ₹5,000(P/F, 12\%, 1) + ₹10,000(P/F, 12\%, 2) + ₹50,000(P/F, 12\%, 3) + ₹40,000(P/A, 12\%, 7)(P/F, 12\%, 3)$$

$$P_0 = ₹5,000(0.8929) + ₹10,000(0.7972) + ₹50,000(0.7118) + ₹40,000(4.5638)(0.7118)$$

$$P_0 = ₹1,77,967.014$$

$$F = P(F/P, 12\%, 10)$$

$$= ₹1,77,967.014(3.1058)$$

$$F = ₹5,52,729.95$$

$$A = P(A/P, 12\%, 10)$$

$$= ₹1,77,967.014(0.17698)$$

$$= ₹31,496.60$$

4.15 CALCULATIONS OF EQUIVALENT PRESENT WORTH AND EQUIVALENT ANNUAL WORTH FOR SHIFTED GRADIENTS

It has been shown in section 4.10 that conventional arithmetic gradient series begins from the end of period 2 and the present worth of this series is located two periods before the gradient starts. A gradient starting at any other time is called shifted gradient. In order to calculate equivalent present worth of the shifted arithmetic gradient series, use the following steps:

1. Calculate present worth of the gradient using the factor $(P/G, i\%, n)$. This present worth will be located two periods before the gradient starts.
2. Consider the present worth calculated at step 1 as future worth F and calculate the equivalent present worth at time 0 using the factor $(P/F, i\%, n)$.

To find the equivalent A series of a shifted gradient through all periods, use the equivalent present worth obtained at step 2 and apply the factor $(A/P, i\%, n)$. Examples 4.28 and 4.29 illustrate the calculation of equivalent present worth P and equivalent annual worth A for shifted arithmetic gradient series.

Example 4.28

The production manager of a manufacturing firm has tracked the average repair and maintenance cost of a machine for 10 years. Cost averages were steady at ₹20,000 for the first five years, but have increased consistently by ₹5,000 for each of the last 5 years. Calculate the equivalent present worth and annual worth of the costs, if the rate of interest is 6% per year.

Solution

The cash flow diagram is shown in Fig. 4.41.

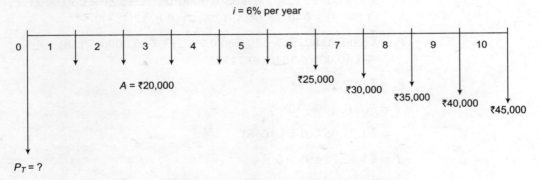

Figure 4.41   Cash flow diagram, Example 4.28

It can be seen from Fig. 4.41 that the base amount is ₹ 20,000 and the arithmetic gradient $G = ₹ 5,000$ starting from the end of year 6. Fig. 4.42 (a) and (b) partitions these two series.

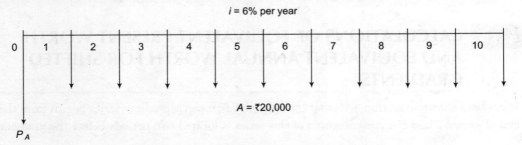

Figure 4.42(a)   Cash flow diagram with base amount

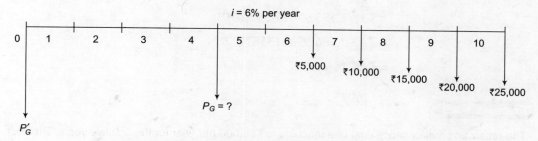

Figure 4.42(b)   Cash flow diagram with shifted arithmetic gradient

The present worth of the cash flow diagram shown in Fig. 4.42(a) i.e. P_A is obtained as:

$$P_A = A(P/A, i\%, n)$$
$$= ₹20,000(P/A, 6\%, 10)$$
$$= ₹20,000(7.3601)$$
$$= ₹1,47,202$$

In order to calculate the present worth of the shifted arithmetic gradient series shown in Fig. 4.42(b), first present worth of the arithmetic gradient $G = ₹5,000$ is calculated as:

$$P_G = G(P/G, i\%, 6)$$
$$= ₹5,000(P/G, 6\%, 6)$$
$$= ₹5,000(11.4594)$$
$$= ₹57,297$$

The P_G will be located two periods before gradient starts i.e. at the end of period 4 (Fig. 4.42(b)). Considering P_G calculated above as future worth, the present worth P_G' is calculated as;

$$P_G' = F(P/F, i\%, n)$$
$$= ₹57,297(P/F, 6\%, 4)$$
$$= ₹57,297(0.7921)$$
$$= ₹45,384.95$$

The equivalent present worth of the cash flow diagram shown in Fig. 4.41 i.e. P_T is obtained as:

$$P_T = P_A + P_G'$$
$$= ₹1,47,202 + ₹45,384.95$$
$$= ₹1,92,586.95$$

The equivalent annual worth A_T for the above calculated P_T is obtained as:

$$A_T = P_T(A/P, i\%, n)$$

$$= ₹1,92,586.95(A/P, 6\%, 10)$$

$$= ₹1,92,586.95(0.13587)$$

$$= ₹26,166.79$$

Example 4.29

The repair and maintenance cost of a machine is ₹50,000 per year for the first five years. The cost increases by 10% per year thereafter for the next five years. Use $i = 12\%$ per year to determine the equivalent total present worth and annual worth for all these cash flows.

Solution

The cash flow diagram is shown in Fig. 4.43.

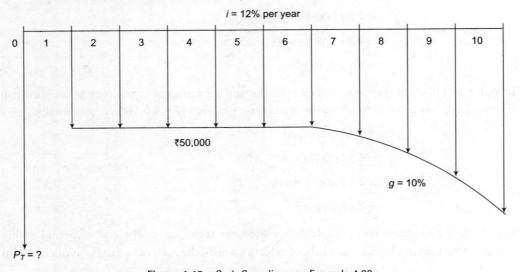

Figure 4.43   Cash flow diagram, Example 4.29

Fig. 4.43 shows that there is a uniform series, $A = ₹50,000$, from year 1 through year 4 and from year 5 geometric gradient series starts with $A_1 = ₹50,000$. The equivalent total present worth P_T and annual worth A_T is calculated as:

$$P_T = A(P/A, 12\%, 4) + A_1(P/A, 10\%, 12\%, 6)(P/F, 12\%, 4)$$

$$= 50,000(3.0373) + 50,000\left[\frac{1 - \left(\frac{1+g}{1+i}\right)^6}{0.12 - 0.10}\right](0.5718)$$

$$= 1,51,865 + 50,000 \left[\frac{1 - \left(\dfrac{1.1}{1.12} \right)^6}{0.12 - 0.10} \right] (0.5718)$$

$$= ₹1,51,865 + ₹1,46,483.816$$

$$= ₹2,98,348.816$$

Now,

$$A_T = P_T(A/P, 12\%, 10)$$

$$= ₹2,98,348.816(0.17698)$$

$$= ₹52,801.773$$

4.16 CALCULATIONS OF EQUIVALENT PRESENT WORTH AND EQUIVALENT ANNUAL WORTH FOR SHIFTED DECREASING ARITHMETIC GRADIENTS

Examples 4.30 and 4.31 illustrate the calculation of equivalent present worth and equivalent annual worth for problems that involve shifted decreasing arithmetic gradients.

Example 4.30

Determine the equivalent present worth and annual worth for the problem for which the cash flow diagram is shown in Fig. 4.44.

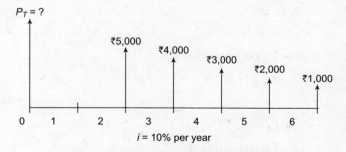

Figure 4.44 Cash flow diagram, Example 4.30

Solution

The partitioned cash flow diagrams of the cash flow diagram shown in Fig. 4.44 are shown in Fig. 4.45(a) and (b).

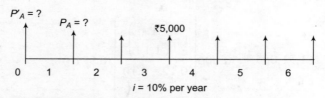

Figure 4.45(a)   Cash flow diagram of shifted uniform series

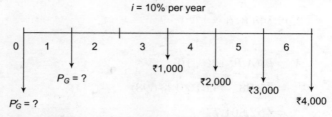

Figure 4.45(b)   Cash flow diagram of shifted arithmetic gradient

For Fig. 4.45(a):

$$P_A = A(P/A, 10\%, 5)$$

$$= ₹5,000(3.7908)$$

$$= ₹18,954$$

Now,

$$P_A' = P_A(P/F, 10\%, 1)$$

$$= ₹18,954(0.9091)$$

$$= ₹17,231.08$$

For Fig. 4.45(b)

$$P_G = G(P/G, 10\%, 5)$$

$$= ₹1,000(6.8618)$$

$$= ₹6,861.8$$

Now,

$$P_G' = P_G(P/F, 10\%, 1)$$

$$= ₹6,861.8(0.9091)$$

$$= ₹6,238.06$$

Here,

$$P_T = P_A' - P_G'$$

$$= ₹17,231.08 - ₹6,238.06$$

$$= ₹10,993.02$$

Now,

$$A_T = P_T(A/P, 10\%, 6)$$

$$= ₹\,10,993.02(0.22961)$$

$$= ₹\,2,524.11$$

Example 4.31

Determine the equivalent present worth and annual worth for the problem for which the cash flow diagram is shown in Fig. 4.46.

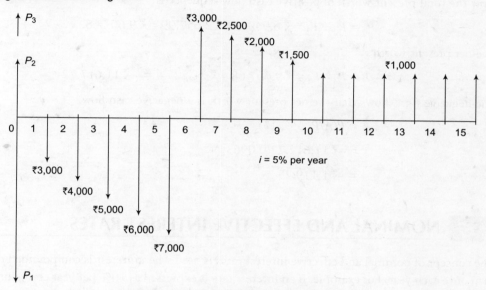

Figure 4.46   Cash flow diagram, Example 4.31

Solution

For the negative cash flow sequence from year 1 to 5, the base amount is ₹ 3,000, $G = ₹\,1,000$ and $n = 5$. For the positive cash flow sequence from year 6 to 10, the base amount is ₹ 3,000, $G = -₹\,500$ and $n = 5$. In addition, there is a 5-year annual series with $A = ₹\,1,000$ from year 11 to 15.

For the negative cash flow sequence from year 1 to 5, let P_1 be the present worth at $t = 0$.

$$P_1 = ₹\,3,000(P/A, 5\%, 5) + ₹\,1,000(P/G, 5\%, 5)$$

$$= ₹\,3,000(4.3295) + ₹\,1,000(8.2369)$$

$$= ₹\,21,225.4$$

For the positive cash flow series from year 6 to 10, let P_2 be the present worth at $t = 0$

$$P_2 = [₹\,3,000(P/A, 5\%, 5) - ₹\,500(P/G, 5\%, 5)](P/F, 5\%, 5)$$

$$= [₹\, 3,000(4.3295) - ₹\, 500(8.2369)](0.7835)$$

$$= ₹\, 6,949.68$$

For the positive cash flow series from year 11 through 15, let P_3 be the present worth at $t = 0$

$$P_3 = ₹\, 1,000(P/A, 5\%, 5)(P/F, 5\%, 10)$$

$$= ₹\, 1,000(4.3295)(0.6139)$$

$$= ₹\, 2,657.88$$

Now the total present worth of positive cash flow sequence is

$$P_4 = P_2 + P_3 = ₹\, 6,949.8 + ₹\, 2,657.99 = ₹\, 9,607.68.$$

The net present worth

$$P = P_4 - P_1 = ₹\, 9,607.68 - ₹\, 21,225.4 = -₹\, 11,617.72$$

The negative sign shows that the net present worth is a negative cash flow.

$$A = P(A/P, 5\%, 15)$$

$$= -₹\, 11,617.72(0.09634)$$

$$= -₹\, 1,119.25$$

4.17 NOMINAL AND EFFECTIVE INTEREST RATES

The concept of nominal and effective interest rates is used when interest is compounded more than once each year. For example, if an interest rate is expressed as 10% per year compounded quarterly, the terms: nominal and effective interest rates must be considered.

Nominal interest rate r is an interest rate that does not include any consideration of compounding. It is given as:

$$r = \text{interest rate per period} \times \text{number of periods} \qquad (4.37)$$

For example, the nominal rate of $r = 2\%$ per month is same as each of the following rates:

$$r = 2\% \text{ per month} \times 24 \text{ months}$$

$$= 48\% \text{ per 2-year period}$$

$$= 2\% \text{ per month} \times 12 \text{ months}$$

$$= 24\% \text{ per year}$$

$$= 2\% \text{ per month} \times 6 \text{ months}$$

$$= 12\% \text{ per semi-annual period}$$

$$= 2\% \text{ per month} \times 3 \text{ months}$$

$$= 6\% \text{ per quarter}$$

$$= 2\% \text{ per month} \times 1 \text{ month}$$

$$= 2\% \text{ per month}$$

$$= 2\% \text{ per month} \times 0.231 \text{ months}$$

$$= 0.462\% \text{ per week}$$

Effective interest rate is the actual rate that applies for a stated period of time. The compounding of interest during the time period of the corresponding nominal rate is accounted for by the effective interest rate. It is commonly expressed on an annual basis as the effective rate i_a, but any time basis can be used.

When compounding frequency is attached to the nominal rate statement, it becomes the effective interest rate. If the compounding frequency is not stated, it is assumed to be the same as the time period of r, in which case the nominal and effective rates have the same value. The following are nominal rate statements. However, they will not have the same effective interest rate value over all time periods, due to the different compounding frequencies.

12% per year, compounded monthly   (Compounding more often than time period)

12% per year, compounded quarterly   (Compounding more often than time period)

4% per quarter, compounded monthly   (Compounding more often than time period)

6% per quarter, compounded daily   (Compounding more often than time period)

All the above rates have the format "*r*% per time period *t*, compounded m/y." In this format, the *m* is a month, a quarter, a week, or some other time unit. The following three time-based units are always associated with an interest rate statement:

➢ *Time period*: It is the period over which the interest is expressed. This is the *t* in the statement of *r*% per time period *t*, for example, 2% per month. One year is the most common time period and it is assumed when not stated otherwise.

➢ *Compounding period (CP)*: It is shortest time unit over which interest is charged or earned. This is defined by the compounding term in the interest rate statement. For example, 10% per year compounded quarterly. Here, the compounding period is one quarter i.e. 3 months. If not stated, it is assumed to be one year.

➢ *Compounding frequency*: It represents the number of times *m* compounding occurs within the time period *t*. If the compounding period CP and the time period *t* are the same, the compounding frequency is 1, for example 2% per month compounded monthly.

Consider the rate 10% per year compounded quarterly. It has a time period *t* of 1 year, a compounding period CP of 1 quarter, and a compounding frequency *m* of 4 times per year. A rate of 12% per year compounded monthly, has *t* = 1 year, CP = 1 month, and *m* = 12.

The effective interest rate per compounding period CP is determined by using the following relation:

$$\text{Effective interest rate per CP} = \frac{r\% \text{ per time period } t}{m \text{ compounding periods per } t} = \frac{r\%}{m} \tag{4.38}$$

Example 4.32 illustrates the determination of effective rate per CP.

Example 4.32

Determine the effective rate per compounding period CP for each of the following:
(a) 12% per year, compounded monthly
(b) 12% per year, compounded quarterly
(c) 5% per 6-month, compounded weekly

Solution

Apply Equation (4.37) to determine effective rate per CP.

(a) $r = 12\%$ per year

CP = 1 month

$m = 12$

Effective interest rate per month $= \dfrac{r\%}{m}$

$$= \frac{12}{12} = 1\% \text{ per month}$$

(b) $r = 12\%$ per year

CP = 1 quarter

$m = 4$

Effective interest rate per quarter $= \dfrac{r\%}{m}$

$$= \frac{12}{4} = 3\% \text{ per quarter}$$

(c) $r = 5\%$ per 6-month

CP = 1 week

$m = 26$

Effective interest rate per week $= \dfrac{r\%}{m}$

$$= \frac{5}{26} = 0.192\% \text{ per week}$$

Sometimes, it becomes really difficult to identify whether a stated rate is a nominal or an effective rate. Basically the following three ways are used to express the interest rates:

1. Nominal rate stated, compounding period stated. For example, 12% per year, compounded quarterly.
2. Effective rate stated. For example, effective 10% per year, compounded quarterly.
3. Interest rate stated, no compounding period stated. For example, 12% per year or 3% per quarter.

The statement given at (1) does not include statement of either nominal or effective rate. However, the compounding frequency is stated. Thus, 12% is the nominal interest rate per year and compounding will take place four times in a year and therefore effective interest rate needs to be calculated.

In the second statement given at (2), the stated rate is identified as effective and therefore, 10% is the effective rate per year.

In the third format given at (3), no compounding frequency is identified. This is effective only over the compounding period. Thus 12% is effective per year and 3% is effective per quarter. The effective rate for any other time period must be calculated.

For a given nominal interest rate, effective annual interest rate is calculated from the following relationship:

$$i_a = \left(1 + \frac{r}{m}\right)^m - 1 \tag{4.39}$$

where,

r = nominal interest rate per year

m = number of compounding periods per year

i_a = effective interest rate per year

If i is the effective interest rate per compounding period then from Equation (4.38), $i = \dfrac{r}{m}$. Equation (4.39) can be written as:

$$i_a = (1 + i)^m - 1 \tag{4.40}$$

If the effective annual rate i_a and compounding frequency m are known, Equation (4.40) can be solved for i to determine the effective interest rate per compounding period.

$$i = (1 + i_a)^{1/m} - 1 \tag{4.41}$$

It is also possible to determine the nominal annual rate r using the above stated definition i.e. $i = \dfrac{r}{m}$.

r% per year = (i% per CP) (number of compounding periods per year) = (i)(m) (4.42)

Example 4.33

Determine the effective annual interest rate for each of the following nominal interest rates:
(a) 12% per year, compounded monthly
(b) 12% per year, compounded quarterly

Solution

(a) Here $r = 12\%$ per year; $m = 12$; $i_a = ?$ Using Equation (4.39), we get

$$i_a = \left(1 + \frac{0.12}{12}\right)^{12} - 1 = 0.1268 = 12.68\% \text{ per year}$$

(b) Here $r = 12\%$ per year; $m = 4$; $i_a = ?$ Using Equation (4.39), we get

$$i_a = \left(1 + \frac{0.12}{4}\right)^{4} - 1 = 0.1255 = 12.55\% \text{ per year}$$

Table 4.4 shows the effective annual interest rates for various nominal interest rates and compounding periods.

Table 4.4   Effective annual interest rates for various nominal rates and compounding periods

Compounding Period	Number of Periods per year, m	Effective annual rate (%) for nominal rate of		
		5%	10%	15%
Annually	1	5.00	10.00	15.00
Semi-annually	2	5.06	10.25	15.56
Quarterly	4	5.09	10.38	15.87
Bimonthly	6	5.11	10.43	15.97
Monthly	12	5.12	10.47	16.08
Weekly	52	5.12	10.51	16.16
Continuously	∞	5.13	10.52	16.18

Consider a situation in which you deposit a certain amount each month into your savings account that pays a nominal interest rate of 12% per year, compounded quarterly. In this situation, the payment period (PP) is 1 month while the compounding period (CP) is 1 quarter i.e. 3 months. Similarly, if a person deposits money each year into his or her savings account which compounds interest monthly, the PP is 1 year while CP is 1 month. The following relationship is used to calculate effective interest rate for any time period or payment period:

$$\text{Effective } i = \left(1 + \frac{r}{m}\right)^{m} - 1 \qquad (4.43)$$

where,

r = nominal interest rate per payment period (PP)
m = number of compounding periods per payment period (PP)

Example 4.34

A company makes payments on a semi-annual basis only. Determine the effective interest rate per payment period for each of the following interest rate:
(a) 12% per year, compounded monthly
(b) 4% per quarter, compounded quarterly

Solution

(a) Here, $PP = 6$ months; $r = 12\%$ per year $= 12/2 = 6\%$ per 6-months; $m = 6$ months per 6-months

$$\text{Effective } i\% \text{ per 6-months} = \left(1 + \frac{0.06}{6}\right)^6 - 1 = 6.15\% \text{ per 6-months}$$

(b) Here, $PP = 6$ months; $r = 4\%$ per quarter $= 4*2 = 8\%$ per 6-months; $m = 2$ quarters per 6-months

$$\text{Effective } i\% \text{ per 6-months} = \left(1 + \frac{0.08}{2}\right)^2 - 1 = 8.16\% \text{ per 6-months}$$

4.18 INTEREST PROBLEMS WITH COMPOUNDING MORE-OFTEN-THAN-ONCE PER YEAR

4.18.1 Single Amounts

If the nominal interest rate and the number of compounding periods per year and the number of years are known, the future worth or present worth for a problem can easily be calculated using Equations (4.8) and (4.10) respectively.

Example 4.35

A person invests ₹25,000 for 15 years at 8% compounded quarterly. How much will it worth at the end of the 15th year?

Solution

The following methods can be used to solve the problem:

Method I

There are four compounding periods per year or a total of $4 \times 15 = 60$ periods.

The interest rate per interest period is $(8\%)/4 = 2\%$.

Now, the future worth is calculated as:

$$F = P(F/P, 2\%, 60)$$

$$= ₹25,000(3.2810)$$

$$= ₹82,025$$

Method II

The effective annual interest rate is calculated as:

$$i_a = \left(1 + \frac{0.08}{4}\right)^4 - 1 = 0.0824 = 8.24\% \text{ per year}$$

Now $F = P(F/P, 8.24\%, 15)$

$$F = ₹\,25,000(F/P, 8.24\%, 15)$$

The value of the factor F/P for an interest rate of 8.24% for 15 years is not available in the interest tables. However, it can be found by using linear interpolation as described below:

The first step in linear interpolation is to set up the known (values 1 and 2) and unknown factors, as shown in Table 4.5. A ratio equation is then set up and solved for c, as follows:

$$\frac{a}{b} = \frac{c}{d} \text{ or } c = \frac{a}{b}d \tag{4.44}$$

where a, b, c and d represent the differences between the numbers shown in the interest tables.

Table 4.5　Linear interpolation setup

		i or n	Factor		
	a	tabulated	value 1	c	
b		desired	Unlisted		d
		tabulated	value 2		

Using the procedure, we can find the value of $(F/P, 8.24\%, 15)$ as follows:

The value of $(F/P, 8\%, 15)$ is 3.1722 and the value of $(F/P, 9\%, 15)$ is 3.6425.

	a	8%	3.1722	c	
b		8.24%	X		d
		9%	3.6425		

The unknown X is the desired factor value. From Equation (4.44):

$$c = \frac{(8.24 - 8)}{(9 - 8)}(3.6425 - 3.1722)$$

$$(X - 3.1722) = \frac{(8.24 - 8)}{(9 - 8)}(3.6425 - 3.1722)$$

$$(X - 3.1722) = (0.24)(0.4703)$$

$$X = 3.1722 + (0.24)(0.4703)$$

$$X = 3.2851$$

Thus, the value of $(F/P, 8.24\%, 15) = 3.2851$

Therefore, $F = ₹ 25,000(3.2851) = ₹ 82,127.5$.

The difference in the value of F from Method I and Method II is due to the interpolated value of the factor $(F/P, 8.24\%, 15)$.

4.18.2   Uniform Series and Gradient Series

When there is more than one compounded interest period per year, the formulae and tables for uniform series and gradient series can be used as long as there is a cash flow at the end of each interest period, as shown in Figures 4.15 and 4.23 for uniform series and arithmetic gradient series, respectively.

Example 4.36

You have borrowed ₹ 15,000 now and it is to be repaid by equal end-of-quarter installments for 5 years with interest at 8% compounded quarterly. What is the amount of each payment?

Solution

The number of installment payments is $5 \times 4 = 20$ and the interest rate per quarter is $(8\%)/4 = 2\%$. When these values are used in Equation (4.27), we find that

$$A = P(A/P, 2\%, 20)$$

$$= ₹ 15,000(0.06116)$$

$$= ₹ 917.4 \text{ per quarter}$$

Example 4.37

The repair and maintenance cost of a machine is zero at the end of first quarter, ₹ 5,000 at the end of second quarter and then it increases by ₹ 5,000 at the end of each quarter thereafter for a total of 3 years. Determine the equivalent uniform payment at the end of each quarter if interest rate is 12% compounded quarterly.

Solution

The cash flow diagram is shown in Fig. 4.47. Here each period represents one quarter. The number of quarters in 3 years are $3 \times 4 = 12$ and the interest rate per quarter is $(12\%)/4 = 3\%$. When these values are used in Equation (4.27), we find that:

$$A = G(A/G, 3\%, 12)$$

$$= ₹ 5,000(5.1485)$$

$$= ₹ 25,742.5 \text{ per quarter}$$

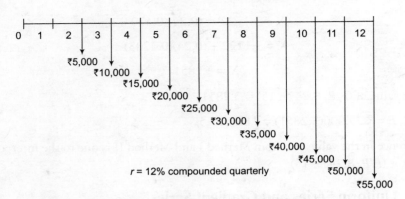

Figure 4.47   Cash flow diagram, Example 4.37

4.18.3   Interest Problems with Uniform Cash Flows Less-Often-Than Compounding Periods

It refers to problems in which uniform cash flow occurs less number of times than the compounding period. For example, the uniform cash flow occurs only once in a year at the end of each year but the compounding of interest takes place quarterly i.e. 4 times in a year. The equivalent future worth can be calculated by using Equation (4.20).

Example 4.38

There exists a series of 5 end-of-year receipts of ₹5,000 each. Determine its equivalent future worth if interest is 8% compounded quarterly.

Solution

The cash flow diagram is shown in Fig. 4.48.

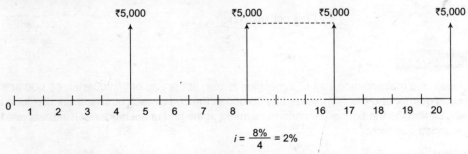

Figure 4.48   Uniform series with cash flow occurring less often than compounding periods, Example 4.38

This problem can be solved by the following methods:

Method I

Interest is (8%)/4 = 2% per quarter, but the uniform series cash flows are not at the end of each quarter. To get the uniform cash flow at the end of each quarter, the cash flow at the end of year 1 i.e. ₹5,000 is considered as future worth F. The uniform cash flow at the end of each quarter of year 1, A is obtained as:

$$A = F(A/F, 2\%, 4)$$

$$= ₹5,000(0.2426)$$

$$= ₹1,213.00$$

Thus, the uniform cash flow at the end of each quarter of year 1 is ₹1,213.0. This is true not only for the first year but also for each of the 5 years under consideration. Hence the original series of 5 end-of-year receipts of ₹5,000 each can be converted to a problem involving 20 end-of-quarter receipts of ₹1,213.0 each as shown in Fig. 4.49.

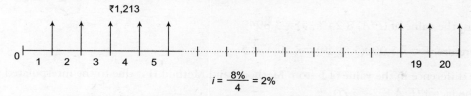

Figure 4.49   Uniform series with cash flow at the end of each quarter

The future worth at the end of fifth year (20th quarter) can then be calculated as:

$$F = A(F/A, 2\%, 20)$$

$$= ₹1,213(24.2974)$$

$$= ₹29,472.75$$

Method II

The effective annual interest rate is calculated as:

$$i_a = \left(1 + \frac{0.08}{4}\right)^4 - 1$$

$$= 0.0824 = 8.24\% \text{ per year}$$

Now,

$$F = A(F/A, 8.24\%, 5)$$

$$F = ₹5,000(F/A, 8.24\%, 5)$$

The value of $(F/A, 8\%, 5)$ is 5.8666 and the value of $(F/A, 9\%, 5)$ is 5.9847.

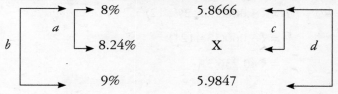

The unknown X is the desired factor value. From Equation (4.44):

$$c = \frac{(8.24 - 8)}{(9 - 8)}(5.9847 - 5.8666)$$

$$(X - 5.8666) = \frac{(8.24 - 8)}{(9 - 8)}(5.9847 - 5.8666)$$

$$(X - 5.8666) = (0.24)(0.1181)$$

$$X = 5.8666 + (0.24)(0.1181)$$

$$X = 5.8949$$

Thus, the value of $(F/A, 8.24\%, 5) = 5.8949$

Therefore, $F = ₹ 5,000(5.8949) = ₹ 29,474.5$.

The difference in the value of F from Method I and Method II is due to the interpolated value of the factor $(F/A\ 8.24\%, 5)$.

4.18.4 Interest Problems with Uniform Cash Flows More-Often-Than Compounding Periods

It refers to problems in which uniform cash flow occurs more number of times than the compounding period. For example, the uniform cash flow occurs monthly, in a year at the end of each month, but the compounding of interest takes place quarterly i.e. 4 times in a year. The equivalent future worth can be calculated by using Equation (4.18).

Example 4.39

You deposit ₹ 1,000 every month in your savings account for 3 years. What is the equivalent worth at the end of 3 years if interest is 8% compounded quarterly?

Solution

This problem can be solved by the following methods:

Method I

It is assumed that there is no interest compounded on any amount except that which is in the account by the end of each quarter. Thus, the amount deposited each quarter totals ₹ 1,000 × 3 = ₹ 3,000. The interest per quarter is $(8\%)/4 = 2\%$ and the number of quarter in 3 years is $3 \times 4 = 12$. The compounded amount at the end of 3 years is:

$$F = ₹ 3,000(F/A, 2\%, 12)$$

$$= ₹ 3,000(13.4121)$$

$$= ₹ 40,236.3$$

Method II

First, the appropriate nominal interest rate per month that compounds to 2% each quarter is calculated as:

$$(1 + r_{12})^3 - 1 = 0.02$$

$$r_{12} = (1.02)^{1/3} - 1$$

$$r_{12} = 0.006623$$

Therefore, the future amount at the end of 3 years with monthly compounding at r_{12} per month is

$$F = ₹ 1,000(F/A, 0.6623, 36)$$

$$F = ₹ 40,503.543$$

Out of the above two methods, the first one is the more realistic and is recommended for solving this type of problem.

PROBLEMS

1. Alphageo Ltd. is considering to procure a new helicopter to ferry its personnel for its drilling operations at Bombay High rig locations. A similar helicopter was purchased 5 years ago at ₹6,50,00,000. At an interest rate of 8% per year, what would be the equivalent value as on today of ₹6,50,00,000 expenditure?

2. Hindustan Explorations Ltd. is planning to set aside ₹75,00,000 for possibly replacing its large gate valves in its oil transportation pipeline network whenever and wherever it becomes necessary. If the rate of return is 21% and the replacement isn't needed for 5 years from now, how much will the company have in its investment account which it has set-aside?

3. Zylog Systems Ltd. manufactures security systems for industrial and defense applications. It wishes to upgrade the technology for some of its units. It is investigating whether it should upgrade it now or consider upgrading later. If the cost of upgradation as on today is ₹50,00,00,000 what will be the equivalent investment 4 years later at an interest rate of 9%?

4. What is the present worth of a future cost of ₹80,00,000 to Sriram Piston and Rings Ltd., 7 years from now at a rate of interest of 11%?

5. Food Corporation of India wants money to construct a new granary. If the granary will cost ₹2,00,00,000 to the company, how much should it set aside each year for 5 years if the rate of interest is 10% per year?

6. Maruti Udyog Ltd, has a budget of ₹7,00,000 per year to pay for seating system components over the next 8 years. If the company expects the cost of the components to increase uniformly according to an arithmetic gradient of ₹10,000 per year, what would be the cost of the components in year 1, if the interest rate is 8% per year?

7. VST Systems, a manufacturer of farm equipment wish to examine its cash flow requirements for the next 7 years. In order to adopt new technology it wishes to replace its manufacturing machines at various times over the next 7 years. It has a plan to invest ₹18,00,000 two years

from now, ₹25,00,000 four years from now and ₹65,00,000 seven years from now. What is the present worth of the planned expenditure at an interest rate of 9%?

8.   A bank employee passed the ICWA examination and his salary was raised by ₹5,000 starting from the end of year 1. At an interest rate of 6% per year, what is the present value of ₹5,000 per year over his remaining 25 years of service?

9.   The Heart Centre, New Delhi wants to purchase an autoclave for its operation theatre (OT). It expects to spend ₹1,20,000 per year for the OT technician and ₹75,000 per year for the consumables for the autoclave. At an interest rate of 12% per year, what is the total equivalent future amount at the end of 5 years?

10. Mahindra Renault has signed a contract worth ₹5,00,00,000 with Motherson Sumi for the supply of electronic control systems of its new upcoming passenger cars. The supply of the systems is to begin after 2 years from now and the payment would be made at the commencement of supply. What would be the present worth of the contract at 15% per year interest?

11. Transport Corporation is planning to install GIS systems and temperature loggers in its entire fleet of 1,500 refrigerated trucks for the efficient monitoring during transit. If the installation of such system is expected to reduce the insurance claims by ₹4,50,00,000 three years from now, how much should the company spend now if it uses an interest rate of 11% per year?

12. Precision Pipes and Profiles uses SS-304 to manufacture flyover railings. It is considering to install a new press in order to reduce its manufacturing cost. If the new press will cost ₹2,20,00,000 now, how much must the company save each year to recover the investment in 5 years at an interest rate of 15% per year?

13. Innovision Advertising expects to invest ₹35,00,000 to launch a campaign for a cosmetic product in the first year of its advertising with amount decreasing by ₹2,50,000 each year subsequently. The company expects an income of ₹1,00,00,000 in first year increasing by ₹15,00,000 each year. Determine the equivalent worth in years 1 through 5 of the company's net cash flow at an interest rate of 15% per year.

14. Mahavir Hanuman Developers expects a revenue of ₹25,00,000 per year over a period of five years for the development of a real estate site. The expenses involved in the project is ₹10,00,000 per year. Assume the interest rate of 10% per year, estimate how much money the company should borrow to finance the project.

15. A savings account was started with an initial deposit of ₹25,000 in year 1. If the annual deposits in the account were to increase 10% each year subsequently, estimate how much money would be in the account over a period of 7 years. Use an interest rate of 6% per year.

16. A software professional is planning for her retirement and wishes to invest 12% of her salary each year into Large Cap Infrastructure Fund. If her current year salary is ₹7,20,000 and if she expects her salary to increase by 6% each year, what will be the present worth of the fund after 10 years if it earns 8% per year?

17. Over a period of 12 years, the future worth of a geometric gradient series was found to be ₹10,00,000. If the rate of interest was 12% per year and the annual rate of increase was 10% per year, what was the cash flow amount in year 1?

18. Determine the value of the present worth in year 0 of the following series of expenditures. Assume a rate of interest of 8% per year:

Year	Expenditures, ₹	Year	Expenditures, ₹
0	0	5	25,000
1	15,000	6	25,000
2	15,000	7	25,000
3	15,000	8	25,000
4	15,000		

19. Mirza Extrusions, Faridabad produces heavy aluminum sections of 13 m standard length. The company extrudes 100 extruded lengths at ₹1,50,000 per length in each of the first 2 years, after which the company expects to extrude 125 lengths at ₹1,75,000 per length through 7 years. If the company's minimum attractive rate of return is 15% per year, what is the present worth of the expected income?

20. A doctor decides to set aside money for his newborn son's university education. He estimates that his needs will be ₹3,50,000 on his 18th, 19th, 20th and 21st birthdays. If he plans to make uniform deposits starting 4 years from now and continue through year 17, what should be the size of each deposit, if the rate of interest that the account earns is 9% per year?

21. OK-Play wishes to purchase an automatic system for ejecting the die-cast products manufactured at its facility. The use of automated system reduces the production cost by ₹25,00,000 per year in each of the first 3 years and by ₹45,00,000 per year in the next 3 years. What is the present worth of the cost savings if the company uses an interest rate of 12% per year?

22. Calculate the future worth of the following income and expenses, if the interest rate is 10% per year:

Year	Income, ₹	Expenses, ₹
0	5,000	3,000
1–6	20,000	8,000
7–12	25,000	10,000

23. Calculate the annual worth in years 1 through 12 of the following series of incomes and expenses. Consider rate of interest to be equal to 10% per year.

Year	Income, ₹	Expenses, ₹
0	5,000	3,000
1–6	20,000	8,000
7–12	25,000	10,000

24. A freelance media person is planning for her retirement 30 years from now. She wishes to invest ₹1,50,000 every year starting now. If she plans to start withdrawing the money 1 year after she makes her last deposit (i.e. 31st year from now), what uniform amount could she withdraw each year for 40 years, if the account earns 8% interest per year?

25. Toyo Rolls expects to increase its income beyond 5 years through 15 years by making suitable investments. In order to achieve this, it plans to invest ₹25,00,000 now and ₹75,00,000 four years from now. If the rate of interest is 10% per year, how much extra income per year would be needed in years 5 through 15 to recover the investment?

26. Mangalore Refineries and Petrochemicals Ltd. wishes to introduce Employees Provident Fund Scheme (EPFS) by converting Employees Welfare Scheme (EWS) into EPFS. It has already deposited ₹1,00,00,000 for each of the five years under EWS corpus. How much should be deposited now in order for the corpus to have ₹18,00,00,000 three years from now, if the corpus grows at a rate of 18% per year?

27. Find the value of "y" in the diagram shown below that will make the equivalent present worth of the cash flow equal to ₹50,000 if the interest rate is 10% per year.

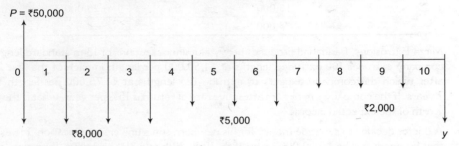

28. Calculate the value of "y" for the cash flows as shown in the table below such that the equivalent total value in 10 years is ₹12,00,000 with an interest rate of 12% per year.

Year	Cash Flow, ₹	Year	Cash Flow, ₹
0	25,000	6	y
1	25,000	7	y
2	y	8	y
3	y	9	15,000
4	y	10	15,000
5	y	11	15,000

29. The receipts and disbursements of Gati Logistics are shown in the following table. Estimate the future worth in 10 years at an interest rate of 10% per year.

Year	Cash Flow, ₹	Year	Cash Flow, ₹
0	−75,00,000	4	35,00,000
1	20,00,000	5	−8,00,000
2	22,00,000	6	50,00,000
3	17,50,000	7	60,00,000

30. Satnam Overseas is planning to install a captive power plant. The company expects that the maintenance cost of the power plant will be ₹3,00,000 per year after the captive power plant is put to service. Beginning 3 years from now, however, the maintenance cost is likely to increase by 5% per year into a foreseeable future. The estimated cost of commissioning the plant, as of now, is ₹4,00,00,000. Once the power plant is put to service, the company will not only save on its own consumed power but sale of excess available power would bring ₹18,00,000 worth of revenue per year. If the company uses a 7-year study period and an interest rate of 12% per year, determine whether the company should go for the captive power plant.

31. Find the present worth at time $t = 0$ of the following cash flow diagram. Assume rate of interest of $i = 10\%$ per year.

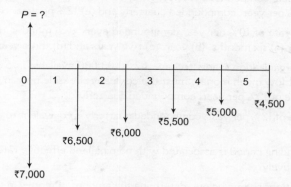

32. Determine the future worth in year 8 at $i = 8\%$ per year for the cash flow diagram shown below:

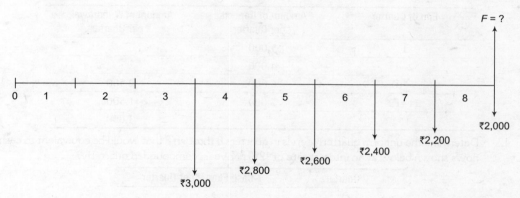

33. Calculate the present worth in year 0 of a lease that requires a payment of ₹1,00,000 now and amounts increasing by 4% per year through year 10. Use an interest rate of 12% per year.

34. For the cash flows shown below find the value of "y" that makes the present worth in year 0 equal to ₹5,00,000 at an interest of 10% per year.

Year	Cash Flow, ₹	Year	Cash Flow, ₹
0	1,000	6	7,000
1	2,000	7	8,000
2	3,000	8	9,000
3	4,000	9	10,000
4	y	10	11,000
5	6,000		

35. A machine has an initial cost of ₹1,50,000, a life of 10 years, and an annual operating cost of ₹10,000 for the first 3 years, increasing by 8% per year thereafter. Use an interest rate of 10% per year and calculate the present worth for the machine.

36. Identify the following interest rate statements as either nominal or effective. (a) 2% per month, (b) 1.5% per week, compounded weekly, (c) effective 5% per quarter, compounded monthly, (d) nominal 12% per year, compounded quarterly and (e) 12% per year compounded monthly.

37. For an interest rate of 10% per year compounded every two months, determine the nominal interest rate per (a) six months, (b) year, (c) two years and (d) three years.

38. A large BPO organization wants to have ₹2,50,00,000 available in 5 years to pay perks to its employees. How much money, must the company set aside now in an account that earns interest at a rate of 12% per year, compounded quarterly?

39. An interest rate of 12% per year, compounded quarterly is equivalent to what effective interest rate per year?

40. What compounding period is associated with nominal and effective rates of 20% and 21.55% per year respectively?

41. For the transactions shown below, determine the amount of money in the account at the end of year 4 if the interest rate is 8% per year compounded semi-annually. Assume no inter-period compounding.

End of Quarter	Amount of Deposit, ₹ per Quarter	Amount of Withdrawal, ₹ per Quarter
1	5,000	-
2–4	4,000	-
7–10	6,000	2,500
11–12	7,000	1,500
15	-	3,000

42. Determine the uniform quarterly series in quarters 0 through 8 that would be equivalent to cash flows shown below at an interest rate of 12% per year, compounded quarterly.

Quarter	Cash Flow, ₹ per Quarter
1	₹5,000
2–3	₹6,000
5–8	₹7,000

43. A present sum of ₹50,000 at an interest rate of 10% per year, compounded quarterly, is equivalent to how much money 5 years ago?

44. Steel Authority of India Ltd. is considering to introduce SAP (an MRP system). The SAP is available for ₹2,00,00,000. If the company wants to recover the cost in 3 years, what is the equivalent amount of new income that must be realized every quarter, with the interest rate is 12% per year, compounded monthly?

45. A person deposits ₹5,000 per month into a savings account that pays interest at a rate of 8% per year, compounded semi-annually. How much will be in the account at the end of 10 years? Assume no inter-period compounding.

5

Methods for Making Economy Studies

5.1 INTRODUCTION

Engineering or business projects require huge capital investments. Economy studies are necessary to be conducted to establish whether a proposed capital investment and its associated expenditures can be recovered over time in addition to a return on the capital that is attractive in view of risks involved and opportunity costs of the limited funds. The concepts of interest and money-time relationships of Chapter 4 are quite useful in arriving at the investment decision.

Since different projects involve different patterns of capital investment, revenue or savings cash flows and expenditure or disbursement cash flows, no single method is perfect for making economy studies of all types of projects. As a result, several methods for making economy studies are commonly used in practice. All methods will produce equally satisfactory results and will lead to the same decision, provided the inherent assumptions of each method are enforced.

This chapter explains the working mechanism of six basic methods for making economy studies and also describes the assumptions and interrelationships of these methods. In making economy studies of the proposed project, the appropriate interest rate to be used for discounting purpose is taken to be equal to the minimum attractive rate of return (*M.A.R.R.*) expected by the fund provider. The value of *M.A.R.R.* is established in view of the opportunity cost of capital which reflects the return forgone as it is invested in one particular project.

5.2 BASIC METHODS

The following six methods are commonly used for making economy studies:

Equivalent worth:
1. Present worth (*P.W.*)
2. Future worth (*F.W.*)
3. Annual worth (*A.W.*)

Rate of return:

1. Internal rate of return (*I.R.R.*)
2. External rate of return (*E.R.R.*)
3. Explicit reinvestment rate of return (*E.R.R.R.*)

5.3   PRESENT WORTH (P.W.) METHOD

In this method, equivalent worth of all cash flows relative to some point in time called present worth i.e. *P.W.* is computed. All cash inflows and outflows are discounted to the present point in time at an interest rate that is generally *M.A.R.R.* using appropriate interest factor. The following steps are used to calculate *P.W.*:

Step 1: Draw the cash flow diagram for the given problem.

Step 2: Determine the *P.W.* of the given series of cash receipts by discounting these future amounts to the present at an interest rate *i* equal to *M.A.R.R.* in the following manner:

$$P.W. = F_0(1+i)^0 + F_1(1+i)^{-1} + F_2(1+i)^{-2} + \ldots + F_k(1+i)^{-k} + \ldots + F_n(1+i)^{-n} \qquad (5.1)$$

Step 3: Determine the *P.W.* of the given series of cash disbursement by discounting these future amounts to the present at an interest rate *i* equal to *M.A.R.R.* using Equation (5.1).

Step 4: Determine the net present worth, (*N.P.W.*) as:

$$N.P.W. = P.W. \text{ at Step } 2 - P.W. \text{ at Step } 3$$

 (i) If *N.P.W.* > 0, the investment in the project is economically justified.
 (ii) If *N.P.W.* < 0, the investment in the project is economically not justified.
(iii) If *N.P.W.* = 0, the investment in the project is barely economically justified.

Examples 5.1 and 5.2 illustrate the *P.W.* method.

Example 5.1

Production engineers of a manufacturing firm have proposed a new equipment to increase productivity of a manual gas-cutting operation. The initial investment (first cost) is ₹5,00,000 and the equipment will have a salvage value of ₹1,00,000 at the end of its expected life of 5 years. Increased productivity will yield an annual revenue of ₹2,00,000 per year. If the firm's minimum attractive rate of return is 15%, is the procurement of the new equipment economically justified? Use *P.W.* method.

Solution

Step 1: The cash flow diagram for this problem is shown in Fig. 5.1.

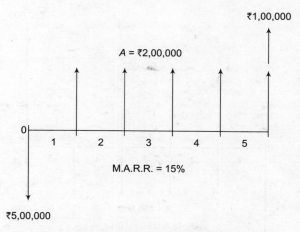

Figure 5.1   Cash flow diagram, Example 5.1

Step 2: The *P.W.* of the series of cash receipts is computed as:

$$P.W. = ₹2,00,000(P/A, 15\%, 5) + ₹1,00,000(P/F, 15\%, 5)$$

$$= ₹2,00,000(3.3522) + ₹1,00,000(0.4972)$$

$$P.W. = ₹7,20,160$$

Step 3: The *P.W.* of the series of cash disbursements is computed as:

$$P.W. = ₹5,00,000$$

Step 4: $N.P.W. = ₹7,20,160 - ₹5,00,000$

$$N.P.W. = ₹2,20,160$$

Since *N.P.W.* > 0, the purchase of this equipment is economically justified.

Example 5.2

An investment of ₹1,05,815.4 can be made in a project that will produce a uniform annual revenue of ₹53,000 for 5 years and then have a salvage value of ₹30,000. Annual disbursements will be ₹30,000 each year for operation and maintenance costs. The company's minimum attractive rate of return is 10%. Show whether this is a desirable investment by using the present worth method.

Solution

Step 1: The cash flow diagram for this problem is shown in Fig. 5.2.

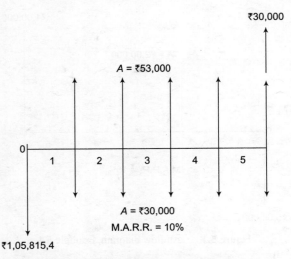

Figure 5.2   Cash flow diagram, Example 5.2

Step 2: The $P.W.$ of the series of cash receipts is computed as:

$$P.W. = ₹53,000(P/A, 10\%, 5) + ₹30,000(P/F, 10\%, 5)$$

$$= ₹53,000(3.7908) + ₹30,000(0.6209)$$

$$P.W. = ₹2,19,539.4$$

Step 3: The $P.W.$ of the series of cash disbursements is computed as:

$$P.W. = ₹1,05,815.4 + ₹30,000(P/A, 10\%, 5)$$

$$= ₹1,05,815.4 + ₹30,000(3.7908)$$

$$P.W. = ₹2,19,539.4$$

Step 4: $N.P.W. = ₹2,19,539.4 - ₹2,19,539.4$

$$N.P.W. = 0$$

Since, $N.P.W. = 0$, the project is shown to be barely justified.

5.4   FUTURE WORTH (F.W.) METHOD

This method involves computation of equivalent worth of all cash flows relative to some point in time called future worth i.e. $F.W.$ All cash inflows and outflows are discounted to the future point in time at an interest rate that is generally $M.A.R.R.$ using appropriate interest factor. Use the following steps are used to calculate $F.W.$

Step 1: Draw the cash flow diagram for the given problem.

Step 2: Determine the *F.W.* of the given series of cash receipts by discounting these present amounts to the future at an interest rate *i* equal to *M.A.R.R.* in the following manner:

$$F.W. = P_0(1+i)^n + P_1(1+i)^{n-1} + P_2(1+i)^{n-2} + \cdots + P_k(1+i)^{n-k} + \ldots + P_n(1+i)^0 \qquad (5.2)$$

Step 3: Determine the *F.W.* of the given series of cash disbursement by discounting these present amounts to the future at an interest rate, i equal to *M.A.R.R.* using Equation (5.2)

Step 4: Determine the net future worth (*N.F.W.*) as:

$$N.F.W. = F.W. \text{ at Step 2} - F.W. \text{ at Step 3}$$

  (i) If $N.F.W. > 0$, the investment in the project is economically justified
  (ii) If $N.F.W. < 0$, the investment in the project is economically not justified
  (iii) If $N.F.W. = 0$, the investment in the project is barely economically justified

Examples 5.3 and 5.4 illustrate the *F.W.* method.

Example 5.3

Solve the problem given in Example 5.1 by *F.W.* method.

Solution

Step 1: The cash flow diagram for this problem is shown in Fig. 5.1.

Step 2: The *F.W.* of the series of cash receipts is computed as:

$$F.W. = ₹\,2,00,000(F/A, 15\%, 5) + ₹\,1,00,000$$

$$= ₹\,2,00,000(6.7424) + ₹\,1,00,000$$

$$F.W. = ₹\,14,48,480$$

Step 3: The *F.W.* of the series of cash disbursements is computed as:

$$F.W. = ₹\,5,00,000(F/P, 15\%, 5)$$

$$= ₹\,5,00,000(2.0114)$$

$$F.W. = ₹\,10,05,700$$

Step 4: $N.F.W. = ₹\,14,48,480 - ₹\,10,05,700$

$$N.F.W. = ₹\,4,42,780$$

Since $N.F.W. > 0$, this equipment is economically justified.

Example 5.4

Solve the problem given in Example 5.2 by *F.W.* method.

Solution

Step 1: The cash flow diagram for this problem is shown in Fig. 5.2.

Step 2: The *F.W.* of the series of cash receipts is computed as:

$$F.W. = ₹53,000(F/A, 10\%, 5) + ₹30,000$$

$$= ₹53,000(6.1051) + ₹30,000$$

$$F.W. = ₹3,53,570.3$$

Step 3: The *F.W.* of the series of cash disbursements is computed as:

$$F.W. = ₹1,05,815.4(F/P, 10\%, 5) + ₹30,000(F/A, 10\%, 5)$$

$$= ₹1,05,815.4(1.6105) + ₹30,000(6.1051)$$

$$F.W. = ₹3,53,568.7$$

Step 4: $N.F.W. = ₹3,53,570.3 - ₹3,53,568.7$

$N.F.W. = ₹1.60$ which is almost equal to 0

Since, $N.F.W. = 0$, this equipment is economically barely justified.

5.5 ANNUAL WORTH (A.W.) METHOD

The term Annual Worth (*A.W.*) refers to a uniform annual series of rupees amounts for a certain period of time that is equivalent to a particular schedule of cash inflows i.e. receipts or savings and/or cash outflows i.e. disbursements. The following procedures can be used to determine the *A.W.* of a particular schedule of cash flows:

Procedure 1:   Use the following steps to calculate *A.W.*

Step 1: Draw the cash flow diagram for the given problem.

Step 2: Determine the *P.W.* of the given series of cash receipts by discounting these future amounts to the present at an interest rate *i* equal to *M.A.R.R.* using Equation (5.1).

Step 3: Determine the *P.W.* of the given series of cash disbursement by discounting these future amounts to the present at an interest rate *i* equal to *M.A.R.R.* using Equation (5.1).

Step 4: Determine the net present worth, *N.P.W.* as:

$$N.P.W. = P.W. \text{ at Step 2} - P.W. \text{ at Step 3}$$

Step 5: Consider *N.P.W.* obtained at Step 4 as present worth *P* and determine its *A.W.* as:

$$A.W. = P(A/P, i\%, n) \tag{5.3}$$

(i) If *A.W.* > 0, the investment in the project is economically justified

(ii) If *A.W.* < 0, the investment in the project is economically not justified

(iii) If *A.W.* = 0, the investment in the project is economically barely justified

Procedure 2: Use the following steps to calculate *A.W.*

Step 1: Draw the cash flow diagram for the given problem.

Step 2: Determine the *F.W.* of the given series of cash receipts by discounting these present amounts to the future at an interest rate *i* equal to *M.A.R.R.* using Equation (5.2).

Step 3: Determine the *F.W.* of the given series of cash disbursement by discounting these present amounts to the future at an interest rate *i* equal to *M.A.R.R.* using Equation (5.2).

Step 4: Determine the net future worth, *N.F.W.* as:

$$N.F.W. = F.W. \text{ at Step 2} - F.W. \text{ at Step 3}$$

Step 5: Consider *N.F.W.* obtained at Step 4 as future worth *F* and determine its *A.W.* as:

$$A.W. = F(A/F, i\%, n) \tag{5.4}$$

(i) If *A.W.* > 0, the investment in the project is economically justified

(ii) If *A.W.* < 0, the investment in the project is economically not justified

(iii) If *A.W.* = 0, the investment in the project is economically barely justified

Procedure 3: Use the following steps to calculate *A.W.*

Step 1: Draw the cash flow diagram for the given problem .

Step 2: Determine the *P.W.* of the given series of cash receipts by discounting these future amounts to the present at an interest rate *i* equal to *M.A.R.R.* using Equation (5.1).

Step 3: Consider *P.W.* obtained at Step 2 as present worth *P* and determine its *A.W.* using Equation (5.3).

Step 4: Determine the *P.W.* of the given series of cash disbursement by discounting these future amounts to the present at an interest rate *i* equal to *M.A.R.R.* using Equation (5.1).

Step 5: Consider *P.W.* obtained at Step 4 as present worth *P* and determine its *A.W.* using Equation (5.3).

Step 6: Determine the annual worth *A.W.* as:

$$A.W. = A.W. \text{ at Step 3} - A.W. \text{ at Step 5}$$

(i) If $A.W. > 0$, the investment in the project is economically justified

(ii) If $A.W. < 0$, the investment in the project is not economically justified

(iii) If $A.W. = 0$, the investment in the project is economically barely justified

Procedure 4: Use the following steps to calculate $A.W.$

Step 1: Draw the cash flow diagram for the given problem.

Step 2: Determine the $F.W.$ of the given series of cash receipts by discounting these present amounts to the future at an interest rate i equal to $M.A.R.R.$ using Equation (5.2).

Step 3: Consider $F.W.$ obtained at Step 2 as future worth F and determine its $A.W.$ using Equation (5.4).

Step 4: Determine the $F.W.$ of the given series of cash disbursement by discounting these present amounts to the future at an interest rate i equal to $M.A.R.R.$ using Equation (5.2).

Step 5: Consider $F.W.$ obtained at Step 4 as future worth F and determine its $A.W.$ using Equation (5.4).

Step 6: Determine the annual worth $A.W.$ as:

$$A.W. = A.W. \text{ at Step } 3 - A.W. \text{ at Step } 5$$

(i) If $A.W. > 0$, the investment in the project is economically justified

(ii) If $A.W. < 0$, the investment in the project is economically not justified

(iii) If $A.W. = 0$, the investment in the project is economically barely justified

Procedure 5: Use the following steps to calculate $A.W.$

Step 1: Draw the cash flow diagram for the given problem.

Step 2: Identify the equivalent annual receipts R from the cash flow diagram drawn at Step 1.

Step 3: Identify the equivalent annual expenses E from the cash flow diagram drawn at Step 1.

Step 4: Identify the initial investment (expenditure at point 0) and S is the salvage value (cash inflow) at the end of useful life.

Step 5: Determine the equivalent annual capital recovery amount, $C.R.$ by using any one of the following formulas:

(i) $C.R. = P(A/P, i\%, n) - S(A/F, i\%, n)$

(ii) $C.R. = (P - S)(A/F, i\%, n) + P(i\%)$

(iii) $C.R. = (P - S)(A/P, i\%, n) + S(i\%)$

Step 6: Determine $A.W.$ as:

$$A.W. = R - E - C.R. \tag{5.5}$$

(i) If $A.W. > 0$, the investment in the project is economically justified

(ii) If $A.W. < 0$, the investment in the project is economically not justified

(iii) If $A.W. = 0$, the investment in the project is economically barely justified

Example 5.5

Solve the problem given in Example 5.1 by *A.W.* method.

Solution

Procedure 1: The *N.P.W.* has already been calculated in Example 5.1 and its value is ₹2,20,160. Considering ₹2,20,160 as *P*, its *A.W.* is calculated by using Equation (5.3) as:

$$A.W. = ₹2,20,160(A/P, 15\%, 5)$$

$$= ₹2,20,160(0.2983)$$

$$A.W. = ₹65,673.73$$

Since $A.W. > 0$, this equipment is economically justified.

Procedure 2: The *N.F.W.* has already been calculated in Example 5.3 and its value is ₹4,42,780. Considering ₹4,42,780 as *F*, its *A.W.* is calculated by using Equation (5.4) as:

$$A.W. = ₹4,42,780 \ (A/F, 15\%, 5)$$

$$= ₹4,42,780 \ (0.1483)$$

$$A.W. = ₹65,664.27$$

Since $A.W. > 0$, this equipment is economically justified.

Procedure 3:

Step 1: The cash flow diagram for the given problem is shown in Fig. 5.1.

Step 2: The *P.W.* of the given series of cash receipts has already been calculated in Example 5.1 and its value is ₹7,20,160.

Step 3: Consider ₹7,20,160 as present worth *P* and determine its *A.W.* using Equation (5.3) as:

$$A.W. = ₹7,20,160(A/P, 15\%, 5)$$

$$= ₹7,20,160(0.2983)$$

$$A.W. = ₹2,14,823.7$$

Step 4: The *P.W.* of the given series of cash disbursements has already been calculated in Example 5.1 and its value is ₹5,00,000.

Step 5: Consider ₹5,00,000 as present worth *P* and determine its *A.W.* using Equation (5.3) as:

$$A.W. = ₹5,00,000(A/P, 15\%, 5)$$
$$= ₹5,00,000(0.2983)$$
$$A.W. = ₹1,49,150$$

Step 6: Determine the annual worth $A.W.$ as:

$$A.W. = A.W. \text{ at Step 3} - A.W. \text{ at Step 5}$$
$$= ₹2,14,823.7 - ₹1,49,150$$
$$A.W. = 65,673.7$$

Since $A.W. > 0$, this equipment is economically justified.

Procedure 4:

Step 1: The cash flow diagram for the given problem is shown in Fig. 5.1.

Step 2: The $F.W.$ of the given series of cash receipts has already been calculated in Example 5.3 and its value is ₹14,48,480.

Step 3: Consider ₹14,48,480 as future worth F and determine its $A.W.$ using Equation (5.4) as:

$$A.W. = ₹14,48,480(A/F, 15\%, 5)$$
$$= ₹14,48,480(0.1483)$$
$$A.W. = ₹2,14,809.6$$

Step 4: The $F.W.$ of the given series of cash disbursements has already been calculated in Example 5.3 and its value is ₹10,05,700.

Step 5: Consider ₹10,05,700 as future worth, F and determine its $A.W.$ using Equation (5.4) as:

$$A.W. = ₹10,05,700(A/F, 15\%, 5)$$
$$= ₹10,05,700(0.1483)$$
$$A.W. = ₹1,49,145.3$$

Step 6: Determine the annual worth $A.W.$ as:

$$A.W. = A.W. \text{ at Step 3} - A.W. \text{ at Step 5}$$
$$= ₹2,14,809.6 - ₹1,49,145.3$$
$$A.W. = ₹65,664.3$$

Since $A.W. > 0$, this equipment is economically justified.

Procedure 5: The following steps are used to calculate $A.W.$:

Step 1: The cash flow diagram for the given problem is shown in Fig. 5.1.

Step 2: From the cash flow diagram shown in Fig. 5.1, it is clear that the equivalent annual receipts, $R = ₹2,00,000$.

Step 3: From the cash flow diagram shown in Fig. 5.1, it is clear that the equivalent annual expenses, $E = 0$.

Step 4: From the cash flow diagram shown in Fig. 5.1, it is clear that the initial investment, $P = ₹5,00,000$ and salvage value, $S = ₹1,00,000$.

Step 5: The equivalent annual capital recovery amount $(C.R.)$ is calculated by using the following formula:

$$C.R. = P(A/P, i\%, n) - S(A/F, i\%, n)$$

$$C.R. = ₹5,00,000(A/P, 15\%, 5) - ₹1,00,000(A/F, 15\%, 5)$$

$$= ₹5,00,000(0.2983) - ₹1,00,000(0.1483)$$

$$= ₹1,49,150 - ₹14,830$$

$$C.R. = ₹1,34,320$$

Step 6: $A.W.$ is calculated as:

$$A.W. = R - E - C.R.$$

$$A.W. = ₹2,00,000 - ₹0 - ₹1,34,320$$

$$A.W. = ₹65,680$$

Since $A.W. > 0$, this equipment is economically justified.

Note: It may be noted that the small difference in the values of $A.W.$ obtained from the above procedures is due to rounding-off the values of factors in the interest table. However, the result by all methods is same.

Example 5.6

Solve the problem given in Example 5.2 by $A.W.$ method.

Solution

Let us solve this problem by Procedure 5 discussed in Example 5.5.

Procedure: The following steps are used to calculate $A.W.$:

Step 1: The cash flow diagram for the given problem is shown in Fig. 5.2.

Step 2: From the cash flow diagram shown in Fig. 5.2, it is clear that the equivalent annual receipts $R = ₹53,000$.

Step 3: From the cash flow diagram shown in Fig. 5.2, it is clear that the equivalent annual expenses $E = ₹ 30,000$.

Step 4: From the cash flow diagram shown in Fig. 5.2, it is clear that the initial investment $P = ₹ 1,05,815.4$ and salvage value $S = ₹ 30,000$.

Step 5: The equivalent annual capital recovery amount $(C.R.)$ is calculated by using the following formula:

$$C.R. = P(A/P, i\%, n) - S(A/F, i\%, n)$$

$$C.R. = ₹ 1,05,815.4(A/P, 10\%, 5) - ₹ 30,000(A/F, 10\%, 5)$$

$$= ₹ 1,05,815.4(0.2638) - ₹ 30,000(0.1638)$$

$$= ₹ 27,914.10 - ₹ 4,914$$

$$C.R. = ₹ 23,000.1$$

Step 6: $A.W.$ is calculated as:

$$A.W. = R - E - C.R.$$

$$A.W. = ₹ 53,000 - ₹ 30,000 - ₹ 23,000.1$$

$$A.W. = -₹ 0.1 \text{ which is almost equal to } 0.$$

Since $A.W. = 0$, this equipment is economically barely justified.

5.6 INTERNAL RATE OF RETURN (I.R.R.) METHOD

Out of all the rate of return methods, this method is widely used for making economy studies. This method is also known by several other names such as investor's method, discounted cash flow method, receipts versus disbursements method and profitability index. In this method an interest rate i' called *I.R.R.* is computed and it is compared with *M.A.R.R.* to take decision on the economic viability of the project. This method can be used only when both positive and negative cash flows are present in the problem. *I.R.R.* is also defined as an interest rate at which net present worth (*N.P.W.*) is 0. The following steps are used to compute *I.R.R.*:

Step 1: Draw the cash flow diagram for the given problem.

Step 2: Determine the *P.W.* of the net receipts at an interest rate of i' in the following manner:

$$\sum_{k=0}^{n} R_k(P/F, i'\%, k) \qquad (5.6)$$

where R_k = net receipts or savings for the kth year.

n = project life

Step 3: Determine the *P.W.* of the net expenditures at an interest rate i' in the following manner:

$$\sum_{k=0}^{n} E_k(P/F, i'\%, k) \qquad (5.7)$$

where, R_k = net expenditures including investments for the kth year

Step 4: Determine the net present worth (*N.P.W.*) as:

$$N.P.W. = \sum_{k=0}^{n} R_k(P/F, i'\%, k) - \sum_{k=0}^{n} E_k(P/F, i'\%, k) \qquad (5.8)$$

Step 5: Set *N.P.W.* = 0 and determine the value of $i'\%$

Step 6: Compare the value of $i'\%$ with *M.A.R.R.*
   (i) If $i'\% > M.A.R.R.$, the investment in the project is economically justified
   (ii) If $i'\% < M.A.R.R.$, the investment in the project is economically not justified
   (iii) If $i'\% = M.A.R.R.$, the investment in the project is economically barely justified

Note: The value of $i'\%$ can also be determined as the interest rate at which net *F.W.* = 0 or at which net *A.W.* = 0.

Examples 5.7 and 5.8 illustrates the *I.R.R.* method.

Example 5.7

Solve the problem given in Example 5.1 by *I.R.R.* method.

Solution

Step 1: The cash flow diagram for the given problem is shown in Fig. 5.1.

Step 2: The *P.W.* of the net receipts at an interest rate of i' is calculated as:

$$P.W. = ₹\,2,00,000(P/A, i'\%, 5) + ₹\,1,00,000(P/F, i'\%, 5)$$

Step 3: The *P.W.* of the net expenditures at an interest rate of i' is calculated as:

$$P.W. = ₹\,5,00,000$$

Step 4: The net present worth, *N.P.W.* is obtained as:

$$N.P.W. = ₹\,2,00,000(P/A, i'\%, 5) + ₹\,1,00,000(P/F, i'\%, 5) - ₹\,5,00,000$$

Step 5: $0 = ₹\,2,00,000(P/A, i'\%, 5) + ₹\,1,00,000(P/F, i'\%, 5) - ₹\,5,00,000$

The equation given at Step 5 normally involves trial-and-error calculations until the $i'\%$ is found. However, since we do not know the exact value of $i'\%$, we will probably try a relatively low $i'\%$, such as 5%, and also a relatively high $i'\%$, such as 40%.

At $i'\% = 5\%$:

$$₹2,00,000(P/A, 5\%, 5) + ₹1,00,000(P/F, 5\%, 5) - ₹5,00,000$$

$$₹2,00,000(4.3295) + ₹1,00,000(0.7835) - ₹5,00,000 = +₹4,44,250$$

At $i'\% = 40\%$:

$$₹2,00,000(P/A, 40\%, 5) + ₹1,00,000(P/F, 40\%, 5) - ₹5,00,000$$

$$₹2,00,000(2.035) + ₹1,00,000(0.1859) - ₹5,00,000 = -₹74,410$$

Since we have both a positive and a negative *P.W.* of net cash flows, linear interpolation can be used as given below to find an approximate value of $i'\%$

$$\frac{40\% - 5\%}{₹4,44,250 - (-₹74,410)} = \frac{i'\% - 5\%}{₹4,44,250 - ₹0}$$

$$i'\% = 5\% + \frac{₹4,44,250}{₹4,44,250 - (-₹74,410)}(40\% - 5\%)$$

Hence, $i'\% = 34.98\%$

Step 6: Since the value of $i'\% = 34.98\% > M.A.R.R. = 15\%$, the investment in the project is economically justified.

Example 5.8

Solve the problem given in Example 5.2 by the *I.R.R.* method.

Solution

Step 1: The cash flow diagram for the given problem is shown in Fig. 5.2.

Step 2: The *P.W.* of the net receipts at an interest rate of i' is calculated as:

$$P.W. = ₹53,000(P/A, i'\%, 5) + ₹30,000(P/F, i'\%, 5)$$

Step 3: The *P.W.* of the net expenditures at an interest rate of i' is calculated as:

$$P.W. = ₹1,05,815.4 + ₹30,000(P/A, i'\%, 5)$$

Step 4: The net present worth *N.P.W.* is obtained as:

$$N.P.W. = ₹53,000(P/A, i'\%, 5) + ₹30,000(P/F, i'\%, 5)$$
$$- ₹1,05,815.4 - ₹30,000(P/A, i'\%, 5)$$

Step 5: $0 = ₹53,000(P/A, i'\%, 5) + ₹30,000(P/F, i'\%, 5) - ₹1,05,815.4 - ₹30,000(P/A, i'\%, 5)$

The equation given at Step 5 normally involves trial-and-error calculations until the $i'\%$ is found. However, since we do not know the exact value of $i'\%$, we will probably try a relatively low $i'\%$, such as 5%, and a relatively high $i'\%$, such as 12%.

At $i'\% = 5\%$:

₹53,000(P/A, 5%, 5) + ₹30,000(P/F, 5%, 5) − ₹1,05,815.4 − ₹30,000(P/A, 5%, 5)

₹53,000(4.3295) + ₹30,000(0.7835) − ₹1,05,815.4 − ₹30,000(4.3295) = + ₹17,268.1

At $i'\% = 12\%$:

₹53,000(P/A, 25%, 5) + ₹30,000(P/F, 25%, 5) − ₹1,05,815.4 − ₹30,000(P/A, 25%, 5)

₹53,000(3.6048) + ₹30,000(0.5674) − ₹1,05,815.4 − ₹30,000(3.6048) = − ₹5,883

Since we have both a positive and a negative *P.W.* of net cash flows, linear interpolation can be used as given below to find an approximate value of $i'\%$:

$$\frac{12\% - 5\%}{₹17,268.1 - (-₹5,883)} = \frac{i'\% - 5\%}{₹17,268.1 - ₹0}$$

$$i'\% = 5\% + \frac{₹17,268.1}{₹17,268.1 - (-₹5,883)}(12\% - 5\%)$$

$i'\% = 10.22\%$, which is approximately equal to 10%.

Step 6: Since the value of $i'\% = M.A.R.R.$, the investment in the project is economically barely justified.

Note: Let us check whether the value of *N.P.W.* at $i' = 10\%$ is 0.

$N.P.W. = ₹53,000(P/A, i'\%, 5) + ₹30,000(P/F, i'\%, 5) − ₹1,05,815.4 − ₹30,000(P/A, i'\%, 5)$

At $i'\% = 10\%$:

$N.P.W. = ₹53,000(P/A, 10\%, 5) + ₹30,000(P/F, 10\%, 5) − ₹1,05,815.4 − ₹30,000(P/A, 10\%, 5)$

$$= ₹53,000(3.7908) + ₹30,000(0.6209) − ₹1,05,815.4 − ₹30,000(3.7908)$$

$N.P.W. = 0$

Thus $i' = 10\%$ which is equal to the given *M.A.R.R.* and therefore, the investment in the project is economically barely justified.

5.7　EXTERNAL RATE OF RETURN (E.R.R.) METHOD

The following steps are involved in the *E.R.R.* method:

Step 1: Draw the cash flow diagram for the given problem.

Step 2: Consider an external interest rate *e*, equal to given *M.A.R.R.* However, if a specific value of *e* is given then take *e* equal to that value. *e* is, indeed, an external interest rate at which net cash flows generated or required by a project over its life can be reinvested or borrowed outside the firm. If this external reinvestment rate happens to be equal to the project's *I.R.R.*, then *E.R.R.* method produces results same as that of *I.R.R.* method.

Step 3: Discount all cash outflows (negative cash flows/expenditures) to period 0 (the present) at $e\%$ in the following manner:

$$\sum_{k=0}^{n} E_k(P/F, e\%, k) \tag{5.9}$$

where E_k is excess of expenditures over receipts in period k
n is project life or number of periods for the study

Step 4: Compound all cash inflows (positive cash flows/receipts) to period n (the future) at $e\%$ in the following manner:

$$\sum_{k=0}^{n} R_k(F/P, e\%, n-k) \tag{5.10}$$

where R_k is excess of receipts over disbursements in period k

Step 5: Equate Equations (5.9) and Equations (5.10) as:

$$\sum_{k=0}^{n} E_k(P/F, e\%, k) = \sum_{k=0}^{n} R_k(F/P, e\%, n-k) \tag{5.11}$$

Step 6: The left hand side (L.H.S.) of Equation (5.11) represents present worth whereas, its right hand side (R.H.S.) represents future worth. Therefore, the two sides cannot be equal. Introduce the factor $(F/P, i'\%, n)$ on to the L.H.S. to bring a balance between two sides of equation in the following manner:

$$\sum_{k=0}^{n} E_k(P/F, e\%, k)\,(F/P, i'\%, n) = \sum_{k=0}^{n} R_k(F/P, e\%, n-k) \tag{5.12}$$

where $i'\%$ is the external rate of return (*E.R.R.*)

Step 7: Determine the value of $i'\%$ by solving Equation (5.12) and compare it with the *M.A.R.R.*
   (i) If $i'\% > M.A.R.R.$, the investment in the project is economically justified
  (ii) If $i'\% < M.A.R.R.$, the investment in the project is economically not justified
 (iii) If $i'\% = M.A.R.R.$, the investment in the project is economically barely justified

Examples 5.9 and 5.10 illustrate the *E.R.R.* method.

Example 5.9

Solve the problem given in Example 5.1 by *E.R.R.* method.

Solution

Step 1: The cash flow diagram for the given problem is shown in Fig. 5.1.

Step 2: Since a specific value of e is not given take e = M.A.R.R. = 15%.

Step 3: In this problem ₹ 5,00,000 is the only cash outflow and it occurs at period 0. Therefore, the result of this step is ₹ 5,00,000.

Step 4: All cash inflows (positive cash flows/receipts) are compounded to period n (the future) at e% in the following manner:

$$₹ 2,00,000(F/A, 15\%, 5) + ₹ 1,00,000$$

Step 5: The result of this step is given as:

$$₹ 5,00,000 = ₹ 2,00,000(F/A, 15\%, 5) + ₹ 1,00,000$$

Step 6: The L.H.S. of the equation given at Step 5 represents present worth whereas, its R.H.S. represents future worth. Therefore, the two sides cannot be equal. Introduce the factor $(F/P, i'\%, n)$ on to the L.H.S. to bring a balance between two sides of equation in the following manner:

$$₹ 5,00,000(F/P, i'\%, 5) = ₹ 2,00,000(F/A, 15\%, 5) + ₹ 1,00,000$$

Step 7: The equation given at Step 6 is solved as:

$$₹ 5,00,000(F/P, i'\%, 5) = ₹ 2,00,000(6.7424) + ₹ 1,00,000$$

$$₹ 5,00,000(F/P, i'\%, 5) = ₹ 14,48,480$$

$$(F/P, i'\%, 5) = ₹ 14,48,480/₹ 5,00,000$$

$$(F/P, i'\%, 5) = 2.897$$

$$(1 + i')^5 = 2.897$$

$$1 + i' = 1.237$$

$$i' = 1.237 - 1 = 0.237 = 23.7\%$$

Since the value of i'% = 23.7% > M.A.R.R. = 15%, the investment in the project is economically justified.

Example 5.10

Solve the problem given in Example 5.2 by *E.R.R.* method.

Solution

Step 1: The cash flow diagram for the given problem is shown in Fig. 5.2.

Step 2: Since a specific value of e is not given, take $e = M.A.R.R. = 10\%$.

Step 3: All cash outflows (negative cash flows/expenditures) are discounted to period 0 (the present) at $e\%$ in the following manner:

$$₹1,05,815.4 + ₹30,000(P/A, 10\%, 5)$$

$$= ₹1,05,815.4 + ₹30,000(3.7908)$$

$$= ₹2,19,539.4$$

Step 4: All cash inflows (positive cash flows/receipts) are compounded to period n (the future) at $e\%$ in the following manner:

$$₹53,000(F/A, 10\%, 5) + ₹30,000$$

$$= ₹53,000(6.1051) + ₹30,000$$

$$= ₹3,53,570.3$$

Step 5: The result of this step is given as:

$$₹2,19,539.4 = ₹3,53,570.3$$

Step 6: The L.H.S. of the equation given in Step 5 represents present worth whereas its R.H.S. represents future worth. Therefore, the two sides cannot be equal. Introduce the factor $(F/P, i'\%, n)$ on to the L.H.S. to bring a balance between two sides of equation in the following manner:

$$₹2,19,539.4(F/P, i' \%, 5) = ₹3,53,570.3$$

Step 7: The equation given at Step 6 is solved as:

$$₹2,19,539.4(F/P, i' \%, 5) = ₹3,53,570.3$$

$$(F/P, i'\%, 5) = ₹3,53,570.3/₹2,19,539.4$$

$$(F/P, i'\%, 5) = 1.61$$

$$(1 + i')^5 = 1.61$$

$$1 + i' = 1.0999$$

$$i' = 1.0999 - 1 = 0.0999 = 9.99\%, \text{which is almost equal to } 10\%.$$

Since the value of $i' \% = 9.99\% = M.A.R.R. = 10\%$, the investment in the project is economically barely justified.

5.8 EXPLICIT REINVESTMENT RATE OF RETURN (E.R.R.R.) METHOD

The following steps are involved in the *E.R.R.R.* method:

Step 1: Draw the cash flow diagram for the given problem.

Step 2: Determine the value of *E.R.R.R.* from the following relationship:

$$E.R.R.R. = \frac{(R - E) - (P - S)(A/F, e\%, n)}{P} \tag{5.13}$$

where, *R* is equivalent annual receipts

E is equivalent annual expenditures

P is the initial investment (expenditure at point 0)

S is the salvage value (cash inflow) at the end of useful life

e% is the external interest rate which is taken equal to *M.A.R.R.* if not given specifically

Step 3: Compare the value of *E.R.R.R.* with *M.A.R.R.*

   (i)   If *E.R.R.R.* > *M.A.R.R.*, the investment in the project is economically justified

  (ii)   If *E.R.R.R.* < *M.A.R.R.*, the investment in the project is economically not justified

 (iii)   If *E.R.R.R.* = *M.A.R.R.*, the investment in the project is economically barely justified

Examples 5.11 and 5.12 illustrate the *E.R.R.R.* method.

Example 5.11

Solve the problem given in Example 5.1 by *E.R.R.R.* method.

Solution

Step 1: The cash flow diagram for the given problem is shown in Fig. 5.1.

Step 2: From the cash flow diagram shown in Fig. 5.1, it is clear that the equivalent annual receipts, $R = ₹\,2,00,000$; the equivalent annual expenses, $E = 0$; the initial investment, $P = ₹\,5,00,000$; salvage value, $S = ₹\,1,00,000$; $n = 5$ and $e\% = M.A.R.R. = 15\%$. Thus,

$$E.R.R.R. = \frac{(₹\,2,00,000 - ₹0) - (₹\,5,00,000 - ₹\,1,00,000)(A/F, 15\%, 5)}{₹\,5,00,000}$$

$$= \frac{(₹\,2,00,000) - (₹\,4,00,000)(0.1483)}{₹\,5,00,000}$$

$$= 0.28136 = 28.14\%$$

Step 3: Since *E.R.R.R.* > *M.A.R.R.*, the investment in the project is economically justified.

Example 5.12

Solve the problem given in Example 5.2 by *E.R.R.R.* method.

Solution

Step 1: The cash flow diagram for the given problem is shown in Fig. 5.2.

Step 2: From the cash flow diagram given at Step 1, it is clear that the equivalent annual receipts, $R = ₹53,000$; the equivalent annual expenses, $E = ₹30,000$; the initial investment, $P = ₹1,05,815.40$; salvage value, $S = ₹30,000$; $n = 5$ and $e\% = M.A.R.R.$ $= 10\%$. Thus,

$$\text{E.R.R.R.} = \frac{(₹53,000 - ₹30,000) - (₹1,05,815.4 - ₹30,000)(A/F, 10\%, 5)}{₹1,05,815.4}$$

$$= \frac{(₹23,000) - (₹75,815.4)(0.1638)}{₹1,05,815.4}$$

$$= 0.0999 = 9.99\% \text{ which is almost equal to } 10\%$$

Step 3: Since *E.R.R.R.* = *M.A.R.R.*, the investment in the project is economically barely justified.

5.9 CAPITALIZED COST CALCULATION AND ANALYSIS

Capitalized cost (*CC*) is defined as the present worth of an alternative or project with infinite or very long life. Public sector projects such as bridges, dams, irrigation systems and rail roads are examples of such projects. In addition, permanent and charitable organization endowments are evaluated by capitalized cost analysis.

The formula to calculate *CC* is derived from the relation $P = A(P/A, i\%, n)$, where $n = \infty$. The equation for P using P/A factor formula is:

$$P = A\left[\frac{(1+i)^n - 1}{i(1+i)^n}\right]$$

Divide the numerator and denominator by $(1 + i)^n$

$$P = A\left[\frac{1 - \dfrac{1}{(1+i)^n}}{i}\right]$$

As n approaches ∞, the term in the bracket becomes $1/i$, and the symbol *CC* replaces P.W. and P.

$$CC = P = \frac{A}{i} \tag{5.14}$$

Thus,

$$A = Pi = CC(i) \tag{5.15}$$

If A value is an annual worth ($A.W.$) determined through equivalence calculations of cash flows over n years, the CC value is:

$$CC = \frac{A.W.}{i} \tag{5.16}$$

The following steps are used in calculating CC for an infinite sequence of cash flows:

Step 1: Draw the cash flow diagram for the given problem showing all non-recurring (one-time) cash flows and at least two cycles of all recurring (periodic) cash flows.

Step 2: Find the present worth of all non-recurring amounts. This is their CC value.

Step 3: Find the equivalent uniform annual worth (A value) through one life cycle of all recurring amounts. Add this to all other uniform amounts occurring in years 1 through infinity and the result is the total equivalent uniform annual worth ($A.W.$).

Step 4: Divide the $A.W.$ obtained in Step 3 by the interest rate i to obtain a CC value.

Step 5: Add the CC values obtained in Steps 2 and 4 to obtain total CC value.

Examples 5.13 and 5.14 illustrate the calculation of CC.

Example 5.13

Determine the capitalized cost of an expenditure of ₹2,00,000 at time 0, ₹25,000 in years 2 through 5, and ₹40,000 per year from year 6 onward. Use an interest rate of 12% per year.

Solution

Step 1: The cash flow diagram for the two cycles is shown in Fig. 5.3.

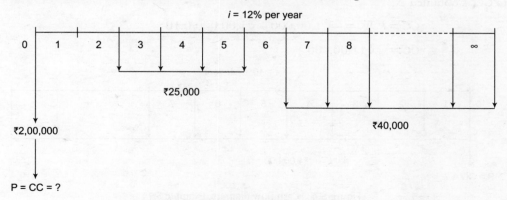

Figure 5.3   Cash flow diagram, Example 5.13

Step 2: The present worth of all non-recurring amount is ₹2,00,000. Thus its *CC* value – ₹2,00,000.

Step 3: The equivalent uniform annual worth (*A.W.*) through one life cycle of all recurring amounts is obtained as:

$$A.W. = -[₹25,000(P/A, 12\%, 4)(P/F, 12\%, 1)](0.12) - [(₹40,000/0.12)(P/F, 12\%, 5)](0.12)$$

$$= -[₹25,000(3.0373)(0.8929)](0.12) - [₹3,33,333.3(0.5674)](0.12)$$

$$= -₹8,136.02 - ₹22,695.99$$

$$= -₹30,832.01$$

Step 4: Divide the *A.W.* obtained in Step 3 by the interest rate *i* to obtain a *CC* value.

Thus, $CC = -₹30,832.01/0.12 = -₹2,56,933.42$

Step 5: Add the *CC* values obtained in Steps 2 and 4 to obtain total *CC* value.

Thus, $CC = -₹2,00,000 - ₹2,56,933.42$

$$CC = -₹4,56,933.42$$

Another way to solve the problem is as follows:

$$CC = -₹2,00,000 - ₹25,000(P/A, 12\%, 4)(P/F, 12\%, 1) - [₹40,000/0.12](P/F, 12\%, 5)$$

$$= -₹2,00,000 - ₹25,000(3.0373)(0.8929) - [₹40,000/0.12])(0.5674)$$

$$CC = -₹4,56,933.$$

Example 5.14

An alumni of Jamia Millia Islamia wanted to set up an endowment fund that would award scholarships to engineering students totaling ₹1,00,000 per year for ever. The first scholarships are to be granted now and continue each year for ever. How much should the alumnus donate now, if the endowment fund is expected to earn interest at a rate of 10% per year?

Solution

The cash flow diagram for this problem is shown in Fig. 5.4. The present worth which is equal to *CC* is obtained as:

$$CC = P.W. = -₹1,00,000 - ₹1,00,000/0.10$$

$$CC = -₹11,00,000$$

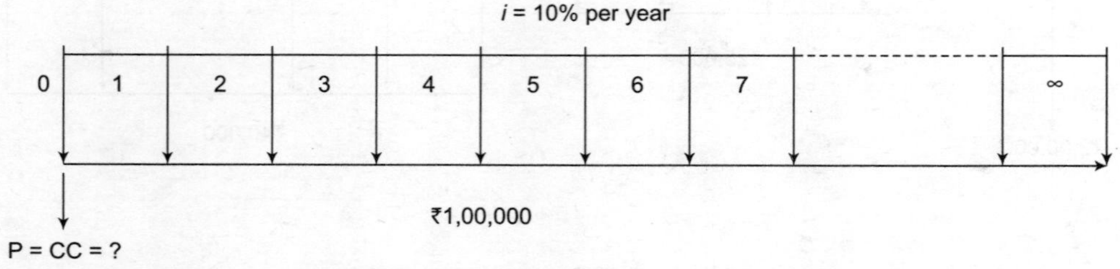

Figure 5.4   Cash flow diagram, Example 5.14

5.10 PAYBACK (PAYOUT) METHOD

This method is an extension of the present worth method. Payback can take two forms: one for $i > 0$ (also called discounted payback method) and the other for $i = 0$ (also called no return payback or simple payback). Payback period, n_p is defined as the estimated time it will take for the estimated revenues and other economic benefits to recover the initial investment and a specified rate of return. It is important to note that the payback period should never be used as the primary measure of worth to select an alternative. However, it should be determined to provide supplemental information in conjunction with an analysis performed using any of the above six methods.

To determine the discounted payback period at a stated rate $i > 0$, calculate the years n_p that make the following expression correct:

$$0 = -P + \sum_{t=1}^{t=n_p} NCF_t(P/F,i,t)$$

(5.17)

where, P is the initial investment or first cost

NCF is the estimated net cash flow for each year t

NCF = Receipt – Disbursements

After n_p years, the cash flows will recover the investment and a return of i %. However, if the alternative is used for more than n_p years, a larger return may result. If the useful life of the alternative is less than n_p years, then there is not enough time to recover the initial investment and i % return.

The n_p value obtained for $i = 0$ i.e. simple payback or no-return payback is used as an indicator to know whether a proposal is a viable alternative worthy of full economic evaluation. Use $i = 0\%$ in Equation (5.17) and find n_p.

$$0 = -P + \sum_{t=1}^{t=n_p} NCF_t$$

(5.18)

It should be noted that the final selection of the alternative on the basis of no-return payback or simple payback would be incorrect.

Examples 5.15 and 5.16 illustrate the calculation of payback period.

Example 5.15

Determine the payback period for an asset that has a first cost of ₹50,000, a salvage value of ₹10,000 any time within 10 years of its purchase, and generates income of ₹8,000 per year. The required return is 8% per year.

Solution

The payback period is determined as:

$$0 = -₹50,000 + ₹8,000(P/A, 8\%, n) + ₹10,000(P/F, 8\%, n)$$

Try $n = 8$: $0 \neq +₹1,375.8$

Try $n = 7$: $0 \neq -₹2,513.8$

Hence, n is between 7 and 8 years.

Example 5.16

A company purchased a small equipment for ₹70,000. Annual maintenance costs are expected to be ₹1,850, but extra income will be ₹14,000 per year. How long will it take for the company to recover its investment at an interest rate of 10% per year?

Solution

$$0 = -₹70,000 + (₹14,000 - ₹1,850)(P/A, 10\%, n)$$

$$(P/A, 10\%, n) = 5.76132$$

n is between 9 and 10;

therefore, it would take 10 years for the company to recover its investment.

PROBLEMS

1. A blender in a pharmaceutical unit costs ₹1,50,000 and has an estimated life of 6 years. If auxiliary equipment is added to it at the time of its initial installation, an annual saving of ₹15,000 would be obtained and its life could be doubled. If the salvage value is negligible and if the effective annual interest rate is 8%, what present expenditure can be justified for the auxiliary equipment? Use a study period of 12 years.

2. An Electric Discharge Machine (EDM) is to be evaluated based on present worth method taking minimum attractive rate of return as 12%. The relevant data is given in the following table will you recommend the machine?

Data	EDM
First Cost	₹7,00,000
Useful Life	15 Years
Salvage Value	₹2,00,000
Annual Operating Cost	₹7,000
Overhaul at the end of fifth year	₹15,000
Overhaul cost at the end of tenth year	₹1,50,000

Also, determine the capital recovery cost of the EDM by all three formulas.

3. An air-conditioning unit is serviced for an estimated service life of 10 years has a first cost of ₹4,00,000, it has no salvage value and annual net receipts of ₹1,25,000. Assuming a minimum attractive rate of return of 18% before tax, find the annual worth of this process and specify whether you would recommend it.

4. A pump installed in a gasoline filling station costs ₹1,25,000, it has a salvage value of ₹50,000 for a useful life of 5 years. At a nominal, interest rate of 12%, compounded quarterly, estimate the annual capital recovery cost of the pump.

5. (a) Find the *P.W.* and *F.W.* of the following proposal A with the *M.A.R.R.* as 15%:

Data	Proposal A
First Cost	₹5,00,000
Useful Life	5 Years
Salvage Value	– ₹25,000
Annual Receipts	₹3,00,000
Annual disbursements	₹1,50,000

   (b) Determine the *I.R.R.* for the proposal. Is the proposal acceptable?

   (c) Determine the *E.R.R.R.* for the proposal when the external reinvestment rate is 12%.

6. M/s Crescent Telesystems Pvt. Ltd. has procured a switching instrument at a present cost of ₹20,00,000. The company will lose on the instrument a sum of ₹80,000 per year for the first 4 years. The company made another investment of ₹6,00,000 during the 4th year and this will result in an annual profit of ₹20,00,000 from 5th year through 12th year. At the end of 12th year the instrument can be sold for ₹8,00,000.

   (a) Determine the *I.R.R*

   (b) Determine the *E.R.R.* when e = 10%.

   (c) Calculate the future worth if *M.A.R.R.* = 14%.

7. Mirza International is considers investing in a CNC lathe and has two alternatives: CNC lathe A and CNC lathe B. Pertinent data with regard to the machines is given in the following table:

Data	CNC Lathe A	CNC Lathe B
Initial Cost	₹6,00,000	₹4,50,000
Annual Net Cash Flow	+ ₹2,00,000	+ ₹1,00,000
Life	6 Years	6 Years
Salvage Value	0	0

   (a) If *M.A.R.R.* = 10%, determine whether the CNC lathes are acceptable alternatives with the three rates of return methods? The reinvestment rate is equal to the *M.A.R.R.*

   (b) Use the equivalent worth methods when *M.A.R.R.* = 10% to determine which of the two CNC lathes is acceptable.

8. Evaluate the lathes A and B in problem 7 by using the simple payback method. Is either of the lathe acceptable?

9. As a project engineer in ABG Heavy Engineering Ltd., you were presented with the summary of a prospective project as per the details given in the following table:

Year Ending	Net Cash Flow
0	− ₹1,36,00,000
1	− ₹13,60,000
2	+ ₹27,50,000
3	+ ₹96,00,000
4	+ ₹2,25,00,000
5	− ₹84,00,000

You were asked to analyze the before-tax *I.R.R.* and evaluate the project. Present the details of your analysis. Also elaborate on the results of your presentation.

10. If you were to rework the problem 9 by *E.R.R.* method and an external reinvestment rate equal to M.A.R.R. 15%, how would you explain the difference in your results?

11. M/s Advanced Valves Ltd. considers investing in a high-pressure vessel testing facility at a cost of ₹6,00,000 and it has a salvage value of ₹1,00,000 at an expected life of 10 years. Once installed, the facility is expected to earn ₹1,00,000 in the first year which will increase by ₹15,000 each year thereafter. Calculate:

    (a) Payout period, if $i = 10\%$.

    (b) The *I.R.R.* for the facility.

12. M/s Technofab Engineering Ltd is considering commissioning a small capacity power plant. The details of the investments to be made in the project are given in the following table:

Title	Capital Expenditure
Land	₹7,50,00,000
Building	₹15,00,00,000
Equipment Cost	₹5,00,00,000
Working Capital	₹1,50,00,000

The company expects to earn ₹2,50,00,000 annually by sale of power for 10 years. By this time company could sell the entire plant for ₹35,00,00,000 and all the working capital recovered. The annual outgoing expenses towards material, labor, services etc. are estimated to total ₹90,00,000. If the company requires a minimum return of 30% on projects of comparable risk, determine, using *I.R.R.* method, if the company should invest in the project?

13. A famous mobile telephone manufacturer wishes to launch a new model. In order to launch the new model, the company has to create an additional facility at a fresh investment of ₹45,00,000 and an additional investment of ₹7,50,000 at the end of 1 year and another ₹4,50,000 at the end of 2 years. A comprehensive market survey has been done for assessing the sales potential and has predicted an year end before tax cash flow as per the following detail:

Year	Cash Flow	Year	Cash Flow
1	− ₹7,00,000	6	+ ₹22,00,000
2	+ ₹3,50,000	7	+ ₹22,00,000
3	+ ₹3,50,000	8	+ ₹15,00,000
4	+ ₹15,00,000	9	+ ₹12,00,000
5	+ ₹18,00,000	10	+ ₹9,00,000

The market research has also revealed that the demand of the model would cease to justify production after 10 years. After 10 years the additional facility would have a salvage value of ₹10,00,000. If the capital is worth not less than 15% before tax evaluate the feasibility of launching the model based on (a) future worth method.

14. A soap manufacture makes an annual expenditure of ₹6,00,000 on the cartons used in packaging. It plans to commission a corrugated-paper box unit to eliminate this expenditure. The unit requires an investment of ₹4,00,000 and has an expected life of 10 years and a salvage value of ₹1,50,000. The unit would, however, require an annual outlay of ₹30,000 towards its up-keep and maintenance and so on. However, the unit would save ₹1,50,000 per year to the company. If the company expects to earn at least 12% nominal interest, compounded annually on the unit determine if it is worth for the company to go ahead with the unit based on (a) present worth method (b) *I.R.R.* method and (c) discounted payback method.

15. A boiler installed in a textile unit consumes ₹1,00,000 worth of fuel annually. The company considers installing an economizer which is expected to cut down the cost of fuel by 20%. If the economizer costs the company ₹90,000 with a life-span of 8 years after which its worth is zero and a minimum return of 12% is desired. Using *I.R.R.* method, determine if it is feasible for the company to install the economizer.

❏❏❏

6

Selection among Alternatives

6.1 INTRODUCTION

Engineering or business projects can be accomplished by more than one method or alternative. The alternatives are developed from project proposals to achieve a stated purpose. The alternatives may be classified as one of the following:

Mutually exclusive When the selection of one alternative excludes the choice of any other alternative being considered, the alternatives are called mutually exclusive. A mutually exclusive alternative selection takes place, for example, when a person must select the one best diesel car from several competing models.

Independent When more than one economically viable alternative can be selected, the alternatives are classified as independent. Independent alternatives do not compete with one another in the evaluation. Each alternative is evaluated separately and thus the comparison is between one alternative at a time and the do-nothing alternative.

Do-nothing (DN) The "do-nothing" refers to a situation when none of the alternatives being considered is selected, as they are not economically viable. DN option is usually understood to be an alternative when the evaluation is performed. If it is required that one of the defined alternatives must necessarily be selected, do nothing is considered an option.

The alternatives available require the investment of different amounts of capital. The disbursement or outflow will be different in each case. Sometimes, the alternatives may produce different revenues, and frequently the useful life of the alternatives will be different. Because in such situations different levels of investment produce different economic outcomes, it is necessary to perform a study to determine economical viability of the alternatives. This chapter describes the methods used to compare mutually exclusive alternatives as well as independent alternatives.

6.2 ALTERNATIVES HAVING IDENTICAL DISBURSEMENTS AND LIVES

This is a simple case of comparison of mutually exclusive alternatives in which revenues from the various alternatives do not exist or may be assumed to be identical, and useful lives of all alternatives being considered are the same. Each of the six methods, i.e., P.W., F.W., A.W., I.R.R., E.R.R. and E.R.R.R discussed in Chapter 5 are used in Example 6.1 for this frequently encountered problem.

Example 6.1

Bongaigaon Refineries wishes to install an air pollution control system in its facility situated in Assam. The system comprises of four types of equipments that involve capital investment and annual recurring costs as given in Table 6.1. Assume a useful life of 10 years for each type of equipment, no salvage value and that the company wants a minimum return of 10% on its capital. Which equipment should be chosen?

Table 6.1 Costs for four equipments

	Equipments			
	A	**B**	**C**	**D**
Investment (₹)	2,00,000	2,76,000	3,00,000	3,25,000
Annual recurring costs:				
Power (₹)	64,000	59,000	36,000	28,000
Labor (₹)	1,20,000	1,36,000	1,84,000	1,96,000
Maintenance (₹)	96,000	64,000	22,000	10,000
Taxes and insurance (₹)	50,000	54,000	68,000	72,000
Total annual recurring costs (₹)	3,30,000	3,13,000	3,10,000	3,06,000

Solution of Example 6.1 by the Present Worth Cost (P.W.C.) Method In Example 6.1, revenues (positive cash flows) for all the four equipments (alternatives) are not known. The present worth (P.W.) of each equipment is calculated which in fact is present worth cost and hence the method is more commonly and descriptively called present worth cost (P.W.C.) method. The alternative that has the minimum P.W.C. is judged to be more desirable and hence selected. The calculation of P.W.C. for each equipment is given below:

Equipment A
The cash flow diagram for this equipment is shown in Fig. 6.1.

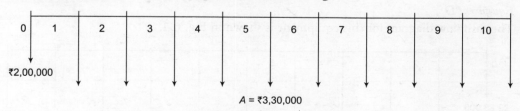

Figure 6.1   Cash flow diagram for equipment A, Example 6.1

$$\text{P.W.C.} = ₹2,00,000 + ₹3,30,000(P/A, 10\%, 10)$$

$$= ₹2,00,000 + ₹3,30,000(6.1446)$$

P.W.C. = ₹22,27,718

Equipment B

The cash flow diagram for this equipment is shown in Fig. 6.2.

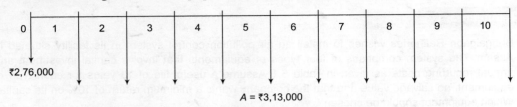

Figure 6.2   Cash flow diagram for equipment *B*, Example 6.1

$$P.W.C. = ₹2,76,000 + ₹3,13,000(P/A, 10\%, 10)$$

$$= ₹2,76,000 + ₹3,13,000(6.1446)$$

$$P.W.C. = ₹21,99,259.8$$

Equipment C

The cash flow diagram for this equipment is shown in Fig. 6.3.

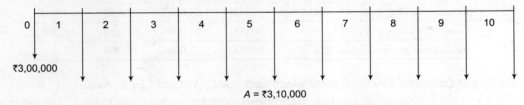

Figure 6.3   Cash flow diagram for equipment *C*, Example 6.1

$$P.W.C. = ₹3,00,000 + ₹3,10,000(P/A, 10\%, 10)$$

$$= ₹3,00,000 + ₹3,10,000(6.1446)$$

$$P.W.C. = ₹22,04,826$$

Equipment D

The cash flow diagram for this equipment is shown in Fig. 6.4.

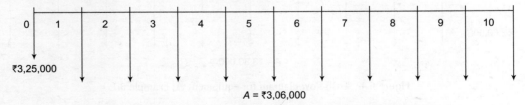

Figure 6.4   Cash flow diagram for equipment *D*, Example 6.1

$$\text{P.W.C.} = ₹3,25,000 + ₹3,06,000(P/A, 10\%, 10)$$

$$= ₹3,25,000 + ₹3,06,000(6.1446)$$

$$\text{P.W.C.} = ₹22,05,247.6$$

The economic criterion is to choose the alternative with the minimum P.W.C., which is Equipment B. The order of desirability among the alternatives in decreasing order is Equipment B, Equipment C, Equipment D and Equipment A.

Solution of Example 6.1 by the Future Worth Cost (F.W.C.) Method   In this method, future worth cost of each alternative is calculated and the alternative with minimum F.W.C. is selected.

The calculation of F.W.C. for each equipment is given below:

Equipment A

The cash flow diagram for this equipment is shown in Fig. 6.1.

$$\text{F.W.C.} = ₹2,00,000(F/P, 10\%, 10) + ₹3,30,000(F/A, 10\%, 10)$$

$$= ₹2,00,000(2.5937) + ₹3,30,000(15.9374)$$

$$\text{F.W.C.} = ₹57,78,082$$

Equipment B

The cash flow diagram for this equipment is shown in Fig. 6.2.

$$\text{F.W.C.} = ₹2,76,000(F/P, 10\%, 10) + ₹3,13,000(F/A, 10\%, 10)$$

$$= ₹2,76,000(2.5937) + ₹3,13,000(15.9374)$$

$$\text{F.W.C.} = ₹57,04,267.4$$

Equipment C

The cash flow diagram for this equipment is shown in Fig. 6.3.

$$\text{F.W.C.} = ₹3,00,000(F/P, 10\%, 10) + ₹3,10,000(F/A, 10\%, 10)$$

$$= ₹3,00,000(2.5937) + ₹3,10,000(15.9374)$$

$$\text{F.W.C.} = ₹57,18,704$$

Equipment D

The cash flow diagram for this equipment is shown in Fig. 6.4.

$$\text{F.W.C.} = ₹3,25000(F/P, 10\%, 10) + ₹3,06,000(F/A, 10\%, 10)$$

$$= ₹3,25,000(2.5937) + ₹3,06,000(15.9374)$$

$$\text{F.W.C.} = ₹57,19,796.9$$

The F.W.C. of Equipment B is minimum and hence should be selected. Once again, the order of desirability among the alternatives in decreasing order is Equipment B, Equipment C, Equipment D and Equipment A.

Solution of Example 6.1 by the Annual Cost (A.C.) Method
In this method, annual cost of each alternative is calculated and the alternative with minimum A.C. is selected.

The calculation of A.C. for each equipment is as follows:

Equipment A
The cash flow diagram for this equipment is shown in Fig. 6.1.

$$\text{A.C.} = ₹2,00,000(A/P, 10\%, 10) + ₹3,30,000$$

$$= ₹2,00,000(0.1627) + ₹3,30,000$$

$$\text{A.C.} = ₹3,62,540$$

Equipment B
The cash flow diagram for this equipment is shown in Fig. 6.2.

$$\text{A.C.} = ₹2,76,000(A/P, 10\%, 10) + ₹3,13,000$$

$$= ₹2,76,000(0.1627) + ₹3,13,000$$

$$\text{A.C.} = ₹3,57,905.2$$

Equipment C
The cash flow diagram for this equipment is shown in Fig. 6.3.

$$\text{A.C.} = ₹3,00,000(A/P, 10\%, 10) + ₹3,10,000$$

$$= ₹3,00,000(0.1627) + ₹3,10,000$$

$$\text{A.C.} = ₹3,58,810$$

Equipment D
The cash flow diagram for this equipment is shown in Fig. 6.4.

$$\text{A.C.} = ₹3,25000(A/P, 10\%, 10) + ₹3,06,000$$

$$= ₹3,25,000(0.1627) + ₹3,06,000$$

$$\text{A.C.} = ₹3,58,877.5$$

The A.C. of Equipment B is minimum and hence should be selected. Once again, the order of desirability among the alternatives in decreasing order is Equipment B, Equipment C, Equipment D and Equipment A.

Solution of Example 6.1 by the Internal Rate of Return (I.R.R.) Method   In this method, alternatives are ordered according to increasing amounts of investment so as to facilitate step-by-step consideration of each increment of investment. Pair-wise comparison of alternatives is done by taking into consideration increment of investment and increment of other disbursements. I.R.R. is calculated on increment considered and compared with M.A.R.R. If I.R.R. is greater than the M.A.R.R., then increment of investment is considered to be economically justified and therefore, the alternative ordered at number 2 is selected and it is compared with alternative ordered at number 3. If I.R.R. is less than or equal to the M.A.R.R., then increment of investment is not considered to be economically justified and therefore the alternative ordered at number 1 is selected and it is compared with alternative ordered at number 3. This process is continued till only one alternative is selected. Table 6.2 shows the cash flows, increment considered and I.R.R. for each increment of investment considered.

Table 6.2 Comparison of four equipments by I.R.R. method

	Equipment			
	A	**B**	**C**	**D**
Investment (₹)	2,00,000	2,76,000	3,00,000	3,25,000
Annual disbursements (₹)	3,30,000	3,13,000	3,10,000	3,06,000
Useful life (years)	10	10	10	10
Increment considered:		A→B	B→C	B→D
Δ Investment (₹)		76,000	24,000	49,000
Δ Annual disbursements (savings) (₹)		17,000	3,000	7,000
I.R.R. on Δ Investment (%)		18.15	4.28	7.08
Is the increment justified?		Yes	No	No

The first increment subject to analysis is the (₹2,76,000 − ₹2,00,000 = ₹76,000) extra investment required for equipment B compared to equipment A. For this increment, annual disbursements are reduced or saved by ₹3,30,000 − ₹3,13,000 = ₹17,000. I.R.R. on the incremental investment is the interest rate at which the present worth of the incremental net cash flow is zero. Thus,

$$- ₹76,000 + ₹17,000(P/A, i', 10) = 0$$

$$(P/A, i', 10) = ₹76,000/ ₹17,000 = 4.4706$$

Interpolating between the tabled factors for the two closest interest rates, we can find that i' = I.R.R. = 18.15%. Since 18.15% > 10% (the M.A.R.R.) the increment A→B is justified.

The next increment subject to analysis is the (₹3,00,000 − ₹2,76,000 = ₹24,000) extra investment required for equipment C compared to equipment B. This increment results in reduction of annual disbursement by ₹3,13,000 − ₹3,10,000 = ₹3,000. The I.R.R. on increment B→C can then be determined by finding the i' at which:

$$- ₹24,000 + ₹3,000(P/A, i', 10) = 0$$

$$(P/A, i', 10) = ₹24,000/ ₹3,000 = 8.0$$

Thus, i' = I.R.R. can be found to be 4.28%. Since 4.28% < 10%, the increment B→C is not justified and hence we can say that equipment C itself is not justified.

The next increment that should be analyzed is B→D and not C→D. This is because equipment C has already been shown to be unsatisfactory and hence it can no longer be a valid basis for comparison with other alternatives.

For $B \rightarrow D$, the incremental investment is ₹3,25,000 – ₹2,76,000 = ₹49,000 and the incremental saving in annual disbursement is ₹3,13,000 – ₹3,06,000 = ₹7,000. The I.R.R. on increment $B \rightarrow D$ can then be determined by finding the i' at which:

$$- ₹49,000 + ₹7,000(P/A, i', 10) = 0$$

$$(P/A, i', 10) = ₹49,000/ ₹7,000 = 7.0$$

Thus, i' can be found to be 7.08%. Since 7.08% < 10%, the increment $B \rightarrow D$ is not justified.

Based on the above analysis, equipment B would be the choice and should be selected.

Solution of Example 6.1 by the External Rate of Return (E.R.R.) Method with e = 10% The procedure and criteria for using the E.R.R. method to compare alternatives are the same as that of I.R.R. method. The only difference is in the calculation methodology. Table 6.3 shows the calculation and acceptability of each increment of investment considered.

Table 6.3 Comparison of four equipment by E.R.R. method

	Equipment			
	A	B	C	D
Investment (₹)	2,00,000	2,76,000	3,00,000	3,25,000
Annual disbursements (₹)	3,30,000	3,13,000	3,10,000	3,06,000
Useful life (years)	10	10	10	10
Increment considered:		A→B	B→C	B→D
Δ Investment (₹)		76,000	24,000	49,000
Δ Annual disbursements (savings) (₹)		17,000	3,000	7,000
E.R.R. on Δ Investment (%)		13.51	7.12	8.58
Is the increment justified?		Yes	No	No

The first increment that should be analyzed is $A \rightarrow B$. For $A \rightarrow B$, the incremental investment is ₹2,76,000 – ₹2,00,000 = ₹76,000 and the incremental saving in annual disbursement is ₹3,30,000 – ₹3,13,000 = ₹17,000. The E.R.R. on increment $A \rightarrow B$ can then be determined as:

$$₹76,000(F/P, i', 10) = ₹17,000(F/A, e\%, 10)$$

Since, the value of external interest rate e is not given, it is taken as equal to the M.A.R.R. i.e., e = M.A.R.R. = 10%.

$$₹76,000(F/P, i', 10) = ₹17,000(F/A, 10\%, 10)$$

$$₹76,000(F/P, i', 10) = ₹17,000(15.9374)$$

$$₹76,000(F/P, i', 10) = ₹2,70,935.8$$

$$(F/P, i', 10) = ₹2,70,935.8/ ₹76,000$$

$$(F/P, i', 10) = 3.56$$

Linear interpolation between 12% and 14% results in $i' = 13.51\%$, which is the E.R.R. Since 13.51% > 10%, the increment $A{\rightarrow}B$ is justified.

The next increment that should be analyzed is $B{\rightarrow}C$. For $B{\rightarrow}C$, the incremental investment is ₹3,00,000 – ₹2,760,000 = ₹24,000 and the incremental saving in annual disbursement is ₹3,13,000 – ₹3,10,000 = ₹3,000. The E.R.R. on increment $B{\rightarrow}C$ can then be determined as:

$$₹24,000(F/P, i', 10) = ₹3,000(F/A, e\%, 10)$$

$$₹24,000(F/P, i', 10) = ₹3,000(F/A, 10\%, 10)$$

$$₹24,000(F/P, i', 10) = ₹3,000(15.9374)$$

$$₹24,000(F/P, i', 10) = ₹47,812.2$$

$$(F/P, i', 10) = ₹47,812.2/ ₹24,000$$

$$(F/P, i', 10) = 1.99$$

Linear interpolation between 7% and 8% results in $i' = 7.12\%$, which is the E.R.R. Since 7.12% < 10%, the increment $B{\rightarrow}C$ is not justified.

The next increment that should be analyzed is $B{\rightarrow}D$. For $B{\rightarrow}D$, the incremental investment is ₹3,25,000 – ₹2,760,000 = ₹49,000 and the incremental saving in annual disbursement is ₹3,13,000 – ₹3,06,000 = ₹7,000. The E.R.R. on increment $B{\rightarrow}D$ can then be determined as:

$$₹49,000(F/P, i', 10) = ₹7,000(F/A, e\%, 10)$$

$$₹49,000(F/P, i', 10) = ₹7,000(F/A, 10\%, 10)$$

$$₹49,000(F/P, i', 10) = ₹7,000(15.9374)$$

$$₹49,000(F/P, i', 10) = ₹1,11,561.8$$

$$(F/P, i', 10) = ₹1,11,561.8/ ₹49,000$$

$$(F/P, i', 10) = 2.28$$

Linear interpolation between 8% and 9% results in $i' = 8.58\%$, which is the E.R.R. Since 8.58% < 10%, the increment $B{\rightarrow}D$ is not justified.

Based on the above analysis, equipment B would be the choice and should be selected.

Solution of Example 6.1 by the Explicit Reinvestment Rate of Return (E.R.R.R.) Method with $e = 10\%$ The rationale and criteria for using the E.R.R.R. method to compare alternatives are the same as that of the I.R.R. method. The only difference is in the calculation methodology. Table 6.4 shows the calculation and acceptability of each increment of investment considered.

Table 6.4   Comparison of four equipments by E.R.R.R. method

	Equipment			
	A	**B**	**C**	**D**
Investment (₹)	2,00,000	2,76,000	3,00,000	3,25,000
Annual disbursements (₹)	3,30,000	3,13,000	3,10,000	3,06,000
Useful life (years)	10	10	10	10
Increment considered:		A→B	B→C	B→D
Δ Investment (₹)		76,000	24,000	49,000
Δ Annual disbursements (savings) (₹)		17,000	3,000	7,000
E.R.R.R. on Δ Investment (%)		16.09	6.23	8.01
Is the increment justified?		Yes	No	No

The first increment that should be analyzed is $A \rightarrow B$. For $A \rightarrow B$, the incremental investment is ₹2,76,000 − ₹2,00,000 = ₹76,000 and the incremental saving in annual disbursement is ₹3,30,000 − ₹3,13,000 = ₹17,000. The E.R.R.R. on increment $A \rightarrow B$ can then be determined as:

$$E.R.R.R. = \frac{(₹17,000 - 0) - (₹76,000 - 0)(A/F, 10\%, 10)}{₹76,000}$$

$$E.R.R.R. = \frac{₹17,000 - ₹76,000(0.0627)}{₹76,000}$$

$$E.R.R.R. = 0.1609 = 16.09\%$$

Since 16.09% > 10%, the increment $A \rightarrow B$ is justified.

The next increment that should be analyzed is $B \rightarrow C$. For $B \rightarrow C$, the incremental investment is ₹3,00,000 − ₹2,76,000 = ₹24,000 and the incremental saving in annual disbursement is ₹3,13,000 − ₹3,10,000 = ₹3,000. The E.R.R.R. on increment $B \rightarrow C$ can then be determined as:

$$E.R.R.R. = \frac{(₹3,000 - 0) - (₹24,000 - 0)(A/F, 10\%, 10)}{₹24,000}$$

$$E.R.R.R. = \frac{₹3,000 - ₹24,000(0.0627)}{₹24,000}$$

$$E.R.R.R. = 0.0623 = 6.23\%$$

Since 6.23% < 10%, the increment $B \rightarrow C$ is not justified.

The next increment that should be analyzed is $B \rightarrow D$. For $B \rightarrow D$, the incremental investment is ₹3,25,000 − ₹2,76,000 = ₹49,000 and the incremental saving in annual disbursement is

₹3,13,000 − ₹3,06,000 = ₹7,000. The E.R.R.R. on increment $B \rightarrow D$ can then be determined as:

$$\text{E.R.R.R.} = \frac{(₹7,000 - 0) - (₹49,000 - 0)(A/F, 10\%, 10)}{₹49,000}$$

$$\text{E.R.R.R.} = \frac{₹7,000 - ₹49,000(0.0627)}{₹49,000}$$

$$\text{E.R.R.R.} = 0.0801 = 8.01\%$$

Since 8.01% < 10%, the increment $B \rightarrow D$ is not justified.

Based on the above analysis, Equipment B would be the choice and should be selected.

6.3 ALTERNATIVES HAVING IDENTICAL REVENUES AND DIFFERENT LIVES

In situations where alternatives that are to be compared have identical revenues or their revenues are not known but their lives are different, the problem becomes a little bit complicated. To make economic studies of such cases, it is necessary to adopt some procedure that will put the alternatives on a comparable basis. Two types of assumptions are commonly employed in such a situation: (1) the repeatability assumption and (2) the coterminated assumption.

The *repeatability assumptions* are employed when the following two conditions prevail:
1. The period for which the alternatives are being compared is either infinitely long or a length of time equal to a lowest common multiple (LCM) of the lives of the alternatives.
2. What is estimated to happen in an alternative's initial life span will happen also in all succeeding life spans, if any, for each alternative.

The *coterminated assumption* involves the use of a finite study period for all alternatives. This time span may be the period of needed service or any arbitrarily specified length of time such as:
1. The life of the shorter-lived alternative
2. The life of the longer-lived alternative
3. Organization's planning horizon

6.3.1 Comparisons Using the Repeatability Assumption

Consider Example 6.2 for which economics study is to be made using the repeatability assumption.

Example 6.2

Consider the following two alternatives

	Alternative A	Alternative B
First cost (₹)	60,000	2,00,000
Annual maintenance & operation cost (₹/year)	11,000	5,000
Salvage value at the end of life (₹)	0	50,000
Useful life (years)	10	25

Determine which alternative is better if the M.A.R.R. is 15% using repeatability assumption.

Solution of Example 6.2 by the Present Worth Cost (P.W.C.) Method   In order to compare the two alternatives by P.W.C. method, it is necessary to compare them over the same length of time. By use of the repeatability assumption, the length of time is chosen to be the lowest common multiple (LCM) of the lives for the two alternatives. In this case, the LCM of 10 and 25 is 50. Thus, the life of both alternatives should be taken as 50 years. Since the life of alternative A is 10 years, the data of alternative A will be repeated 4 times as shown in Fig. 6.5.

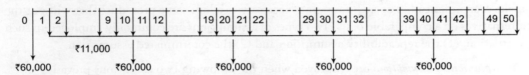

Figure 6.5   Cash flow diagram for alternative A, Example 6.2

P.W.C. of alternative A is determined as:

$$(P.W.C.)_A = -₹60,000 - ₹11,000(P/A, 15\%, 50) - ₹60,000(P/F, 15\%, 10)$$
$$- ₹60,000(P/F, 15\%, 20) - ₹60,000(P/F, 15\%, 30)$$
$$- ₹60,000(P/F, 15\%, 40)$$

$$(P.W.C.)_A = -₹60,000 - ₹11,000(6.6605) - ₹60,000(0.2472) - ₹60,000(0.0611)$$
$$- ₹60,000(0.0151) - ₹60,000(0.0037)$$

$$(P.W.C.)_A = -₹1,52,891.5$$

The life of alternative B is 25 years and therefore, the data of alternative B will be repeated 1 time as shown in Fig. 6.6.

P.W.C. of alternative B is determined as:

$$(P.W.C.)_B = -₹2,00,000 - ₹5,000(P/A, 15\%, 50) - ₹2,00,000(P/F, 15\%, 25)$$
$$+ ₹50,000(P/F, 15\%, 25) + ₹50,000(P/F, 15\%, 50)$$

$$(P.W.C.)_B = -₹2,00,000 - ₹5,000(6.6605) - ₹2,00,000(0.0304)$$
$$+ ₹50,000(0.0304) + ₹50,000(0.0009)$$

$$(P.W.C.)_B = -₹2,37,817.5$$

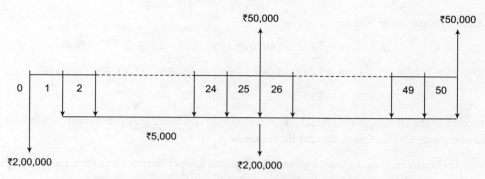

Figure 6.6 Cash flow diagram for alternative B, Example 6.2

On comparing $(P.W.C.)_A$ with $(P.W.C.)_B$, it is found that alternative A is better than alternative B and therefore, alternative A should be chosen.

Solution of Example 6.2 by the Future Worth Cost (F.W.C.) Method F.W.C. of alternative A is determined as:

$$(F.W.C.)_A = -₹60,000(F/P, 15\%, 50) - ₹11,000(F/A, 15\%, 50)$$
$$- ₹60,000(F/P, 15\%, 40) - ₹60,000(F/P, 15\%, 30)$$
$$- ₹60,000(F/P, 15\%, 20) - ₹60,000(F/P, 15\%, 10)$$

$$(F.W.C.)_A = -₹60,000(1083.66) - ₹11,000(7217.72) - ₹60,000(267.8635)$$
$$- ₹60,000(66.2118) - ₹60,000(16.3665) - ₹60,000(4.0456)$$

$$(F.W.C.)_A = -₹16,56,83,763$$

F.W.C. of alternative B is determined as:

$$(F.W.C.)_B = -₹2,00,000(F/P, 15\%, 50) - ₹5,000(F/A, 15\%, 50)$$
$$- ₹2,00,000(F/P, 15\%, 25) + ₹50,000(F/P, 15\%, 25) + ₹50,000$$

$$(F.W.C.)_B = -₹2,00,000(1083.66) - ₹5,000(7217.72) - ₹2,00,000(32.9190)$$
$$+ ₹50,000(32.9190) + ₹50,000$$

$$(F.W.C.)_B = -₹25,77,08,450$$

On comparing $(F.W.C.)_A$ with $(F.W.C.)_B$, it is found that alternative A is economically better than alternative B and, therefore, alternative A should be chosen.

Solution of Example 6.2 by the Annual Cost (A.C.) Method The A.C. of alternative A is calculated over 10 years only as it will be same over 50 years. Similarly, the A.C. of alternative B is calculated over 25 years only as it will be same over 50 years.

A.C. of alternative A is determined as:

$$(A.C.)_A = -₹60,000(A/P, 15\%, 10) - ₹11,000$$
$$(A.C.)_A = -₹60,000(0.19925) - ₹11,000$$
$$(A.C.)_A = -₹22,955$$

A.C. of alternative B is determined as:

$$(\text{A.C.})_B = -₹2,00,000(A/P, 15\%, 25) - ₹5,000 + ₹50,000(A/F, 15\%, 25)$$

$$(\text{A.C.})_B = -₹2,00,000(0.15470) - ₹5,000 + ₹50,000(0.00470)$$

$$(\text{A.C.})_B = -₹35,705$$

On comparing $(\text{A.C.})_A$ with $(\text{A.C.})_B$, it is found that alternative A is better than alternative B and therefore, alternative A should be chosen.

Note: It should be noted that when the alternatives have different lives then comparing them by A.C. method requires less calculations and this method is easy too as compared to P.W.C. and F.W.C. methods.

Solution of Example 6.2 by the Internal Rate of Return (I.R.R.) Method   In the I.R.R. method, the I.R.R. on the incremental investment of alternative B over alternative A is required. The I.R.R. on the increment can be computed by finding the interest rate at which the equivalent worths (P.W. or F.W. or A.W.) or equivalent costs (P.W.C. or F.W.C. or A.C.) of the two alternatives are equal.

Let us compute I.R.R. by finding A.C. for the two alternatives at an unknown i' and equating them.

$$-₹60,000(A/P, i', 10) - ₹11,000 = -₹2,00,000(A/P, i', 25) - ₹5,000 + ₹50,000(A/F, i', 25)$$

Linear interpolation between 5% and 6% results in $i' = 5.5\%$, which is the I.R.R. Since 5.5% < 15%, the increment $A \rightarrow B$ is not justified. Thus, alternative A is better than alternative B and it should be selected.

Solution of Example 6.2 by the External Rate of Return (E.R.R.) Method with e = 15%

	Alternative A	Alternative B
Investment (₹)	79,260[*]	2,06,080[#]
FW of annual disbursements (₹): ₹11,000(F/A, 15%, 50) ₹5,000(F/A, 15%, 50)	7,93,94,920	3,60,88,600
FW of Salvage (₹):		50,000(F/P, 15%, 25) + 50,000 = 16,95,950
Increment considered:		
Δ Investment (₹)		1,26,460
Δ FW of savings (₹)		(7,93,94,920 − 3,60,88,600) + 16,95,950 = 4,50,02,270
E.R.R. on Δ Investment		1,26,460(F/P, i', 50) = 4,50,02,270 which yields i' = 12.46%

[*]$60,000 + 60,000(P/F, 15\%, 10) + 60,000(P/F, 15\%, 20) + 60,000(P/F, 15\%, 30) + 60,000(P/F, 15\%, 40)$
$= 79,260$
[#]$2,00,000 + 2,00,000(P/F, 15\%, 25) = 2,06,080$

Since E.R.R. (12.46%) < 15%, the increment $A \rightarrow B$ is not justified. Thus, alternative A is better than alternative B and it should be selected.

Solution of Example 6.2 by the Explicit Reinvestment Rate of Return (E.R.R.R.) Method with e = 15%

	Alternative A	Alternative B
Annual expenses (₹): Disbursements	11,000	5,000
Depreciation (₹): ₹60,000(A/F, 15%, 10) ₹(2,00,000 – 50,000)(A/F, 15%, 25)	2,955	705
Total annual expenses (₹):	11,000 + 2955 = 13,955	5,000 + 705 = 5,705
Δ Annual expenses (annual savings) (₹)		13,955 – 5705 = 8,250
Δ Investment (₹):		2,00,000 – 60.000 = 1,40,000
E.R.R.R. on Δ Investment		8,250/1,40,000 = 5.89%

Since E.R.R.R. (5.89%) < 15%, the increment $A \rightarrow B$ is not justified. Thus, alternative A is better than alternative B and it should be selected.

6.3.2 Comparisons Using the Coterminated Assumption

The coterminated assumption is used in engineering economy studies if the period of needed service is less than a common multiple of the lives of alternatives. In this case, the specified period of needed service should be used as study period. Even if the period of needed service is not known, it is convenient to use some arbitrarily specified study period for all alternatives. In such cases, an important concern involves the salvage value to be assigned to any alternative that will not have reached its useful life at the end of the study period. A cotermination point commonly used in these situations is the life of the shortest-lived alternative. Example 6.3 illustrates the coterminated assumption.

Example 6.3

Consider the same two alternatives of Example 6.2 as given below:

	Alternative A	Alternative B
First cost (₹)	60,000	2,00,000
Annual maintenance & operation cost (₹/year)	11,000	5,000
Salvage value at the end of life (₹)	0	50,000
Useful life (years)	10	25
M.A.R.R.	15%	15%

Determine which alternative is better by terminating the study period at the end of 10 years and assuming a salvage value for alternative B at that time as ₹1,25,000.

Solution of Example 6.3 by the Present Worth Cost (P.W.C.) Method   Since the study period is specified as 10 years, the lives of both alternatives will be taken as 10 years. The P.W.C. of the alternatives is calculated as:

P.W.C. of alternative A is determined as:

$$(P.W.C.)_A = -₹60,000 - ₹11,000(P/A, 15\%, 10)$$

$$(P.W.C.)_A = -₹60,000 - ₹11,000(5.0188)$$

$$(P.W.C.)_A = -₹1,15,206.8$$

P.W.C. of alternative B is determined as:

$$(P.W.C.)_B = -₹2,00,000 - ₹5,000(P/A, 15\%, 10) + ₹1,25,000(P/F, 15\%, 10)$$

$$(P.W.C.)_B = -₹2,00,000 - ₹5,000(5.0118) + ₹1,25,000(0.2472)$$

$$(P.W.C.)_B = -₹1,94,194$$

On comparing $(P.W.C.)_A$ with $(P.W.C.)_B$, it is found that $(P.W.C.)_A$ is smaller than $(P.W.C.)_B$ and, therefore, alternative A should be chosen.

Solution of Example 6.3 by the Future Worth Cost (F.W.C.) Method   F.W.C. of alternative A is determined as:

$$(F.W.C.)_A = -₹60,000(F/P, 15\%, 10) - ₹11,000(F/A, 15\%, 10)$$

$$(F.W.C.)_A = -₹60,000(4.0456) - ₹11,000(20.3037)$$

$$(F.W.C.)_A = -₹4,66,076.7$$

F.W.C. of alternative B is determined as:

$$(F.W.C.)_B = -₹2,00,000(F/P, 15\%, 10) - ₹5,000(F/A, 15\%, 10) + ₹1,25,000$$

$$(F.W.C.)_B = -₹2,00,000(4.0456) - ₹5,000(20.3037) + ₹1,25,000$$

$$(F.W.C.)_B = -₹7,85,638.5$$

On comparing $(F.W.C.)_A$ with $(F.W.C.)_B$, it is found that $(F.W.C.)_A$ is smaller than $(F.W.C.)_B$ and therefore, alternative A should be chosen.

Solution of Example 6.3 by the Annual Cost (A.C.) Method   A.C. of alternative A is determined as:

$$(A.C.)_A = -₹60,000(A/P, 15\%, 10) - ₹11,000$$

$$(A.C.)_A = -₹60,000(0.19925) - ₹11,000$$

$$(A.C.)_A = -₹22,955$$

A.C. of alternative B is determined as:

$$(A.C.)_B = -₹2,00,000(A/P, 15\%, 10) - ₹5,000 + ₹1,25,000(A/F, 15\%, 10)$$

$$(A.C.)_B = -₹2,00,000(0.1993) - ₹5,000 + ₹1,25,000(0.0493)$$

$$(A.C.)_B = -₹51,022.5$$

On comparing $(A.C.)_A$ with $(A.C.)_B$, it is found that $(A.C.)_A$ is smaller than $(A.C.)_B$ and therefore, alternative A should be chosen.

Solution of Example 6.3 by the Internal Rate of Return (I.R.R.) Method Let us compute I.R.R. by finding A.C. for the two alternatives at an unknown i' and equating them.

$$-₹60,000(A/P, i', 10) - ₹11,000 = -₹2,00,000(A/P, i', 10) - ₹5,000 + ₹1,25,000(A/F, i', 10)$$

Linear interpolation between 3% and 4% results in $i' = 3.37\%$, which is the I.R.R. Since $3.37\% < 15\%$, the increment $A{\rightarrow}B$ is not justified. Thus, alternative A is better than alternative B and it should be selected.

Solution of Example 6.3 by the External Rate of Return (E.R.R.) Method with e = 15%

	Alternative A	Alternative B
Investment (₹)	60,000	2,00,000
FW of annual disbursements (₹): ₹11,000(F/A, 15%, 10) ₹5,000(F/A, 15%, 10)	2,23,340.7	1,01,518.5
FW of Salvage (₹):		1,25,000
Increment considered:		
Δ Investment (₹)		1,40,000
Δ FW of savings (₹)		$(2,23,340.7 - 1,01,518.5) + 1,25,000 = 2,46,822.2$
E.R.R. on Δ Investment		$1,40,000(F/P, i', 10) = 2,46,822.2$ which yields $i' = 5.83\%$

Since E.R.R. (5.83%) < 15%, the increment $A{\rightarrow}B$ is not justified. Thus, alternative A is better than alternative B and it should be selected.

Solution of Example 6.3 by the Explicit Reinvestment Rate of Return (E.R.R.R.) Method with e = 15%

	Alternative A	Alternative B
Annual expenses (₹): Disbursements	11,000	5,000
Depreciation (₹): ₹60,000(A/F, 15%, 10) ₹(2,00,000 − 1,25,000)(A/F, 15%, 10)	2,955	3,697.5
Total annual expenses (₹):	$11,000 + 2955 = 13,955$	$5,000 + 3,697.5 = 8,697.5$
Δ Annual expenses (annual savings) (₹)		$13,955 - 8,697.5 = 5,257.5$
Δ Investment (₹):		$2,00,000 - 60,000 = 1,40,000$
E.R.R.R. on Δ Investment		$5,257.5/1,40,000 = 3.75\%$

Since E.R.R.R. (3.75%) < 15%, the increment $A \rightarrow B$ is not justified. Thus, alternative A is better than alternative B and it should be selected.

6.4   ALTERNATIVES HAVING DIFFERENT REVENUES AND IDENTICAL LIVES

Example 6.4 illustrates the use of basic methods for making economy studies to compare the alternatives having different revenues and identical lives:

Example 6.4

Consider the following alternatives:

	Alternative 1	Alternative 2	Alternative 3
First cost (₹)	1,40,000	90,000	1,20,000
Net cash flow per year (₹/year)	+ 27,500	+ 12,000	+ 24,000
Salvage value (₹)	+ 7,500	0	+ 2,500
Useful life (years)	10	10	10

Assume that the interest rate (M.A.R.R.) is 15%. Determine which alternative should be selected.

Solution of Example 6.4 by the Present Worth (P.W.) Method   P.W. of alternative 1 is determined as:

$$(P.W.)_1 = -₹1,40,000 + ₹27,500(P/A, 15\%, 10) + ₹7,500(P/F, 15\%, 10)$$

$$(P.W.)_1 = -₹1,40,000 + ₹27,500(5.0188) + ₹7,500(0.2472)$$

$$(P.W.)_1 = -₹129$$

P.W. of alternative 2 is determined as:

$$(P.W.)_2 = -₹90,000 + ₹12,000(P/A, 15\%, 10)$$

$$(P.W.)_2 = -₹90,000 + ₹12,000(5.0188)$$

$$(P.W.)_2 = -₹29,774.4$$

P.W. of alternative 3 is determined as:

$$(P.W.)_3 = -₹1,20,000 + ₹24,000(P/A, 15\%, 10) + ₹2,500(P/F, 15\%, 10)$$

$$(P.W.)_3 = -₹1,20,000 + ₹24,000(5.0188) + ₹2,500(0.2472)$$

$$(P.W.)_3 = +₹1,069.2$$

On comparing P.W. of the three alternatives, it is found that P.W. of alternative 3 is maximum and therefore, alternative 3 should be chosen.

Solution of Example 6.4 by the Future Worth (F.W.) Method F.W. of alternative 1 is determined as:

$$(\text{F.W.})_1 = -₹1,40,000(F/P, 15\%, 10) + ₹27,500(F/A, 15\%, 10) + ₹7,500$$

$$(\text{F.W.})_1 = -₹1,40,000(4.0456) + ₹27,500(20.3037) + ₹7,500$$

$$(\text{F.W.})_1 = -₹532.25$$

F.W. of alternative 2 is determined as:

$$(\text{F.W.})_2 = -₹90,000(F/P, 15\%, 10) + ₹12,000(F/A, 15\%, 10)$$

$$(\text{F.W.})_2 = -₹90,000(4.0456) + ₹12,000(20.3037)$$

$$(\text{F.W.})_2 = -₹1,20,459.6$$

F.W. of alternative 3 is determined as:

$$(\text{F.W.})_3 = -₹1,20,000(F/P, 15\%, 10) + ₹24,000(F/A, 15\%, 10) + ₹2,500$$

$$(\text{F.W.})_3 = -₹1,20,000(4.0456) + ₹24,000(20.3037) + ₹2,500$$

$$(\text{F.W.})_3 = +₹4,316.8$$

On comparing F.W. of the three alternatives, it is found that F.W. of alternative 3 is maximum and therefore, alternative 3 should be chosen.

Solution of Example 6.4 by the Annual Worth (A.W.) Method A.W. of alternative 1 is determined as:

$$(\text{A.W.})_1 = -₹1,40,000(A/P, 15\%, 10) + ₹27,500 + ₹7,500(A/F, 15\%, 10)$$

$$(\text{A.W.})_1 = -₹1,40,000(0.1993) + ₹27,500 + ₹7,500(0.0493)$$

$$(\text{A.W.})_1 = -₹32.25$$

A.W. of alternative 2 is determined as:

$$(\text{A.W.})_2 = -₹90,000(A/P, 15\%, 10) + ₹12,000$$

$$(\text{A.W.})_2 = -₹90,000(0.1993) + ₹12,000$$

$$(\text{A.W.})_2 = -₹5,937$$

A.W. of alternative 3 is determined as:

$$(\text{A.W.})_3 = -₹1,20,000(A/P, 15\%, 10) + ₹24,000 + ₹2,5000(A/F, 15\%, 10)$$

$$(\text{A.W.})_3 = -₹1,20,000(0.1993) + ₹24,000 + ₹2,500(0.0493)$$

$$(\text{A.W.})_3 = +₹207.25$$

On comparing A.W. of the three alternatives, it is found that A.W. of alternative 3 is maximum and therefore, alternative 3 should be chosen.

Solution of Example 6.4 by the Internal Rate of Return (I.R.R.) Method Let us consider the increment 2→3 and compute I.R.R. by finding A.W. for the two alternatives at an unknown i' and equating them.

$$- ₹90,000(A/P, i'\%, 10) + ₹12,000 = - ₹1,20,000(A/P, i'\%, 10) + ₹24,000 + ₹2,500(A/F, i'\%, 10)$$

Linear interpolation between 35% and 40% results in $i' = 38.58\%$, which is the I.R.R. Since 38.58% > 15%, the increment 2→3 is justified. Thus, alternative 3 is better than alternative 2 and it should be selected.

Next increment considered is 3→1. The I.R.R. is obtained by equating A.W. of the alternatives at an unknown i'.

$$- ₹1,20,000(A/P, i'\%, 10) + ₹24,000 + ₹2,500(A/F, i'\%, 10) = - ₹1,40,000(A/P, i'\%, 10) + ₹27,500 + ₹7,500(A/F, i'\%, 10)$$

Linear interpolation between 13% and 14% results in $i' = 13.53\%$, which is the I.R.R. Since 13.53% < 15%, the increment 3→1 is not justified. Thus, alternative 3 is better than alternative 1 and it should be selected.

Solution of Example 6.4 by the External Rate of Return (E.R.R.) Method with e = 15%

	Alternative 2	Alternative 3	Alternative 1
Investment (₹)	90,000	1,20,000	1,40,000
FW of annual cash flows (₹): ₹12,000(F/A, 15%, 10) ₹24,000(F/A, 15%, 10) ₹27,500(F/A, 15%, 10)	2,43,644.4	4,87,288.8	5,58,351.75
FW of Salvage (₹):	0	2,500	7,500
Increment considered:		2→3	3→1
Δ Investment (₹)		30,000	20,000
Δ FW of savings (₹)		(4,87,288.8 – 2,43,644.4) + 2,500 = 2,46,144.4	(5,58,351.75 – 4,87,288.8) + 7,500 = 78,562.95
E.R.R. on Δ Investment		30,000(F/P, i', 10) = 2,46,144.4 which yields $i' = 23.42\%$	20,000(F/P, i', 10) = 78,562.95 which yields $i' = 14.6\%$
Is the increment justified		Yes	No

Thus, alternative 3 is the best and it should be selected.

Solution of Example 6.4 by the Explicit Reinvestment Rate of Return (E.R.R.R.) Method with e = 15%

	Alternative 2	Alternative 3	Alternative 1
Investment (₹)	90,000	1,20,000	1,40,000
Annual cash flows (₹):	+ 12,000	+ 24,000	+ 27,500
Depreciation (₹): ₹90,000(A/F, 15%, 10) ₹(1,20,000 − 2,500)(A/F, 15%, 10) ₹(1,40,000 − 7,500)(A/F, 15%, 10)	4,437	5,792.75	6,532.25
Total annual cash flows (₹):	12,000 − 4,437 = 7,563	24,000 − 5,792.75 = 18,207.25	27,500 − 6,532.25 = 20,967.75
Increment considered		2→3	3→1
Δ Annual cash flows (₹)		18,207.25 − 7,563 = 10,644.25	20,967.75 − 18,207.25 = 2,760.5
Δ Investment (₹)		1,20,000 − 90,000 = 30,000	1,40,000 − 1,20,000 = 20,000
E.R.R.R. on Δ Investment		10,644.25/30,000 = 35.48%	2,760.5/20,000 = 13.80%
Is increment justified		Yes	No

Thus, alternative 3 is the best and it should be selected.

6.5 ALTERNATIVES HAVING DIFFERENT REVENUES AND DIFFERENT LIVES

Example 6.5 illustrates the use of basic methods for making economy studies to compare the alternatives having different revenues and different lives.

Example 6.5

Consider the following two alternatives:

	Alternative A	Alternative B
Investment (₹)	17,500	25,000
Annual revenue (₹/year)	9,500	12,500
Annual disbursements (₹/year)	3,225	6,915
Useful life (years)	4	8
Net salvage value (₹)	0	0

Determine which alternative is better using repeatability assumption, if the M.A.R.R. is 10%.

Solution of Example 6.5 by the Present Worth (P.W.) Method   P.W. of alternative A is determined as:

$$(P.W.)_A = -₹17,500 + ₹9,500(P/A, 10\%, 8) - ₹3,225(P/A, 10\%, 8)$$
$$\qquad - ₹17,500(P/F, 10\%, 4)$$

$$(P.W.)_A = -₹17,500 + ₹9,500(5.3349) - ₹3,225(5.3349) - ₹17,500(0.6830)$$

$$(P.W.)_A = ₹4,023.99$$

P.W. of alternative B is determined as:

$$(P.W.)_B = -₹25,000 + ₹12,500(P/A, 10\%, 8) - ₹6,915(P/A, 10\%, 8)$$

$$(P.W.)_B = -₹25,000 + ₹12,500(5.3349) - ₹6,915(5.3349)$$

$$(P.W.)_B = ₹4,795.42$$

On comparing $(P.W.)_A$ with $(P.W.)_B$, it is found that alternative B is better than alternative A and therefore, alternative B should be chosen.

Solution of Example 6.5 by the Future Worth (F.W.) Method   F.W. of alternative A is determined as:

$$(F.W.)_A = -₹17,500(F/P, 10\%, 8) + ₹9,500(F/A, 10\%, 8) - ₹3,225(F/A, 10\%, 8)$$
$$\qquad - ₹17,500(F/P, 10\%, 4)$$

$$(F.W.)_A = -₹17,500(2.1436) + ₹9,500(11.4359) - ₹3,225(11.4359)$$
$$\qquad - ₹17,500(1.4641)$$

$$(F.W.)_A = ₹8,625.52$$

F.W. of alternative B is determined as:

$$(F.W.)_B = -₹25,000(F/P, 10\%, 8) + ₹12,500(F/A, 10\%, 8) - ₹6,915(F/A, 10\%, 8)$$

$$(F.W.)_B = -₹25,000(2.1436) + ₹12,500(11.4359) - ₹6,915(11.4359)$$

$$(F.W.)_B = ₹10,279.50$$

On comparing $(F.W.)_A$ with $(F.W.)_B$, it is found that alternative B is better than alternative A and therefore, alternative B should be chosen.

Solution of Example 6.5 by the Annual Worth (A.W.) Method   A.W. of alternative A is determined as:

$$(A.W.)_A = -₹17,500(A/P, 10\%, 4) + ₹9,500 - ₹3,225$$

$$(A.W.)_A = -₹17,500(0.3155) + ₹9,500 - ₹3,225$$

$$(A.W.)_A = ₹753.75$$

A.W. of alternative B is determined as:

$$(A.W.)_B = -₹25,000(A/P, 10\%, 8) + ₹12,500 - ₹6,915$$

$$(\text{A.W.})_B = -₹25,000(0.1874) + ₹12,500 - ₹6,915$$

$$(\text{A.W.})_B = ₹900$$

On comparing $(\text{A.W.})_A$ with $(\text{A.W.})_B$, it is found that alternative B is better than alternative A and therefore, alternative B should be chosen.

Solution of Example 6.5 by the Internal Rate of Return (I.R.R.) Method   Let us compute I.R.R. by finding A.W. for the two alternatives at an unknown i' and equating them.

$$-₹17,500(A/P, i\%, 4) + ₹9,500 - ₹3,225 = -₹25,000(A/P, i\%, 8) + ₹12,500 - ₹6,915$$

Linear interpolation between 12% and 15% results in $i' = 12.68\%$, which is the I.R.R. Since $12.68\% > 10\%$, the increment $A{\rightarrow}B$ is justified. Thus, alternative B is better than alternative A and it should be selected.

Solution of Example 6.5 by the External Rate of Return (E.R.R.) Method with e = 10%

	Alternative A	Alternative B
Investment (₹)	17,500	25,000
FW of net annual cash flows (₹):		
₹(9,500 – 3,225)(F/A, 10%, 8)	71,760.27	63,869.50
₹(12,500 – 6,915)(F/A, 15%, 8)		
FW of Salvage (₹):	-	-
Increment considered:		
Δ Investment (₹)		7,500
Δ FW of savings (₹)		(71,760.27 + 63,869.50)
		= 1,35,629.77
E.R.R. on Δ Investment		7,500(F/P, i', 8) = 1,35,629.77
		which yields $i' = 43.6\%$

Since E.R.R. (43.6%) >10%, the increment $A{\rightarrow}B$ is justified. Thus, alternative B is better than alternative A and it should be selected.

Solution of Example 6.5 by the Explicit Reinvestment Rate of Return (E.R.R.R.) Method with e = 10%

	Alternative A	Alternative B
Net annual cash flows (₹):	(9,500 – 3,225) = 6,275	(12,500 – 6,915) = 5,585
Depreciation (₹):		
₹17,500(A/F, 10%, 8)	1,529.5	2,185
₹25,000(A/F, 10%, 8)		
Total annual cash flows (₹):	6,275 – 1,529.5 = 4,745.5	5,585 – 2,185 = 3,400
Δ Annual expenses (annual savings) (₹)		4,745.5 + 3,400 = 1,345.5
Δ Investment (₹):		25,000 – 17,500 = 7,500
E.R.R.R. on Δ Investment		1,345.5/7,500= 17.93%

Since E.R.R.R. (17.93%) >10%, the increment $A{\rightarrow}B$ is justified. Thus, alternative B is better than alternative A and it should be selected.

6.6   COMPARISON OF ALTERNATIVES BY THE CAPITALIZED WORTH METHOD

Capitalized worth method is an effective technique to compare mutually exclusive alternatives when the period of needed service is indefinitely long or when the common multiple of the lives is very long and the repeatability assumption is applicable. This method involves determination of the Present Worth of all receipts and/or disbursements over an infinitely-long length of time. If disbursements only are considered, results obtained by this method can be expressed as capitalized cost (CC) and the procedure for calculating CC is discussed in Chapter 5 (section 5.9).

Example 6.6 illustrates the use of capitalized worth method for comparing mutually exclusive alternatives:

Example 6.6

Compare alternatives A and B given in Example 6.2 by the capitalized worth (cost) method.

Solution

	Capitalized Cost	
	Alternative A	**Alternative B**
First cost (₹)	60,000	2,00,000
Replacements (₹/year):		
60,000(A/F, 15%, 10)/0.15	19,700	
(2,00,000 – 50,000) (A/F, 15%, 25)/0.15		4,700
Annual disbursements (₹/year)		
11,000/0.15	73,333.33	
5,000/0.15		33,333.33
Total Capitalized Cost (₹)	60,000 + 19,700 + 73,333.33 = 1,53,033.33	2,00,000 + 4,700 + 33,333.33 = 2,38,033.33

The capitalized cost for alternative A is smaller than alternative B. Thus, alternative A is better than alternative B and therefore, alternative A should be chosen. The result of capitalized cost method is consistent with the results for Example 6.2 by the other methods of comparison.

6.7 SELECTION AMONG INDEPENDENT ALTERNATIVES

The examples that have been considered so far in this chapter involved mutually exclusive alternatives, i.e., the selection of one alternative excludes the selection of any other and therefore, only one alternative is chosen. This section deals with the selection among independent alternatives/projects. Independent alternatives mean that the selection of one alternative does not affect the selection of any other, and any number of alternatives may be selected as long as sufficient capital is available. Independent project alternatives are often descriptively called opportunities. Any of the six basic methods of study of economics can also be used for the comparison of independent alternatives.

Example 6.7

Consider the following three independent projects. Determine which should be selected, if the minimum attractive rate of return (M.A.R.R.) is 10% and there is no limitation on total investment funds available.

Project	Investment, P	Life, n	Salvage Value, S	Net Annual Cash Flow, A
A	₹5,00,000	5 years	₹5,00,000	+ ₹1,15,000
B	₹6,00,000	5 years	0	+ ₹1,40,000
C	₹7,50,000	5 years	0	+ ₹2,00,000

Solution of Example 6.7 by the Present Worth (P.W.) Method

Project	P.W. of P	P.W. of S	P.W. of A	Net P.W.
A	− ₹5,00,000	+ ₹5,00,000(P/F, 10%, 5) = + ₹3,10,450	+ ₹1,15,000(P/A, 10%, 5) = + ₹4,35,942	+ ₹2,46,392
B	− ₹6,00,000	0	+ ₹1,40,000(P/A, 10%, 5) = + ₹5,30,712	− ₹69,288
C	− ₹7,50,000	0	+ ₹2,00,000(P/A, 10%, 5) = + ₹7,58,160	+ ₹8,160

From the above calculations it is clear that projects A and B, having positive net P.W., would be satisfactory for investment, but project B would not be satisfactory.

Solution of Example 6.7 by the Future Worth (F.W.) Method

Project	F.W. of P	F.W. of S	F.W. of A	Net F.W.
A	− ₹5,00,000(F/P, 10%, 5) = − ₹8,05,250	+ ₹5,00,000	+ ₹1,15,000(F/A, 10%, 5) = + ₹7,02,086.5	+ ₹396836.5
B	− ₹6,00,000(F/P, 10%, 5) = − ₹9,66,300	0	+ ₹1,40,000(F/A, 10%, 5) = ₹8,54,714	− ₹1,11,586
C	− ₹7,50,000(F/P, 10%, 5) = − ₹12,07,875	0	+ ₹2,00,000(F/A, 10%, 5) = ₹12,21,020	+ ₹13,145

From the above calculations, it is clear that projects A and B, having positive net F.W., would be satisfactory for investment, but project B would not be satisfactory.

Solution of Example 6.7 by the Internal Rate of Return (I.R.R.) Method

Project A

The equation for Net Present Worth (N.P.W.) at $i'\%$ is written as follows:

$$\text{N.P.W.} = -₹5,00,000 + ₹5,00,000(P/F, i'\%, 5) + ₹1,15,000(P/A, i'\%, 5)$$

We have to find i' at which N.P.W. is zero. Linear interpolation between 20% and 25% results in $i' = 23.13\%$, which is the I.R.R. Since $23.13\% > 10\%$, investment in project A is justified. Thus, project A should be selected.

Project B

The equation for Net Present Worth (N.P.W.) at $i'\%$ is written as follows:

$$\text{N.P.W.} = -₹6,00,000 + ₹1,40,000(P/A, i'\%, 5)$$

We have to find i' at which N.P.W. is zero. Linear interpolation between 5% and 8% results in $i' = 5.39\%$, which is the I.R.R. Since $5.39\% < 10\%$, investment in project B is not justified. Thus, project B should not be selected.

Project C

The equation for Net Present Worth (N.P.W.) at $i'\%$ is written as follows:

$$\text{N.P.W.} = -₹7,50,000 + ₹2,00,000(P/A, i'\%, 5)$$

We have to find i' at which N.P.W. is zero. Linear interpolation between 10% and 12% results in $i' = 11.92\%$, which is the I.R.R. Since $11.929\% > 10\%$, investment in project C is justified. Thus, project C should be selected.

From the above, we conclude that independent projects A and C should be selected, but project B should not be selected.

Solution of Example 6.7 by the External Rate of Return (E.R.R.) Method

Project A

$$₹5,00,000(F/P, i'\%, 5) = ₹5,00,000 + ₹1,15,000(F/A, 10\%, 5)$$

Hit and trial method results in $i' = 19.18\%$, which is the value of E.R.R. Since $19.18\% > 10\%$, investment in project A is justified. Thus, project A should be selected.

Project B

$$₹6,00,000(F/P, i'\%, 5) = ₹1,40,000(F/A, 10\%, 5)$$

Hit and trial method results in $i' = 7.33\%$, which is the value of E.R.R. Since $7.33\% < 10\%$, investment in project B is not justified. Thus, project B should not be selected.

Project C

$$₹7,50,000(F/P, i'\%, 5) = ₹2,00,000(F/A, 10\%, 5)$$

Hit and trial method results in $i' = 10.23\%$, which is the value of E.R.R. Since $10.23\% > 10\%$, investment in project C is justified. Thus, project C should be selected.

From the above, we conclude that independent projects A and C should be selected, whereas, project B should not be selected.

Solution of Example 6.7 by the Explicit Reinvestment Rate of Return (E.R.R.R.) Method

Project A

$$\text{E.R.R.R.} = \frac{(₹1,15,000 - 0) - (₹5,00,000 - 5,00,000)(A/F,10\%,10)}{₹5,00,000}$$

$$\text{E.R.R.R.} = 23\%$$

Since $23\% > 10\%$, investment in project A is justified. Thus, project A should be selected.

Project B

$$\text{E.R.R.R.} = \frac{(₹1,40,000 - 0) - (₹6,00,000 - 0)(A/F,10\%,10)}{₹6,00,000}$$

$$\text{E.R.R.R.} = 6.95\%$$

Since $6.95\% < 10\%$, investment in project B is not justified. Thus, project B should not be selected.

Project C

$$\text{E.R.R.R.} = \frac{(₹2,00,000 - 0) - (₹7,50,000 - 0)(A/F,10\%,10)}{₹7,50,000}$$

$$\text{E.R.R.R.} = 10.29\%$$

Since $10.29\% > 10\%$, investment in project C is justified economically. Thus, project C should be selected.

From the above, we conclude that independent projects A and C should be selected whereas, project B should not be selected.

Example 6.8

Consider the following three independent projects. Determine which should be selected if the minimum attractive rate of return (M.A.R.R.) is 10% and total funds available for such projects is ₹2 crores. Use P.W. method.

Project	Investment, P	Life, n	Net Annual Cash Flow, A
A	₹90,00,000	20 years	+ ₹12,00,000
B	₹60,00,000	15 years	+ ₹9,00,000
C	₹80,00,000	25 years	+ ₹10,00,000

Solution

Let us use P.W. method to see investment in which project(s) is economically justified. If each project is acceptable, we must next determine which combination of projects maximizes net present worth without exceeding ₹2 crores, the limit of available funds.

Project	P.W. of P	P.W. of A	Net P.W.
A	− ₹90,00,000	+ ₹12,00,000(P/A, 10%, 20) = + ₹1,02,16,320	+ ₹12,16,320
B	− ₹60,00,000	+ ₹9,00,000(P/A, 10%, 15) = + ₹68,45,490	+ ₹8,45,490
C	− ₹80,00,000	+ ₹10,00,000(P/A, 10%, 25) = + ₹90,77,000	+ ₹10,77,000

Investment in all three projects is economically justified. However, all three projects cannot be selected because of limitation on the availability of funds, which is ₹2 crores. Projects A and C should be recommended because this combination maximizes net P.W. (= ₹12,16,320 + ₹10,77,000 = ₹22,93,320), with a total investment of ₹90,00,000 + ₹80,00,000 = ₹1,70,00,000, i.e., ₹1.7 crores. The leftover funds (₹2 − ₹1.7 = ₹0.3 crores or ₹30 lakhs) would be invested elsewhere at the M.A.R.R. of 10%.

PROBLEMS

1. Economic data pertaining to four mutually exclusive alternatives of a major project are given in the following table. (i) If M.A.R.R. is 15% per year, use P.W. method and suggest which alternative is economically viable and should be selected, (ii) Recommend the alternative when a total capital investment budget of ₹1,00,00,000 is available.

	Alternatives			
	I	II	III	IV
Capital Investment (₹)	− 60,00,000	− 95,00,000	− 1,12,00,000	− 1,35,00,000
Net Annual Revenue (₹)	5,50,000	9,65,000	11,65,000	13,95,000
Salvage Value (₹)	3,99,000	50,000	5,00,000	7,50,000
Useful Life (Years)	10	10	10	10

2. Economic data pertaining to three mutually exclusive layout designs of a small facility are given in the following table. Suggest, for an M.A.R.R. is 15% per year, the best alternative using (i) A.W. method and (ii) F.W. method.

	Layout Designs		
	I	II	III
Capital Investment (₹)	− 13,00,000	− 6,50,000	− 10,00,000
Net Annual Revenue (₹)	4,50,000	2,65,000	3,84,000
Salvage Value (₹)	99,900	15,000	25,000
Useful Life (Years)	10	10	10

3. Municipal Corporation of a mega metropolitan city invited expressions of interest for green development of all its sewer lines. It has received four bids the financial details of which are

given in the following table. Recommend which bid should be selected considering M.A.R.R. of 15% and an analysis period of 10 years. Use all three worth methods.

	Bid			
	I	II	III	IV
Capital Investment (₹)	– 3,60,00,000	– 5,35,00,000	– 7,12,00,000	– 8,35,00,000
Annual Expenses (₹)				
Power	2,65,000	2,95,000	4,75,000	4,95,000
Labor	5,65,000	5,96,000	6,75,000	6,00,000
Interests and Taxes	84,00,000	72,50,000	64,00,000	48,00,000
Salvage Value (₹)	– 5,00,000	– 3,00,000	0	0

4. A small FMCG brand considers expansion of its business, for which it has two alternatives. It can launch a new product or undertake vertically backward integration. The estimated new cash flow for each of the alternatives is given in the following table. If the M.A.R.R. is 12% per year, show that the same alternative would be made with proper application of (a) the P.W. method and (b) the I.R.R. method.

	Alternatives	
Year Ending	I	II
0	–3,15,00,000	–7,50,00,000
1	2,50,000	4,65,000
2	4,50,000	11,50,000
3	8,50,000	14,50,000
4	10,50,000	18,50,000

5. A company is evaluating three mutually exclusive alternatives. The investment data pertaining to the alternatives are shown in the following table. Recommend the best alternative if M.A.R.R. is 15% per year and salvage value is zero, using (i) P.W. method and (ii) A.W. method.

	Alternatives		
	I	II	III
Capital Investment (₹)	–5,00,000	–20,50,000	–60,00,000
Net Annual Revenue (₹)	50,000	2,25,000	3,84,000
Useful Life (Years)	5	5	10

6. A large telecom service provider has offered, to one of its major industrial customers two alternative group GSM service plans P and Q. Plan P has a first cost ₹5,00,000, service lock-in period of 10 years, zero end-of-service market value and net annual revenue ₹1,60,000. The corresponding data for plan Q are ₹9,50,000, 20 years, zero market value and revenue ₹3,00,000. Using M.A.R.R. of 15% per year before tax, find the F.W. of each of the plan and recommend the best plan.

7. A company is evaluating two alternatives for modernization of its facility. The capital investment data for the two alternatives is given in the following table. Evaluate and recommend the best alternative. Use M.A.R.R. of 12% per year and an analysis period of 10 years. Evaluation basis of recommendation are: (a) A.W. Method and (b) P.W. Method. Confirm the recommendation using I.R.R. method.

	Alternatives	
	I	II
Capital Investment (₹)	– 7,00,000	– 11,00,000
Annual Expenses (₹)	1,50,000	2,95,000
Annual Revenues (₹)	4,50,000	7,50,000
End of life Market Values (₹)	1,50,000	50,000
Useful Life (Years)	5	10

8. A district administration department is planning to build a small entertainment-cum-recreation facility. It is evaluating two alternative proposals "A" and "B". Plan "A" has a first cost of ₹25,00,000, a life of 25 years, ₹2,50,000 market value and annual operation and maintenance cost of ₹1,96,000. The corresponding data for plan "B" are ₹45,00,000, a life of 50 years, zero market value; the annual operation and maintenance cost for the first 15 years is estimated to be ₹2,75,000 and is expected to increase by ₹20,000 per year from 16 to 50 years. Assuming interest at 12% per year, compare the two alternatives by the Capitalized Cost method.

9. A dye and chemicals firm is evaluating two alternatives for the disposal of its waste. The comparative estimates prepared by the company are given in the following table. An M.A.R.R. of 15% per year and analysis period of 10 years is used. For the alternative A, a contract from the service provider has to be made for 8–10 years at an estimated lost of ₹28,00,000.

	Alternatives	
	A	B
Capital Investment (₹)	– 8,50,000	– 10,00,000
Annual Expenses (₹)	1,95,000	1,80,000
Useful Life (Years)	5	10
Value at the end of useful life (₹)	1,15,000	1,40,000

Recommend the preferred alternative using the I.R.R. method and confirm the recommendation using the E.R.R. method.

□□□

7

Replacement and Retention Decisions

7.1 INTRODUCTION

Business organizations use physical assets in the form of machines and equipments to produce goods and services. After a particular period of service, there might be a need for the replacement of the existing assets and therefore, the firms must constantly monitor the performance of the assets so as to decide whether they should be continued in service or they should be replaced with new assets. Replacement study is carried out to make an economic decision whether to retain or replace an existing asset at a point of time. If the decision is to replace, the study is complete. However, if the decision is to retain, the cost estimates and decision will be reconsidered at periodic intervals to ensure that the decision to retain is still economically correct.

A replacement study involves an application of the A.W. method of comparing alternatives having different useful lives. In situations where a replacement study is required to be performed for no specified study period, the A.W. values are determined by a technique of cost evaluation called the economic service life (ESL) analysis. However, there is a different procedure for conducting replacement study for cases where study period is specified. All these procedures are discussed in this chapter.

7.2 REASONS FOR REPLACEMENT

Replacement study can be required for the following reasons:

Reduced Performance   Due to physical impairment deterioration or accident, the ability of the asset to reliably perform its intended functions reduces. This leads to increased costs of operation, higher wastage and rework costs, lost sales, reduced quality and higher maintenance costs.

Altered Requirements   In situations where the existing asset cannot meet the new requirements of accuracy, speed and other specifications.

Obsolescence   In the presence of a technologically advanced machine or equipment, the existing asset becomes obsolete. The product development cycle time to bring new products to market is getting shorter and this necessitates the replacement study even before the estimated useful life of the asset is reached.

7.3   TERMINOLOGIES USED IN REPLACEMENT STUDY

Before the procedure for replacement study is explained, it is important to understand the following terms:

Defender and Challenger   Defender and challenger are two mutually exclusive alternatives and a replacement study compares these two alternatives. The existing asset is called the defender whereas its potential replacement is known as the challenger.

A.W. Values   The defender and challenger are compared on the basis of annual worth (A.W.) values. Since often, only costs are included in the computation of A.W. values, the term equivalent uniform annual cost (EUAC) may be used in lieu of A.W.

Economic Service Life (ESL)   The number of years at which the lowest A.W. of cost of an asset occurs is called its economic service life (ESL). The equivalency calculations to determine ESL establish the life n for the best challenger, and it also establishes the lowest cost life for the defender in a replacement study.

Defender First Cost   It represents the investment amount P used for the defender. In replacement study the value of P is correctly estimated by the current market value (MV) of the defender. The estimated salvage value of the defender at the end of a particular year becomes its MV at the beginning of the next year, provided the estimates remain correct as the years pass. If the defender must be upgraded or augmented to make it equivalent to the challenger, this cost is added to the MV to obtain the estimate of defender first cost.

Challenger First Cost   This represents the amount of capital that must be recovered when replacing a defender with a challenger. It is equal to P, the first cost of challenger. There may be occasions when a high trade-in value (TIV) may be offered for the defender. In this event, the correct amount to recover and use in economic analysis for the challenger is $P - (TIV - MV)$. It is obvious that when the trade-in value and market value are the same, P is used as challenger first cost.

7.4   ECONOMIC SERVICE LIFE

The economic service life (ESL) is defined as the number of years n at which the equivalent annual worth (A.W.) of costs is the minimum, considering the most current cost estimates over all possible years during which the asset may provide the needed service. ESL represents the estimated life of an asset and this life should be used in engineering economy studies. When n years have passed, the ESL indicates that the asset should be replaced to minimize overall costs. In replacement study, it is required to determine ESL of both defender and challenger.

While an asset is in service, its ESL is determined by calculating the total A.W. of costs in 1 year, 2 years, 3 years, and so on of its service. Total A.W. of costs is the sum of capital recovery (CR), which is the A.W. of the initial investment and any salvage value and the A.W. of the estimated annual operating cost (A.O.C.), that is,

Total A.W. = – Capital recovery – A.W. of annual operating costs

$$= -\ CR - A.W.\ of\ A.O.C. \tag{7.1}$$

CR is calculated from the following formula:

$$CR = -P(A/P,\ i,\ n) + S(A/F,\ i,\ n) \tag{7.2}$$

Figure 7.1 shows the variation of CR, A.W. of A.O.C. and total A.W. of costs with the life of the asset. It can be seen from this figure that CR decreases with each year of life of the asset. However, A.W. of A.O.C. increases with the life of the asset. The figure also shows that the total A.W. of costs decreases in the early life of the asset, attains a minimum value and then increases as the asset gets older. The life of the asset at which total A.W. of costs is the minimum is its ESL. In order to calculate A.W. of A.O.C., determine the present worth of each A.O.C value with the P/F factor, then redistribute this P value over the entire life of the asset using the A/P factor.

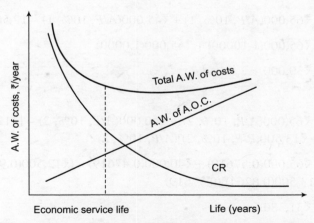

Figure 7.1   A.W. curves of Costs and Economic Service Life (ESL)

The following equation is used to calculate A.W. of costs over k years:

$$\text{Total A.W.}_k = -P(A/P, i, k) + S_k (A/F, i, k) - \left[\sum_{j=1}^{j=k} \text{A.O.C.}_j (P/F, i, j) \right] (A/P, i, k) \tag{7.3}$$

where, P = Initial investment or current market value

S_k = Salvage value or market value after k years

A.O.C.$_j$ = Annual operating cost for year j (j = 1 to k)

The current MV is used for P when the asset is the defender and the estimated future MV values are substituted for the S values in years 1, 2, 3, . . .

Example 7.1 illustrates the determination of ESL.

Example 7.1

A small manufacturing asset, which is 3 years old, is being considered for early replacement. Its current market value is ₹65,000. Estimated future market values and annual operating costs for the next 5 years are given below. What is the economic service life of this defender if the interest rate is 10% per year?

Year, j	Market value, MV$_j$ (₹)	Annual operating costs, A.O.C.$_j$ (₹)
1	45,000	– 12,500
2	40,000	– 13,500
3	30,000	– 15,000
4	10,000	– 17,500
5	0	– 22,500

Solution

Equation (7.3) is used to calculate total A.W.$_k$ for k = 1, 2, 3, 4 and 5.

Year 1

Total A.W.$_1$ = – ₹65,000(A/P, 10%, 1) + ₹45,000(A/F, 10%, 1) – 12,500

         = – ₹65,000(1.10000) + ₹45,000(1.0000)

Total A.W.$_1$ = – ₹39,000

Year 2

Total A.W.$_2$ = – ₹65,000(A/P, 10%, 2) + ₹40,000(A/F, 10%, 2) – [₹12,500(P/F, 10%, 1)
         + ₹13,500(P/F, 10%, 2)](A/P, 10%, 2)

         = – ₹65,000(0.57619) + ₹40,000(0.47619) – [₹12,500(0.9091) +
         ₹13,500(0.8264)](0.57619)

Total A.W.$_2$ = – ₹31,380.6

Year 3

Total A.W.$_3$ = – ₹65,000(A/P, 10%, 3) + ₹30,000(A/F, 10%, 3) – [₹12,500(P/F, 10%, 1)
         + ₹13,500(P/F, 10%, 2) + ₹15,000(P/F, 10, 3)](A/P, 10%, 3)

         = – ₹65,000(0.40211) + ₹30,000(0.30211) – [₹12,500(0.9091) +
         ₹13,500(0.8264) + ₹15,000(0.7513)](0.40211)

Total A.W.$_{.3}$ = $-$ ₹30,661.01

Year 4

Total A.W.$_{.4}$ = $-$ ₹65,000(A/P, 10%, 4) + ₹10,000(A/F, 10%, 4) $-$ [₹12,500(P/F, 10%, 1) + ₹13,500(P/F, 10%, 2) + ₹15,000(P/F, 10, 3) + ₹17,500(P/F, 10%, 4)](A/P, 10%, 4)

$\quad$ = $-$ ₹65,000(0.31547) + ₹10,000(0.21547) $-$ [₹12,500(0.9091) + ₹13,500(0.8264) + ₹15,000(0.7513) + ₹17,500(0.6830)](0.31547)

Total A.W.$_{.4}$ = $-$ ₹32,781.13

Year 5

Total A.W.$_{.5}$ = $-$ ₹65,000(A/P, 10%, 5) + ₹0(A/F, 10%, 5) $-$ [₹12,500(P/F, 10%, 1) + ₹13,500(P/F, 10%, 2) + ₹15,000(P/F, 10, 3) + ₹17,500(P/F, 10%, 4) + ₹22,500(P/F, 10%, 5)](A/P, 10%, 5)

$\quad$ = $-$₹65,000(0.26380) $-$ [₹12,500(0.9091) + ₹13,500(0.8264) + ₹15,000(0.7513) + ₹17,500(0.6830) + ₹22,500(0.6209)](0.26380)

Total A.W.$_{.5}$ = $-$ ₹32,899.13

Thus, the ESL of the asset is 3 years as the total A.W. is the minimum at that time.

The problems discussed in the previous chapters had an estimated life of *n* years with associated other estimates such as first cost in year 0, possibly salvage value in year *n* and an A.O.C. that remained constant or varied each year. For all such problems, the calculation of A.W. using these estimates determined the A.W. over *n* years. This is also the ESL when *n* is fixed. In all previous problems, there were no year-by-year MV estimates applicable over the years. Thus, we can conclude that when the expected life is known for the challenger or defender, determine its A.W. over *n* years, using the first cost or current market value, estimated salvage value after *n* years and A.O.C. estimates. This A.W. is used in the replacement study.

The following points should be remembered about *n* and A.W. values to be used in a replacement study:

> ➤ In situations where year-by-year market value estimates are made, use them to perform an ESL analysis and determine the *n* value with the lowest total A.W. of costs. These are the best *n* and A.W. values for the replacement study.
> ➤ In cases where yearly market value estimates are not made, use it to calculate the A.W. over *n* years and these values of *n* and A.W. are used in the replacement study.

7.5 PROCEDURE FOR PERFORMING REPLACEMENT STUDY

Replacement studies are performed in one of the two ways: without a study period specified or with a defined study period. This section describes the procedure for making replacement study when no study period (planning horizon) is specified. The procedure for making replacement study when study period is specified is discussed in the subsequent section.

A replacement study determines when a challenger replaces the in-use defender. If the challenger (C) is selected to replace the defender (D) now, then the replacement study is completed. However, if the defender is retained now, the study may extend over a number of years, equal to the life of the defender n_D, after which a challenger replaces the defender. Use the annual worth and life values for C and D determined in the ESL analysis to apply the following replacement study procedure. It assumes that the services provided by the defender could be obtained at the A.W.$_D$ amount.

Step 1: Determine A.W.$_C$ and A.W.$_D$ values using the estimates of challenger and defender respectively. Based on the better A.W.$_C$ or A.W.$_D$ value, select the challenger or defender. When the challenger is selected, replace the defender now and keep the challenger for n_C years. This replacement study is complete. If the defender is selected, plan to retain it for up to n_D more years. Next year, perform the following steps.

Step 2: After one year, look at the estimates, especially first cost, market value and AOC of both challenger and defender. If these estimates are not current estimates, proceed to Step 3. However, if the estimates are current estimates and this is year n_D, replace the defender. If this is not year n_D, retain the defender for another year and repeat the same step. This step may be repeated several times.

Step 3: Whenever the estimates have changed, update them and determine new A.W.$_C$ and A.W.$_D$ values. Initiate a new replacement study (Step 1).

If the defender is selected initially (Step 1), estimates may need updating after 1 year of retention (Step 2). It may be possible that a new best challenger is available and therefore it should be compared with the defender. Either significant changes in defender estimates or availability of a new challenger indicates that a new replacement study needs to be performed. In actual sense, replacement study is performed each year to take decisions about replacing or retaining any defender, provided a competitive challenger is available.

Example 7.2 illustrates the procedure discussed above for performing replacement study.

Example 7.2

Five years ago, a company purchased a small welding machine for ₹2,25,000. It is expected to have the market values and annual operating costs shown for the rest of its useful life of up to 3 years. It could be traded now at an appraised market value of ₹40,000.

Year	Market value at the end of year (₹)	A.O.C. (₹)
1	30,000	– 2,50,000
2	20,000	– 2,65,000
3	5,000	– 3,00,000

A replacement welding machine costing ₹6,25,000 has an estimated ₹50,000 salvage value after its 5-year life and an A.O.C. of ₹1,55,000 per year. At an interest rate of 15% per year, determine how many years the company should retain the present welding machine.

Solution

Find ESL of the defender; compare with $A.W._C$ over 5 years.

For n = 1: $A.W._D = -40,000(A/P, 15\%, 1) - 2,50,000 + 30,000(A/F, 15\%, 1)$

$$= -40,000(1.15) - 2,50,000 + 30,000(1.0000)$$

$$= -₹2,66,000$$

For n = 2: $A.W._D = -40,000(A/P, 15\%, 2) - 2,50,000 + (-15,000 + 20,000)(A/F, 15\%, 2)$

$$= -40,000 (0.61512) - 2,50,000 + 5,000(0.46512)$$

$$= -₹2,72,279$$

For n = 3: $A.W._D = -40,000(A/P, 15\%, 3) - [2,50,000(P/F, 15\%, 1)$
$$+ 2,65,000(P/F, 15\%, 2)](A/P, 15\%, 3)$$
$$+ (-3,00,000 + 5,000)(A/F, 15\%, 3)$$

$$= -40,000 \quad (0.43798) \quad - \quad [2,50,000(0.8696) \quad + \quad 2,65,000(0.7561)]$$
$$(0.43798) - 2,95,000(0.28798)$$

$$= -₹2,85,447$$

The ESL is now 1 year with $A.W._D = -₹2,66,000$

$$A.W._C = -6,25,000(A/P, 15\%, 5) - 1,55,000 + 50,000(A/F, 15\%, 5)$$

$$= -6,25,000(0.29832) - 1,55,000 + 50,000(0.14832)$$

$$= -₹3,34,034$$

Since the ESL A.W. value of defender is lower than the challenger A.W., the company should keep the defender now and replace it after 1 year.

7.6 REPLACEMENT STUDY OVER A SPECIFIED STUDY PERIOD

In cases where replacement study is performed over a specified study period, for example, 5 years, the determination of A.W. values for the challenger and for the remaining life of the defender are usually not based on the economic service life. It is assumed that the services of the alternatives are not needed beyond the study period and therefore, what happens to them after the study period is not considered in the replacement analysis.

When performing a replacement study over a fixed study period, it is important that the estimates used to determine the A.W. values be accurate. When the defender's remaining life is shorter than the study period, the cost of providing the defender's services from the end of its expected remaining life to the end of the study period must be estimated as accurately as possible and included in the replacement study.

In order to perform replacement study over a specified study period, the P.W. or A.W. for each option, i.e., defender and challenger, is calculated over the study period. The option with the lowest cost, or highest income if revenues are estimated, is then selected.

Examples 7.3 and 7.4 illustrate the procedure for performing replacement study over a specified period.

Example 7.3

Three years ago, Delhi Fire Service purchased a new fire-truck. Because of increase in fire cases, new fire-fighting capacity is needed once again. An additional fire-truck of the same capacity can be purchased now, or a double capacity truck can replace the current fire-truck. Estimates are given below. Compare the options at 10% per year, using a 10-year study period.

	Currently Owned	New Purchase	Double Capacity
First cost P, (₹)	– 37,75,000	– 43,75,000	– 47,50,000
AOC (₹)	– 37,500	– 37,500	– 62,500
Market value (₹)	17,50,000	-	-
Salvage value (₹)	3,77,500	5,25,000	4,75,000
Life in years	10	10	10

Solution

Let us identify Option 1 as retention of the presently owned truck and augmentation with a new same-capacity truck and Option 2 as replacement with the double-capacity truck.

	Option 1		Option 2
	Currently Owned	Augmentation	Double Capacity
P (₹)	– 17,50,000	– 43,75,000	– 47,50,000
A.O.C. (₹)	– 37,500	– 37,500	– 62,500
S (₹)	3,77,500	5,25,000	4,75,000
n, years	7	10	10

For a full-life 10-year study period of Option 1:

$$\text{A.W.}_1 = (\text{A.W. of currently owned}) + (\text{A.W. of augmentation})$$
$$= [-17,50,000(A/P, 10\%, 7) + 37,75,000(A/F, 10\%, 7) - 37,500]$$
$$+ [-43,75,000(A/P, 10\%, 10) + 5,25,000(A/F, 10\%, 10) - 37,500]$$
$$= [-17,50,000(0.20541) + 37,75,000(0.10541) - 37,500]$$
$$+ [-43,75,000(0.16275) + 5,25,000(0.06275) - 37,500]$$
$$= 955.25 - 7,16,588$$

$$A.W._{-1} = - ₹7,15,632$$

$$A.W._{-2} = - 47,50,000(A/P, 10\%, 10) + 4,75,000(A/F, 10\%, 10) - 62,500$$

$$= - 47,50,000(0.16275) + 4,75,000(0.06275) - 62,500$$

$$A.W._{-2} = - ₹8,05,756$$

So the result is: retain the currently owned fire-truck and perform the replacement study again next year.

Example 7.4

An oil exploring company placed an equipment into service 5 years ago for which a replacement study is required. It has been decided that the current equipment will have to serve for either 2, 3, or 4 more years before replacement. The equipment has a current market value of ₹5,00,000, which is expected to decrease by ₹1,25,000 per year. The A.O.C. is expected to remain constant at ₹1,25,000 per year. The replacement challenger is a fixed-price contract to provide the same services at ₹3,00,000 per year for a minimum of 2 years and a maximum of 5 years. Use M.A.R.R. of 10% per year to perform replacement study over a 6-year period to determine when to sell the current equipment and purchase the contract services.

Solution

Since the defender will be retained for 2, 3, or 4 years, there are 3 feasible options (A, B and C).

	Defender Retained	Challenger Serves
A	2 years	4 years
B	3 years	3 years
C	4 years	2 years

The A.W. values of the defender for 2, 3 and 4 years are:

$$A.W._{.D2} = - 5,00,000(A/P, 10\%, 2) + 2,50,000(A/F, 10\%, 2) - 1,25,000$$

$$= - 5,00,000(0.57619) + 2,50,000(0.47619) - 1,25,000$$

$$= - ₹2,94,048$$

$$A.W._{.D3} = - 5,00,000(A/P, 10\%, 3) + 1,25,000(A/F, 10\%, 3) - 1,25,000$$

$$= - 5,00,000(0.40211) + 1,25,000(0.30211) - 1,25,000$$

$$= - ₹2,50,528$$

$$A.W._{.D4} = - 5,00,000(A/P, 10\%, 4) - 1,25,000$$

$$= - 5,00,000(0.31547) - 1,25,000$$

$$= - ₹2,82,735$$

For all options, the challenger has A.W. of:

$$A.W._C = -₹3,00,000$$

The table below shows the cash flows and P.W. values for each option over the 6-year study period. P.W. computation for options A, B and C are shown below:

$$P.W._A = -2,94,048(P/A, 10\%, 2) - 3,00,000(F/A, 10\%, 4)(P/F, 10\%, 6)$$

$$= -2,94,048(1.7355) - 3,00,000(4.6410)(0.5645)$$

$$P.W._A = -₹12,96,274$$

$$P.W._B = -2,50,528(P/A, 10\%, 3) - 3,00,000(F/A, 10\%, 3)(P/F, 10\%, 6)$$

$$= -2,50,528(2.4869) - 3,00,000(3.3100)(0.5645)$$

$$P.W._B = -₹11,83,587$$

$$P.W._C = -2,82,735(P/A, 10\%, 4) - 3,00,000(F/A, 10\%, 2)(P/F, 10\%, 6)$$

$$= -2,82,735(3.1699) - 3,00,000(2.1000)(0.5645)$$

$$P.W._C = -₹12,51,877$$

Option B has the lowest cost P.W. value (– ₹11,83,587). Keep the defender for 3 years, then replace it. It should be noted that the same answer will result if the annual worth, or future worth, of each option is calculated at the given M.A.R.R.

Time in Service, Years		A.W. Cash Flows for Each Option, ₹/Year						Option P.W.,
Defender	Challenger	1	2	3	4	5	6	(₹)
A 2	4	– 2,94,048	– 2,94,048	– 3,00,000	– 3,00,000	– 3,00,000	– 3,00,000	– 12,96,274
B 3	3	– 2,50,528	– 2,50,528	– 2,50,528	– 3,00,000	– 3,00,000	– 3,00,000	– 11,83,587
C 4	2	– 2,82,735	– 2,82,735	– 2,82,735	– 2,82,735	– 3,00,000	– 3,00,000	– 12,51,877

PROBLEMS

1. Anas Khan wishes to replace his Skoda car, which he bought for ₹8,00,000 two years ago, with a new Toyota car. The Toyota costs (MRP) ₹12,00,000. He expects ₹3,00,000 for his Skoda. He approaches a Toyota showroom which gives him a replacement offer of ₹3,85,000 for the Skoda which is ₹25,000 more than the currently owned Skoda model and a discount of ₹1,50,000 on the MRP of the new Toyota brand. The showroom further offers free insurance and accessories with ₹65,000. In case Anas wishes to consider the replacement offer, what is the correct first cost for (a) the defender and (b) the challenger?

2. Eveready Industries Limited bought a testing and inspection machine for ₹80,00,000 two years ago. At the time of its purchase, the company expected to use it for 5 years. The company also expected that the machine could be salvaged for ₹9,00,000 after 5 years. The fast developments in its business and high competition forced the company to procure a new machine for ₹1.0 Cr. In case the company wishes to sell the machine now, it would fetch the company ₹42,00,000. The company may also retain the machine for two more years at which time the salvage value

of the machine is estimated at ₹80,000. If the company retains the machine for these two more years, it expects that it would have to spend ₹1,20,000 annually as maintenance and overhead costs on the machine (other than operating cost). Determine the values of P, n, S and A.O.C. for this defender if a replacement analysis were to be performed today.

3. Zoya purchased a keyboard one year ago for ₹2,50,000 which consumes more battery than expected. She purchased the keyboard with a view to use it for 7 years with annual battery consumption worth ₹2,500 and a salvage value of ₹20,000. However, she had to spend ₹12,000 towards some maintenance during the previous year. She expects that the keyboard now requires periodic maintenance with the present year's maintenance cost of ₹13,500 and an increase of ₹1,500 each year. The salvage value is estimated to follow the relationship $S = $ ₹20,000 − 3,500C where S is the salvage value and C is the number of years since the keyboard was purchased. It is now estimated that the keyboard would be useful for a maximum of 3 more years. Determine the values of P, A.O.C., n and S for a replacement study performed now.

4. Toyama Electric commissioned a new SMT system in its manufacturing unit two years ago for ₹2.5 Cr and an expected life of 5 years. At the time of its installation, its market value was described as ₹$(2,10,00,000 - 15,00,000C^{1.3})$, where C was the number of years since purchase. It was estimated that O&M cost of the SMT system follows a relationship: ₹$25,000+100C^{2.5}$. Determine the values of P, S and A.O.C. for this defender if a replacement analysis is performed (a) now with a study period of 3 years specified and (b) 2 years from now with no study period specified.

5. Wockhardt Ltd. wishes to replace one of the incubators installed in its plant at an estimated first cost of ₹45,00,000, an annual operating cost of ₹6,40,000 and a maximum useful life of 5 years. The salvage value of the incubator any time it is sold is expected to be ₹4,80,000. At an interest rate of 12% per year, determine its economic service life and corresponding A.W. value.

6. A heavy-mining equipment with a first cost of ₹9,00,00,000 is expected to have a useful life of 10 years and a market value that decreases by ₹9,00,000 every year. The annual operating cost of the equipment is to remain constant at ₹9,00,000 for 5 years and to increase 20% per year every year thereafter. Unicom Limited, which owns the equipment, manages to get finance at a low interest rate of 7% per year. (a) Verify that the ESL is 5 years. Is the ESL sensitive to the changing market value and A.O.C. estimates? (b) During a replacement analysis, the company finds that equipment would have an ESL of 10 years when it is evaluated against any challenger. If the estimated A.O.L. series has proved to be correct, determine the minimum market value that will make ESL equal to 10 years.

7. Emad Consulting estimated the following A.W. values for a precision inspection system owned by its client Sameer Aerospace:

Number of Years (to be retained)	A.W. Value ₹/Year
1	− 6,40,000
2	− 5,00,000
3	− 4,40,000
4	− 5,50,000
5	− 7,50,000

A challenger has ESL = 2 years and A.W.$_c$ = −₹4,80,000 per year. If the consultant must recommend a replacement/retain decision today, should the company purchase the challenger?

The M.A.R.R. is 15% per year.

8. Sachin Sethi Movers Pvt. Ltd. is considering replacing a heavy-duty crane, which was purchased 2 years ago for ₹5,00,00,000, with a new one. The existing crane is valued at present at ₹10,00,000. If the company decides to upgrade the existing crane, it would cost ₹3,50,00,000 and would be worthy for use for another 3 years of its operation, after which the crane would have to be sold for an estimated value of ₹8,50,000. The challenger can be purchased at a cost of ₹6,00,00,000, has an expected life of 10 years and has a salvage value of ₹10,00,000. Determine whether the company should upgrade or replace at a M.A.R.R. of 12% per year. Assume the A.O.C. estimates are the same for both the options.

9. Kitply Ltd. manufactures different types of wooden board. It is evaluating the replacement alternative for one of its chemical treatment facility. The cost estimates for this possible replacement is given in the following table:

	Current System	New Chemical Treatment Facility
First Cost 8 Years ago (₹)	– 2,00,00,000	
First Cost (₹)		– 3,85,00,000
Remaining Life (Years)	5	10
Current Value (₹)	24,00,000	
A.O.C. (₹/year)	– 65,00,000	– 57,00,000
Future Salvage (₹)	0	24,00,000

For the replacement proposition (a) perform the replacement analysis and (b) find the minimum resale value required to make the challenger replacement choice now. Is this a reasonable amount to expect for the current system? Use an interest rate of 12% per year.

10. Rishi Laser Cutting Ltd. purchased a CNC Wire EDM for ₹6,50,000 five years ago and its A.W. data for rest of its useful life of up to 3 years is shown in the following table:

Year	Year-end Market Value (₹)	A.O.C. (₹)
1	1,20,000	– 87,500
2	96,000	– 99,000
3	56,000	– 1,25,000

It is known that the machine can now be traded for ₹45,000. A new EDM with latest software and pulse control power source costs ₹11,00,000 with a ₹3,60,000 salvage value after 5 years and an A.O.C. of ₹55,000 per year. Find out for how many more years, the company should retain the present machine if interest rate is 10% per year.

11. A replacement analysis is being performed for an in-situ machining center with an equivalent annual worth of – ₹6,00,000 for each year of its maximum remaining useful life of 2 years. A suitable replacement is determined to have equivalent annual worth values of – ₹8,25,000, – ₹7,15,000 and – ₹7,65,000 per year for 1, 2 or 3 years respectively. If the company uses a fixed 3-year planning period, when should the company replace the machine? Assume an interest rate of 15%.

12. Two years ago, Cipla Ltd. installed an emission control system for its medicine plant for ₹4,50,00,000 with an estimated salvage value of ₹3,00,000 after 10 years. Currently, the expected service life is 8 years with an A.O.C. of ₹4,50,000 per year. Thermax Ltd. has recently developed a challenger which costs ₹9,60,00,000 with an A.O.C. of ₹2,50,000 and has an estimated useful life of 15 years with a salvage value of ₹2,80,000. If the M.A.R.R. is 15% find (a) the minimum trade-in value necessary now to make the challenger economically viable,

(b) number of years to retain the defender up to break-even if the trade-in offer is ₹4,50,00,000. Assume that the salvage value can be realized for all retention periods up to 8 years.

13. Mahindra & Mahindra Ltd. purchased a multi-axis CNC machining center for ₹5,00,00,000 three years ago. To meet the demand and to catch-up with competition, the company finds that either the center be upgraded with new accessories or be replaced with a new machining center. The cost of upgradation is ₹2,00,00,000, which enables the machine to be good for operation for 3 more years. In case the company goes for replacement, the existing center may be traded for ₹1,10,00,000. The annual operating cost of the centre is ₹2,20,00,000. The replacement machine would cost ₹7,50,00,000 and would satisfactorily serve for at least 8 more years. The salvage value of the replaced machine would be ₹1,50,00,000 for year 1 through 5, ₹60,00,000 after 6 years and ₹6,00,000 thereafter. It will have an estimated A.O.C. of ₹5,50,000 per year. Using a 5-year time horizon, perform an economic analysis at 20% per year. Also recommend if the company should replace the existing machine now or do so after 3 years from now. What are the A.W. values?

14. Three years ago, The Heart Centre, New Delhi, had installed a Japanese robotic heart surgery facility for ₹3,50,00,000. The system can be satisfactorily used for 3 more years. The hospital is considering a replacement option. If replaced, the hospital can realize ₹50,00,000. The market value and operating costs of the present system and replacement system are tabulated below:

Year	Present System		Replacement Alternative		
	Market Value (₹)	A.O.C. (₹)	Market Value (₹)	A.O.C. (₹)	Inspection Cost (₹)
1	25,00,000	– 5,20,000	80,00,000	– 2,50,000	
2	15,00,000	– 6,00,000	50,00,000	– 3,00,000	
3	6,00,000	– 7,20,000	32,00,000	– 3,60,000	
4	5,00,000	– 8,50,000	6,00,000	– 4,00,000	
5			3,00,000	– 4,50,000	
6			0	– 5,00,000	– 12,50,000
7			0	– 5,50,000	– 12,50,000

The replacement option, a Taiwan make, is cheaper at ₹27,500,000, but has a considerably higher O&M costs in later years of service. It has a maximum life of 7 years. It is also anticipated that the new system would incur a recurring cost, after 4 years, for periodic inspection by a service personnel from Taiwan, costing ₹1,250,000 per year. Using the tabulated financial data and i=15%, estimate ESL and A.W. values for the defender and challenger and also find in what year the current system should be replaced.

❑❑❑

8

Depreciation

8.1 INTRODUCTION

Depreciation is defined as decrease in the value of a physical property or asset with the passage of time. A physical asset has value because it provides monetary benefits to its owner. These benefits are in the form of future cash flows resulting from (i) the use of the asset to produce saleable goods or services, or (ii) the ultimate sale of the asset at the end of its useful life. Depreciation, thus, represents decrease in the value due to lessening in the ability to produce these future cash flows, as a result of several causes such as wear and tear and obsolescence.

Although, depreciation does occur, it is really difficult to determine its magnitude well in advance. Depreciation is not an actual cash flow but it is a tax-allowed deduction included in tax calculations in almost all industrialized countries. Depreciation lowers income tax as per the relation:

$$\text{Tax} = (\text{income} - \text{deductions}) \, (\text{Tax rate})$$

8.2 DEPRECIATION TERMINOLOGY

The definitions of the terms that are used in calculation are given below:

Book Depreciation   Book depreciation is used by a corporation or business firm for internal financial accounting. It indicates the reduced investment in an asset based upon the usage pattern and expected useful life of the asset.

Tax Depreciation   This is used in tax calculations as per government regulations. A company subtracts an annual tax of depreciation from its annual income to calculate the amount of taxes due each year. However, the tax depreciation amount must be calculated using a government approved method.

First Cost or Unadjusted Basis   This represents the installed cost of the asset. It includes purchase price, delivery and installation fees, and other depreciable direct costs incurred to prepare the asset for use. The term unadjusted basis, B or simply basis, is used when the asset is new, whereas the term adjusted basis is used after some depreciation has been charged.

Book value This represents the remaining, un-depreciated capital investment on the books or record of the company after the total amount of depreciation charges to date have been subtracted from the basis. The book value, BV_t is usually determined at the end of each year.

Recovery period This represents the depreciable life n of the asset in years.

Market value This is defined as the estimated amount that can be realized if the asset was sold in the open market. The book value and market value of different assets may be substantially different. For example, the market value of real estate tends to increase, but the book value will decrease as depreciation charges are taken. On the other hand, a computer may have a market value much lower than its book value due to changing technology.

Salvage value This is defined as the estimated value of the asset at the end of its useful life. The salvage value S of the asset may be positive, zero, or negative depending upon the type of asset and the situation in which it is being sold.

Depreciation rate or recovery rate This represents the fraction of the first cost removed by depreciation each year. This rate, d_t may be the same each year or different for each year of the recovery period.

8.3　METHODS OF DEPRECIATION

Although, there are several methods for calculation of depreciation, the following methods are commonly used and therefore, only these methods are discussed in detail here.

1. Straight Line (SL) Method
2. Declining Balance (DB) Method
3. Sum-of-the-Year-Digits (SYD) Method
4. Sinking Fund Method
5. Service Output Method

8.3.1　Straight Line (SL) Method

This method assumes that the decrease in the value of the asset is directly proportional to its age. This method derives its name from the fact that the book value of the asset decreases linearly with time. This method is widely used for computing depreciation costs owing to the fact that it is simple and gives uniform annual depreciation charge. If

n = depreciable life of the asset or recovery period in years

B = first cost or original cost or unadjusted basis in rupees

S = estimated salvage value at the end of the life of the asset in rupees

d_k = annual cost of depreciation in the kth year ($1 \le k \le n$) in rupees

D_k = cumulative depreciation cost through the kth year in rupees

BV_k = book value of the asset at the end of k years, in rupees

Then:

$$d_k = \left(\frac{B-S}{n}\right)$$ (8.1)

$$D_k = \frac{K(B-S)}{n}$$ (8.2)

$$BV_k = B - D_k$$

$$BV_k = B - \frac{K(B-S)}{n}$$ (8.3)

The term $(B-S)$ is referred to as the depreciable value of an asset. Examples 8.1 and 8.2 illustrate depreciation calculation by SL method.

Example 8.1

A small equipment has first cost of ₹20,000 and a 10–year estimated life. The estimated salvage value of the equipment is zero at the end of 10 years. Calculate (a) depreciation cost during the fifth year, (b) cumulative depreciation cost through the fifth year, and (c) book value at the end of fifth year.

Solution

Given B = ₹20,000

   $n = 10$ years

   $S = 0$

   (a) $d_5 = ?$ (b) $D_5 = ?$ (c) $BV_5 = ?$

(a) Using Equation (8.1), we get:

$$d_5 = \left(\frac{₹20,000 - 0}{10}\right)$$

$$d_5 = ₹2,000$$

(b) Using Equation (8.2), we get:

$$D_5 = \frac{5(₹20,000 - 0)}{10}$$

$$D_5 = ₹10,000$$

(c) Using Equation (8.3), we get:

$$BV_5 = ₹20,000 - \frac{5(₹20,000 - 0)}{10}$$

$$BV_5 = ₹10,000$$

Example 8.2

A company purchased a small machine for ₹1,00,000. It paid sales tax and shipping costs of ₹ 10,000. The installation cost of the machine is ₹5,000 and its estimated useful life is 5 years. The estimated salvage value of the machine at the end of its useful life is ₹10,000. Calculate (a) depreciation cost during the third year, (b) cumulative depreciation cost through the third year and (c) book value at the end of third year.

Solution

Given $B = ₹1,00,000 + ₹10,000 + ₹5,000 = ₹1,15,000$

$$n = 5 \text{ years}$$

$$S = ₹10,000$$

(a) $d_3 = ?$ (b) $D_3 = ?$ (c) $BV_3 = ?$

(a) $d_3 = \left(\dfrac{₹1,15,000 - ₹10,000}{5} \right)$

$D_3 = ₹21,000$

(b) $D_3 = \dfrac{3(₹1,15,000 - ₹10,000)}{5}$

$D_3 = ₹63,000$

(c) $BV_3 = B - D_3$

$= ₹1,15,000 - ₹63,000$

$= ₹52,000$

8.3.2 The Declining Balance (DB) Method

This method is also known by other names such as Matheson formula and the constant percentage method. In this method, it is assumed that the annual cost of depreciation is a fixed percentage of the book value at the beginning of the year. The ratio of the depreciation in any one year to the book value at the beginning of the year is constant throughout the life of the asset and is designated by R ($0 \leq R \leq 1$). R is taken equal to $2/n$ when a 200% declining balance is used. This gives the rate of recovery twice the straight line rate of $1/n$ and the declining balance method is known as double declining balance (DDB) method. Similarly, $R = 1.5/n$ if 150% declining balance method is specified. The following relationships are used for calculating depreciation costs by DB method:

For the first year

$$R = \frac{d_1}{BV_1}$$

Since $BV_1 = B$

$$d_1 = B(R) \qquad\qquad (8.4)$$

Similarly, for the second year

$$R = \frac{d_2}{BV_2}$$

Here $BV_2 = B - d_1$

Therefore, $d_2 = (B - d_1)\,R$

$$= (B - BR)\,R$$

$$= B(1 - R)\,R$$

For the third year

$$R = \frac{d_3}{BV_3}$$

Here $BV_3 = B - d_1 - d_2$

Therefore, $d_3 = (B - BR - BR + BR^2)\,R$

$$= B(1 - R)^2\,R$$

In general, depreciation cost in the kth year, i.e., d_k is:

$$d_k = B(1 - R)^{k-1}\,R \qquad\qquad (8.5)$$

The relationship for cumulative depreciation cost through kth year is obtained as:

$$D_k = d_1 + d_2 + d_3 + \ldots + d_k$$

$$= BR + BR(1 - R) + BR(1 - R)^2 + \ldots + BR(1 - R)^{k-1}$$

$$= BR + BR[(1 - R) + (1 - R)^2 + \ldots + (1 - R)^{k-1}]$$

$(1 - R) + (1 - R)^2 + \ldots + (1 - R)^{k-1}$ forms a geometric series and its sum is obtained as

$$(1 - R)\left[\frac{1 - (1 - R)^{k-1}}{1 - (1 - R)}\right]$$

$$\left[\frac{(1 - R) - (1 - R)^k}{R}\right]$$

Thus,

$$D_k = BR + BR\left[\frac{(1 - R) - (1 - R)^k}{R}\right]$$

$$D_k = BR + B(1 - R) - B(1 - R)^k$$

$$= BR + B - BR - B(1 - R)^k$$

$$D_k = B[1 - (1 - R)^k]$$

(8.6)

$$BV_k = B - D_k$$

$$= B - B[1 - (1 - R)^k]$$

$$BV_k = B(1 - R)^k$$

(8.7)

Similarly, book value at the end of useful life of the asset is given as:

$$BV_n = B(1 - R)^n$$

(8.8)

Example 8.3

Solve Example 8.1 by the double declining balance (DDB) method.

Solution

Given, $B = ₹20,000$

$n = 10$ years

$S = 0$

(a) $d_5 = ?$ (b) $D_5 = ?$ (c) $BV_5 = ?$

(a) Using Equation (8.5), we get:

$$d_5 = ₹20,000(1 - R)^{5-1}R$$

For DDB,

$$R = \frac{2}{n} = \frac{2}{10} = 0.2$$

$$d_5 = ₹20,000(1 - 0.2)^4 (0.2)$$

$$= ₹1,638.40$$

(b) Using Equation (8.6), we get

$$D_5 = ₹20,000[1 - (1 - 0.2)^5]$$

$$= ₹13,446.40$$

(c) Using Equation (8.7), we get

$$BV_5 = ₹20,000 (1 - 0.2)^5$$

$$= ₹6,553.60$$

8.3.3 Sum-of-the-Years'-Digits (SYD) Method

The annual depreciation cost for a given year by the SYD method is obtained by taking the product of SYD depreciation factor for that year and the difference between unadjusted basis (B) and the salvage value of the asset at the end of its useful life. In order to compute the SYD depreciation factor for any year, use the following steps:

  (i) List the digits corresponding to the number for each year of life in reverse order.
 (ii) Compute the sum of these digits.
(iii)   Divide each number of the reverse-ordered list by the sum to obtain SYD depreciation factor for each year of life of the asset.

For example, for an asset having useful life of 6 years, SYD depreciation factors are as follows:

Year	Digits in Reverse Order	SYD Depreciation Factor
1	6	6/21
2	5	5/21
3	4	4/21
4	3	3/21
5	2	2/21
6	1	1/21
Sum of digits	21	

The annual cost of depreciation for any year k is given as

$$d_k = (B - S)\left[\frac{2(n - k + 1)}{n(n + 1)}\right] \quad (8.9)$$

The book value at the end of year k is given as:

$$BV_k = B - \left[\frac{2(B - S)}{n}\right]k + \left[\frac{(B - S)}{n(n + 1)}\right]k(k + 1) \quad (8.10)$$

The cumulative depreciation through the kth year is given as:

$$D_k = B - BV_k \quad (8.11)$$

Example 8.4

Solve Example 8.2 by the SYD method.

Solution

Given, $B = ₹1,15,000$

$$n = 5 \text{ years}$$

$$S = ₹10,000$$

(a) $d_3 = ?$ (b) $D_3 = ?$ (c) $BV_3 = ?$

(a) Using Equation (8.9), we get

$$d_3 = (₹1,15,000 - ₹10,000)\left[\frac{2(5 - 3 + 1)}{5(5 + 1)}\right]$$

$$= ₹21,000$$

(b) Using Equation (8.10), we get

$$BV_3 = ₹1,15,000 - \left[\frac{2(₹1,15,000 - ₹10,000)}{5}\right](3) + \left[\frac{(₹1,15,000 - ₹10,000)}{5(5 + 1)}\right]3(3 + 1)$$

$$BV_3 = ₹31,000$$

(c) Using Equation (8.11), we get

$$D_3 = B - BV_3$$

$$= ₹1,15,000 - ₹31,000$$

$$= ₹84,000$$

8.3.4   The Sinking Fund Method

In this method, it is assumed that the funds required to replace the asset accumulates in a sinking fund. By sinking fund we mean that each year depreciation is so charged that the future worth of all depreciation and salvage value (S) becomes equal to the unadjusted basis (B). The total fund accumulated in the sinking fund plus interest on it up to any given time is assumed to be equal to the total depreciation cost up to that time. With this method, a uniform yearly deposit (d) can easily be calculated, if the estimated life, salvage value and interest on the sinking fund are known. The depreciation cost for any year is subsequently obtained by adding annual deposits and accumulated interest for that year. For an interest rate of $i\%$, the following relationships are used to calculate depreciation costs and book value of the asset:

$$B = S + \sum (\text{Future worth of depreciation till nth year})$$

$$B = S + d + d(1+i) + d(1+i)^2 + \ldots + d(1+i)^{n-1}$$

$$B = S + d[1 + (1+i) + (1+i)^2 \ldots + (1+i)^{n-1}]$$

$$B = S + d\left[\frac{(1+i)^n - 1}{(1+i) - 1}\right] = S + d\left[\frac{(1+i)^n - 1}{i}\right]$$

$$d = (B - S)\left[\frac{i}{(1+i)^n - 1}\right]$$

$$d = (B - S)(A/F, i\%, n) \tag{8.12}$$

Depreciation in 1st year $= d$

Depreciation in 2nd year $= d(1+i)$

Depreciation in 3rd year $= d(1+i)^2$

Depreciation in kth year $= d(1+i)^{k-1}$

$$d_k = d(F/P, i\%, k-1) \tag{8.13}$$

Total depreciation through kth year is obtained as:

$$D_k = d + d(1+i) + d(1+i)^2 + \ldots + d(1+i)^{k-1}$$

$$D_k = d[1 + (1+i) + (1+i)^2 + \ldots + (1+i)^{k-1}]$$

$$D_k = d\left[\frac{(1+i)^k}{i}\right] = d(F/A, i\%, k)$$

Substituting the value of d from Eqn. (8.12) we get

$$D_k = (B - S)(A/F, i\%, n)(F/A, i\%, k) \qquad (8.14)$$

$$BV_k = B - D_k$$

$$BV_k = B - [(B - S)(A/F, i\%, n)(F/A, i\%, k)] \qquad (8.15)$$

$$D = B - BV_k \qquad (8.16)$$

Example 8.5

Solve Example 8.1 by sinking fund method taking $i = 8\%$.

Solution

Given, B = ₹20,000

     $n = 10$ years

     $S = 0$

     $i = 8\%$

(a) $d_5 = ?$ (b) $D_5 = ?$ (c) $BV_5 = ?$

$$d = (₹20,000 - 0)(A/F, 8\%, 10)$$

$$= (₹20,000)(0.0690)$$

$$d = ₹1,380$$

(a) $d_5 = (₹13,800)(F/P, 8\%, 4)$

     $= (₹1,380)(1.3605)$

     $= ₹1,877.49$

(b) $D_5 = (₹20,000 - 0)(A/F, 8\%, 10)(F/A, 8\%, 5)$

     $= (₹20,000)(0.0690)(5.8666)$

     $= ₹8,095.91$

(c) $BV_5 = ₹20,000 - [(₹20,000 - 0)(A/F, i\%, n)(F/A, i\%, k)]$

     $= ₹20,000 - ₹8,095.91$

     $= ₹11,904.09$

8.3.5   The Service Output Method

This method assumes that the depreciation cost of an asset is based on the service that it will render during the useful life. Depreciation for any period is computed on the basis of service that has been rendered during that period. If,

B = first cost or original cost or unadjusted basis in rupees.

S = estimated salvage value at the end of the life of the asset in rupees.

Y = maximum capacity of service of the asset during its useful life.

y = quantity of service rendered in a period.

Then, the depreciation is computed per unit of service rendered as given below:

Depreciation per unit of service $= (B - S)/Y$

Depreciation for y units of service in a period $= \dfrac{(B - S)}{Y}\,(y)$ (8.17)

Example 8.6

A small equipment has a first cost of ₹20,000 and a 10-year estimated life. The estimated salvage value of the equipment is zero at the end of 10 years. It is expected that the equipment will be used for a total of 10,000 hours over a period of 10 years. In the fifth year of operation, the estimated usage is 600 hours and the cumulative usage by the end of year 5 is 6,000 hours. If the depreciation is based on hours of use, determine (a) depreciation cost during the fifth year, (b) cumulative depreciation cost through the fifth year and (c) book value at the end of fifth year.

Solution

Solution: Depreciation per hour $= \dfrac{(₹20,000 - 0)}{10,000 \text{ hours}} = ₹2.0$ per hour

$$d_5 = 600 \text{ hours } (₹2.0 \text{ per hour}) = ₹1,200$$

$$D_5 = 6,000 \text{ hours } (₹2.0 \text{ per hour}) = ₹12,000$$

$$BV_5 = ₹20,000 - ₹12,000 = ₹8,000$$

PROBLEMS

1. Define depreciation. What are the reasons for computing depreciation?
2. Distinguish between:
    (a) Book depreciation and tax depreciation.
    (b) Book Value and Market Value.

3. Commonwealth Infra Ltd. has just purchased a semi-automatic concrete mixer for ₹22,00,000. Its expected life is 15 years and the salvage value at the end of its useful life is ₹3,50,000. Using the double declining balance (DDB) method, compute (a) depreciation cost during the sixth year, (b) cumulative depreciation cost through the sixth year and (c) book value at the end of sixth year.

4. Kailash Automobiles has purchased a biometric system for its employees in ₹12,00,000. The expected life of the system is 15 years. The salvage value of the system at the end of its useful life is ₹20,000. Find the following, using the 8% sinking fund method of depreciation:

   (a) Depreciation at the end of fifth and eighth year.
   (b) Book value at the end of third and sixth year.

5. By each of the methods indicated below, calculate the book value of a book-binding machine at the end of 5 years if the machine originally costs ₹1,50,000 and had an estimated salvage value of ₹10,000. The useful life of the machine is 8 years.

   (a) Straight line method of depreciation.
   (b) 6% sinking fund method.

6. Ranbaxy has purchased a bottle-filling machine for ₹5,00,000. The plant engineer estimates that the machine has a useful life of 5 years and a salvage value of ₹10,000 at the end of its useful life. Using the straight line method, compute (a) depreciation cost during the third year, (b) cumulative depreciation cost through the third year and (c) book value at the end of third year.

7. Alpha Chemicals has procured a blending unit for ₹5,00,000. Its salvage value at the end of its useful life is almost negligible and can be taken as zero. Its output during the useful life is equivalent to ₹1,20,00,000. During the fifth year of its operation it gave an output equivalent to ₹10,00,000 and the cumulative output worth ₹50,00,000. Determine the depreciation for the fifth year of the operation of the unit and book value at the end of year 5 using the service output method.

8. A machine costing ₹5,00,000 is estimated to be usable for 1,000 units and then have no salvage value.

   (a) What would be the depreciation charge for a year in which 400 units were produced?
   (b) What would be the total depreciation charged after 800 units were produced?

9. An asset costs ₹2,00,000, has a useful life of 10 years and a salvage value of ₹20,000. Determine the depreciation charge for the third year and the book value at the end of third year using (a) double declining balance (DDB) method and (b) sum-of-year-digits (SYD) method.

10. Nova Auto Ltd. has purchased a car-painting plant for ₹20,00,000. Its expected life is 10 years and the salvage value at the end of its useful life is ₹1,00,000. Using the sum-of-the-year-digits (SYD) method, compute (a) depreciation cost during the third year, (b) cumulative depreciation cost through the fifth year and (c) book value at the end of sixth year.

9

Economic Evaluation of Public Sector Projects

9.1 INTRODUCTION

The projects that are owned, used and financed by the government (and hence the citizens) are classified as public sector projects. The primary purpose of such projects is to provide services to the people at no profit. Examples of projects related to public sector include hospitals and clinics, educational institutions, transportation, police and fire service, public housing, postal services, etc. It has been observed that such projects often require large initial investments, possibly distributed over several years. Public sector projects do not yield direct profit, but they require huge investments that are paid by the appropriate government unit. The successful implementation of the public sector projects provides benefits to the citizens. It is found that certain undesirable consequences, for example impact on environment, are often associated with the public sector projects. Therefore, the economic analysis of such projects must not only consider these negative consequences but also should quantify them with great accuracy. To perform an economic analysis of public sector projects, the costs (initial and annual), the benefits, and the disbenefits, if considered, must be estimated as accurately as possible in monetary units.

Costs refer to the estimated expenditures to the government body for construction, operation and maintenance of the project, less any expected salvage value.

Benefits refer to the advantages expected by the owners, i.e., the public.

Disbenefits refer to any expected undesirable consequences to the owners if the project is implemented.

Economic analysis of public sector projects is different from those of private sector projects. For public sector projects, the initial costs are usually large, the expected life is long and the sources for capital are usually a combination of taxes levied on the citizens, user fees, bond issues and private lenders. It is very difficult to make accurate estimates of benefits and disbenefits for a public sector project. The interest rates for the public sector are lower than those for private sector. The methods that are used to perform economic analysis of public sector projects are discussed in the following sections.

9.2   BENEFIT/COST ANALYSIS OF A SINGLE PROJECT

The benefit/cost (B/C) ratio is a fundamental analysis method for public sector projects. In this method, all costs and benefits estimates are converted to a common equivalent monetary unit (P.W., A.W., or F.W.) at the given interest rate. The B/C ratio is then calculated from one of the following relations:

$$B/C = \frac{P.W. \text{ of benefits}}{P.W. \text{ of costs}} = \frac{A.W. \text{ of benefits}}{A.W. \text{ of costs}} = \frac{F.W. \text{ of benefits}}{F.W. \text{ of costs}} \tag{9.1}$$

The P.W. and A.W. equivalencies are more used than F.W. values. It is important to note that the sign convention for B/C analysis is a positive sign, so the costs are preceded by a + sign. If salvage values are also estimated, then they are subtracted from costs. If disbenefit values are also estimated then they are subtracted from benefits and placed in the numerator of Equation (9.1). The decision guidelines are given below:

If B/C ≥ 1.0, accept the project as economically acceptable for the estimates and interest rate applied.

If B/C < 1.0, the project is not economically acceptable.

If the B/C value is exactly 1.0 or very near 1.0, non-economic factors will help make the decision for the best alternative.

Another version of B/C ratio, called conventional B/C ratio is calculated as:

$$B/C = \frac{P.W. \text{ of benefits} - P.W. \text{ of disbenefits}}{P.W. \text{ of costs}} = \frac{A.W. \text{ of benefits} - A.W. \text{ of disbenefits}}{A.W. \text{ of costs}} \tag{9.2}$$

In Equation (9.2) disbenefits are subtracted from benefits and not added to costs. It should be noted that if disbenefits are considered as costs, then B/C value will change considerably. For example, if numbers 12, 10 and 10 represent P.W. of benefits, P.W. of disbenefits and P.W. of costs respectively, then the correct procedure results in $B/C = \dfrac{(12-10)}{10} = 0.2$. If

disbenefits are regarded as costs, then incorrect procedure results in $B/C = \dfrac{12}{(10+10)} = 0.6$,

which is three times the correct B/C value of 0.2. Thus, it is clear that the manner in which disbenefits are handled affects the B/C value. It should be noted that, no matter whether disbenefits are correctly subtracted from the numerator or incorrectly added to costs in the denominator, a B/C ratio of less than 1.0 by the first method will always yield a B/C ratio less than 1.0 by the second method, and vice versa.

There is another way to calculate B/C ratio. The B/C ratio obtained from this method is called modified B/C ratio. The following relationship is used to calculate modified B/C ratio:

$$\text{Modified } B/C = \frac{\text{Benefits} - \text{Disbenefits} - \text{Maintenance and Operation costs}}{\text{Initial investment}} \tag{9.3}$$

In Equation (9.3), benefits, disbenefits, maintenance and operation costs, and initial investment are expressed in terms of either P.W. or A.W. or F.W. Salvage value is included in the denominator as a negative cost. The modified B/C ratio yields a different value than the conventional B/C method, but it does not change the decision to accept or reject the project.

Examples 9.1 and 9.2 illustrate the use of B/C ratio in the evaluation of projects.

Example 9.1

University Grants Commission (UGC) is planning to award a grant of ₹5,00,00,000 to Jamia Millia Islamia to upgrade its engineering education. The grant will extend over a period of 10 years and will create an estimated savings of ₹50,00,000 per year in faculty salaries and student-related expenses. UGC uses a rate of return of 5% per year on all grants award. This grant will share UGC funding with ongoing activities, so an estimated ₹2,00,000 will be removed from other program funding. To make this program successful, ₹5,00,000 per year operating cost will be incurred from the regular maintenance and operation budget. Use the B/C method to determine if the grant is economically justified.

Solution

Let us use annual worth (A.W.) as the common monetary equivalent.

$$\text{A.W. of investment cost} = ₹5,00,00,000(A/P, 5\%, 10)$$

$$= ₹5,00,00,000(0.12950)$$

$$= ₹64,75,000$$

$$\text{A.W. of benefits} = ₹50,00,000$$

$$\text{A.W. of disbenefits} = ₹2,00,000$$

$$\text{A.W. of M\&O costs} = ₹5,00,000$$

Use Equation (9.2) to obtain conventional B/C ratio as:

$$\text{B/C} = \frac{₹50,00,000 - ₹2,00,000}{₹64,75,000 + ₹5,00,000} = \frac{₹48,00,000}{₹69,75,000} = 0.69$$

The project is not justified, since B/C < 1.0.

Use Equation (9.3) to obtain modified B/C ratio as:

$$\text{Modified B/C} = \frac{₹50,00,000 - ₹2,00,000 - ₹5,00,000}{₹64,75,000} = \frac{₹43,00,000}{₹64,75,000} = 0.66$$

The project is also not justified as expected, since B/C < 1.0.

Example 9.2

The fire chief of Delhi has estimated that the initial cost of a new fire station will be ₹2,00,00,000. Annual maintenance and operation costs are estimated at ₹15,00,000. Benefits to citizens of

₹27,50,000 per year and disbenefits of ₹4,50,000 have also been identified. The discount rate is estimated to be 5% per year. Use the B/C method to determine if the fire station is economically justified.

Solution

Let us use annual worth (A.W.) as the common monetary equivalent.

A.W. of investment cost = (₹2,00,00,000)(0.05) = ₹10,00,000

A.W. of benefits = ₹27,50,000

A.W. of disbenefits = ₹4,50,000

A.W. of M&O costs = ₹15,00,000

Use Equation (9.2) to obtain conventional B/C ratio as:

$$B/C = \frac{₹27,50,000 - ₹4,50,000}{₹10,00,000 + ₹15,00,000} = \frac{₹23,00,000}{₹25,00,000} = 0.92$$

The project is not justified, since B/C < 1.0.

Use Equation (9.3) to obtain modified B/C ratio as:

$$\text{Modified B/C} = \frac{₹27,50,000 - ₹4,50,000 - ₹15,00,000}{₹10,00,000} = \frac{₹8,00,000}{₹10,00,000} = 0.8$$

The project is also not justified as expected, since B/C < 1.0.

9.3 SELECTION BETWEEN TWO MUTUALLY EXCLUSIVE ALTERNATIVES USING INCREMENTAL B/C ANALYSIS

Follow the following steps to compare two mutually exclusive alternatives by benefit/cost analysis:

Step 1: Determine the total equivalent costs for both alternatives (use either P.W. or A.W. or F.W. equivalencies).

Step 2: Order the alternatives by total equivalent cost: smaller first, then larger. Calculate the incremental cost (ΔC) for the larger-cost alternative.

Step 3: Calculate the total equivalent benefits and disbenefits estimated for both alternatives (use either P.W. or A.W. or F.W. equivalencies). Calculate the incremental benefits (ΔB) for the larger cost alternative. ΔB is calculated as:

$$\Delta B = \Delta(B - D) \tag{9.4}$$

where, B is either P.W. or A.W. or F.W. of benefits

D is either P.W. or A.W. or F.W. of disbenefits

Step 4: Calculate the incremental B/C ratio using the following relationship:

$$\Delta(B/C) = \Delta(B - D)/\Delta C \qquad (9.5)$$

Step 5: Select the higher-cost alternative, if B/C $\geq$ 1.0, otherwise select the lower-cost alternative.

Example 9.3 Illustrates the use of incremental B/C analysis to compare two mutually exclusive alternatives:

Example 9.3

Use incremental B/C analysis at an interest rate of 5% per year to determine which alternative should be selected. Use a 25-year study period.

	Alternative A	Alternative B
Initial cost (₹)	30,00,000	40,00,000
Annual M&O costs (₹/year)	2,50,000	3,50,000
Usage cost (₹/year)	3,50,000	1,50,000

Solution

Since most of the cash flows are already annualized, the incremental B/C ratio will use A.W. values. No disbenefits estimates are considered. Follow the steps of the procedure explained above.

Step 1: The A.W. of costs is the sum of initial cost and annual M&O costs.

$$A.W._A = ₹30,00,000(A/P, 5\%, 25) + ₹2,50,000 = ₹4,62,850$$

$$A.W._B = ₹40,00,000(A/P, 5\%, 25) + ₹3,50,000 = ₹6,33,800$$

Step 2: Alternative B has the larger A.W. of costs, so it is the alternative to be incrementally justified. The incremental cost value is:

$$\Delta C = A.W._B - A.W._A = ₹6,33,800 - ₹4,62,850 = ₹1,70,950$$

Step 3: The A.W. benefits are derived from the usage cost. The benefits for the B/C analysis are not the costs themselves, but the difference, if alternative B is selected. The lower usage cost each year is a positive benefit for alternative B.

$$\Delta B = Usage_A - Usage_B = ₹3,50,000 - ₹1,50,000 = ₹2,00,000$$

Step 4: The incremental B/C ratio is calculated by Equation (9.5)

$$B/C = ₹2,00,000/ ₹1,70,950 = 1.17$$

Step 5: The B/C ratio is greater than 1, indicating that the extra costs associated with alternative *B* are justified. Therefore, select alternative *B*.

9.4 SELECTION AMONG MULTIPLE MUTUALLY EXCLUSIVE ALTERNATIVES USING INCREMENTAL B/C ANALYSIS

The procedure for selecting one from more than two mutually exclusive alternatives by incremental B/C analysis is essentially the same as that of the Section 9.3. Take the following steps to compare more than two mutually exclusive alternatives by benefit/cost analysis:

Step 1: Determine the total equivalent costs for all alternatives (use either P.W. or A.W. or F.W. equivalencies for equal lives; use A.W. for unequal lives).

Step 2: Order the alternatives by total equivalent cost; smallest first.

Step 3: Calculate the total equivalent benefits (and any disbenefits estimated) for each alternative. (Use either P.W. or A.W. or F.W. equivalencies for equal lives; use A.W. for unequal lives).

Step 4: Calculate the B/C for the first ordered alternative. This makes the "Do Nothing" the defender and the first alternative the challenger. If B/C < 1.0, eliminate the challenger and go to the next challenger. Repeat this until B/C ≥ 1.0. The defender is eliminated and the next alternative is now the challenger.

Step 5: Calculate incremental costs (ΔC) and benefits (ΔB) using the following relations:

$$\Delta C = \text{Challenger cost} - \text{Defender cost} \tag{9.6}$$

$$\Delta B = \text{Challenger benefits} - \text{Defender benefits} \tag{9.7}$$

If relative usage costs are estimated for each alternative, rather than direct benefits, ΔB may be found using the relation:

$$\Delta B = \text{Defender usage costs} - \text{Challenger usage costs} \tag{9.8}$$

Step 6: Calculate the incremental B/C for the first challenger compared to the defender.

$$\text{B/C} = \Delta B / \Delta C \tag{9.9}$$

If incremental B/C ≥ 1.0 in Equation (9.9), the challenger becomes the defender and the previous defender is eliminated. Conversely, if B/C < 1.0, remove the challenger and the defender remains against the next challenger.

Step 7: Repeat Steps 5 and 6 until only one alternative remains. This is the selected alternative.

In all the steps above, incremental disbenefits may be considered by replacing ΔB with $\Delta(B - D)$, as the conventional B/C ratio.

Example 9.4 illustrates the use of incremental B/C analysis to compare more than two mutually exclusive alternatives.

Example 9.4

A pharmaceutical company is considering using any one of the following four methods to control chemical fume leakage into the surrounding air from a mixing machine. The estimated costs and benefits, in terms of reduced employee health costs, are given for each method. Assuming that all methods have a 10-year life and no salvage value, determine which one should be selected using a M.A.R.R. of 10% per year and the B/C method.

	Method 1	Method 2	Method 3	Method 4
Installed cost (₹)	75,000	95,000	1,25,000	1,65,000
Annual M&O costs (₹/year)	50,000	60,000	45,000	55,000
Benefits (₹/year)	75,000	1,00,000	95,000	1,10,000

Solution

Step 1: Find the A.W. of costs for each method, order them and determine the incremental B/C values.

$$\text{A.W. of costs} = \text{installed cost}(A/P, 15\%, 10) + \text{A.O.C.}$$

Method	A.W. of cost calculation
1	₹75,000(A/P, 10%,10) + ₹50,000 = ₹75,000(0.16275) + ₹50,000 = ₹62,206.25
2	₹95,000(A/P, 10%,10) + ₹60,000 = ₹95,000(0.16275) + ₹60,000 = ₹75,461.25
3	₹1,25,000(A/P, 10%,10) + ₹45,000 = ₹1,25,000(0.16275) + ₹45,000 = ₹65,343.75
4	₹1,65,000(A/P, 10%,10) + ₹55,000 = ₹1,65,000(0.16275) + ₹55,000 = ₹81,853.75

Step 2: Order of incremental analysis is: "Do Nothing", i.e., DN, Method 1, Method 3, Method 2, Method 4.

Step 3: A.W. of benefits of Method 1, 2, 3 and 4 is given in the problem as ₹75,000, ₹1,00,000, ₹95,000 and ₹1,10,000 respectively.

Step 4, Step 5, Step 6:

Method 1 vs. DN

$$B/C = ₹75,000/₹62,206.25$$

$$= 1.21 > 1.0 \qquad \text{Eliminate DN, keep Method 1.}$$

Method 3 vs. 1

$$\Delta C = ₹65,343.75 - ₹62,206.25 = ₹3,137.5$$

$$\Delta B = ₹95,000 - ₹75,000 = ₹20,000$$

$$\Delta B/C = ₹20,000/₹3,137.5$$

$$= 6.37 > 1 \qquad \text{Eliminate Method 1, keep 3.}$$

Method 2 vs. 3

$$\Delta C = ₹75,461.25 - ₹65,343.75 = ₹10,117.5$$

$$\Delta B = ₹1,00,000 - ₹95,000 = ₹5,000$$

$$\Delta B/C = ₹5,000/₹10,117.5$$

$$= 0.49 < 1.0 \qquad \text{Eliminate Method 2, keep 3.}$$

Method 4 vs. 3

$$\Delta C = ₹81,853.75 - ₹65,343.75 = ₹16,510$$

$$\Delta B = ₹1,10,000 - ₹95,000 = ₹15,000$$

$$\Delta B/C = ₹15,000/₹16,510$$

$$= 0.91 < 1.0 \qquad \text{Eliminate Method 4, keep 3}$$

Step 7:

Select Method 3.

PROBLEMS

1. Haryana government is considering extending Bhakra canal into a desert area for irrigation. The initial cost of the project is expected to be ₹3,75,00,000 with annual maintenance costs of ₹6,25,000 per year. (a) If agricultural revenue is expected to be ₹47,00,000 per year, do a B/C analysis to determine whether the project should be undertaken, using a 15-year study period and a discount rate of 5% per year. (b) Rework the problem, using the modified B/C ratio.

2. The annual cash flows (C) for a public project is estimated to be ₹1,12,50,000 per year, benefits (B) of ₹1,50,00,000 per year and disbenefits (D) of ₹25,00,000 per year. Determine the (a) B/C ratio and (b) value of B – C.

3. The B/C ratio for new flood-control project along the banks of Yamuna river is required to be 1.2. If the benefit is estimated at ₹1,50,00,000 per year and the maintenance cost is expected to be ₹75,00,000 per year, what is the allowed maximum initial cost of the project? The discount rate is 6% per year and a project life of 50 years is expected.

4. The modified B/C ratio for a municipal corporation project is 1.5. If the initial cost is ₹2,50,00,000 and the annual benefits are ₹37,50,000, what is the amount of the annual M&O costs used in the calculation, if a discount rate of 5% is considered and the estimated life is 25 years?

5. Delhi Government is planning a hydroelectric project for Yamuna river basin primarily for the generation of electric power. However, this project will also provide flood control, irrigation and recreation benefits. The estimated costs and benefits associated with the project are given below. Apply (a) conventional B/C ratio and (b) modified B/C ratio to determine whether the project is economically justified. Use a discount rate of 10% per year.

Item	Cash Flow
Initial cost (₹)	54,00,00,000
Annual operating and maintenance costs (₹/year)	1,50,00,000
Annual power sales (₹/year)	3,60,00,000
Annual flood control savings (₹/year)	1,50,00,000
Annual irrigation benefits (₹/year)	2,40,00,000
Annual recreation benefits (₹/year)	1,20,00,000
Project life (years)	30

6. Calculate the B/C ratio for the following cash flow estimates at a discount rate of 5% per year:

Item	Cash Flow
P.W. of benefits (₹)	9,50,00,000
A.W. of disbenefits (₹/year)	11,25,000
First cost (₹)	5,50,00,000
Maintenance & Operation costs (₹/year)	75,00,000
Project life (years)	20

7. Two mutually exclusive alternatives are being considered for investment. The cost data and benefit values of these alternatives are given below. Use incremental B/C analysis at an interest rate of 8% per year to determine which alternative should be selected. Use a 20-year study period.

	Alternative A	Alternative B
Initial cost (₹)	1,50,00,000	2,00,00,000
Annual M&O costs (₹/year)	12,50,000	15,50,000
Benefits (₹/year)	17,50,000	7,50,000

8. Solar and conventional alternatives are being considered for providing energy to a far-off site. The data pertaining to these alternatives is given below. Use the incremental B/C method to determine which should be selected at a discount rate of 1% per month over a 5-year study period.

	Solar	Conventional
Initial cost (₹)	11,25,00,000	5,00,00,000
M&O costs (₹/month)	2,50,000	12,50,000
Salvage value (₹)	37,50,000	0

9. The Central Government is considering three sites for mineral extraction. The cash flows associated with each site are given below. Use the B/C method to determine which site, if any, is best, if the extraction period is limited to 5 years and the interest rate is 8% per year.

	Site A	Site B	Site C
Initial cost (₹)	1,25,00,00,000	2,25,00,00,000	5,00,00,00,000
Annual cost (₹/year)	7,50,00,000	10,00,00,000	15,00,00,000
Annual benefits (₹/year)	50,00,00,000	72,50,00,000	1,52,50,00,000
Annual disbenefits (₹/year)	1,25,00,000	3,75,00,000	5,25,00,000

10. A state government is planning a hydroelectric project for Yamuna river basin primarily for the generation of electric power. However, this project will also provide various other benefits. The estimated costs and benefits associated with three alternatives under consideration are given below. If the interest rate is 8% per year and the life of alternatives is estimated to be 40 years, determine which alternative should be selected using B/C ratio analysis.

	Alternative A	Alternative B	Alternative C
Initial cost (₹)	30,00,00,000	50,00,00,000	80,00,00,000
Annual O&M cost (₹/year)	40,00,000	50,00,000	70,00,000
Annual power sales (₹/year)	2,00,00,000	2,40,00,000	3,60,00,000
Annual benefits (₹/year)	1,40,00,000	2,00,00,000	2,90,00,000

❑❑❑

10

Economics Study Considering Inflation

10.1 INTRODUCTION

Inflation is observed nearly every day in personal and professional life. The government, units, business and industrial corporations closely watch and analyze the impact of inflation rate. When inflation is considered, an engineering economics study yields different results compared to one in which it is not considered. In the last few years, inflation has been a major concern in India and in most industrialized nations. Inflation rate is sensitive to real as well as perceived factors of the economy. Factors such as the cost of energy, interest rates, availability and cost of skilled people, scarcity of materials, political instability and money supply have significant short-term and long-term impacts on the inflation rate. Owing to the significance of inflation, it is necessary to incorporate the effects of inflation into an economic analysis. This chapter covers the basic techniques that integrate inflation into economic analysis.

10.2 EFFECTS OF INFLATION

We understand very well that ₹1,000 now does not have the same purchasing power as it had few years ago. In other words, ₹1,000 now cannot purchase the same amount of a commodity as ₹1,000 did a few years ago. This happens due to inflation. Inflation is defined as an increase in the amount of money necessary to obtain the same amount of a commodity or service before the inflated price was present. Inflation occurs because the value of currency has gone down and, consequently, it takes more money to purchase the same amount of goods or services. In order to compare the monetary amounts that occur in different time-periods, the different-valued currency must first be converted to constant-value currency so that they represent the same purchasing power over time.

Money in one period of time t_1 can be brought to the same value as money in another period of time t_2 by using the following equation:

$$\text{Rupees in period } t_1 = \frac{\text{Rupees in period } t_2}{\text{Inflation rate between } t_1 \text{ and } t_2} \tag{10.1}$$

Rupees in period t_1 are called constant-value rupees or today's rupees. Rupees in period t_2 are called future rupees. If f represents the inflation rate per period (years) and n is the number of time periods (years) between t_1 and t_2, Equation (10.1) is

$$\text{Constant value rupees} = \text{Today's rupees} = \frac{\text{Future rupees}}{(1+f)^n} \qquad (10.2)$$

$$\text{Future rupees} = \text{Today's rupees } (1+f)^n \qquad (10.3)$$

As an illustration, consider the price of a pizza as:

₹25   on March 2009

If inflation averaged 5% during the previous year, this cost is last year's equivalent of:

₹25/(1.05) = ₹23.81   on March 2008

A predicted price in 2010 is:

₹25(1.05) = ₹26.25   on March 2010

If inflation averages 5% over the next 5 years, Equation (10.3) is used to predict the price of pizza in 2014:

₹25(1.05)^5 = ₹31.91   on March 2014

This is a 27.64% increase over the 2009 price at 5% inflation. If inflation averages 8% per year, the pizza cost in 5 years will be ₹36.73, an increase of 46.92%. In some parts of the world, hyperinflation may average 50% per year. In such an unfortunate economy, the pizza price in 5 years rises from ₹25 to ₹189.84. This is why countries experiencing hyperinflation must devalue their currency by factors of 100 and 1,000 when unacceptable inflation rates persist.

In industrial or business environment, at a reasonably low inflation rate averaging 5% per year, equipment or services with a first cost of ₹5,00,000 will increase by 27.64% to ₹6,38,200 over a 5-year-span.

It is important to understand the meaning of the following three different rates:

Real or Inflation-free Interest rate i   This represents the rate at which interest is earned when the effects of inflation have been removed. The real interest rate presents an actual gain in purchasing power.

Inflation-adjusted Interest rate i_f   This is the interest rate that has been adjusted to take inflation into account. The market interest rate, which we hear every day, is an inflation-adjusted rate. This rate is, in fact, a combination of the real interest rate i and the inflation rate f and, therefore, it changes as the inflation rate changes. It is also known as the inflated interest rate.

Inflation rate f   It is a measure of the rate of change in the value of currency.

Deflation is the opposite of inflation. In the presence of deflation, the purchasing power of the monetary unit is greater in the future than at present and therefore, it will take fewer rupees in the future to buy the same amount of goods or services as it does today. Computations considering deflation can be done by using Equations (10.2) and (10.3), but the deflation rate is used as $-f$ in these equations. For example, if deflation is estimated to be 3% per year, an asset that costs ₹5,00,000 today would have a first cost 5 years from now as:

$$₹5,00,000(1-f)^n = ₹5,00,000(1-0.03)^5 = ₹4,29,367.$$

10.3   PRESENT WORTH CALCULATIONS ADJUSTED FOR INFLATION

We have seen in Chapter 3 that the relationship between present worth (P) and future worth (F) is:

$$P = F\frac{1}{(1+i)^n}$$

In the above relation, F is a future-rupees amount with inflation built in. F can be converted into today's rupees by using Equation (10.2).

$$P = \frac{F}{(1+f)^n}\frac{1}{(1+i)^n}$$

$$= F\frac{1}{(1+i+f+if)^n} \tag{10.4}$$

If the term $i+f+if$ is defined as i_f, the equation becomes:

$$P = F\frac{1}{(1+i_f)^n} = F(P/F, i_f, n) \tag{10.5}$$

The symbol i_f is called the inflation-adjusted interest rate and is defined as

$$i_f = i + f + if \tag{10.6}$$

where i = real interest rate

f = inflation rate

For a real interest rate of 10% per year and inflation rate of 5% per year, Equation (10.6) yields an inflated interest rate of 15.5%.

$$i_f = 0.10 + 0.05 + (0.10)(0.05) = 0.155$$

Example 10.1

An engineering alumni of Jamia Millia Islamia wishes to donate to scholarship fund of the faculty of Engineering and Technology. The following three options are available:
Plan 1   ₹15,00,000 now.
Plan 2   ₹3,75,000 per year for 10 years beginning 1 year from now.
Plan 3   ₹12,50,000 five years from now and another ₹20,00,000 five years from now.
The Dean of the faculty wants to select the plan that maximizes the buying power of the rupees received. If the donation earns a real interest rate of 10% per year and the inflation rate is expected to average 5% per year, which plan should be accepted? Evaluate the plans by taking into account the inflation.

Solution

The inflation-adjusted interest rate i_f is calculated as:

$$i_f = 0.10 + 0.05 + (0.10)(0.05) = 0.155$$

$$(P.W.)_{\text{Plan 1}} = ₹15,00,000$$

$$(P.W.)_{\text{Plan 2}} = ₹3,75,000(P/A, 15.5\%, 10)$$

$$= ₹3,75,000(4.926)$$

$$(P.W.)_{\text{Plan 2}} = ₹18,47,250$$

$$(P.W.)_{\text{Plan 3}} = ₹12,50,000(P/F, 15.5\%, 5) + ₹20,00,000(P/F, 15.5\%, 5)$$

$$= ₹12,50,000(0.4867) + ₹20,00,000(0.4867)$$

$$(P.W.)_{\text{Plan 3}} = ₹15,81,775$$

Since $(P.W.)_{\text{Plan 2}}$ is the largest in today's rupees, select Plan 2.

Example 10.2

A 15-year bond of ₹5,00,000 that has a dividend rate of 5% per year, payable semiannually, is currently for sale. If the expected rate of return of the purchaser is 8% per year, compounded semiannually and if the inflation rate is expected to be 1.5% for each 6-month period, what is the bond worth now (a) without an adjustment for inflation and (b) when inflation is considered?

Solution

(a) Without inflation adjustment: The semi-annual dividend is $I = [(₹5,00,000)(0.05)]/2 = ₹12,500$. At a nominal 4% per 6 months for 30 periods, the P.W. is:

$$P.W. = ₹12,500(P/A, 4\%, 30) + ₹5,00,000(P/F, 4\%, 30)$$

$$= ₹12,500(17.2920) + ₹5,00,000(0.3083)$$

$$P.W. = ₹3,70,300$$

(b) With inflation: Use the inflated rate i_f:

$$i_f = 0.04 + 0.015 + (0.04)(0.015) = 0.0556$$

$$P.W. = ₹12,500(P/A, 5.6\%, 30) + ₹5,00,000(P/F, 5.6\%, 30)$$

$$= ₹12,500(14.4079) + ₹5,00,000(0.1970)$$

$$P.W. = ₹2,78,598.80$$

Example 10.3

The manager of a company wishes to calculate present worth (P.W.) of a small project with estimated costs of ₹1,75,000 now and ₹35,000 per year for 5 years beginning 1 year from now, with increase of 10% per year thereafter for the next 10 years. The real interest rate is 12% per year. Calculate the present worth (a) without an adjustment for inflation and (b) considering inflation at a rate of 8% per year.

Solution

(a) Figure 10.1 shows the cash flow diagram. The P.W. without an adjustment for inflation is found using $i = 12\%$ and $g = 10\%$.

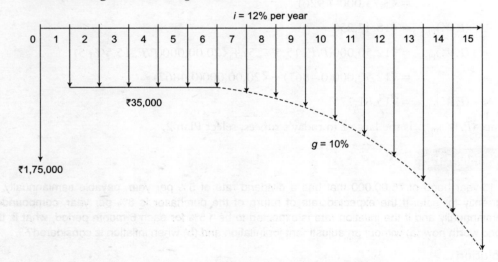

Figure 10.1   Cash flow diagram, Example 10.3

$$P.W. = -₹1,75,000 - ₹35,000(P/A, 12\%, 4) - \left\{ \dfrac{₹35,000\left[1 - \left(\dfrac{1.10}{1.12}\right)^{11}\right]}{0.12 - 0.10} \right\}(P/F, 12\%, 4)$$

$$= -₹1,75,000 - ₹35,000(3.0373) - (₹3,14,645)(0.6355)$$

$$P.W. = -₹4,81,262$$

To adjust for inflation, calculate the inflated interest rate i_f:

$$i_f = i + f + if = 0.12 + 0.08 + (0.12)(0.08) = 0.2096$$

$$P.W. = -₹1,75,000 - ₹35,000(P/A, 20.96\%, 4)$$

$$-\left\{ \frac{₹35,000\left[1 - \left(\dfrac{1.10}{1.2096}\right)^{11}\right]}{0.2096 - 0.10} \right\}(P/F, 20.96\%, 4)$$

$$= -₹1,75,000 - ₹35,000(2.5431) - (₹2,07,007)(0.4675)$$

$$P.W. = -₹3,60,784$$

It is clear from the result of the above example that the present value of future inflated rupees is reasonably less when inflation adjustment is considered. Thus, in a high-inflation economy, it is advantageous for the borrower to use future (inflated) rupees whenever possible to make payments to repay a loan.

10.4 FUTURE WORTH CALCULATIONS ADJUSTED FOR INFLATION

If future worth F is the actual amount of money accumulated at time n, then it is obtained using the inflation adjusted interest rate from Equation (10.7). It should be noted that the inflation adjusted interest rate is also known as market rate.

$$F = P(1 + i_f)^n = P(F/P, i_f, n) \tag{10.7}$$

For example, if market rate (including inflation) is 12%, ₹10,000 invested now for 10 years will accumulate to:

$$F = ₹10,000(F/P, 12\%, 10) = ₹10,000(3.1058) = ₹31,058$$

If F represents the purchasing power of the actual amount accumulated at time n, then it is obtained from the following relation:

$$F = \frac{P(1 + i_f)^n}{(1 + f)^n} = \frac{P(F/P, i_f, n)}{(1 + f)^n} \tag{10.8}$$

For example, consider the same ₹10,000 now, a 12% per year market rate and inflation rate of 5% per year. In 10 years, the purchasing power has risen, but only to ₹19,066.85.

$$F = \frac{₹10,000(F/P, 12\%, 10)}{(1.05)^{10}} = \frac{₹31,058}{1.6289} = ₹19,066.85$$

This is ₹11,991.15 (38.61%) less than the ₹31,058 actually accumulated at 12%. Thus, it can be concluded that 5% inflation over 10 years reduces the purchasing power of money by 38.61%.

Alternately, the future amount of money accumulated with today's buying power is determined by calculating the real interest rate and using it in the F/P factor to compensate for the decreased purchasing power of the rupees. This real interest rate is the i in Equation (10.6).

$$i_f = i + f + if$$
$$= i(1 + f) + f$$
$$i = \frac{i_f - f}{1 + f} \tag{10.9}$$

The real interest rate i represents the rate at which today's rupees expand with their same purchasing power into equivalent future rupees. An inflation rate larger than the market interest rate leads to a negative real interest rate. The use of this interest rate is appropriate for calculating the future worth. For the example of ₹10,000 in today's rupees from Equation (10.9):

$$i = \frac{0.12 - 0.05}{1 + 0.05} = 0.0666, \text{ or } 6.66\%$$

$$F = ₹10,000(F/P, 6.66\%, 10) = ₹10,000(1.9072) = ₹19,072$$

If F is required to be calculated without considering interest rate but considering only the inflation rate, then the following relation is used:

$$F = P(1 + f)^n = P(F/P, f, n) \tag{10.10}$$

Let us once again consider the ₹10,000 used previously. If this amount is rising at the inflation rate of 5% per year, the amount 10 years from now will be:

$$F = ₹10,000(F/P, 5\%, 10) = ₹10,000(1.6289) = ₹16,289$$

If F is to be calculated for a situation in which both inflation rate and real interest rates are considered, then we must use the market (inflation-adjusted) rate. The market rate is determined as:

$$i_f' = i + f + if$$

where $i = \dfrac{i_f - f}{1 + f}$

Thus for the same ₹10,000 amount,

$$i = \frac{0.12 - 0.05}{1 + 0.05} = 0.0666$$

$$i'_f = 0.0666 + 0.05 + (0.0666)(0.05) = 0.1199 = 11.99\% \approx 12\%$$

Thus, $F = P(F/P, i'_f, n) = ₹10,000(F/P, 12\%, 10) = ₹10,000(3.1058) = ₹31,058$

The result shows that ₹31,058 ten years in the future will be equivalent to ₹10,000 now with a real return of $i = 6.66\%$ per year and inflation of $f = 5\%$ per year.

Example 10.4

A company wants to buy an equipment. The management of the company is considering the following two plans:

Plan A   Buy now

Plan B   Buy later

If the company selects plan A, the equipment will be purchased now for ₹10,00,000. However, if the company selects plan B, the purchase will be deferred for 4 years when the cost is expected to ₹17,00,000. The company expects a real M.A.R.R. of 10% per year. The average inflation rate in the country is 8% per year. Determine whether the company should purchase now or later (a) when inflation is not considered and (b) when inflation is considered.

Solution

(a) *Inflation not considered*: Calculate the *F.W.* value for plan A at an interest rate $i = $ M.A.R.R. $= 10\%$ and $n = 4$ years. Compare the calculated *F.W.* value with the expected cost of plan B, i.e., ₹17,00,000 and select the one having the lower cost.

$$(F.W.)_{\text{Plan A}} = -₹10,00,000(F/P, 10\%, 4) = -₹10,00,000(1.4641) = -₹14,64,100$$

$$(F.W.)_{\text{Plan B}} = -₹17,00,000$$

Select plan A, i.e., purchase now.

(b) *Inflation considered*: First calculate the inflation-adjusted M.A.R.R. as:

$$i_f = i + f + if = 0.10 + 0.08 + (0.10)(0.08) = 0.188$$

Use i_f to compute F.W. for Plan A:

$$(F.W.)_{\text{Plan A}} = -₹10,00,000(F/P, 18.8\%, 4) = -₹10,00,000(1.9927) = -₹19,92,700$$

$$(F.W.)_{\text{Plan B}} = -₹17,00,000$$

Select plan B, i.e., purchase later.

10.5 CAPITAL RECOVERY CALCULATIONS ADJUSTED FOR INFLATION

Due to inflation, future rupees will have less purchasing power than today's rupees and, therefore, more rupees will be required to recover the present investment. Thus, when inflation is considered, it is necessary to use inflated interest rate in the A/P formula. For example, if ₹10,000 is invested today at a real interest rate of 12% per year when the inflation rate is 7% per year, the equivalent amount that must be recovered each year for 5 years in future rupees is calculated as:

$$i_f = i + f + if = 0.12 + 0.07 + (0.12)(0.07) = 0.1984$$

$$A = P(A/P, i_f, 5) = ₹10,000(A/P, 19.84\%, 5) = ₹10,000(0.33321) = ₹3332.10$$

It should also be noted that one can invest fewer rupees in the present to accumulate a specified amount of future (inflated rupees). This suggests the use of a higher interest rate, i.e., i_f rate, to produce a lower value in the A/F formula. For example, the annual equivalent (inflation adjusted) of $F = ₹10,000$ five years from now in future rupees is:

$$A = ₹10,000(A/F, 19.84\%, 5) = ₹10,000(0.13481) = ₹1348.10$$

Example 10.5

What annual deposit is required for 5 years to accumulate an amount of money with the same purchasing power as ₹1,000 today, if the market interest rate is 12% per year and inflation is 7% per year?

Solution

First, find the actual number of future (inflated) rupees required 5 years from now.

$$F = (\text{present buying power})(1 + f)^5 = ₹1,000(1.07)^5 = ₹1,402.55$$

The actual amount of the annual deposit is calculated using the market (inflated) interest rate of 12%:

$$A = ₹1,000(A/F, 12\%, 5) = ₹1,402.55(0.15741) = ₹220.77$$

PROBLEMS

1. Convert ₹50,000 future rupees in year 8 into constant-value rupees of today if the inflation-adjusted (market) interest rate is 12% per year and the real interest rate is 5% per year.

2. Convert ₹50,000 future rupees in year 8 into constant-value rupees of today if the inflation-adjusted (market) interest rate is 12% per year and the inflation rate is 6% per year.

3. Tata Motors announced that the price of Nano car is going to increase by only the inflation rate for the next 3 years. If the current price of a car is ₹1,20,000 and the inflation rate is expected to average 5% per year, what is the expected price 3 years from now?

4. The cost of a machine was ₹2,50,000 five years ago. If a similar machine costs ₹3,00,000 now and its price increased only by the inflation rate, what was the annual inflation rate over that 5-year period?

5. What annual inflation rate is implied from a market interest rate of 12% per year when the real interest rate is 5% per year?

6. What is the real interest rate per quarter when the market interest rate is 24% per year, compounded quarterly and the inflation rate is 2.5% per quarter?

7. Compare the following two alternatives on the basis of their present worth values, using a real interest rate of 8% per year and an inflation rate of 5% per year, (a) without any adjustment for inflation and (b) with inflation considered.

	Alternative 1	Alternative 2
First cost, ₹	−1,50,000	−2,40,000
Annual operating cost, ₹/year	−1,40,000	−1,00,000
Salvage value, ₹	25,000	40,000
Life, Years	5	5

8. A company is considering to purchase a machine for its operations. The following two alternatives are available. The company has a M.A.R.R. of a real 12% per year and it expects inflation to be 5% per year. Use *P.W.* analysis to determine which machine should be purchased.

	Machine 1, Future rupees	Machine 2, Today's rupees
First cost, ₹	− 3,00,000	− 4,75,000
Annual operating cost, ₹/year	− 2,75,000	− 1,75,000
Life, Years	10	10

9. A company wants to set aside money now so that it will be able to purchase new equipment 5 years from now. The price of the equipment is expected to increase only by the inflation rate of 5% per year for each of the next 5 years. If the total cost of the equipment now is ₹2,25,000, determine (a) the expected cost of the equipment after 5 years and (b) how much the company will have to set aside now, if it earns interest at a rate of 10% per year.

10. If you make an investment in real estate that is guaranteed to net you ₹75 lakhs, 20 years from now, what will be the buying power of that money with respect to today's rupees if the market interest rate is 10% per year and the inflation rate is expected to be 4% over the time period?

11. What is the annual worth in then-current rupees in years 1 through 5 years of a receipt of ₹37,50,000 now, if the market interest rate is 8% per year and the inflation rate is 4% per year?

12. Compare the following two alternatives on the basis of annual worth analysis, using M.A.R.R. equal to a real interest rate of 10% per year and an inflation rate of 5% per year, if the estimates are (a) constant-value rupees and (b) future rupees.

	Alternative 1	Alternative 2
First cost, ₹	−7,50,000	−51,25,000
Annual operating cost, ₹/year	−3,50,000	−25,000
Salvage value, ₹	2,00,000	10,00,000
Life, Years	5	∞

□□□

11

Make or Buy Decision

INTRODUCTION

Every business organization engaged in the manufacturing of tangible products faces the problem to decide whether it is more advantageous to make a particular item in-house, or to buy it from a supplier. The choice involves aspects which are both qualitative such as quality control, and quantitative such as the relative cost factors. In order to decide whether to make or buy a product, the company generally analyzes the capacity of its facilities and the cost at which the product can be bought. Subsequently, the option that involves the least cost is selected by the company. Buying option often appears to be feasible and beneficial for a company which manufactures low volumes of product. In situations where the total cost of buying a product is more than or equal to the total cost of making it, it is recommended to make the product within the company.

11.2 ## FEASIBLE ALTERNATIVES FOR LAUNCHING NEW PRODUCTS

It is a natural tendency of a business organization to thrive and grow in order to earn higher profits. Some of the most efficient ways in which this can be done include (i) backward integration and (ii) forward integration. In both these approaches new product(s) is introduced by the organization. To implement its new preposition, the company may adopt any one of the following strategies:

- Products may be manufactured in-house with the existing facility.
- Products may be purchased from external suppliers or vendors.
- A few components of the products may be made in-house, a few others may be bought from the outside vendors and finally the product may be assembled by the company.

Although the alternatives are limited in number and appear simple, but the decision is strategically critical because it has far-reaching effects on the ultimate success or failure of the new preposition. In case the company decides to make the product (entire product or part of it) in-house, it involves the following:

- Fresh investments in terms of new machines, construction of new buildings, jigs and fixtures, raw materials, and manpower.

- Involvement of more number of work hours of their existing machines, equipment, and manpower that may be critical to the company's existing product line.
- The decision needs to be weighed in terms of profits gained by the new product viz-a-viz the fresh expenditures. If the decision has not been weighed properly, it may not only lead to failure of the new preposition, but may also have undesirable effects on its existing line of products.

The decision to make or buy may be made by carrying out a systematic analysis of many factors that include:

- Nature of demand of the product, i.e., whether seasonal, fluctuating, temporary or permanent and continuous.
- The annual expected product volume.
- Whether the tools, jigs and fixtures, equipment required for its production, inspection and assembly are readily available or not.
- Is there any need for new manufacturing techniques?
- Does the company have past experience in all the manufacturing techniques and skilled personnel who are needed in the new product?
- Does it demand special production personnel or machines and are these already available in the company? If already available, then can these be engaged more profitably and fruitfully elsewhere?
- Does the product need frequent design and drawing changes?
- Is there any involvement of any copyright items, patented design or methods of manufacturing?
- Is the product or some of its components proprietary items?
- Are the vendors available who have all the facilities, machines, raw materials, equipment and personnel required for the product?
- What is the reliability of the vendor in terms of the delivery of the product in required lead time and quantity?

11.3 DECISIVE FACTORS FOR MAKE OR BUY DECISION

The criteria that need to be considered to arrive at a suitable decision to either make or buy product(s) are discussed in the following section.

11.3.1 Criteria for Make Decision

The following conditions favor the decision to make the entire or part of the product in-house:

- All the facilities required for making the product in-house including machines, tools, jigs and fixtures and personnel exists in the company and the product can be made in-house cheaper.
- The product is proprietary and/or requires patented technology and requires very strict quality control.
- There is a long-term demand for the product.

11.3.2   Criteria for Buy Decision

Buy decision is opted under the following conditions:
- Non availability of the facilities that are required to make the product(s).
- Non availability of sufficient capital to buy machines, equipment and other resources to be used in the manufacturing of the product(s).
- Non availability of skilled workers who can make the product(s).
- Better and economical utilization of the existing facilities in the manufacturing of other parts and/or products.
- Low demand for the product(s).
- Existence of a patent or other legal barriers that prevent the company from making product(s).

11.4   TECHNIQUES USED TO ARRIVE AT MAKE OR BUY DECISION

The following techniques are generally used to arrive at either make or buy decision:
- Simple cost Analysis
- Economic Analysis
- Break-even Analysis

11.4.1   Simple Cost Analysis

This analysis involves determination of cost associated with making a product and the cost of buying it from an external vendor. Subsequently, the option that results in the minimum cost is selected and implemented. Example 11.1 illustrates the concept of this approach.

Example 11.1

A Faridabad-based automobile manufacturing unit buys gears from one of its vendors for ₹2,800 per unit. The company has realized that it has enough extra capacity to produce the gears in-house. The cost estimates associated with producing the gears in-house are given below:

Cost of raw material - ₹175 per unit

Labor cost - ₹250 per unit

Variable overhead cost - ₹75

The annual fixed cost associated with the unused capacity is ₹15,00,000. The demand for gears over the next year is estimated to be 25,000 units.

(a) Is it profitable for the company to make the gears?

(b) If the existing capacity of the company is used to produce some other parts, then this would cover its fixed and variable cost and contribute ₹3,00,000 to profit. What would be more profitable: gear production or production of other parts?

Solution

(a) Let us assume that the unused capacity has alternative use.

Cost to make:

Variable cost/unit = Raw material cost + Labor cost + Variable overhead cost

$$= ₹175 + ₹250 + ₹75 = ₹500$$

Total variable cost = (25,000 units) × (₹500/unit) = ₹1,25,00,000

Total cost = Fixed cost + Total variable cost

$$= ₹15,00,000 + ₹1,25,00,000$$

$$= ₹1,40,00,000$$

Cost to buy:

Purchase cost = (25,000 units) × (₹2,800) = ₹7,00,00,000

Total cost = Fixed cost + Purchase cost

$$= ₹15,00,000 + ₹7,00,00,000$$

$$= ₹7,15,00,000$$

From the above calculations, it is clear that the cost of making the gears is much less than that of buying them from outside and, therefore, making of gears in-house is advantageous.

(b) Here the company has the following options:

- Make the gears in-house
- Buy the gears from external vendor and utilize the existing unused capacity to manufacture other products

The computation of cost for each of the above options is shown in the following table:

	Make Gears	Buy Gears and Make Other Products
Total variable cost (₹)	1,25,00,000	7,00,00,000
Fixed cost (₹)	15,00,000	0
Total cost (₹)	1,25,00,000 + 15,00,000 = 1,40,00,000	7,00,00,000 + 0 = 7,00,00,000
Contribution to profit (₹)	0	3,00,000
Net cost (₹)	1,40,00,000 − 0 = 1,40,00,000	7,00,00,000 − 3,00,000 = 6,97,00,000

From the above calculations, it is clear that the net cost of make option is less than that of buy option and therefore, it is advisable to make gears in-house.

11.4.2   Economic Analysis

Economic analysis involves computation of the economic order quantity (EOQ) and total cost for make and buy alternatives based on the inventory control concept. The alternative with the minimum cost is selected and implemented. The following formulae are used to calculate EOQ and to find the total cost for make and buy alternatives:

Formula for EOQ and total cost for make option:

$$Q_P = \sqrt{\frac{2C_s D}{C_c (1 - D/r)}} \tag{11.1}$$

$$TC_P = D \times P + \frac{DC_o}{Q_P} + C_c(r - D)\frac{Q_P}{2 \times r} \tag{11.2}$$

Formula for EOQ and total cost for buy option:

$$\theta_0 = \sqrt{\frac{2C_o D}{C_c}} \tag{11.3}$$

$$TC_0 = D \times P + \frac{DC_o}{Q_o} + \frac{Q_o \times C_c}{2} \tag{11.4}$$

where, D = Demand/year

   P = Purchase price/unit

   C_c = Carrying cost/unit/year

   C_o = Ordering cost/unit

   C_s = Setup cost/setup

   r = Production rate/unit per year

   Q_P = Economic production size

   Q_o = Economic order size

   TC_P = Total cost of making economic production size/year

   TC_o = Total cost of economic order size/year

Example 11.2

OK Play Ltd. manufactures an article from a polymer, the annual demand for which is 20,000 units. Various cost elements in respect of make or buy option of the article are as follows. Recommend the best option.

	Buy	Make
Unit Cost of Item (₹)	85	52
Procurement cost/order (₹)	750	
Set-up cost/set-up (₹)		1,000
Carrying cost/per item/per year (₹)	25	12
Annual rate of Production		35,000 units

Solution

Make option:

$C_s = ₹1,000/\text{set-up}$

$C_c = ₹12/\text{unit/year}$

$D = 20,000$ units/year

$r = 35,000$ units/year

$P = ₹52/\text{unit}$

$$Q_P = \sqrt{\frac{2C_s D}{C_c(1 - D/r)}}$$

$$Q_P = \sqrt{\frac{2 \times 1,000 \times 20,000}{12(1 - 20,000/35,000)}} = 2,788.867 \approx 2,789 \text{ units (approximately)}$$

$$TC_P = D \times P + \frac{DC_0}{Q_P} + C_c(r - D)\frac{Q_P}{2 \times r}$$

$$= 20,000 \times 52 + \frac{20,000 \times 12}{2,789} + 12 \times (35,000 - 20,000) \times \frac{2,789}{2 \times 35,000}$$

$$= ₹10,47,257.77$$

Buy option:

$C_0 = ₹750/\text{order}$

$C_c = ₹25/\text{unit/year}$

$D = 20,000$ units/year

$P = ₹85/\text{unit}$

$$Q_0 = \sqrt{\frac{2C_0 D}{C_c}}$$

$$Q_o = \sqrt{\frac{2 \times 750 \times 20{,}000}{25}} = 1{,}095.45 \approx 1{,}096 \text{ units (approximately)}$$

$$TC_o = D \times P + \frac{DC_o}{Q_o} + \frac{Q_o \times C_c}{2}$$

$$= 20{,}000 \times 85 + \frac{20{,}000 \times 750}{1{,}096} + \frac{1{,}096 \times 25}{2}$$

$$= ₹17{,}27{,}386.131$$

From the above calculations, it can be seen that the cost of making is less than the cost of buying. Therefore, the company should opt for making the item in-house rather than buying it from an outside vendor.

11.4.3   Break-Even Analysis

Break-even analysis is also known as cost-volume-profit analysis. This analysis is used to study the relationship between the total revenue, total costs and total profits of the firm at various levels of output. This analysis is often used by business executives to determine the cut-off production or sales volume from where the firm will make profit. The analysis uses a break-even chart in which the total revenue (TR) and the total cost (TC) curves are represented by straight lines as shown in Figure 11.1.

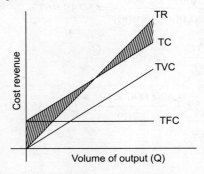

Figure 11.1   Break-even or cost-volume-profit chart

In Figure 11.1, the TR line intersects the TC line at a point which is known as the break-even point. The production or sales volume corresponding to break-even point is called break-even output or sales. At the break-even point, TR is equal to TC and the firm neither earns profit nor incurs loss when it produces break-even quantity. However, the firm earns profit when it produces more than break-even quantity and incurs loss when it produces less than break-even quantity. The break-even quantity can be found algebraically as follows:

At break-even point, the total revenue (TR) is equal to the total costs (TC)

TR is equal to the selling price (*P*) per unit times the quantity of output or sales (*Q*), i.e.,

$$TR = (P)(Q) \qquad (11.5)$$

TC equal total fixed costs (*TFC*) plus total variable costs (*TVC*). Since *TVC* is equal to the average or per-unit variable costs (*v*) times the quantity of output or sales, we have:

$$TC = TFC + (v)(Q) \qquad (11.6)$$

Setting *TR* equal to *TC* and substituting Q_B (the break-even output) for *Q*, we have:

$$TR = TC \qquad (11.7)$$

$$(P)(Q_B) = TFC + (v)(Q_B) \qquad (11.8)$$

Solving Equation (11.8) for the break-even output Q_B, we get:

$$(P)(Q_B) - (v)(Q_B) = TFC$$

$$(Q_B)(P - v) = TFC$$

$$Q_B = \frac{TFC}{(P - v)} \qquad (11.9)$$

The denominator in Equation (11.9), i.e., (*P* − *v*) is called the contribution margin per unit as it represents the portion of selling price that can be applied to cover the fixed costs of the firm and to provide for profits.

More generally, suppose that the firm wishes to earn a specific profit and wants to estimate the quantity that it must sell to earn that profit. Break-even analysis can be used to determine the target output (Q_T) at which target profit (S_T) can be achieved. To do so, we simply add S_T to the numerator of Equation (11.9) and have:

$$Q_T = \frac{TFC + S_T}{(P - v)} \qquad (11.10)$$

The margin of safety (*M.S.*) is defined as the sales over and above the break-even sales and it is computed from the following formula:

$$M.S. = \text{Actual sales} - \text{break-even sales} \qquad (11.11)$$

In terms of percentage of sales, *M.S.* can be expressed as:

$$M.S. \text{ as a percent of sales} = (M.S./\text{Sales}) \times 100 \qquad (11.12)$$

The concept of break-even analysis discussed above can be used in make or buy decision. Examples 11.3 and 11.4 illustrate the application of break-even analysis in make or buy decision.

Example 11.3

AIWA Ltd. requires 7,500 cabinets for one of its niche products which it buys from a vendor for ₹350 per unit. If the company starts making the cabinet in its manufacturing facility, it has to bear ₹175 per unit towards variable cost and ₹10,00,000 as fixed costs. Should the company buy or purchase the cabinets?

Solution

Selling price per unit $(P) = ₹350$

Variable cost per unit $(v) = ₹175$

Total fixed costs $(TFC) = ₹10,00,000$

Using the above data and Equation (11.9), let us calculate break-even output Q_B

$$Q_B = \frac{10,00,000}{(350 - 175)} = 5,714.29 \approx 5,715 \text{ units (approximately)}$$

Since the demand (7,500 units) is greater than break-even quantity, the firm should make the cabinets.

Example 11.4

An auto-parts manufacturing industry has the required facility to make a component either by Laser Welding or by GTAW or it can buy the component from an outside vendor. The annual demand of the component is about 12,000 units. Various cost elements associated with the component are given below:

	Laser Welding	GTAW	Buy
Fixed annual cost, TFC (₹)	15,00,000	9,00,000	
Unit variable cost, v (₹)	350	500	
Unit purchase cost, P (₹)			400

Should the company make the component by Laser Welding or by GTAW or buy it?

Solution

Annual cost of Laser Welding $= TFC + (v)(Q)$

$$= 15,00,000 + (350)(12,000)$$

$$= ₹57,00,000$$

Annual cost of GTAW $= TFC + (v)(Q)$

$$= 9,00,000 + (500)(12,000)$$

$$= ₹69,00,000$$

Annual cost of buy $= (P)(Q)$

$$= (400)(12,000)$$

$$= ₹48,00,000$$

From the above calculations, it can be seen that the cost of buy option is the minimum among all possible alternatives. Therefore, the company should opt for buying the component from outside vendor.

PROBLEMS

1. Enumerate the criteria for make as well as for buy decisions.

2. List the factors that are considered by a firm whenever it decides to make a given product in-house.

3. What is break-even analysis? How does this analysis help in the process of make or buy decision?

4. An automobile ancillary unit has extra capacity that can be used to produce a pin that is required in one of its products. Currently the company is buying pins for ₹250 each. If it decides to make the pin in-house utilizing its available capacity, its estimated costs of raw material, labor and overhead variable costs is ₹70, ₹150 and ₹45 per pin respectively. The annual fixed costs associated with the unused capacity are ₹9, 50,000. An expected demand for the pins for next year is estimated to be 7,500 units.

   (a) Recommend, if it is economically viable for the company to make pins in-house.

   (b) The company can also alternatively use its extra capacity to produce bushes. It is expected that this would cover the fixed and variable costs associated with the extra capacity and also contribute ₹1,50,000 towards profit. Recommend whether the company should make pins or bushes.

5. An auto ancillary manufacturing firm found that it can produce cam shafts using two different processes: Forging or Machining. The firm also has an option of buying the cam shafts from an external vendor. The estimated costs that are incurred during all the three alternatives are given below:

Cost elements	Buy	Forging	Machining
Fixed cost/year (₹)		24,00,000	29,00,000
Variable cost/unit (₹)		475	425
Purchase cost/unit (₹)	700		

The annual demand of the cam shaft is 15,000 units.

   (a) Should the company make the cam shafts by Forging or Machining, or Buy them from the external vendor?

   (b) What is the annual production that will force the firm to switch from Forging to Machining?

   (c) At what annual volume should the company switch from manufacturing the cam shafts by Forging to Buying?

6. A company is an official vendor for the supply of fuel injection systems of various brands of vehicles. An actuator used in the assembly of the systems costs ₹560, which the company buys

from outside. In case it produces it in-house, the estimated fixed and variable costs would be ₹45,00,000 and ₹210 respectively.

Annual requirement of the actuator is 12,000. Should the company make it in-house or buy it from outside?

7. The annual demand of a particular product is 10,000 units. The different costs pertaining to make or buy are given below in the table. Recommend whether the products should be made or purchased.

	Make	Buy
Product cost/unit (₹)	145	165
Procurement cost/order (₹)		600
Inventory carrying cost/unit/year (₹)	12	18
Set-up cost/set-up (₹)	800	
Production rate/year	25,000	

12

Project Management

12.1 INTRODUCTION

A project consists of a set of activities that are interrelated in a logical sequence and they need to be executed in a certain order for timely completion of the project. Most realistic projects undertaken by organizations such as Infosys, Honda Motors, or the Indian Defense Department, etc., are large and complex. Delhi Development Authority (DDA) building residential houses, for example, must complete thousands of activities costing crores of rupees. Indian Space Research Organization (ISRO) must inspect countless components before it launches a satellite. Almost every industry worries about how to manage similar large scale, complicated projects effectively. It is a difficult problem, and the stakes are high. Crores of rupees in cost overruns have been wasted due to poor planning of projects. Unnecessary delays have occurred due to poor scheduling. How can such problems be solved? The tools and techniques of Project Management (PM) can provide a solution to such problems. PM is concerned with planning, scheduling, monitoring and control of the projects. It aims at timely and cost effective completion of the projects and, to achieve this objective, it performs functions in phases which are discussed in the next section.

12.2 PHASES OF PROJECT MANAGEMENT

Effective management of a project requires appropriate actions to be taken by the management at different stages of the project, from its planning to completion. Thus, project management involves the following phases:

1. Planning
2. Scheduling
3. Monitoring and Control

12.2.1 Planning

Planning is essential for economical and timely execution of projects. The various stages of the planning process are:

12.2.1.1   Pre-planning

The planning undertaken before the decision has been made to take up the project is called pre-planning. This phase includes formulation of the project, assessment about its necessity, financial and/or economic analysis and decision making about whether or not to take up the project.

12.2.1.2   Detailed Planning

Once the decision has been made to take up the project, detailed planning is carried out, which includes the following phases:
  (a) Preparation of detailed designs and detailed working plans.
  (b) Preparation of specifications, detailed bill of quantities, etc.
  (c) Working out of plan-of-action for carrying out the work. This involves the following:
      (i) Project breakdown, i.e., breaking up the project into small component jobs.
      (ii) Establishment of sequence of various operations and their interrelationship.

12.2.2   Scheduling

Once the planning is over, one would like to know whether the work would be taken in hand, what the detailed time-table of various operations would be, and when the task would get completed. This requires estimation of duration of the different operations involved in execution of the task and calculation of the project duration. The scheduling phase involves preparation of a detailed time-table of various operations and includes estimation of duration of the different operations, preparation of a detailed schedule of work, etc.

12.2.3   Monitoring and Control

Once the work-plan and schedule of various operations have been evolved, actual execution of the project starts. Monitoring of progress and updating of schedule must go on, through the entire execution stage, if the project is to be completed. It is an important phase of PM and includes:
  (a) Monitoring of progress, vis-à-vis the proposed schedule.
  (b) Updating of the schedule to take into account the actual progress of different operations with reference to the original schedule and revised forecasts of availability of various resources.

12.3   BAR OR GANTT CHARTS

At the beginning of the twentieth century, Henry L. Gantt developed a pictorial method of showing the time-table of various operations required to be performed for execution of a project. In this method, the elapsed time of each of the operations is shown in the form of bar, against a time-scale. These work-versus-time graphs are commonly known as Bar charts or Gantt charts. These charts, indeed, represent the first successful attempt of engineers in systematic planning.

Bar charts are simple in concept and are easy to prepare and understand. In these charts, time is represented in terms of hours/days/weeks/months/years along one of the axis (usually the horizontal axis) and the various operations/jobs are represented along the other axis (vertical axis, if the time has been represented along the horizontal axis). However, it would not be wrong if time is represented along the vertical axis and jobs are shown along the horizontal axis. The duration of each of the operations or jobs is represented by bars drawn against each of the jobs. The length of the bar is proportional to the duration of the activity: the beginning of the bar indicates the time of start and its end indicates the finishing of the concerned activity. A typical Bar chart for a simple work described in Example 12.1 is shown in Fig. 12.1.

Example 12.1

Draw a Gantt chart schedule for a task, the start/finish time of component jobs of which are given below:

Job	Duration (Days)	Start	Finish
A	6	0	6
B	8	0	8
C	12	0	12
D	5	7	12
E	8	10	18

Solution

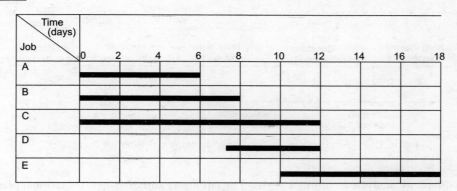

Figure 12.1   Bar or Gantt chart for Example 12.1

Bar charts provide an excellent method of both the representation of the work-schedule and monitoring of the progress. They are simple in concept and easy to make and understand. However, despite all these merits, bar charts suffer from some shortcomings: they are unable to show the interrelationships between the various component jobs, and therefore, it becomes difficult to appreciate how delay in completion of some of the jobs is likely to affect the overall completion of the project. Due to these shortcomings, bar charts have very limited use in the area of PM.

Efforts were made in USA and other western countries to develop new techniques of project planning in order to get over the shortcomings of the bar charts. This led to the development of two new techniques known as Critical Path Method (CPM) and Program Evaluation and Review Technique (PERT). These techniques belong to the family of network analysis technique and they are discussed in detail in the subsequent sections.

12.4 NETWORK ANALYSIS TECHNIQUE

Network analysis refers to a term used to represent techniques that use networks and the basic Critical Path concepts for solving project planning and scheduling problems. CPM and PERT are two well-known techniques that belong to this family. Network analysis techniques are graphical-numeric in nature as they depend, basically, upon graphical portrayal of the work-plan in the form of networks and then their numerical analysis to generate information which is subsequently used by the management.

Before going into further details, it is necessary to explain the meaning of a network. A network is a pictorial statement of the project or task which, with the help of arrows and nodes, depicts graphically the planned sequence of the various operations required to be performed for achievement of project objectives and to show interdependencies between them. The network for the installation of a complex air filter on the smokestack of a metal-working firm is shown in Fig. 12.2.

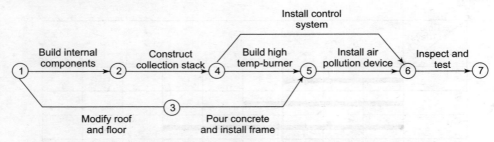

Figure 12.2 A network for installation of a complex air filter on smokestack

Network analysis techniques are used by engineers and managers for rational planning and effective management of large complex projects. These techniques enable sequencing of the various operations, from a logical standpoint, and help in expeditious and economic completion of projects. As the schedule of operations as well as the logic behind the interrelationship of activities is explicitly indicated, the techniques are ideal for helping in effective control of the progress during the course of execution. This explains why these techniques have been used so extensively in business and industry.

12.5 CRITICAL PATH METHOD (CPM)

As mentioned earlier, CPM stands for Critical Path Method. It belongs to the family of network analysis techniques and, therefore, a network is needed for developing CPM schedule. A network in the form of an arrow diagram in which arrows representing the activities and nodes representing the events is required to be developed for CPM analysis. Various aspects of the arrow diagram have been discussed in detail in the following sections. CPM was developed as a consequence of providing a solution to the difficult scheduling problems of a chemical plant of DuPont company.

CPM originators had assumed that the time-duration of various activities could be estimated reasonably accurately and therefore only one estimate of duration was considered good enough for working out the schedule. Owing to this, CPM is considered to be a deterministic approach to project scheduling. In addition, CPM also takes into account cost-duration relationship in order to arrive at an optimal schedule, i.e., completion of the project in minimum possible time with minimum cost expenditure.

12.5.1 Arrow Diagrams

An arrow diagram can be defined as a network in which arrows represent the activities and the nodes represent the events. A typical arrow diagram is shown in Fig. 12.3. Arrow diagrams were used for the first time with CPM. However, these are also employed extensively with network techniques, for planning and scheduling of projects. Arrow diagrams are indeed, pictorial representation of the project and they show the activities required to be performed, the sequence or order in which these are to be undertaken and the way these activities are interrelated.

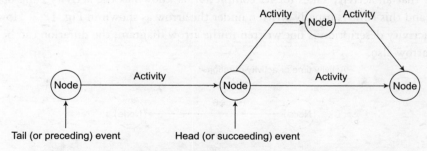

Figure 12.3   A typical arrow diagram which shows activities and events

12.5.1.1 Activity

An activity is that element of the project which consumes resources and time for its execution. In arrow diagrams, an arrow (Fig. 12.4) defines an activity completely. The beginning of the arrow, i.e., its tail represents the start of the activity and the end of the arrow, i.e., its head represents the completion. Each arrow is unique in the sense that an activity cannot be represented by more than one arrow. In other words, there can be only one activity between a pair of nodes.

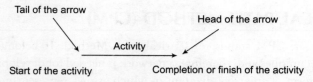

Figure 12.4   An arrow represents an activity completely

Each activity must start from and end in a node (Fig. 12.5).

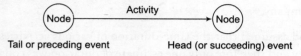

Figure 12.5   True representation of an activity

The name of the activity which the arrow represents may be written over the arrow in full or in the form of an abbreviation such a letter or a symbol over the arrow (Fig. 12.6).

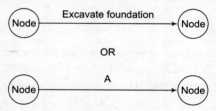

Figure 12.6   Activity description may be indicated over the arrow in full or through abbreviation

The time that an activity takes for its completion is known as the activity time or activity duration and this time is usually written under the arrow as shown in Fig. 12.7. However, in case the activity description is not written in the arrow diagram, the duration can be written over the arrow also.

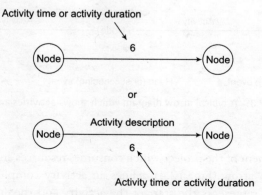

Figure 12.7   Method of showing activity duration

12.5.1.2 *Events or Nodes*

An event is that element of the project which represents a specified point of time during execution of the project to mark start or finish of an activity or a group of activities. Events do not consume resources and time, they simply occur to indicate start or finish of activities. Nodes or events are usually represented by circles. However, other geometrical shapes such as rectangle, squares, ovals, etc., could also be used for this purpose. Events in arrow diagrams are assigned distinct numbers. Event's number is written inside the circle or squares. Some of the common ways representing events in arrow diagrams are shown in Fig. 12.8. Certain specific information such as earliest occurrence time (T_E), latest occurrence time (T_L) pertains to each event in the arrow diagram. Different ways of indicating this information in nodes are shown in Fig. 12.9.

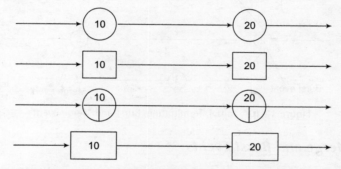

Figure 12.8   Different geometrical shapes used for indicating events

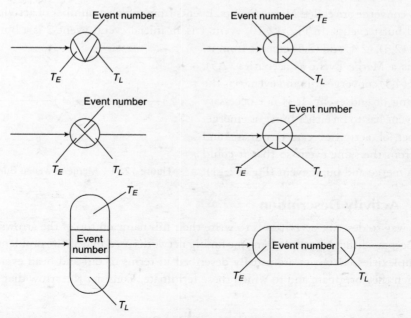

Figure 12.9   Different ways of indicating relevant information pertaining to events

12.5.1.3 Head and Tail Events

An event from which an activity starts is called the tail event or the preceding event. An event at which the activity ends is called head event or the succeeding event. For any activity, therefore, there is a tail event, i.e., from where it starts and a head event where it ends (Fig. 12.5).

12.5.1.4 Initial and Terminating Events

The very first event of an arrow diagram is known as the initial event and the very last one as the terminating event. An initial event has only outgoing arrows, no incoming arrows; a terminating event has only incoming arrows and no outgoing arrows (Fig. 12.10).

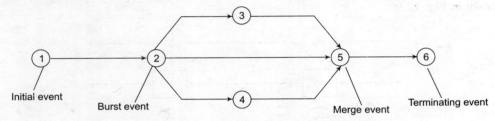

Figure 12.10   Initial, Terminating, Burst and Merge events

12.5.1.5 Merge and Burst Events

In arrow diagrams, there are some events to which a number of activities converge; there may be others from which a number of activities diverge. The events to which a number of activities converge are called Merge Events. Events from which a number of activities emerge are called burst events. In Fig. 12.10, event 1 is an initial event, event 2 is a burst event as

activities (2,3), (2,4) and (2,5) emerge from it, event 5 is a Merge Event as activities (2,5), (3,5) and (4,5) converge at it and event 6 is the terminating or end event. It is not necessary that an event has to be either burst or emerge. A number of activities can terminate and emerge from the same event so that it could be both a merge and burst event (Fig. 12.11).

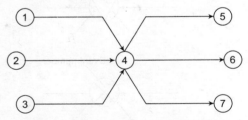

Figure 12.11   Merge as well as Burst Events

12.5.2 Activity Description

The best way to describe activities is to write their full name on top of the arrows. However, this can become difficult and cumbersome for big networks used in real-life problems. To avoid such complexities, activities are usually described in terms of tail and head event numbers from which they originate and to which these terminate. Consider the arrow diagram shown in Fig. 12.12. Here, activity A starts from node 1 and ends up in node 2. So, activity A can be designated as (1,2) or (1–2). Similarly, activity B which originates from event 2 and terminates in event 3 can be designated as (2,3) or (2–3). Similarly, activities C, D, E, F and G can be designated as (3,4), (3,5), (3,6), (4,6) and (5,6) respectively.

By this method of designating of activities, each activity gets defined in a unique way as the nodes in the arrow diagram have distinct numbers. This method of designating the activities is useful for both manual as well as for computer-based computations and is, therefore, universally adopted. In general, an activity P can be designated as (i, j) or $(i–j)$ where "i" is the tail event number and "j" is the head event number (Fig. 12.13). It must be noted that "j" is greater than "i".

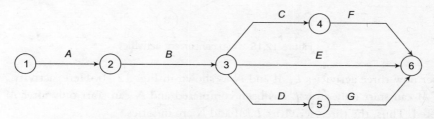

Figure 12.12 Designating activities in terms of Head and Tail event numbers

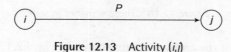

Figure 12.13 Activity (i,j)

12.5.3 Understanding Logic of Arrow Diagrams

Arrow diagrams depict, through a typical arrangement of arrows, the interrelationship between various activities and the sequence in which these activities are to be performed. In order that these diagrams are interpreted in a unique manner, it is necessary to understand clearly the logic involved in the development of arrow diagrams. The dependency rule that "an activity that depends on another activity or activities can start only after all the preceding activities on which it is dependent have been completed" should be followed in order to show interrelationship between various activities in the arrow diagrams. This rule can be explained with the help of a few examples given below:

Consider a portion of the arrow diagram shown in Fig. 12.14. It shows two activities L and M. Here, activity L precedes activity M or activity M follows activity L or activity M succeeds activity L. Activity M cannot start until activity L is completed.

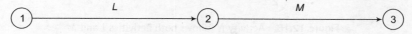

Figure 12.14 Two activities in series

Let us now look at the portion of the arrow diagram shown in Fig. 12.15. Here, the two activities L and M take off from the same event, i.e., event number 1. This implies that both the activities L and M can go on concurrently. Both can start only after the occurrence of their tail event, i.e., event number 1. However, it is not necessary that both must start and end together.

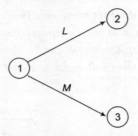

Figure 12.15   Two concurrent activities

Consider now three activities L, M and N as shown in Fig. 12.16. Here, activity M follows L, i.e., M can start only after L has been completed and N can start only after M has been completed. Thus, the three activities L, M and N are in series.

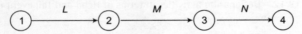

Figure 12.16   Three activities in series

Let us consider a portion of the arrow diagram shown in Fig. 12.17. Here, activity L precedes both the activities M and N. This implies that both the activities M and N can start only after activity L has been completed, thereafter these can go on concurrently.

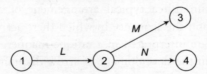

Figure 12.17   Activities M & N follow activity L

Consider a portion of arrow diagram shown in Fig. 12.18. Here, activity N follows both the activities L and M which however, can go on concurrently until they reach event number 3.

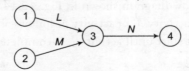

Figure 12.18   Activity N follows both activities L and M

12.5.4   Dummy Activities

Dummy activities are those activities which do not consume either resources or time. Thus, the time consumed by dummy activities is zero. These activities are shown as dashed arrows (Fig. 12.19) in arrow diagrams.

---------------------------➤

Figure 12.19   A dummy activity

Dummy activities are provided in the network diagrams or arrow diagrams in order to (i) show the correct logic of sequential relationship between different activities and (ii) ensure that the activities are uniquely defined, i.e., no two arrows representing two different activities are drawn between the same pair of nodes or events. Examples 12.2 and 12.3 illustrate the use of dummy activities in the arrow diagrams.

Example 12.2

Let there be five activities A, B, C, D and E. It is required to draw an arrow diagram to show the following sequential relationships between them:
(a) B and C follow A
(b) D follows B
(c) E follows C
(d) E is dependent on B
(e) D is independent of C

Solution

The arrow diagram for Example 12.2 is shown in Fig. 12.20.

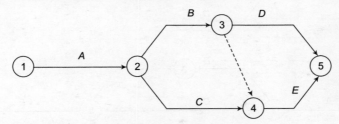

Figure 12.20   Arrow diagram for Example 12.2

Example 12.3

Let there be five activities A, B, C, D and E. It is required to draw an arrow diagram to show the following sequential relationship between them:
(a) B follows A
(b) C follows A
(c) D follows A
(d) E follows B, C and D

Solution

In order to show the sequential relationship between activities given in this example, one may draw the arrow diagram as shown in Fig. 12.21. But this would be incorrect.

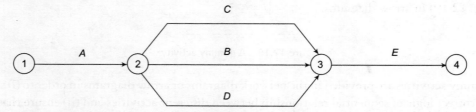

Figure 12.21 Incorrect way of showing relationship between activities in Example 12.3

In Fig. 12.21, three activities namely B, C and D undoubtedly follow activity A, but their tail event and head event are the same, which implies that these activities are not unique, rather they are the same. As per the basic logic of arrow diagrams, there must be only one activity between a pair of events. However, in this case there are three activities B, C and D between event number 2 and 3, which is against the logic.

The unique characteristics of the activities B, C and D can be shown by using dummy activities. The correct arrow diagram for this example is shown in Fig. 12.22.

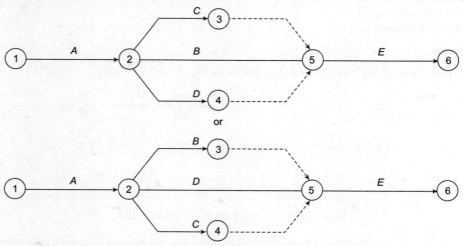

Figure 12.22 Correct way of showing relationship between activities in Example 12.3

12.6 GUIDELINES FOR DRAWING NETWORK DIAGRAMS OR ARROW DIAGRAMS

The following points pertain to the network diagrams or arrow diagrams and should be followed for making them:

(i) Arrows represent activities and nodes represent events.

(ii)   The arrows are drawn in left-right direction.

(iii) Length and orientation of arrows are not important.

(iv)   Events must have a distinct number. Head events have a higher number than the corresponding tail events.

(v)   Each activity must start and end in a node.

(vi)   To be uniquely defined, only one activity can span across a pair of events.

(vii) Activities other than dummy activities consume both resources and time.

(viii) An event cannot occur until all the activities leading into it have been completed.

(ix)   An activity cannot start unless all the preceding activities on which it depends have been completed.

(x)   Dummy activities do not consume either resources or time. However, they follow the same dependency rule as the ordinary activities.

(xi)   Crossing of arrows should be avoided as far as possible. Complete elimination of crossing may not be possible in all cases. However, efforts should be made to reduce the crossing by rearranging the arrows, keeping, of course, the basic logic to be the same. Where the arrows have to cross necessarily, it is important to draw the crossing in such a way that the arrow diagram does not become confusing. Fig. 12.23 shows an improper way of showing crossing of arrows. Where crossing of arrows cannot be avoided, it should be shown in either of the ways depicted in Fig. 12.24 and Fig. 12.25.

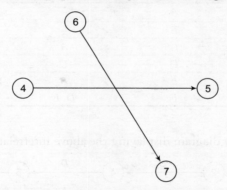

Figure 12.23   An improper way of showing crossing of arrows

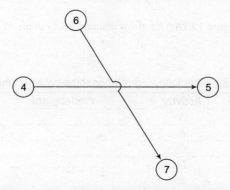

Figure 12.24   One of the ways of showing crossing of arrows

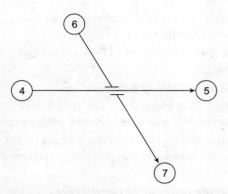

Figure 12.25   Another way of showing crossing of arrows

Examples 12.4, 12.5 and 12.6 illustrate the application of the above mentioned guidelines to make arrow diagrams.

Example 12.4

Draw an arrow diagram to show the following relationships between the various activities:

Activity	Predecessor
A	–
B	–
C	A
D	C
E	B
F	D, E

Solution

Fig. 12.26 shows an arrow diagram displaying the above interrelationship.

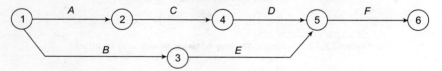

Figure 12.26   An arrow diagram for Example 12.4

Example 12.5

Draw an arrow diagram to show the following relationships between the various activities:

Activity	Predecessor
A	–
B	–
C	–
D	A, C
E	B
F	D, E

Solution

The above interrelationship between the various activities is shown in the arrow diagram depicted in Fig. 12.27.

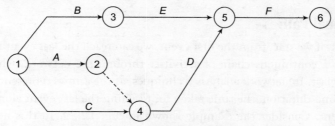

Figure 12.27   An arrow diagram for Example 12.5

Example 12.6

Draw an arrow diagram to show the following relationships between the various activities:

Activity	Predecessor
A	–
B	–
C	A
D	A
E	B, D
F	C
G	C, E
H	F, G

Solution

Fig. 12.28 shows an arrow diagram displaying the above interrelationship.

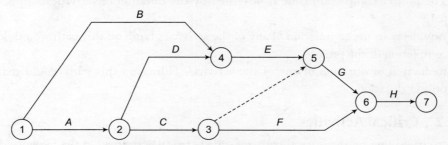

Figure 12.28   An arrow diagram for Example 12.6

12.7 CPM CALCULATIONS

The first step in CPM calculations is to develop an arrow diagram as per the interrelationship between various activities of the given task or project. After developing the arrow diagram, each activity of this diagram is assigned only one time-duration. Calculations are then made to determine Critical Path, Critical Activities, Project Completion Time and Float of the activities. Before explaining the procedure for making CPM calculations, it is important to understand the meaning of the different terms used.

12.7.1   Critical Path

In every network, if we start from the first event, we can reach the last event through different paths. A path is a continuous chain of activities through the network, which connects the first event to the last. In network analysis techniques, as the arrows representing the activities have a definite time-duration, the time taken for reaching the last event along different paths would be different. Consider the example shown in Fig. 12.29. In this network, there are three paths through which the last event, i.e., event 5 can be reached starting from the first event. These are $1 - 2 - 3 - 4 - 5$, $1 - 2 - 4 - 5$ and $1 - 2 - 5$. The path that takes the longest time to traverse through the network, starting from the first to the last event is called the critical path. In this example, path $1 - 2 - 3 - 4 - 5$ takes $2 + 4 + 5 + 2 = 13$ days, path $1 - 2 - 4 - 5$ takes $2 + 6 + 2 = 10$ days and path $1 - 2 - 5$ takes $2 + 7 = 9$ days. Thus, path $1 - 2 - 3 - 4 - 5$ is the critical path as it takes longest time to reach event 5 starting from event 1.

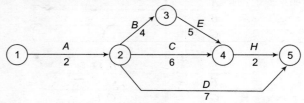

Figure 12.29   A simple network diagram

In project scheduling, the critical path carries a significant importance due to the following reasons:
  (i) The project completion time is governed by the duration of activities falling on this path.
 (ii) Any delay in the completion of any of the activities lying on this path would delay the completion of the project.
(iii) Reduction of duration of any of the activities falling on this path would reduce the project duration.

12.7.2   Critical Activities

The activities lying on the critical path are called critical activities. In the network diagram shown in Fig. 12.29, activities $A\ (1 - 2)$, $B\ (2 - 3)$, $E\ (3 - 4)$ and $H\ (4 - 5)$ are critical activities as they fall on the critical path. Critical activities are important, as they govern the completion of the project. Any delay in their execution would result in delay in the project completion.

12.7.3   Non-critical Activities

In a network diagram, all activities other than critical activities are called non-critical activities. These have a certain amount of spare time or float available. Non-critical activities can be delayed to the extent of their available float and by doing so the project completion time would not be affected.

12.7.4 Earliest Event Time

Earliest event time (T_E) is defined as the earliest time at which an event can occur, without affecting the total project time. It is also known as earliest occurrence time.

12.7.5 Latest Event Time

Latest event time (T_L) is defined as the latest time at which an event can occur, without affecting the total project time. It is also known as latest occurrence time.

12.8 CALCULATION OF THE EARLIEST OCCURRENCE TIME OF EVENTS

Earliest occurrence times of events (T_E) of the various events in a network are computed, starting from the first event. T_E of the first event is taken as zero. The calculations proceed from left to right, until the last event is reached. T_E of a succeeding or head event is obtained by adding activity-duration to the T_E of the preceding or tail event. This method of computation of T_E of head events is referred to as "Forward Pass".

For a merge event at which a number of activities converge, the occurrence times taken for reaching this event through different paths are calculated by adding to T_E of the tail events, the duration of the corresponding activity and then the highest of these is taken as T_E of the merge event. Example 12.7 illustrates the calculation of T_E of the various events in a network.

Example 12.7

Consider a simple network shown in Fig. 12.30. Compute the earliest occurrence time of the events.

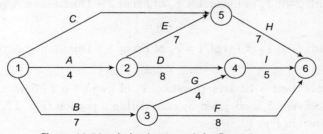

Figure 12.30 A simple network for Example 12.7

Solution

(i) T_E of Event 1 = 0.
(ii) T_E of Event 2 is obtained by considering a portion (Fig. 12.30 a) of the arrow diagram shown in Fig. 12.30.

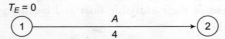

Figure 12.30(a)   Portion of the arrow diagram shown in Figure 12.30

T_E of Event $2 = T_E$ of Event $1 +$ Duration of Activity A.

$$= 0 + 4 = 4 \text{ days}.$$

(iii) T_E of Event 3 is obtained by considering a portion (Fig. 12.30 b) of the arrow diagram shown in Fig. 12.30.

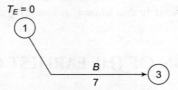

Figure 12.30(b)   Portion of the arrow diagram shown in Figure 12.30

T_E of Event $3 = T_E$ of Event $1 +$ Duration of Activity B

$$= 0 + 7 = 7 \text{ days}.$$

(iv) T_E of Merge Event 4 is obtained by considering a portion (Fig. 12.30 c) of the arrow diagram shown in Fig. 12.30.

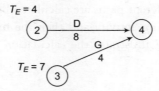

Figure 12.30(c)   Portion of the arrow diagram shown in Figure 12.30

Through path 2–4: T_E of Event $4 = T_E$ of Event $2 +$ Duration of Activity D

$$= 4 + 8 = 12 \text{ days}.$$

Through path 3–4: T_E of Event $4 = T_E$ of Event $3 +$ Duration of Activity G

$$= 7 + 4 = 11 \text{ days}.$$

The highest of these is 12 days, therefore, T_E of Event 4 is 12 days.

(v) T_E of Merge Event 5 is obtained by considering a portion (Fig. 12.30 d) of the arrow diagram shown in Fig. 12.30.

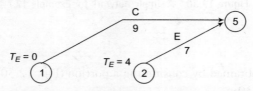

Figure 12.30(d)   Portion of the arrow diagram shown in Figure 12.30

Through path 1–5: T_E of Event 5 = T_E of Event 1 + Duration of Activity C

$$= 0 + 9 = 9 \text{ days.}$$

Through path 2–5: T_E of Event 5 = T_E of Event 2 + Duration of Activity E

$$= 4 + 7 = 11 \text{ days.}$$

The highest of these is 11 days, therefore, T_E of Event 5 is 11 days.

(vi) T_E of Merge Event 6 is obtained by considering a portion (Fig. 12.30 e) of the arrow diagram shown in Fig. 12.30.

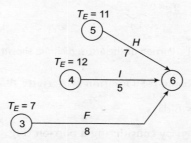

Figure 12.30(e)   Portion of the arrow diagram shown in Figure 12.30

Through path 3–6: T_E of Event 6 = T_E of Event 3 + Duration of Activity F

$$= 7 + 8 = 15 \text{ days.}$$

Through path 4–6: T_E of Event 6 = T_E of Event 4 + Duration of Activity I

$$= 12 + 5 = 17 \text{ days.}$$

Through path 5–6: T_E of Event 6 = T_E of Event 5 + Duration of Activity H

$$= 11 + 7 = 18 \text{ days.}$$

The highest of these is 18 days, therefore, T_E of Event 6 is 18 days.

Example 12.8

Compute the Earliest Event Times for the simple network shown in Fig. 12.31.

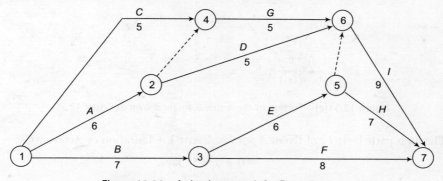

Figure 12.31   A simple network for Example 12.8

Solution

(i) T_E of Event 1 = 0.

(ii) T_E of Event 2 is obtained by considering a portion (Fig. 12.31 a) of the arrow diagram shown in Fig. 12.31.

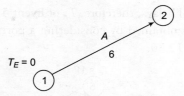

Figure 12.31(a)   Portion of the arrow diagram shown in Figure 12.31

T_E of Event 2 = T_E of Event 1 + Duration of Activity A.

$$= 0 + 6 = 6 \text{ days.}$$

(iii) T_E of Event 3 is obtained by considering a portion (Fig. 12.31 b) of the arrow diagram shown in Fig. 12.31.

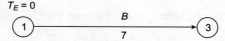

Figure 12.31(b)   Portion of the arrow diagram shown in Figure 12.31

T_E of Event 3 = T_E of Event 1 + Duration of Activity B

$$= 0 + 7 = 7 \text{ days.}$$

(iv) T_E of Merge Event 4 is obtained by considering a portion (Fig. 12.31 c) of the arrow diagram shown in Fig. 12.31.

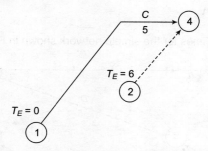

Figure 12.31(c)   Portion of the arrow diagram shown in Figure 12.31

Through path 1–4: T_E of Event 4 = T_E of Event 1 + Duration of Activity C

$$= 0 + 5 = 5 \text{ Days.}$$

Through path 2–4: T_E of Event $4 = T_E$ of Event 2 + Duration of Dummy Activity 2–4

$$= 6 + 0 = 6 \text{ days.}$$

The highest of these is 6 days, therefore, T_E of Event 4 is 6 days.

(v) T_E of Event 5 is obtained by considering a portion (Fig. 12.31 d) of the arrow diagram shown in Fig. 12.31.

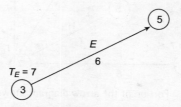

Figure 12.31(d)   Portion of the arrow diagram shown in Figure 12.31

T_E of Event $5 = T_E$ of Event 3 + Duration of Activity E

$$= 7 + 6 = 13 \text{ days.}$$

(vi) T_E of Merge Event 6 is obtained by considering a portion (Fig. 12.31 e) of the arrow diagram shown in Fig. 12.31.

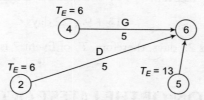

Figure 12.31(e)   Portion of the arrow diagram shown in Figure 12.31

Through path 2–6: T_E of Event $6 = T_E$ of Event 2 + Duration of Activity D

$$= 6 + 5 = 11 \text{ days.}$$

Through path 4–6: T_E of Event $6 = T_E$ of Event 4 + Duration of Activity G

$$= 6 + 5 = 11 \text{ days.}$$

Through path 5–6: T_E of Event $6 = T_E$ of Event 5 + Duration of Dummy Activity 5–6

$$= 13 + 0 = 13 \text{ days.}$$

The highest of these is 13 days, therefore, T_E of Event 6 is 13 days.

(vii) T_E of Merge Event 7 is obtained by considering a portion (Fig. 12.31 f) of the arrow diagram shown in Fig. 12.31.

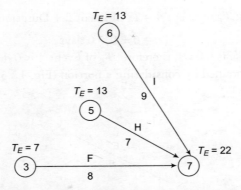

Figure 12.31(f) Portion of the arrow diagram shown in Figure 12.31

Through path 3–7: T_E of Event 7 = T_E of Event 3 + Duration of Activity F

$$= 7 + 8 = 15 \text{ days.}$$

Through path 5–7: T_E of Event 7 = T_E of Event 5 + Duration of Activity H

$$= 13 + 7 = 20 \text{ days.}$$

Through path 6–7: T_E of Event 7 = T_E of Event 6 + Duration of Activity I

$$= 13 + 9 = 22 \text{ days.}$$

The highest of these is 22 days, therefore, T_E of Event 7 is 22 days.

12.9 CALCULATION OF THE LATEST OCCURRENCE TIME OF EVENTS

Latest occurrence times of events (T_L) of the various events in a network are computed, starting from the last event. T_E of the last event is taken as its T_L, unless a time earlier than T_E has been specified. In case a time earlier than T_E is specified, then this specified value is taken as T_L of the last event. The calculations proceed from right to left, i.e., in the backward direction, until the first event is reached. T_L of a preceding or tail event is obtained by subtracting activity-duration from the T_L of the succeeding or head event. This method of computation of T_L of tail events is referred to as "Backward Pass".

For a burst event, from which a number of activities emerge, the occurrence times taken for reaching this event through different paths are calculated by subtracting from T_L of head events, the duration of the corresponding activity and then the lowest of these is taken as T_L of the burst event. Examples 12.9 and 12.10 illustrate the calculation of T_L of the various events in a network.

Example 12.9

Determine the latest occurrence times of the events shown in Fig. 12.32.

Solution

Fig. 12.30 shown earlier has been reproduced and shown in Fig. 12.32. Earliest Occurrence Times (T_E) of its various events calculated in Example 12.7 have been written on top of the respective events. The Latest Occurrence Times (T_L) of the various events of the network shown in Fig. 12.32 is calculated as follows:

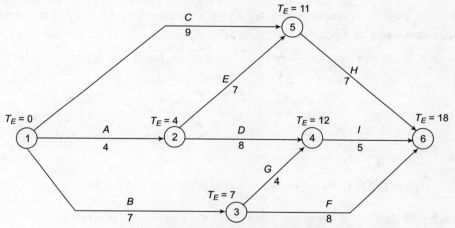

Figure 12.32   A simple network as shown in Figure 12.30

(i) T_L of the last event, i.e. Event 6, is taken as its T_E. Thus, T_L of Event 6 = 18 days.

(ii) T_L of Event 5 is obtained by considering a portion (Fig. 12.32 a) of the arrow diagram shown in Fig. 12.32.

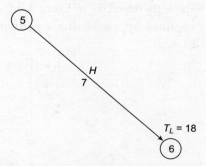

Figure 12.32(a)   Portion of the arrow diagram shown in Figure 12.32

T_L of Event 5 = T_L of Event 6 − Duration of Activity H.

$$= 18 - 7 = 11 \text{ days.}$$

(iii) T_L of Event 4 is obtained by considering a portion (Fig. 12.32 b) of the arrow diagram shown in Fig. 12.32.

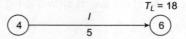

Figure 12.32(b)   Portion of the arrow diagram shown in Figure 12.32

T_L of Event 4 = T_L of Event 6 – Duration of Activity I.

$$= 18 - 5 = 13 \text{ days.}$$

(iv) T_L of Burst Event 3 is obtained by considering a portion (Fig. 12.32 c) of the arrow diagram shown in Fig. 12.32.

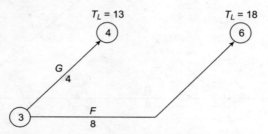

Figure 12.32(c)   Portion of the arrow diagram shown in Figure 12.32

Through path 3–4: T_L of Event 3 = T_L of Event 4 – Duration of Activity G

$$= 13 - 4 = 9 \text{ days.}$$

Through path 3–6: T_L of Event 3 = T_L of Event 6 – Duration of Activity F

$$= 18 - 8 = 10 \text{ days.}$$

The lowest of these is 9 days, therefore, T_L of Event 3 is 9 days.

(v) T_L of Burst Event 2 is obtained by considering a portion (Fig. 12.32 d) of the arrow diagram shown in Fig. 12.32.

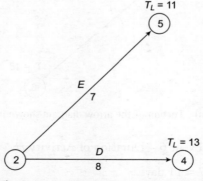

Figure 12.32(d)   Portion of the arrow diagram shown in Figure 12.32

Through path 2–4: T_L of Event 2 = T_L of Event 4 – Duration of Activity D

$$= 13 - 8 = 5 \text{ days.}$$

Through path 2–5: T_L of Event 2 = T_L of Event 5 – Duration of Activity E

$$= 11 - 7 = 4 \text{ days.}$$

The lowest of these is 4 days, therefore, T_L of Event 2 is 4 days.

(vi) T_L of Burst Event 1 is obtained by considering a portion (Fig. 12.32 e) of the arrow diagram shown in Fig. 12.32.

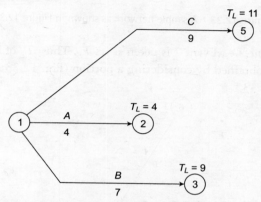

Figure 12.32(e) Portion of the arrow diagram shown in Figure 12.32

Through path 1–2: T_L of Event 1 = T_L of Event 2 – Duration of Activity A

$$= 4 - 4 = 0 \text{ days.}$$

Through path 1–3: T_L of Event 1 = T_L of Event 3 – Duration of Activity B

$$= 9 - 7 = 2 \text{ days.}$$

Through path 1–5: T_L of Event 1 = T_L of Event 5 – Duration of Activity C

$$= 11 - 9 = 2 \text{ days.}$$

The lowest of these is 0 days, therefore, T_L of Event 1 is 0 days.

Example 12.10

Compute the Latest Occurrence Times of the events shown in Fig. 12.33.

Solution

Fig. 12.31 shown earlier has been reproduced and shown in Fig. 12.33. Earliest Occurrence Times (T_E) of its various events calculated in Example 12.8 have been written on top of the respective events. The Latest Occurrence Times (T_L) of the various events of the network shown in Fig. 12.33 is calculated as follows:

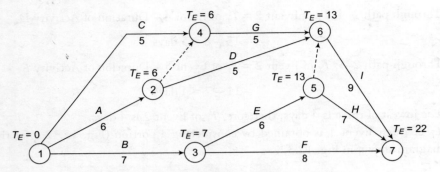

Figure 12.33   A simple network as shown in Figure 12.31

(i) T_L of the last event, i.e., Event 7 is taken as its T_E. Thus, T_L of Event 7 = 22 days.

(ii) T_L of Event 6 is obtained by considering a portion (Fig. 12.33 a) of the arrow diagram shown in Fig. 12.33.

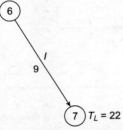

Figure 12.33(a)   Portion of the arrow diagram shown in Figure 12.33

T_L of Event 6 = T_L of Event 7 – Duration of Activity I.

$$= 22 - 9 = 13 \text{ days.}$$

(iii) T_L of Burst Event 5 is obtained by considering a portion (Fig. 12.33 b) of the arrow diagram shown in Fig. 12.33.

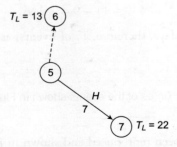

Figure 12.33(b)   Portion of the arrow diagram shown in Figure 12.33

Through path 5–6: T_L of Event 5 = T_L of Event 6 – Duration of the Dummy Activity

$$= 13 - 0 = 13 \text{ days.}$$

Through path 5–7: T_L of Event 5 = T_L of Event 7 – Duration of Activity H

$$= 22 - 7 = 15 \text{ days.}$$

The lowest of these is 13 days, therefore, T_L of Event 5 is 13 days.

(iv) T_L of Event 4 is obtained by considering a portion (Fig. 12.33 c) of the arrow diagram shown in Fig. 12.33.

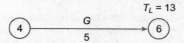

Figure 12.33(c) Portion of the arrow diagram shown in Figure 12.33

T_L of Event 4 = T_L of Event 6 – Duration of Activity G.

$$= 13 - 5 = 8 \text{ Days.}$$

(v) T_L of Burst Event 3 is obtained by considering a portion (Fig. 12.33 d) of the arrow diagram shown in Fig. 12.33.

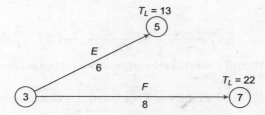

Figure 12.33 d Portion of the arrow diagram shown in Figure 12.33

Through path 3–5: T_L of Event 3 = T_L of Event 5 – Duration of the Activity E

$$= 13 - 6 = 7 \text{ days.}$$

Through path 3-7: T_L of Event 3 = T_L of Event 7 – Duration of Activity F

$$= 22 - 8 = 14 \text{ days.}$$

The lowest of these is 7 days, therefore, T_L of Event 3 is 7 days.

(vi) T_L of Burst Event 2 is obtained by considering a portion (Fig. 12.33 e) of the arrow diagram shown in Fig. 12.33.

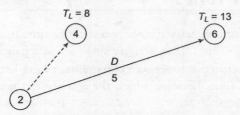

Figure 12.33(e) Portion of the arrow diagram shown in Figure 12.33

Through path 2–4: T_L of Event 2 = T_L of Event 4 – Duration of the Dummy Activity

$$= 8 - 0 = 8 \text{ days.}$$

Through path 2–6: T_L of Event 2 = T_L of Event 6 – Duration of Activity D

$$= 13 - 5 = 8 \text{ days.}$$

The lowest of these is 8 days, therefore, T_L of Event 2 is 8 days.

(vii) T_L of Burst Event 1 is obtained by considering a portion (Fig. 12.33 f) of the arrow diagram shown in Fig. 12.33.

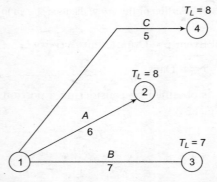

Figure 12.33(f)   Portion of the arrow diagram shown in Figure 12.33

Through path 1–2: T_L of Event 1 = T_L of Event 2 – Duration of Activity A

$$= 8 - 6 = 2 \text{ days.}$$

Through path 1–3: T_L of Event 1 = T_L of Event 3 – Duration of Activity B

$$= 7 - 7 = 0 \text{ days.}$$

Through path 1–4: T_L of Event 1 = T_L of Event 4 – Duration of Activity C

$$= 8 - 5 = 3 \text{ days.}$$

The lowest of these is 0 days, therefore, T_L of Event 1 is 0 days.

12.10 ACTIVITY TIMES

Activities, except the dummy activities, have a definite time duration and, therefore, they have both start and finish times. Each activity spans across a pair of events. Since each event has earliest and latest occurrence times, the two events, i.e., tail and head events provide the boundaries, in terms of time, between which the activity can take place. For each activity, there are four types of times:

(i) Earliest Start Time (E.S.T.)

(ii) Earliest Finish Time (E.F.T.)

(iii) Latest Start Time (L.S.T.)

(iv) Latest Finish Time (L.F.T.)

12.10.1 Earliest Start Time

Earliest Start Time (E.S.T.) is the earliest time at which the activity can start without affecting the project time. An activity can start only after its tail event has occurred and, tail event will occur only when all activities leading into it have been completed. Therefore, E.S.T. of an activity is same as the T_E of its tail event, i.e., E.S.T. of an activity = T_E of its tail event.

Fig. 12.34 shows an activity $(i - j)$ between tail event 'i' and head event 'j'. The activity-duration is $d_{(i-j)}$. For this activity:

$$(\text{E.S.T.})_{(i-j)} = T_{Ei}$$

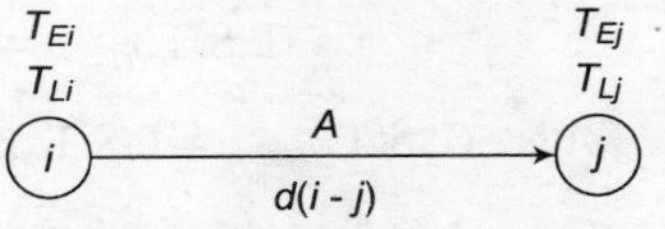

where $(\text{E.S.T.})_{(i-j)}$ = E.S.T. of activity $(i - j)$ and

T_{Ei} = Earliest occurrence time of tail event 'i'.

Figure 12.34   An activity (i-j)

12.10.2 Earliest Finish Time

Earliest Finish Time (E.F.T.) is the earliest possible time at which an activity can finish without affecting the total project time. E.F.T. of an activity is obtained by adding its E.S.T. to its duration.

E.F.T. of an activity = E.S.T. of the activity + duration of the activity.

or   E.F.T. of an activity = T_E of the tail event + duration of the activity.

In general, for the activity shown in Fig. 12.34:

$$(\text{E.F.T.})_{(i-j)} = T_{Ei} + d_{(i-j)}$$

where $(\text{E.F.T.})_{(i-j)}$ = E.F.T. of activity $(i - j)$

12.10.3 Latest Finish Time

Latest Finish Time (L.F.T.) of an activity is the latest time by which the activity must get finished, without delaying the project completion. L.F.T. of an activity is equal to the latest occurrence time (T_L) of its head event.

E.F.T. of an activity = Latest Occurrence Time of the head event.

In general, for the activity shown in Fig. 12.34:

$$(\text{L.F.T.})_{(i-j)} = T_{Lj}$$

where $(\text{L.F.T.})_{(i-j)}$ = L.F.T. of activity $(i - j)$ and

T_{Lj} = Latest occurrence time of head event 'j'.

12.10.4   Latest Start Time

Latest Start Time (L.S.T.) of an activity is the latest time by which an activity must start, without affecting the project duration. L.S.T. of an activity is obtained by subtracting the activity-duration from its L.F.T. Thus:

> L.S.T. = L.F.T. – duration of the activity.

Since   L.F.T. = T_L of head event

> L.S.T. = T_L of head event – duration of activity.

In general, for the activity shown in Fig. 12.34:

$$(\text{L.S.T.})_{(i-j)} = T_{Lj} - d_{(i-j)}$$

where $(\text{L.S.T.})_{(i-j)}$ = L.S.T. of activity $(i - j)$.

Example 12.11 illustrates the computation of E.S.T., E.F.T, L.F.T. and L.S.T. of the various activities in a network diagram.

Example 12.11

Consider the simple network shown in Fig. 12.35. Compute E.S.T., E.F.T., L.F.T. and L.S.T. of the various activities.

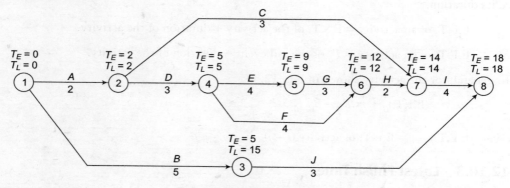

Figure 12.35   A simple network

Solution

The earliest Occurrence Times (T_E) and the Latest Occurrence Times (T_L) of the various events of Fig. 12.35 have been computed as per the procedure described in sections 12.8 and 12.9 respectively.

Earliest Start Times   E.S.T.'s of the activities are same as the Earliest Occurrence Times (T_E) of the tail events of the activities. These have already been worked out and have been tabulated in Table 12.1.

Earliest Finish Times E.F.T.'s of the activities are obtained by adding the duration of the activity to the corresponding E.S.T.'s. Thus, E.F.T. of the activity,

$(1–2) = 0 + 2 = 2$ days

$(1–3) = 0 + 5 = 5$ days

$(2–4) = 2 + 3 = 5$ days

$(2–7) = 2 + 3 = 5$ days

$(3–8) = 5 + 3 = 8$ days

$(4–5) = 5 + 4 = 9$ days

$(4–6) = 5 + 4 = 9$ days

$(5–6) = 9 + 3 = 12$ days

$(6–7) = 12 + 2 = 14$ days

$(7–8) = 14 + 4 = 18$ days

Latest Finish Times L.F.T.'s of activities are given by the Latest Occurrence Times (T_L) of the head events of the activities. These have already been worked out and have been tabulated in Table 12.1.

Latest Start Times L.S.T.'s of the activities are obtained by subtracting the duration of the activity from the corresponding L.F.T.'s. Thus, L.S.T. of the activity,

$(1–2) = 2 – 2 = 0$ day

$(1–3) = 15 – 5 = 10$ days

$(2–4) = 5 – 3 = 2$ days

$(2–7) = 14 – 3 = 11$ days

$(3–8) = 18 – 3 = 15$ days

$(4–5) = 9 – 4 = 5$ days

$(4–6) = 12 – 4 = 8$ days

$(5–6) = 12 – 3 = 9$ days

$(6–7) = 14 – 2 = 12$ days

$(7–8) = 18 – 4 = 14$ days

Table 12.1 Activity-times for network shown in Fig. 12.35

(1) Activity $(i-j)$	(2) Duration $d_{(i-j)}$ (days)	(3) E.S.T. = T_{Ei}	(4) = (3) + (2) E.F.T. = E.S.T. + $d_{(i-j)}$	(5) = (6) − (2) L.S.T. = L.F.T. − $d_{(i-j)}$	(6) L.F.T. = T_{Lj}
1–2	2	0	2	0	2
1–3	5	0	5	10	15
2–4	3	2	5	2	5
2–7	3	2	5	11	14
3–8	3	5	8	15	18
4–5	4	5	9	5	9
4–6	4	5	9	8	12
5–6	3	9	12	9	12
6–7	2	12	14	12	14
7–8	4	14	18	14	18

Example 12.12

Consider the simple network shown in Fig. 12.36. Compute E.S.T., E.F.T., L.F.T. and L.S.T. of its various activities.

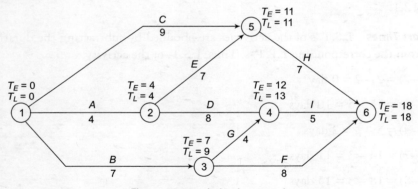

Figure 12.36 A simple network

Solution

The earliest occurrence times (T_E) and the latest occurrence times (T_L) of the various events of Fig. 12.36 have been computed as per the procedure described in sections 12.8 and 12.9 respectively. E.S.T., E.F.T., L.S.T. and L.F.T. of the various activities of the network shown in Fig. 12.36 have been computed as per the procedure given in Example 12.11. These values have been tabulated in Table 12.2.

Table 12.2 Activity-times for network shown in Fig.12.36

(1) Activity $(i-j)$	(2) Duration $d_{(i-j)}$ (days)	(3) E.S.T. $= T_{Ei}$	(4) = (3) + (2) E.F.T. = E.S.T. + $d_{(i-j)}$	(5) = (6) − (2) L.S.T. = L.F.T. − $d_{(i-j)}$	(6) L.F.T. $= T_{Lj}$
1–2	4	0	4	0	4
1–3	7	0	7	2	9
1–5	9	0	9	2	11
2–4	8	4	12	5	13
2–5	7	4	11	4	11
3–4	4	7	11	9	13
3–6	8	7	15	10	18
4–6	5	12	17	13	18
5–6	7	11	18	11	18

Example 12.13

Consider the simple network shown in Fig.12.37. Compute E.S.T., E.F.T., L.F.T. and L.S.T. of its various activities.

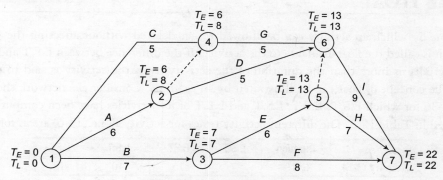

Figure 12.37 A simple network

Solution

The earliest occurrence times (T_E) and the latest occurrence times (T_L) of the various events of Fig. 12.37 have been computed as per the procedure described in sections 12.8 and 12.9 respectively. E.S.T., E.F.T., L.S.T. and L.F.T. of the various activities of the network shown in Fig. 12.37 have been computed as per the procedure given in Example 12.11. These values have been tabulated in Table 12.3.

Table 12.3 Activity-times for network shown in Fig. 12.37

(1) Activity $(i-j)$	(2) Duration $d_{(i-j)}$ (days)	(3) E.S.T. = T_{Ei}	(4) = (3) + (2) E.F.T. = E.S.T. + $d_{(i-j)}$	(5) = (6) − (2) L.S.T. = L.F.T. − $d_{(i-j)}$	(6) L.F.T. = T_{Lj}
1–2	6	0	6	2	8
1–3	7	0	7	0	7
1–4	5	0	5	3	8
2–4	0	6	6	8	8
2–6	5	6	11	8	13
3–5	6	7	13	7	13
3–7	8	7	15	14	22
4–6	5	6	11	8	13
5–6	0	13	13	13	13
5–7	7	13	20	15	22
6–7	9	13	22	13	22

12.11 FLOAT

The time by which an activity can be allowed to get delayed without affecting the project duration is called the float available to the activity. If the difference between L.F.T and E.S.T of an activity is more than the duration of the activity, then the activity is said to possess float. The concept of float can be illustrated by an example. Consider the network shown in Fig. 12.36 for which E.S.T, E.F.T, L.S.T and L.F.T of its activities have been computed and tabulated in Table 12.2. The different activity times for activity F, i.e., (3-6) are as follows:

	Duration	E.S.T.	E.F.T.	L.S.T.	L.F.T.
Activity (3-6)	8	7	15	10	18

The earliest that this activity can start is by the 7th day. As it has duration of 8 days, it would finish on 15th day. However, it can be allowed to finish by the 18th day. So, this activity has a float of 3 days. This is the time available by which it can float. The following alternatives are available for scheduling this activity:

(a) The activity may start on the 7th day and complete on 15th so that the entire float is available towards the end.

   or

(b) The start of the activity may be postponed to the 10th day to coincide with the L.S.T, so that it may get completed by the 18th day, i.e., by L.F.T.

   or

(c) The activity may start on any day between 7th to 10th days, finish during any time between 15th to 18th day.

   or

(d) It may start on 7th day, but may be allowed to absorb the time available, i.e., it may take longer time than the scheduled duration. The increase in duration, in this case, could be anything from zero to three days.

Computation of float of activities of a network enables the project management to identify the activities that are important and those which are not.

Activities that do not possess any float, i.e., those having zero float are known as critical activities. These critical activities, indeed, determine the project duration or project completion time. The activities that possess certain amount of positive float are called non- critical activities.

Float also enables diversion of resources from non-critical activities to critical activities so as to complete the project by its scheduled duration.

12.11.1 Types of Float

There are four different types of float:
   (i) Total float
   (ii) Free float
   (iii) Independent float
   (iv) Interfering float

Total Float   The total float for an activity is obtained by subtracting the activity-duration from the total time available for the performance of the activity. The total time available for the performance of an activity is obtained by subtracting E.S.T. from L.F.T. Thus,

$$\text{Total float} = \text{L.F.T.} - \text{E.S.T.} - d$$

where d is the activity duration.

Since

$$\text{L.F.T.} - d = \text{L.S.T.}$$

$$\text{Total float} = \text{L.S.T.} - \text{E.S.T.}$$

In general, for an activity $(i - j)$

$$F^t_{(i-j)} = (\text{L.F.T.})_{(i-j)} - (\text{E.S.T.})_{(i-j)} - d_{(i-j)}$$

$$= T_{Lj} - T_{Ei} - d_{(i-j)}$$

$$= (\text{L.S.T.})\,(i-j) - (\text{E.S.T.})_{(i-j)}$$

where $F^t_{(i-j)}$ is total float of the activity $(i - j)$

Free Float   This is that part of the total float which does not affect the subsequent activities. This float is obtained when all the activities are started at the earliest. This is obtained from the following formula:

$$F^F_{(i-j)} = T_{Ej} - T_{Ei} - d_{(i-j)}$$

where, $F^F_{(i-j)}$ is the free float of the activity $(i - j)$

Independent float   The part of the float that remains unaffected by utilization of float by the preceding activities and does not affect the succeeding activities is called independent float.

This is obtained by the following formula:

$$F^{I}_{(i-j)} = T_{Ej} - T_{Li} - d_{(i-j)}$$

where, $F^{I}_{(i-j)}$ is the independent float of the activity $(i-j)$

Interfering float   The portion of the total float which affects the start of the subsequent activities is known as interfering float.

This is obtained by following formula:

$$F^{intf}_{(i-j)} = T_{Lj} - T_{Ej}$$

where, $F^{intf}_{(i-j)}$ is the interfering float of the activity $(i-j)$.

Example 12.14 illustrates the computation of different types of floats for the activities.

Example 12.14

Compute different types of floats for the network shown in Fig.12.38.

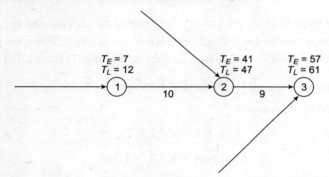

Figure 12.38   A simple network

Solution

For activity (1–2)

Total available time $= T_{L2} - T_{E1} = 47 - 7 = 40$ days.

Time required to perform this activity $= 10$ days.

So, the spare time available to this activity $= 40 - 10 = 30$ days.

This activity can start at the latest on $7 + 30 = 37^{\text{th}}$ day without delaying the project completion time. However, if the entire float, i.e., 30 days is utilized by this activity, event 2 can occur only at its latest time, i.e., 47th day. Consequently, activity (2–3) cannot start till 47th day. Thus, the utilization of full float by activity (1–2) would affect the availability of float of the subsequent activity (2–3).

In order that the floats of subsequent activities are not affected, event 2 must occur at its earliest occurrence time, i.e., 41st day. E.S.T. of activity (1–2) is 7th day and E.S.T. of activity (2-3) is 41st day. So, if both these activities start at their earliest times, then total available time = 41 –7 = 34 days. Duration of activity (1–2) = 10 days. Therefore, free float = 34 – 10 = 24 days. This implies that the float of subsequent activity (2–3) would get curtailed by 6 days, if the float of activity (1–2) were to get utilized. The 6 days is then the interfering float of activity (1–2).

Now, supposing that only 24 days out of the float of 30 days of activity (1–2) is utilized, then there would be no effect at all on the float of the subsequent activities. Thus, out of float of 30 days, 24 days are free float of activity (1–2).

Now, if the activity (1–2) were to start at time $T_L = 12$, it is obvious that all the preceding activities would have been completed by then as per the schedule. So, the time by when activity (1–2) would finish if started at $T_L = 12$, would be (12 +10) day = 22 days. The T_E of event 2 is, however, 41 days. So, the balance time (41– 22 = 19 days) is available for event 2 to occur. This 19 days is the independent float of activity (1–2). Thus, activity (1–2) has a total float of 30 days, a free float of 28 days, an independent float of 19 days and interfering float of 6 days.

For activity (2–3) The various types of float for activity (2–3) can be computed in a similar manner. It has a total float of 61 – 41 – 9 = 11 days and a free float of 57 – 41 – 9 = 7 days. Interfering float is 11 – 7 = 4 days.

The latest start of time of activity (2-3) is 47 days. The time duration of activity is 9 days. The activity can, therefore, finish by 56th day if it was to start on $T_L = 47$. The earliest occurrence time of event 3 is 57 days. Hence this activity possesses an independent float of 1 day. The various types of floats for the two activities (1–2) and (2–3) discussed are tabulated in Table 12.4.

Table 12.4 Calculation of Floats for network shown in Fig. 12.38

Activity (i–j)	Duration (days)	Float (in days)			
		Total	Free	Independent	Interfering
1–2	10	30	24	19	6
2–3	9	11	7	1	4

Example 12.15

Compute different types of floats for the various activities of the network shown in Fig.12.36.

Solution

Fig.12.36 is reproduced below as Fig. 12.39 for the convenience of the readers. The Earliest and the Latest Occurrence Times of the various events of Fig. 12.39 have already been computed and indicated in the figure. The activity times have also been shown in Table 12.2. The various types of floats are computed as:

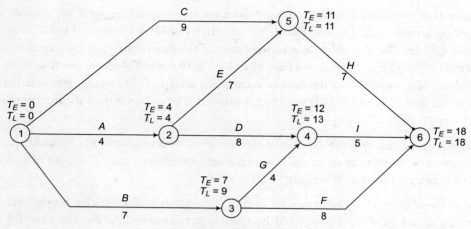

Figure 12.39   A simple network

For Activity (1–2)

Total float $= T_{Lj} - T_{Ei} - d_{(i-j)}$

$\qquad\qquad = T_{L2} - T_{E1} - d_{(1-2)}$

$\qquad\qquad = 4 - 0 - 4 = 0$

Free float $= T_{Ej} - T_{Ei} - d_{(i-j)}$

$\qquad\qquad = T_{E2} - T_{E1} - d_{(1-2)}$

$\qquad\qquad = 4 - 0 - 4 = 0$

Independent float $= T_{Ej} - T_{Li} - d_{(i-j)}$

$\qquad\qquad\qquad = T_{E2} - T_{L1} - d_{(1-2)}$

$\qquad\qquad\qquad = 4 - 0 - 4 = 0$

Interfering float $= T_{Lj} - T_{Ej}$

$\qquad\qquad\qquad = T_{L2} - T_{E2}$

$\qquad\qquad\qquad = 4 - 4 = 0$

For Activity (1–3)

Total float $= T_{Lj} - T_{Ei} - d_{(i-j)}$

$\qquad\qquad = T_{L3} - T_{E1} - d_{(1-3)}$

$\qquad\qquad = 9 - 0 - 7 = 2 \text{ days}$

Free float $= T_{Ej} - T_{Ei} - d_{(i-j)}$

$\qquad\qquad = T_{E3} - T_{E1} - d_{(1-3)}$

$\qquad\qquad = 7 - 0 - 7 = 0$

Independent float $= T_{Ej} - T_{Li} - d_{(i-j)}$

$$= T_{E3} - T_{L1} - d_{(1-3)}$$

$$= 9 - 0 - 7 = 2 \text{ days}$$

Interfering float $= T_{Lj} - T_{Ej}$

$$= T_{L3} - T_{E3}$$

$$= 9 - 7 = 2 \text{ days}$$

For Activity (1–5)

Total float $= T_{Lj} - T_{Ei} - d_{(i-j)}$

$$= T_{L5} - T_{E1} - d_{(1-5)}$$

$$= 11 - 0 - 9 = 2 \text{ days}$$

Free float $= T_{Ej} - T_{Ei} - d_{(i-j)}$

$$= T_{E5} - T_{E1} - d_{(1-5)}$$

$$= 11 - 0 - 9 = 2 \text{ days}$$

Independent float $= T_{Ej} - T_{Li} - d_{(i-j)}$

$$= T_{E5} - T_{L1} - d_{(1-5)}$$

$$= 11 - 0 - 9 = 2 \text{ days}$$

Interfering float $= T_{Lj} - T_{Ej}$

$$= T_{L5} - T_{E5}$$

$$= 11 - 11 = 0$$

For Activity (2–4)

Total float $= T_{Lj} - T_{Ei} - d_{(i-j)}$

$$= T_{L4} - T_{E2} - d_{(2-4)}$$

$$= 13 - 4 - 8 = 1 \text{ day}$$

Free float $= T_{Ej} - T_{Ei} - d_{(i-j)}$

$$= T_{E4} - T_{E2} - d_{(2-4)}$$

$$= 12 - 4 - 8 = 0$$

Independent float $= T_{Ej} - T_{Li} - d_{(i-j)}$

$$= T_{E4} - T_{L2} - d_{(2-4)}$$

$$= 12 - 4 - 8 = 0$$

$$\text{Interfering float} = T_{Lj} - T_{Ej}$$
$$= T_{L4} - T_{E4}$$
$$= 13 - 12 = 1 \text{ day}$$

For Activity (2–5)

$$\text{Total float} = T_{Lj} - T_{Ei} - d_{(i-j)}$$
$$= T_{L5} - T_{E2} - d_{(2-5)}$$
$$= 11 - 4 - 7 = 0$$

$$\text{Free float} = T_{Ej} - T_{Ei} - d_{(i-j)}$$
$$= T_{E5} - T_{E2} - d_{(2-5)}$$
$$= 11 - 4 - 7 = 0$$

$$\text{Independent float} = T_{Ej} - T_{Li} - d_{(i-j)}$$
$$= T_{E5} - T_{L2} - d_{(2-5)}$$
$$= 11 - 4 - 7 = 0$$

$$\text{Interfering float} = T_{Lj} - T_{Ej}$$
$$= T_{L5} - T_{E5}$$
$$= 11 - 11 = 0$$

For Activity (3–4)

$$\text{Total float} = T_{Lj} - T_{Ei} - d_{(i-j)}$$
$$= T_{L4} - T_{E3} - d_{(3-4)}$$
$$= 13 - 7 - 4 = 2 \text{ days}$$

$$\text{Free float} = T_{Ej} - T_{Ei} - d_{(i-j)}$$
$$= T_{E4} - T_{E3} - d_{(3-4)}$$
$$= 12 - 7 - 4 = 1 \text{ day}$$

$$\text{Independent float} = T_{Ej} - T_{Li} - d_{(i-j)}$$
$$= T_{E4} - T_{L3} - d_{(3-4)}$$
$$= 12 - 9 - 4 = -1 \text{ day}$$

$$\text{Interfering float} = T_{Lj} - T_{Ej}$$
$$= T_{L4} - T_{E4}$$
$$= 13 - 12 = 1 \text{ day}$$

For Activity (3–6)

Total float $= T_{Lj} - T_{Ei} - d_{(i-j)}$

$\qquad = T_{L6} - T_{E3} - d_{(3-6)}$

$\qquad = 18 - 7 - 8 = 3$ days

Free float $= T_{Ej} - T_{Ei} - d_{(i-j)}$

$\qquad = T_{E6} - T_{E3} - d_{(3-6)}$

$\qquad = 18 - 7 - 8 = 3$ day

Independent float $= T_{Ej} - T_{Li} - d_{(i-j)}$

$\qquad = T_{E6} - T_{L3} - d_{(3-6)}$

$\qquad = 18 - 9 - 8 = 1$ day

Interfering float $= T_{Lj} - T_{Ej}$

$\qquad = T_{L6} - T_{E6}$

$\qquad = 18 - 18 = 0$

For Activity (4–6)

Total float $= T_{Lj} - T_{Ei} - d_{(i-j)}$

$\qquad = T_{L6} - T_{E4} - d_{(4-6)}$

$\qquad = 18 - 12 - 5 = 1$ day

Free float $= T_{Ej} - T_{Ei} - d_{(i-j)}$

$\qquad = T_{E6} - T_{E4} - d_{(4-6)}$

$\qquad = 18 - 12 - 5 = 1$ day

Independent float $= T_{Ej} - T_{Li} - d_{(i-j)}$

$\qquad = T_{E6} - T_{L4} - d_{(4-6)}$

$\qquad = 18 - 13 - 5 = 0$

Interfering float $= T_{Lj} - T_{Ej}$

$\qquad = T_{L6} - T_{E6}$

$\qquad = 18 - 18 = 0$

For Activity (5–6)

Total float $= T_{Lj} - T_{Ei} - d_{(i-j)}$

$\qquad = T_{L6} - T_{E5} - d_{(5-6)}$

$\qquad = 18 - 11 - 7 = 0$

Free float $= T_{Ej} - T_{Ei} - d_{(i-j)}$

$\qquad\qquad = T_{E6} - T_{E5} - d_{(5-6)}$

$\qquad\qquad = 18 - 11 - 7 = 0$

Independent float $= T_{Ej} - T_{Li} - d_{(i-j)}$

$\qquad\qquad\qquad = T_{E6} - T_{L5} - d_{(5-6)}$

$\qquad\qquad\qquad = 18 - 11 - 7 = 0$

Interfering float $= T_{Lj} - T_{Ej}$

$\qquad\qquad\qquad = T_{L6} - T_{E6}$

$\qquad\qquad\qquad = 18 - 18 = 0$

The different types of float for the various activities of network shown in Fig. 12.39 calculated above are tabulated in Table 12.5.

Table 12.5   Calculation of Floats for network shown in Fig. 12.39

Activity (i–j)	Duration (days)	Float (in days)			
		Total	Free	Independent	Interfering
1–2	4	0	0	0	0
1–3	7	2	0	2	2
1–5	9	2	2	2	0
2–4	8	1	0	0	1
2–5	7	0	0	0	0
3–4	4	2	1	0	1
3–6	8	3	3	1	0
4–6	5	1	1	0	0
5–6	7	0	0	0	0

Example 12.16

Compute different types of floats for the various activities of the network shown in Fig. 12.35.

Solution

The Earliest Occurrence Times (T_E) and the Latest Occurrence Times (T_L) of the various events of Fig. 12.35 have been computed as per the procedure described in sections 12.8 and 12.9 respectively. E.S.T., E.F.T., L.S.T. and L.F.T. of the various activities of the network shown in Fig. 12.35 have been computed as per the procedure given in Example 12.11. These values have been tabulated in Table 12.1. Different types of floats of the various activities of the network shown in Fig. 12.35 have been computed as per the procedure given in Example 12.15. These values have been tabulated in Table 12.6.

Table 12.6   Calculation of Floats for network shown in Fig. 12.35

Activity (i – j)	Duration (days)	Float (in days)			
		Total	Free	Independent	Interfering
1–2	2	0	0	0	0
1–3	5	10	0	0	10
2–4	3	0	0	0	0
2–7	3	9	9	9	0
3–8	3	10	10	0	0
4–5	4	0	0	0	0
5–6	3	0	0	0	0
6–7	2	0	0	0	0
7–8	4	0	0	0	0

Example 12.17

Compute different types of floats for the various activities of the network shown in Fig. 12.37.

Solution

The Earliest Occurrence Times (T_E) and the Latest Occurrence Times (T_L) of the various events of Fig. 12.37 have been computed as per the procedure described in sections 12.8 and 12.9 respectively. E.S.T., E.F.T., L.S.T. and L.F.T. of the various activities of the network shown in Fig. 12.37 have been computed as per the procedure given in Example 12.13. These values have been tabulated in Table 12.3. Different types of floats of the various activities of the network shown in Fig. 12.37 have been computed as per the procedure given in Example 12.15. These values have been tabulated in Table 12.7.

Table 12.7   Calculation of Floats for network shown in Fig. 12.37

Activity (i – j)	Duration (days)	Float (in days)			
		Total	Free	Independent	Interfering
1–2	6	2	0	0	2
1–3	7	0	0	0	0
1–4	5	3	1	1	2
2–4	0	2	0	-2	2
2–6	5	2	2	0	0
3–5	6	0	0	0	0
3–7	8	7	7	7	0
4–6	5	2	2	0	0
5–6	0	0	0	0	0
5–7	7	2	2	2	0
6–7	9	0	0	0	0

12.11.2　Negative Float

While computing the Latest Occurrence Times of the various events in a network, it has been assumed that the Latest Occurrence Time of the terminating event, i.e., the last event is equal to its Earliest Occurrence Time. However, this would be a wrong assumption if a specific target date of completion of the project has been specified. In such cases, the Latest Occurrence Time (T_L) of the last event will have to be taken equal to the specified target date. If the specified target date happens to be earlier than T_L of the last event, then backward pass carried out with reference to this date/time would result in negative float of the activities as illustrated in Example 12.18.

Example 12.18

The project shown in Fig. 12.40 is targeted to be completed in 15 days. Compute E.S.T., E.F.T., L.S.T., L.F.T. and Total Float of its various activities.

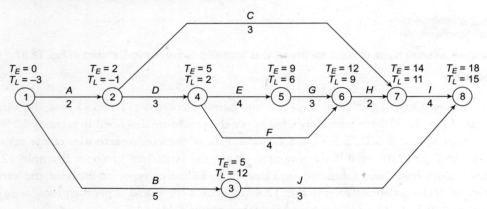

Figure 12.40　A simple network

Solution

The Earliest Occurrence Times (T_E) of the various events of the network shown in Fig. 12.40 have been computed and indicated in the network. The T_E of the last event, i.e., event 8 is 18 days. If T_L has now to be 15 days instead of 18, the process of backward pass will yield a fresh set of Latest Occurrence Times (T_L) of the various events. These are shown in Fig. 12.40. Taking into account the Latest Occurrence Time of the different events, the various activity times and Total Float of the activities can be computed. These are tabulated in Table 12.8.

Table 12.8　Activity-times and Total Float for network shown in Fig. 12.40

Activity $(i-j)$	Duration $d(i-j)$ (days)	E.S.T.	E.F.T.	L.S.T.	L.F.T.	F^t
1–2	2	0	2	−3	−1	−3
1–3	5	0	5	7	12	7
2–4	3	2	5	−1	2	−3
2–7	3	2	5	8	11	6

Activity (i – j)	Duration d(i – j) (days)	E.S.T.	E.F.T.	L.S.T.	L.F.T.	F^t
3–8	3	5	8	12	15	7
4–5	4	5	9	2	6	–3
5–6	3	9	12	6	9	–3
6–7	2	12	14	9	11	–3
7–8	4	14	18	11	15	–3

12.12 IDENTIFICATION OF CRITICAL PATH

Critical path/paths through a network are the path/paths on which critical activities of the network occur. Critical path/paths determine the project completion time. In the networks that do not have target completion date or time constraint, some of the activities possess zero total float. Zero total float of an activity implies that the activity does not have any spare time at all and any delay in completion of the activity would delay the overall completion of the project. These activities having zero total float are termed as critical activities and a continuous chain passing through such activities is known as critical path. There can be more than one critical path through a network.

Example 12.19

Find the critical path for the network shown in Fig. 12.36.

Solution

Total floats of this network have already been worked out in Example 12.15 and Table 12.5.

Activities (1–2), (2–5) and (5–6) have zero floats. So a continuous chain passing through these activities, i.e., (1–2–5–6), would determine the critical path in this network. This is shown in Fig. 12.41.

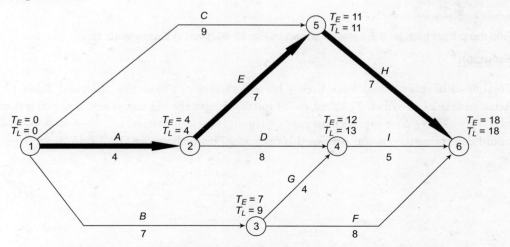

Figure 12.41 Identification of critical path for network shown in Figure 12.36

Example 12.20

Find the critical path for the network shown in Fig. 12.37.

Solution

Total floats of this network have already been worked out in Example 12.17 and Table 12.7. Activities (1–3), (3–5), (5–6) and (6–7) have zero floats. So a continuous chain passing through these activities, i.e., (1-3-5-6-7), would determine the critical path in this network. This is shown in Fig. 12.42.

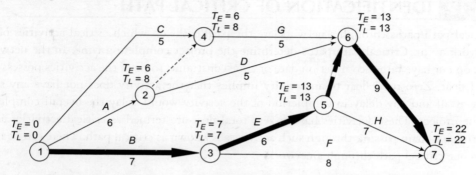

Figure 12.42   Identification of critical path for network shown in Figure 12.37

In cases where the project has to be completed by a target date/time which is earlier than the earliest occurrence time of the terminating, i.e., the last event, some of the activities will have negative floats. The critical path in such cases is obtained by connecting the first event to the last, in a continuous chain, through the events with the highest values of negative floats. Example 12.21 illustrates the identification of critical path of a network in which target date/time is less than the earliest occurrence time of the last event.

Example 12.21

Find the critical path for the network shown in Fig. 12.40 given in Example 12.18.

Solution

Total floats of this network have already been worked out in Example 12.18 and Table 12.7 Activities (1–2), (2–4), (4–5), (5–6), (6–7) and (7–8) have the highest values of negative floats which is − 3. So a continuous chain passing through these activities, i.e., (1–2–4–5–6–7–8), would determine the critical path in this network. This is shown in Fig. 12.43.

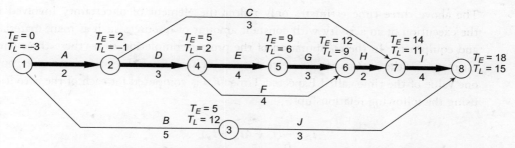

Figure 12.43 Identification of critical path for network shown in Figure 12.40

12.13 PROGRAM EVALUATION AND REVIEW TECHNIQUE (PERT)

As mentioned earlier, PERT stands for Program Evaluation and Review Technique. PERT also belongs to the family of Network Analysis Techniques. Like CPM, a network is needed for developing a PERT schedule. Although CPM and PERT belong to Network Analysis Techniques, they possess the following distinctive features:

1. CPM is an activity oriented technique and thereby, attaches importance to the activities. PERT is event oriented technique and, therefore, it attaches importance to events.
2. CPM uses a deterministic model and therefore, only one time is assigned to each activity of the network. On the other hand, PERT assumes stochastic or probabilistic model to reflect uncertainty in the times of completion of activity. Due to involvement of uncertainty, three times are assigned to each activity of the network.
3. CPM makes use of cost-duration relationships to arrive at an optimal schedule, i.e., a schedule with minimum duration, consistent with economy in cost. PERT on the other hand, does not take into consideration the cost-duration relationship.

12.13.1 PERT Activity Time Estimates

If an activity is performed repeatedly, its performance time would be different on different occasions. This happens owing to the uncertainty in the completion time of the activity. The uncertainty in the completion time is taken care of in PERT by making the following three estimates of the time-duration:

1. **Optimistic Time (t_o):** This is the shortest time that can be expected for completing an activity so as to reach the head event, when everything happens in the desired fashion as per the plan.
2. **Pessimistic Time (t_p):** This is the longest time that can be expected for completing an activity. This signifies a situation in which the activity could not be carried out as per the plan due to reasons beyond control.
3. **Most Likely Time (t_m):** If an activity is repeated many times, then the duration in which the activity is likely to get completed in a very large number of cases is called the most likely time.

The above three time estimates only reflect the element of uncertainty involved in the execution of an activity with constant amount of resources such as man, materials and equipment. In the computation of the project completion time, these three time estimates for the various activities are not used. Using these three time estimates, only one value of the time called Expected Time (T_E) is computed for each of the activities using the following relationships:

$$t_e = \frac{1}{6}(t_o + 4t_m + t_p)$$

The expected times of the activities are used to determine the project completion time.

12.13.2 PERT Computations

PERT computations involve the following steps:

1. Develop a network/arrow diagram to show the logical relationships among the various activities of the project. The procedure for developing the network is same as that of CPM as described in sections 12.5 and 12.6.

2. Using three time estimates, i.e., t_o, t_m and t_p, determine the expected time (t_e) for each of the activities using the following relationship:

$$t_e = \frac{1}{6}(t_o + 4t_m + t_p)$$

The expected time, i.e. t_e, represents the mean of the three time estimates assuming a beta distribution.

3. Reproduce the arrow diagram developed at step 1 and assign only one time value, i.e., t_e (calculated at step 2) to each of the activities.

4. Compute T_E and T_L of all the events following the procedure described in sections 12.8 and 12.9 respectively. Since PERT networks are event oriented, it is not necessary to compute activity times, i.e., E.S.T., E.F.T., L.S.T. and L.F.T.

5. Compute slack for each of the events of the network. Slack is given by the difference between T_L and T_E. Slack represents surplus time available to an event within which it must occur in order that overall completion of the project is not delayed.

6. Join the events that have zero slack through a continuous chain to get Critical Path/ Paths. Events that have zero slack are called critical events.

7. Add the t_e values of all the critical activities, i.e., the activities that fall on the critical path to get project completion time.

Example 12.22 illustrates the use of the above listed steps for PERT computation:

Example 12.22

Compute T_E, T_L and slack of each of the events of the network shown in Fig. 12.44. The three time estimates of each of the activities are given in Table 12.9. Identify the critical events, critical path and determine the project completion time.

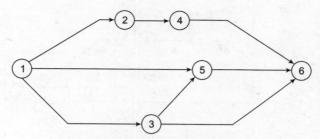

Figure 12.44   PERT network

Table 12.9   Three time estimates of activities

Activity (i − j)	t_o (days)	t_m (days)	t_p (days)
1–2	4	6	8
1–3	3	5	7
1–5	4	6	8
2–4	3	4	5
3–5	4	5	6
3–6	6	8	10
4–6	6	8	10
5–6	6	7	8

Solution

First, we calculate the expected time (t_e) for each of the activities. The three time estimates and computed values of t_e are tabulated in Table 12.10.

Table 12.10   Three time estimates and computed values of t_e

Activity (i − j)	t_o (days)	t_m (days)	t_p (days)	$t_e = \dfrac{1}{6}(t_0 + 4t_m + t_p)$ (days)
1–2	4	6	8	6
1–3	3	5	7	5
1–5	4	6	8	6
2–4	3	4	5	4
3–5	4	5	6	5
3–6	6	8	10	8
4–6	6	8	10	8
5–6	6	7	8	7

The network with the expected time of the activities is shown in Fig. 12.45.

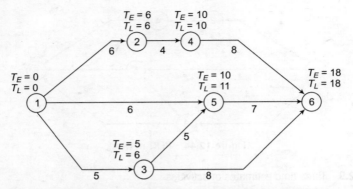

Fig- 12.45　PERT network with expected time of the activities

The T_E of the first event is zero, i.e., $T_{E1} = 0$. The T_E of event 2 can be found by adding the expected time of the activity (1-2) to the T_{E1}. Thus, the T_E of event 2 is $0 + 6 = 6$ days. The T_E of other events can be found in a similar manner. The T_E of various events of this network are tabulated in column 2 of Table 12.11.

Now, assume that T_L of the last event, i.e., event no. 6 is same as its T_E. Working backwards, T_L of event 5 is obtained by subtracting expected time of activity (5-6) from T_L of event 6. So, the T_L of event 5 is $18 - 7 = 11$ days. The T_L of other events can be found in the similar manner. These have been tabulated in column 3 of Table 12.11.

Slack of the event is obtained by taking the difference between T_L and T_E of the event. For example, T_L of event 1 is zero and its T_E is also zero. Therefore, its slack is zero. Slack of other events are obtained in the same manner. These values are listed in column 4 of Table 12.11.

Table 12.11　T_E, T_L and slack of various events of network shown in Fig. 12.45

Event No.	T_E (days)	T_L (days)	Slack (days)
1	0	0	0
2	6	6	0
3	5	6	1
4	10	10	0
5	10	11	1
6	18	18	0

It is clear from Table 12.11 that events 1, 2, 4, and 6 have zero slack and therefore, they are critical events. The critical path is obtained by joining these critical events through a continuous chain in the network, i.e., 1–2–4–6 Te critical path is highlighted in Fig.12.46.

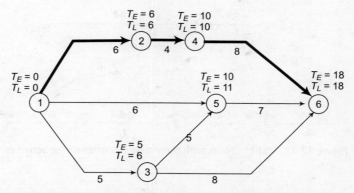

Figure 12.46 Critical path of the PERT network shown in Figure 12.45

Since activities (1–2), (2–4) and (4–6) fall on the critical path, these are the critical activities.

Project completion time is $= t_{e(1-2)} + t_{e(2-4)} + t_{e(4-6)} = 6 + 4 + 8 = 18$ days.

Example 12.23

The three time estimates of various activities of the network shown in Fig. 12.47 are given in Table 12.12. Calculate (i) the earliest and latest occurrence times of various events (ii) Slack of various events (iii) The project duration and (iv) identify the critical events, critical activities and critical paths.

Table 12.12 Three time estimates of various activities of network shown in Fig. 12.47

Activity ($i - j$)	t_o (days)	t_m (days)	t_p (days)
1–2	3	4	5
2–3	4	7	10
2–4	4	5	6
3–5	3	5	7
3–6		0	
4–6		0	
4–7	5	6	7
5–7	5	7	9
6–7	7	8.5	13

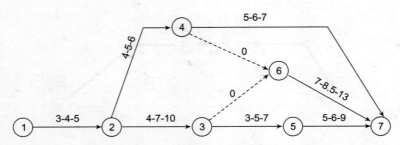

Figure 12.47   PERT network with three time estimates of the activities

Solution

The expected time (t_e) are calculated for each of the activities. These are tabulated in Table 12.13.

Table 12.13   Expected time (t_e) of the various activities of network shown in Fig. 12.47

Activity $(i-j)$	$t_e = \dfrac{1}{6}(t_0 + 4t_m + t_p)$ (days)
1–2	4
2–3	7
2–4	5
3–5	5
3–6	0
4–6	0
4–7	6
5–7	7
6–7	9

The network with expected time (t_e) of each of the activities is shown in Fig. 12.48.

The earliest occurrence times (T_E) and the latest occurrence times (T_L) for the various activities shown in Fig.12.48 can be computed now. These values are shown in Fig.12.48 and also tabulated in the Table 12.14.

Table 12.14   Earliest and Latest event times

Event No.	T_E (days)	T_L (days)
1	0	0
2	4	4
3	11	11
4	9	14
5	16	16
6	11	14
7	23	23

From the earliest and latest event times shown in Table 12.14, the slack of the events can be computed easily and the values of slack are tabulated in Table 12.15.

Table 12.15 T_E, T_L and slack of various events of network shown in Fig. 12.48.

Event No.	T_E (days)	T_L (days)	Slack = $T_L - T_E$ (days)
1	0	0	0
2	4	4	0
3	11	11	0
4	9	14	5
5	16	16	0
6	11	14	3
7	23	23	0

Event Nos. 1, 2, 3, 5 and 7 are critical events since their slack is zero and the critical path is obtained by a continuous chain passing through these events, i.e., the critical path is 1–2–3–5–7. Since activities (1–2), (2–3), (3–5) and (5–7) fall on the critical path, these activities are critical activities. The project duration is obtained by adding the expected time (t_e) of these critical activities, i.e., (1–2), (2–3), (3–5) and (5–7). Thus the project duration = 4 + 7 + 5 + 7 = 23 days.

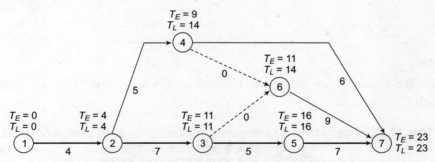

Figure 12.48 PERT network with expected time of the activities

Example 12.24

The Optimistic, Most Likely and Pessimistic time estimates of the various activities of a project shown in Fig. 12.49 are tabulated in Table 12.16. Work out the following:
 (i) The Earliest and Latest Occurrence Times of various events.
 (ii) Slack of various events.
(iii) The project duration.
(iv) Identify the critical events, critical activities and critical paths.

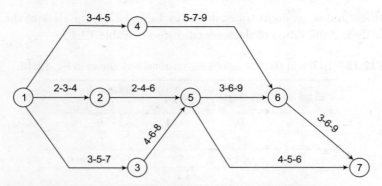

Figure 12.49 PERT network with three time estimates of the activities

Table 12.16 Three time estimates of various activities of network shown in Fig. 12.49

Activity ($i - j$)	t_o (days)	t_m (days)	t_p (days)
1–2	2	3	4
1–3	3	5	7
1–4	3	4	5
2–5	2	4	6
3–5	4	6	8
4–6	5	7	9
5–6	3	6	9
5–7	4	5	6
6–7	3	6	9

Solution

The expected times (t_e) are calculated for each of the activities. These are tabulated in Table 12.17.

Table 12.17 Expected time (t_e) of the various activities of the network shown in Fig. 12.49

Activity ($i - j$)	$t_e = \dfrac{1}{6}(t_0 + 4t_m + t_p)$ (days)
1–2	3
1–3	5
1–4	4
2–5	4
3–5	6
4–6	7
5–6	6
5–7	5
6–7	6

The network with expected time (t_e) of each of the activities is shown in Fig. 12.50.

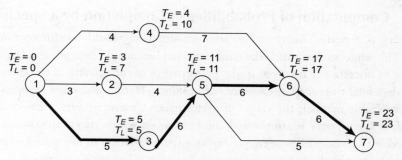

Figure 12.50 PERT network with expected time of the activities

The Earliest Occurrence Times (T_E) and the Latest Occurrence Times (T_L) of the various events of the network shown in Fig. 12.50 can be computed now. These values are shown in Fig. 12.50 and are also tabulated in Table 12.18.

Table 12.18 Earliest and Latest event times

Event No.	T_E (days)	T_L (days)
1	0	0
2	3	7
3	5	5
4	4	10
5	11	11
6	17	17
7	23	23

From the earliest and latest event times shown in Table 12.18, the slack of the events can be computed easily and the values of slack are tabulated in Table 12.19.

Table 12.19 T_E, T_L and slack of the various events of network shown in Fig. 12.50

Event No.	T_E (days)	T_L (days)	Slack = $T_L - T_E$ (days)
1	0	0	0
2	3	7	4
3	5	5	0
4	4	10	6
5	11	11	0
6	17	17	0
7	23	23	0

Event Nos. 1, 3, 5, 6 and 7 are critical events since their slack is zero and the critical path is obtained by a continuous chain passing through these events, i.e., the critical path is 1–3–5–6–7. Since activities (1–3), (3–5), (5–6) and (6–7) fall on the critical path, these activities are critical activities. The project duration is obtained by adding the expected time (t_e) of these critical activities. Thus, the project duration = 5 + 6 + 6 + 6 = 23 days.

12.13.3 Computation of Probabilities of Completion by a Specified Date

If an activity is repeated many times, its time-duration would be different on different occasions. The range within which the time-duration lies, would depend, to a large extent, on the element of uncertainty involved in the performance of the activity. If the execution of the activity is dependent upon a large number of unknown factors which are expected to change in an unpredictable manner, the values of performance times may vary considerably. On the other hand, if the activity is simple and the factors, on which its completion depends, can be predicted with a reasonable degree of certainty, the variation in the values of performance times would be minimal.

It is important for scheduling of projects to have a quantitative idea of the degree of uncertainty involved. The degree of uncertainty for any set of activities can be obtained by computing the probabilities of achieving the specified completion times. These probabilities, if plotted against the corresponding completion times, yield Probability Distribution Curves which are very useful in obtaining the probabilities of finalizing the project in different timings, within an overall range.

The three time estimates of activity duration can be used to calculate the Standard Deviation (S.D.) and the Variance (V). S.D. is a very good measure of spread of the values and, thus, the element of uncertainty involved. A high value of S.D. signifies a high degree of uncertainty. On the other hand, a low value of S.D. implies a low degree of uncertainty. PERT originators had assumed that the three estimates of activity duration lie on a probability distribution curve having a Beta Distribution. However, for ease of calculations, PERT originators suggested the following formula for calculating S.D.

$$\text{Standard Deviation (S.D.)} = \frac{t_p - t_o}{6}$$

$$\text{and Variance (V)} = \left(\frac{t_p - t_o}{6}\right)^2$$

where, t_p is the pessimistic time, and

t_o is the optimistic time

S.D. and variance can be calculated for various activities with the help of the above formulae, making use of the three values of the estimates of the time-duration. These can be used to compute the probability of achieving the scheduled date of completion.

Project duration is determined by the critical activities, as it is the sum of time-duration of the activities on the critical path. If $d_1, d_2, d_3 \ldots d_n$ are the duration of the n activities in a series, on the critical path and D the total project duration, then:

$$D = d_1 + d_2 + d_3 + \ldots + d_n$$

$$= \sum_{m=1}^{n} d_m$$

where d_m is the activity-duration of the mth activity on the critical path.

Let $V_1, V_2, V_3 \ldots V_n$ be the variances of n activities lying on the critical path of a network and V the variance and D the project duration, then:

$$V = V_1 + V_2 + V_3 + \ldots + V_n$$

$$= \sum_{m=1}^{n} V_m$$

where V_m is the variance of the mth activity on the critical path.

From the value of V, we can calculate the S.D. of the total project duration, which would be $\sqrt{V}$.

These values of variance and S.D. can be used to calculate the probability of meeting the scheduled date of completion as explained below:

The shape of the probability distribution curve for the project duration would be nearly normal. From the S.D. calculated above, it is possible to plot the probability distribution curve. The probability of achieving a scheduled date would be given by the area under the distribution curve. Example 12.25 illustrates the computation of probability of achieving a scheduled date of a given project.

Example 12.25

For the network shown in Fig. 12.51, identify the critical path and compute the probability of achieving project duration of (a) 23 days and (b) 26.5 days.

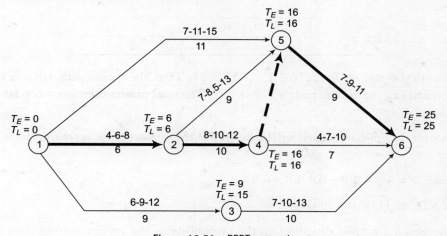

Figure 12.51 PERT network

Solution

The first step would be to calculate the expected time and variance of the various activities. These values are tabulated in Table 12.20.

Table 12.20 t_e and V of various activities of network shown in Fig. 12.51

Activity	t_o	t_m	t_p	$t_e = \dfrac{1}{6}(t_o + 4t_m + t_p)$	$V = \left(\dfrac{t_p - t_o}{6}\right)^2$
1–2	4	6	8	6	4/9
1–3	6	9	12	9	1
1–5	7	11	15	11	16/9
2–4	8	10	12	10	4/9
2–5	7	8.5	13	9	1
3–6	7	10	13	10	1
4–5	0	0	0	0	0
4–6	4	7	10	7	1
5–6	7	9	11	9	4/9

We calculate the Earliest Occurrence Time (T_E), Latest Occurrence Time (T_L) and slack of the various events. These values are tabulated in Table 12.21.

Table 12.21 T_E, T_L and slack of various events of network shown in Fig. 12.51

Event No.	T_E (days)	T_L (days)	Slack = $T_L - T_E$ (days)
1	0	0	0
2	6	6	0
3	9	15	6
4	16	16	0
5	16	16	0
6	25	25	0

So, the critical events are Event Nos. 1, 2, 4, 5 and 6. Thus, the critical path will pass through these events, i.e., the critical path is 1–2–4–5–6. The total project duration = 6 + 10 + 0 + 9 = 25 days.

Variance for the project duration will be the sum of the variance of the activities on the critical path.

Therefore, $V = 4/9 + 4/9 + 0 + 4/9 = 4/3$

So, the S.D. = $\sqrt{4/3} = 1.1547$

Now assuming a normal distribution, the probability of the project getting completed in 23 days will be given by the area under the normal probability curve. This can be found by referring to the standard tables pertaining to normal probability distribution curves (Table 1, Appendix B), the ratio,

$$z = \frac{23 - 25}{1.1547} = -1.73$$

This gives a value 0.0418 or a probability of just 4.18%.

Let us now find the probability of completion of the project in 26.5 days.

$$z = \frac{26.5 - 25}{1.1547} = 1.299 = \cong 1.3$$

Referring to the normal probability distribution curve (Table 1, Appendix B), the area under the curve is 0.9032. So, the probability of achieving this project completion date is 90.32%.

Example 12.26

The three time estimates of the various activities of a project are given in Table 12.22.
(a) Construct a network for this problem.
(b) Determine the expected time and variance for each activity.
(c) Determine the critical path and project time.
(d) Determine the probability that the project will be finished in 70 days.
(e) Determine the probability that the project will be finished in 80 days.

Table 12.22 Three time estimates of the activities given in Example 12.26

Activity	t_o (days)	t_m (days)	t_p (days)	Immediate Predecessor
A	9	11	13	—
B	7	8	9	—
C	3	4	5	—
D	11	21	31	A
E	7	8	9	C
F	10	11	12	B, D, E
G	6	7	7	B, D, E
H	15	16	17	F
I	10	11	12	F
J	7	8	9	G, H
K	4	7	10	I, J
L	1	2	3	G, H

Solution

(a) The network for this problem is shown in Fig. 12.52.

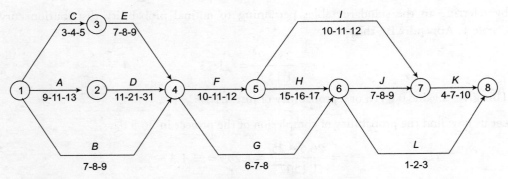

Figure 12.52   Network for Example 12.26

(b) The expected time and variance for each of the activities are tabulated in Table 12.23.

Table 12.23   Expected time (t_e) and Variance (V) of the various activities

Activity (i – j)	$t_e = \frac{1}{6}(t_o + 4t_m + t_p)$	Variance (V) = $\left(\frac{t_p - t_o}{6}\right)^2$
A (1–2)	11	4/9
B (1–4)	8	1/9
C (1–3)	4	1/9
D (2–4)	21	100/9
E (3–4)	8	1/9
F (4–5)	11	1/9
G (4–6)	7	1/9
H (5–6)	16	1/9
I (5–7)	11	1/9
J (6–7)	8	1/9
L (6–8)	2	1/9
K (7–8)	7	1

(c) In order to identify the critical path and to determine the project completion time, Earliest Occurrence Times (T_E), Latest Occurrence Times (T_L) and slack of the various events are required to be computed. These values are tabulated in Table 12.24. T_E and T_L are also shown in Fig. 12.53.

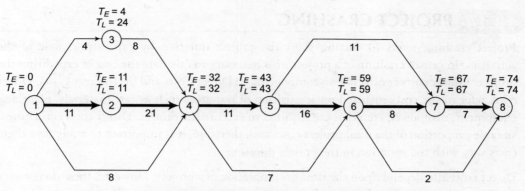

Figure 12.53 Earliest occurrence time and latest occurrence of events

Table 12.24 T_E, T_L and slack of various events of network shown in Fig. 12.52

Event No.	T_E (days)	T_L (days)	Slack = $T_L - T_E$ (days)
1	0	0	0
2	11	11	0
3	4	24	20
4	32	32	0
5	43	43	0
6	59	59	0
7	67	67	0
8	74	74	0

Thus, Event Nos. 1, 2, 4, 5, 6, 7 and 8 are critical events and the critical path is the continuous chain through these events. So, the critical path is 1–2–4–5–6–7–8. Total project duration = 11 + 21 + 11 + 16 + 8 + 7 = 74 days.

Variance for the project duration will be the sum of the variances of the activities on the critical path.

Therefore, $V = 4/9 + 100/9 + 1/9 + 1/9 + 1/9 + 1 = 116/9$

So, the S.D. = $\sqrt{116/9}$ = 3.59

(d) The probability of the project getting completed in 70 days is obtained as given below:

$$z = \frac{70 - 74}{3.59} = -1.11$$

This gives a value 0.1335 or a probability of 13.35%.

(e) Let us now find the probability of completion of the project in 80 days.

$$z = \frac{80 - 74}{3.59} = 1.67$$

Referring to normal probability distribution curve (Appendix B), the area under the curve is 0.9525. So, the probability of achieving this project completion date is 95.25%.

12.14 PROJECT CRASHING

Project crashing refers to cutting short the project duration by expediting some of the activities. In case of crashing of a project, it is necessary to calculate the cost of expediting the project. Project costs consist of two components (i) Direct Costs and (ii) Indirect Costs. Direct costs refer to the costs incurred on acquisition of resources such as raw materials, plant and equipment, manpower, etc., that are required to execute the project. Direct costs constitute a sizeable proportion of the total project costs and, therefore, it is important to study how these costs vary with the variation in the project duration.

Direct costs also depend upon the time of completion of a project. However, these do not vary linearly with the project duration. For a given site and project, normally, there would be an optimum duration at which the direct cost is the minimum. This duration yields optimal utilization of the various resources and, thereby, minimize direct cost. This project duration is sometimes known as *normal duration*. Any change in duration, either increase or decrease, would lower the level of utilization of these resources and direct costs would increase. The variation of direct costs with project duration is shown in Fig. 12.54.

Indirect costs refer to the costs incurred on overheads, administrative expenses and depreciation. Indirect costs are proportional to the duration of project. These increase linearly with time. So, longer the duration, more the expenditure on indirect cost. Conversely, if the project duration can be reduced, the indirect costs would come down. The variation of indirect costs with project duration is also shown in Fig. 12.54.

Total project cost is the sum of direct costs and indirect costs. As direct costs do not have a linear relationship with respect to duration, the total project costs also have a non-linear relationship with respect to time. The variation of total project costs with respect to time is also shown in Fig. 12.54. There is a unique project duration that yields minimum total project costs. This duration is known as *optimal duration*.

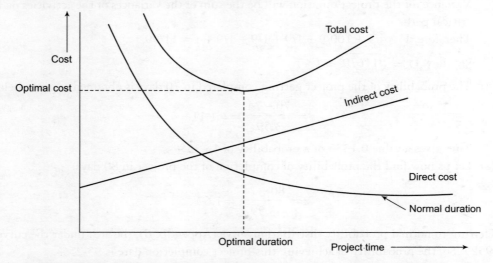

Figure 12.54   Relationship between direct, indirect and total cost and project duration

12.14.1 Cost Slope

The cost slope can be defined as the rate at which the cost either increases or decreases per unit time. Cost slope may be constant or variable, positive or negative. The cost slope with linear variation is shown in Fig. 12.55.

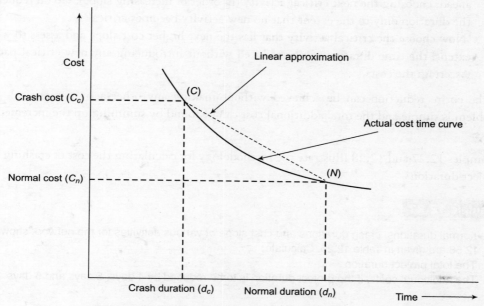

Figure 12.55 Relationship between project cost and duration

If a linear relationship between project cost and duration is assumed, then the cost slope is given by the following relationship:

$$\text{Cost slope} = \frac{C_c - C_n}{d_n - d_c}$$

where, C_c is the project cost for crash duration

C_n is the normal project cost i.e. cost for normal duration

d_n is the normal project duration

d_c is the crash project duration

12.14.2 Cost of Crashing

The procedure to determine the cost of crashing, i.e., the cost of expediting the completion of the project involves the following steps:

1. Work out the optimum project duration based on minimum direct costs.
2. Obtain the cost-duration relationships for the critical and non-critical activities.
3. Calculate the cost slope, i.e., the rate at which the direct costs increase/decrease per unit time.

4. Compute the indirect costs, i.e., the rate at which indirect costs decrease.
5. Choose the critical activity that has the lowest cost slope. Assess to what extent the duration can be reduced and at what costs.
6. Check whether by doing so, any other activity has also become critical. If not, proceed ahead, choosing the next critical activity (in order of increasing slope). Go on reducing the duration only to the extent that no new activity becomes critical.
7. Now choose the critical activity that has the next higher cost slope and assess to what extent, the time duration can be reduced without introducing any new critical paths. Ascertain the costs.

If the entire reduction can be achieved without making any other activities critical, the problem is simple and the total additional cost can be found by summing up the incremental costs.

Examples 12.27 and 12.28 illustrate the methodology for calculating the cost of crashing the project duration.

Example 12.27

The normal durations, crash durations and cost slope of various activities for the network shown in Fig. 12.56 are given in Table 12.25. Calculate:
 (i) The total project duration
(ii) The additional costs, if the project duration is to be reduced by 4 days, 5 days and 6 days

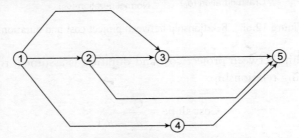

Figure 12.56 Network for Example 12.27

Table 12.25 Normal duration, Crash duration and Cost slope

Activity (i – j)	Normal duration (days)	Crash duration (days)	Cost slope (₹/day)
1–2	6	5	90
1–3	10	8	70
1–4	7	5	80
2–3	9	5	50
2–5	11	8	60
3–5	8	7	110
4–5	9	7	100

Solution

The first step is to identify the critical activities and critical path. In order to identify critical activities, we consider normal durations of each of the activities and compute T_E and T_L of events and E.S.T., E.F.T., L.S.T., L.F.T. and total float (F^t) of each of the activities. The T_E and T_L values of the various events are shown in Fig. 12.57. For each of the activities shown in Fig. 12.57, two time durations are given. The time-duration outside the parentheses is the normal time whereas the one given within parentheses is crash time. The E.S.T., E.F.T., L.S.T., L.F.T. and F^t of the various activities of the network shown in Fig. 12.57 are tabulated in Table 12.26.

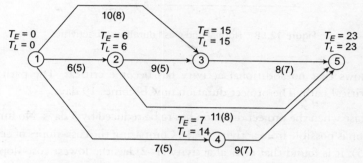

Figure 12.57 T_E and T_L values of events in Example 12.27

Table 12.26 Normal duration, E.S.T., E.F.T., L.S.T., L.F.T. and F^t of activities

Activity (i – j)	Normal durations (days)	E.S.T. (days)	E.F.T. (day)	L.S.T. days)	L.F.T. days)	F^t (days)
1–2	6	0	6	0	6	0
1–3	10	0	10	5	15	5
1–4	7	0	7	7	14	7
2–3	9	6	15	6	15	0
2–5	11	6	17	12	23	6
3–5	8	15	23	15	23	0
4–5	9	7	16	14	23	7

It can be seen from Table 12.26 that activities (1-2), (2-3) and (3-5) are critical activities as their total float is zero. The project completion time is $6 + 9 + 8 = 23$ days. On comparing the cost-slope of these critical activities, it is found that critical activity (2–3) has the lowest cost-slope, i.e., ₹ 50 per day, its normal duration is 9 days and crash, i.e., minimum duration is 5 days. It means that the duration of this activity can be reduced by 4 days. So, the project duration can be reduced by 4 days by reducing the time duration of activity (2–3) alone by 4 days. The additional cost of reduction of time duration by 4 days would therefore, be ₹ 50 * 4 = ₹ 200. Now, after reducing the time duration of the critical activity (2–3), the new time duration of the various activities are shown in Fig. 12.58.

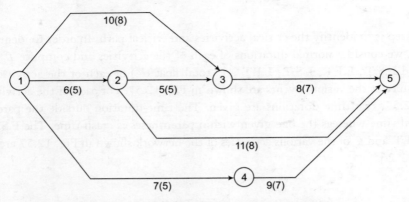

Figure 12.58   Normal and crash durations of activities

Fig. 12.58 shows that no additional activity has become critical. The path 1–2–3–5 still remains the critical path. The project duration now becomes 19 days.

Consider the case when the project duration has to be reduced by 5 days. No further reduction of time duration is possible from activity (2–3). Comparing the cost-slopes of critical activities (1–2) and (3–5), it is found that critical activity (1–2) has the lowest cost-slope, i.e. ₹ 90 per day. So, the additional cost of reduction of project duration by 5 days would be ₹ 200 + ₹ 90 = ₹ 290.

Now, after reducing the time duration of activity (1-2) by 1 day, the new time duration of activities are shown in Fig. 12.59. Fig. 12.59 shows that one more activity, i.e., (1-3) has become critical and in addition to path 1–2–3–5, path 1–3–5 has also become critical. The project duration now becomes 18 days.

Consider the case when the project duration has to be reduced by 6 days. No further reduction of time duration is possible from activities (1–2) and (2–3). Comparing the cost-slopes of critical activities (3–5) and (1–3), it is found that the cost-slope of (1–3) is the lowest, i.e., ₹ 70 per day. If time duration of activity (1–3) alone is reduced by 1 day, then it will not cause a reduction of the project duration by 6 days as the project duration is determined by the path 1–2–3–5, which gives 18 days. Therefore, in order to reduce the project duration by 6 days, it is necessary to reduce the time duration of activity (3–5) by 1 day. Thus, the reduction of project duration by 6 days, is achieved by reducing the time duration of activity (1–2) by 1 day as well as reducing the time duration of activity (3–5) by 1 day. Hence, the additional cost of reducing the project duration by 6 days would be ₹ 290 + ₹ 90 + ₹ 110 = ₹ 490.

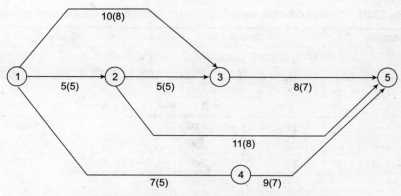

Figure 12.59 Normal and crash durations of activities

Example 12.28

The normal as well as crash costs and durations of the various activities of the network shown in Fig. 12.60 are given in Table 12.27. Calculate:

(i) The normal project cost as well as the duration for completion of the project represented by this network

(ii) The total project cost if the duration of the project has to be cut by 3 days

(iii) The total project cost if the duration of the project has to be cut by 5 days, 6 days and 7 days. Consider the indirect costs to be incurred at the rate of ₹75 per day.

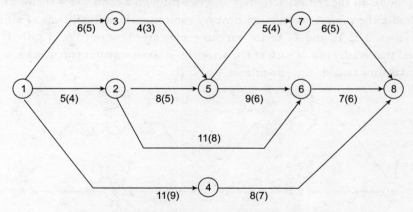

Figure 12.60 Normal and crash durations of activities

Table 12.27   Normal duration, normal cost, crash duration and crash cost of activities

Activity (i – j)	Normal		Crash	
	Duration, d_n (days)	Cost, C_n (₹)	Duration, d_c (days)	Cost, C_c (₹)
1–2	5	410	4	530
1–3	6	440	5	500
1–4	11	610	9	690
2–5	8	290	5	440
2–6	11	510	8	720
3–5	4	280	3	480
4–8	8	360	7	450
5–6	9	410	6	590
5–7	5	490	4	640
6–8	7	370	6	450
7–8	6	460	5	560

Solution

(i) First of all, we would consider normal duration of each of the activities and calculate T_E and T_L of the various events. Then we will calculate E.S.T., E.F.T., L.S.T., L.F.T. and Total Float of each of the activities. Subsequently, the critical path will be identified by joining all the critical activities in series through a continuous chain. Then we will compute the project completion time by summing the normal time of all the critical activities. The T_E and T_L of the various events are shown in Fig. 12.61. It should be noted that crash time of each of the activities is shown within parenthesis, whereas, the normal time outside the parenthesis.

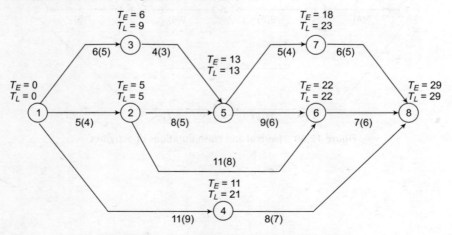

Figure 12.61 T_E and T_L values of the various events

The E.S.T., E.F.T., L.S.T., L.F.T. and F^t of the various activities are tabulated in Table 12.28.

Table 12.28 Normal duration, E.S.T., E.F.T., L.S.T., L.F.T. and F^t of activities

Activity (i – j)	Normal durations (days)	E.S.T. (days)	E.F.T. (day)	L.S.T. days)	L.F.T. days)	F^t (days)	Remarks
1–2	5	0	5	0	5	0	Critical
1–3	6	0	6	3	9	3	
1–4	11	0	11	10	21	10	
2–5	8	5	13	5	13	0	Critical
2–6	11	5	16	11	22	6	
3–5	4	6	10	9	13	3	
4–8	8	11	19	21	29	10	
5–6	9	13	22	13	22	0	Critical
5–7	5	13	18	18	23	5	
6–8	7	22	29	22	29	0	Critical
7–8	6	18	24	23	29	5	

From Table 12.28, it is clear that (1–2), (2–5), (5–6) and (6–8) are critical activities as their F^t values are zero. Thus, the critical path is 1–2–5–6–8. The project completion time is the sum of normal time of activities (1–2), (2–5), (5–6) and (6–8), i.e., $5 + 8 + 9 + 7 = 29$ days.

Total cost of the project when it is completed in normal time, i.e., 29 days, is obtained by adding the direct and indirect costs. Direct costs are obtained by summing the normal cost of each of the activities of the network. Thus, direct costs = 410 + 440 + 610 + 290 + 510 + 280 + 360 + 410 + 490 + 370 + 460 = ₹ 4,630.

Indirect costs are obtained by multiplying the given rate of indirect cost with the project duration. Thus, indirect costs = ₹ 75 * 29 = ₹ 2,175.

Therefore, the total cost of project when it is completed in normal time = ₹ 4,630 + ₹ 2,175 = ₹ 6,805.

(ii) In order to see the effect of reducing the project duration on the project cost, we need to calculate the cost slope of each of the activities. The cost slope of the activity is calculated by the following relationship:

$$\text{Cost slope} = \frac{C_c - C_n}{d_n - d_c}$$

The cost slope values of each of the activities are tabulated in Table 12.29.

Table 12.29   Normal duration, normal cost, crash duration, crash cost and cost slope of activities

Activity (i – j)	Normal		Crash		Cost slope $= \dfrac{C_c - C_n}{d_n - d_c}$ (₹/day)	Remarks
	Duration, d_n (days)	Cost, C_n (₹)	Duration, d_c (days)	Cost, C_c (₹)		
1–2	5	410	4	530	120	Critical
1–3	6	440	5	500	60	
1–4	11	610	9	690	40	
2–5	8	290	5	440	50	Critical
2–6	11	510	8	720	70	
3–5	4	280	3	480	200	
4–8	8	360	7	450	90	
5–6	9	410	6	590	60	Critical
5–7	5	490	4	640	150	
6–8	7	370	6	450	80	Critical
7–8	6	460	5	560	100	

In order to reduce project completion duration by 3 days, we have to reduce the duration of the critical activity that has the lowest cost slope by 3 days. On comparing the cost slopes of the critical activities given in Table 12.29, it is found that activity (2–5) has the lowest cost slope, i.e., ₹ 50 per day, its normal duration is 8 days, crash duration is 5 days and, therefore, 3 days can be reduced from this activity at an additional cost of ₹ 50 * 3 = ₹ 150.

So, the total direct cost would increase to ₹ 4,630 + ₹ 150= ₹ 4,780.

When 3 days are reduced from the critical activity (2–5), the project completion time is also reduced by 3 days and it would now be completed in 26 days. Therefore, indirect costs would be ₹ 75 * 26 = ₹ 1,950.

So, the total project cost when the project completion time is reduced by 3 days = ₹ 4780 + ₹ 1950 = ₹ 6,730. The network with reduced duration of activity (2–5) is shown in Fig. 12.62.

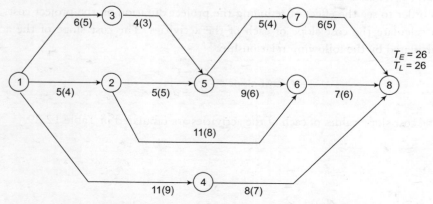

Figure 12.62   Network with reduced duration of activity (2–5)

(iii) It can be seen from Fig. 12.62 that in addition to path 1–2–5–6–8, path 1–3–5–6–8 also becomes critical as the project completion time through this new path is also 26 days. Thus, two more activities, i.e., (1–3) and (3–5) also become critical.

In order to reduce the project completion time by 5 days, 2 more days are required to be reduced from the duration of critical activities. No further reduction is possible from the duration of activity (2-5). Out of the critical activities (1–2), (5–6), (6–8), (1–3) and (3–5), the cost slope of (1–3) and (5–6) is equal and at the same time the lowest also. Activity (5–6) is selected for reduction purpose as it is possible to reduce 2 days from its duration and once duration is reduced both paths, i.e., 1–2–5–6–8 and 1–3–5–6–8 remain critical. The additional cost of reducing the project completion time by 2 days would be ₹ 60 * 2 = ₹ 120. The additional cost of reducing the project completion time by 5 days = additional cost of reducing time by 3 days + additional cost of reducing time by 2 days = ₹ 150 + ₹ 120 = ₹ 270. The direct cost of reducing project completion time by 5 days = ₹ 4,630 + ₹ 270 = ₹ 4,900. Indirect costs when the project completion time is reduced by 5 days = ₹ 75 * 24 = ₹ 1,800. Thus, the total cost when the project completion time is reduced by 5 days = ₹ 4,900 + ₹ 1,800 = ₹ 6,700.

The network with reduced duration of activity (5–6) is shown in Fig. 12.63.

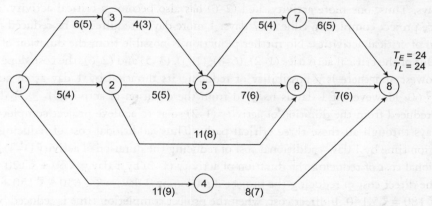

Figure 12.63   Network with reduced duration of activity (5–6)

From Fig. 12.64, it can be seen that paths 1–2–5–6–8 and 1–3–5–6–8 are still critical paths as the project completion time through these paths is 24 days and, therefore, no new activity has become critical. In order to reduce project completion time by 6 days, 1 more days is required to be reduced from the duration of critical activities. There is a further possibility of reducing the duration of activity (5–6) by 1 day. So, 1 day is reduced from the duration of activity (5–6) with an additional cost of ₹ 60. The additional cost of reducing project completion time by 6 days = additional cost of reducing time by 3 days + additional cost of reducing time by 2 days + additional cost of reducing time by 1 day = ₹ 150 + ₹ 120 + ₹ 60 = ₹ 330. The direct cost of reducing project completion time by 6 days = ₹ 4,630 + ₹ 330 = ₹ 4,960. Indirect costs when the project completion time is reduced by 6 days = ₹ 75 * 23 = ₹ 1,725. Thus, the total cost when project completion time is reduced by 6 days = ₹ 4,960 + ₹ 1,725 = ₹ 6,685.

The network with reduced duration of activity (5-6) is shown in Fig. 12.64.

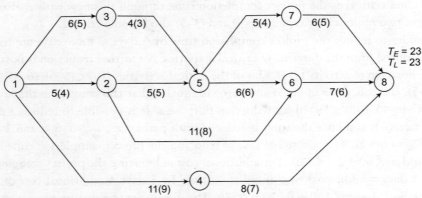

Figure 12.64　Network with reduced duration of activity (5–6)

From Fig. 12.64 it can be seen that in addition to paths 1–2–5–6–8 and 1–3–5–6–8, the path 1–2–6–8 has also become critical as the project completion time through all these three paths is 23 days. Thus, one more activity, i.e., (2–6) has also become a critical activity. In order to reduce project completion time by 7 days, 1 more day is required to be reduced from the duration of critical activities. No further reduction is possible from the duration of activity (5–6). Out of the critical activities (1–2), (6–8), (1–3), (3–5) and (2–6), the cost slope of (1–3) is the lowest and there is a possibility of reducing its duration by 1 day at an additional cost of ₹60. However, if 1 day is reduced from the duration of activity (1–3), 1 day must also be reduced from the duration of activity (1–2) so as to achieve project completion time of 22 days through all these three critical paths. Thus, additional cost of reducing project completion time by 1 day = additional cost of reducing the duration of activity (1–3) by 1 day + additional cost of reducing the duration of activity (1–2) by 1 day = ₹60 + ₹120 = ₹180. Thus, the direct cost of reducing the project duration by 7 days = ₹4,630 + ₹150 + ₹120 + ₹60 + ₹180 = ₹5,140. Indirect cost, when the project completion time is reduced by 7 days = ₹75 * 22 = ₹1650. Thus, the total cost when project completion time is reduced by 7 days = ₹5,140 + ₹1,650 = ₹6,790.

That is one more option that needs to be explored. Although the cost slope of critical activity (6–8) is higher than activity (1–3), but if 1 day is reduced from the duration of activity (6–8), then all three paths, i.e., 1–2–5–6–8, 1–3–5–6–8 and 1–2–6–8 still remain critical. Thus, the additional cost of reducing duration of activity (6-8) by 1 day is ₹80. The additional cost of reducing project completion time by 7 days = ₹150 + ₹120 + ₹60 + ₹80 = ₹410 Direct cost of the project when the project completion time is reduced by 7 days = ₹4,630 + ₹410 = ₹5,040. Indirect cost of project when the project completion time is reduced by 7 days = ₹75 * 22 = ₹1,650. So, the total cost when the project completion time is reduced by 7 days = ₹5,040 + ₹1,650 = ₹6,690. This option gives a relatively lower total cost and, therefore, it should be considered as it provides a better solution. Finally, the total cost when the project completion time is reduced by 7 days is ₹6,690. The total cost when the project completion time when reduced by different durations is summarized in Table 12.30.

Table 12.30 Summary of the project total cost when its completion time is reduced by different days

Project time reduced by	Total cost (₹)
3 days	6,730
5 days	6,700
6 days	6,685
7 days	6,690

From Table 12.30, it is clear that total cost of project when its duration is reduced by 6 days is the minimum and in the event of either increase or decrease in the project duration the total project cost increases. Thus, the optimum solution to this problem is project completion time = 23 days and total project cost= ₹6,685.

PROBLEMS

1. Define the terms "Project" and "Project Management".

2. Explain in brief the various phases of Project Management.

3. Explain construction and use of a Gantt chart for a project of your choice. What are limitations of a Gantt chart?

4. What is Network Analysis Technique?

5. Distinguish between:
   (a) Activity and Event
   (b) Initial and Terminating Events
   (c) Merge and Burst Events

6. What are full forms of CPM and PERT? Differentiate between CPM and PERT. How do these techniques help in decision making?

7. List the guidelines for drawing an arrow diagram.

8. What is a dummy activity? With the help of an example, demonstrate its use in an arrow diagram.

9. What are forward pass and backward pass methods? Explain the determination of Earliest Occurrence Time and Latest Occurrence Time of an event, Earliest Start Time, Latest Start Time, Earliest Finish Time and Latest Finish Time of an activity.

10. Explain the concept of Float and Slack. Distinguish between free, independent, and interfering floats.

11. What is a critical path? Why is it so important in project scheduling and control?

12. Explain the three time estimates that are used in PERT. How are the expected duration of a project and its standard deviation calculated?

13. A project consists of 10 activities from A to J. The relationship between the activities and duration of each activity are given in the table below:

Activity	Immediate predecessor	Duration (in days)
A	–	3
B	–	6
C	–	4
D	A	4
E	D	5
F	D	5
G	E	4
H	F, G	3
I	C, H	5
J	B, I	4

(a) Construct an arrow diagram.

(b) Compute Earliest Start Time, Latest Start Time, Earliest Finish Time and Latest Finish Time of each activity.

(c) Compute Total Float of each activity.

(d) Compute Free, Interfering, and Independent floats.

(e) Determine the Critical Path and Project Completion Time.

14. An overpass construction project consists of 11 activities from A to K. The relationship between the activities and duration of each activity are given in the table below:

Activity	Activity Description	Immediate predecessor	Duration (in weeks)
A	Design the Girder and Column	–	2
B	Design the Foundation	A	1
C	Order raw materials	B	1
D	Excavate Site	B	2
E	Construct foundation reinforcement	D	2
F	Cast Foundation	E	3
G	Cast Columns	F	3
H	Cast Girders	C	3
I	Mount girder supports on columns	G	1
J	Arrange crane	A	3
K	Install girder	H, I, J	1

(a) Construct an arrow diagram.

(b) Compute Earliest Start Time, Latest Start Time, Earliest Finish Time and Latest Finish Time of each activity,

(c) Compute Total Float of each activity.

(d) Determine the Critical Path and Project Completion Time.

15. A project that involves a welding operation for a tanker consists of 8 activities from A to H. The relationship between the activities and duration of each activity are given in the table below:

Activity	Activity Description	Immediate predecessor	Duration (in days)
A	Order the welding consumables, electrodes, gas, etc.	–	2
B	Cut the templates using oxy-gas cutting	A	3
C	Prepare fixtures for welding	–	6
D	Perform the welding operation	B, C	4
E	Inspect the welding by X-Ray	D	1
F	Obtain the antirust covering for tanker	–	15
G	Apply the antirust covering on the tanker	E, F	3
H	Perform final finishing operations	G	1

(a) Construct an arrow diagram.

(b) Compute Earliest Start Time, Latest Start Time, Earliest Finish Time and Latest Finish Time of each activity.

(c) Compute Total Float of each activity.

(d) Determine the Critical Path and Project Completion Time.

16. A project that involves extrusion process for making complex sections consists of 12 activities from A to L. The relationship between activities and duration of each activity are given in the table below:

Activity	Activity Description	Immediate predecessor	Duration (in days)
A	Design die	–	2
B	Manufacture die	A	4
C	Assemble die parts	B	2
D	Heat treat die	C	5
E	Remove previous die from the chamber	–	1
F	Mount mandrel in die	D	1
G	Order billet	–	1
H	Clean the chamber	C	8
I	Mount billet on supply line and cut slug	G	3
J	Arrange take-up line	E, F	3
K	Clean the ram, load slug and heat the chamber	H, I	5
L	Extrude	J, K	1

(a) Construct an arrow diagram.

(b) Compute Earliest Start Time, Latest Start Time, Earliest Finish Time and Latest Finish Time of each activity.

(c) Compute Total Float of each activity.

(d) Determine the Critical Path and Project Completion Time.

17. Consider the project network shown in the figure below:

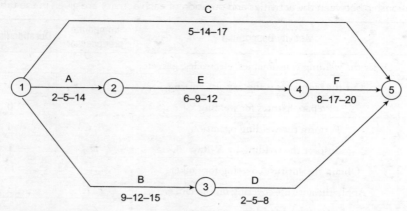

The project executive has made estimates of the optimistic, most likely and pessimistic times (in days) for completion of various activities as:

Activity	Time estimates (in days)		
	Optimistic	Most likely	Pessimistic
A	2	5	14
B	9	12	15
C	5	14	17
D	2	5	8
E	6	9	12
F	8	17	20

(a) Find the critical path.

(b) Determine the expected project completion time and its variance.

(c) What is the probability that the project will be completed in 35 days?

18. Consider a project having the following activities and their time estimates:

Activity	Immediate Predecessor	Time estimates (in days)		
		Optimistic	Most likely	Pessimistic
A	–	3	4	5
B	–	4	8	10
C	B	5	6	8
D	A, C	9	10	15
E	B	4	6	8
F	D, E	3	4	5
G	D, E	5	6	8

Activity	Immediate Predecessor	Time estimates (in days)		
		Optimistic	Most likely	Pessimistic
H	D, E	1	3	4
I	G	2	4	5
J	F, I	7	8	10
K	G	4	5	6
L	H	8	9	13
M	J, K, L	6	7	8

(a) Draw an arrow diagram for the project.

(b) Compute the expected project completion time.

(c) What should be the due date to have 0.9 probability of completion?

(d) Find the total float for all the non-critical activities.

19. The following table gives data on normal time and cost, and crash time and cost for a project.

Activity	Duration (weeks)		Total Cost (₹)	
	Normal	Crash	Normal	Crash
1–2	3	2	300	450
2–3	3	3	75	75
2–4	5	3	200	300
2–5	4	4	120	120
3–4	4	1	100	190
4–6	3	2	90	130
5–6	3	1	60	110

(a) Draw the network, find out the critical path and normal project duration.

(b) Find out the total float associated with each activity.

(c) If the indirect costs are ₹100 per week, find out the optimum duration by crashing and the corresponding project cost.

20. The time and cost estimates for a six-activity project are as follows:

Activity	Preceding Activity	Time estimates (in weeks)		Cost estimates (₹)	
		Normal	Crash	Normal	Crash
A	–	6	4	10000	14000
B	–	4	3	5000	8000
C	A	3	2	4000	5000
D	B	8	3	1000	6000
E	B	14	6	9000	13000
F	C, D	8	4	7000	8000

With the help of CPM technique, crash the project to its minimum length at the lowest possible direct cost, assuming there is no indirect cost.

21. The table below shows jobs, their normal times and costs, and crash time and cost estimates for a project.

Job	Normal Time (days)	Cost (₹)	Crash time (days)	Cost (₹)
1–2	6	1400	4	1900
1–3	8	2000	6	2700
2–3	4	1100	2	1500
2–4	3	800	2	1400
3–4	Dummy	–	–	–
3–5	6	900	3	1800
4–6	10	2500	6	3500
5–6	3	500	2	800

Indirect cost for the project is ₹ 300 per day.

(a) Draw the network of the project.

(b) What is the normal duration and cost of the project?

(c) If all activities are crashed, what will be the project duration and the corresponding cost?

(d) Find the optimum duration and minimum project cost.

□□□

13

Value Engineering

13.1 INTRODUCTION

To thrive in today's competitive markets, organizations are looking for smart ways of doing business and work. They cannot afford extensive time delays and increased costs. Value Engineering (VE) is a methodology developed to assist individuals as well as enterprises that will allow them to make changes to augment their processes or services. Several establishments discovered that the methods being followed in the past were insufficient and, therefore, it was essential for them to bring changes to their working style. Using the VE practices, small changes along with hard work and common sense can improve things for organization.

13.2 CONCEPT OF VALUE ENGINEERING

Larry Miles, a procurement engineer for General Electric, introduced the VE methodology during World War II. The problem he was facing was that he was not able to obtain material to manufacture the turbo-supercharger for the B-17 and P-47 airplanes. He developed a technique which did not solve the problem of procuring the material, but addressed the functionality of each material. The group determined the functionality of every component in the turbo-supercharger. Then they established the more economical constituents to attain the desired function.

VE can be defined as a systematized or methodical tactic, focused on analyzing the functionality of systems, apparatus, amenities, services and supplies with the objective of realizing their necessary functionality at the lowermost life-cycle cost, without affecting requisite capability, consistency, eminence and safety. Since more than seven decades, many organizations have been using, analyzing and refining the VE procedures. The basic aspect of VE is that it makes you acquire the ability to think in different ways. The basic features of VE are shown in Fig. 13.1.

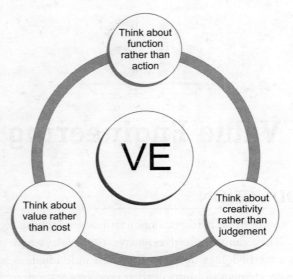

Figure 13.1   Basic aspects of VE

Value is the goal, not the cost. Sometimes the highest price items have the best value. Delivering value to the customer is the aim of the organization. Some people think that making things better means reducing the cost of the product and, in that process, sometimes the functionality of the item is reduced, which reduces the value. However, this is incorrect. VE is not against reducing cost, but not at the cost of reducing the functionality or value of the product. VE also emphasizes on cost reduction, but without reducing the functionality.

Generally, VE is achieved with the help of multidisciplinary teams. Earlier, organizations succeeded by giving a problem or a question to an individual. However, with so much competition now and to gain competitive advantage, an organization has to take into account several disciplines and perspectives. Therefore, it is necessary to work with multidisciplinary teams. The components of VE are shown in Fig. 13.2.

Figure 13.2   Components of VE

VE utilizes cogent logic and the evaluation of functionality to recognize relations that improve value. It is a quantifiable technique, analogous to the logical technique, which concentrates on hypothesis-conclusion methods to assess correlation and also to operations research, which uses mathematical modelling to find predictive interactions. Fig. 13.3 shows the difference between conventional methods and VE approach.

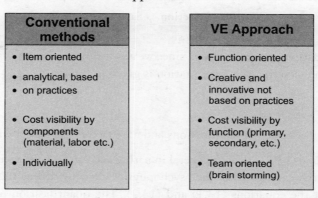

Figure 13.3   Difference between conventional and VE approach

VE utilizes an organized procedure to assess alternatives as depicted in Table 13.1.

Table 13.1   VE Structured process

Process	Action	Sub-action
Collect Information	1. How are the things done presently?	- Who is responsible to do it? - What might it be able to do? - What should not be done?
Quantify	2. How will the options be quantified?	- What are the different choices for achieving the specifications? - How can the required function be achieved alternatively?
Inspect	3. What action should be taken?	- What will it cost?
Produce	4. What are the other alternatives?	
Assess	5. What are the pre-eminent proposals?	
	6. Improve and make detailed plans	- What are the influences? - What will it cost? - What is the capability?
	7. Present the concept	- Promote options

13.3 NATURE AND MEASUREMENT OF VALUE

Measuring value could be an informal managerial decision procedure or a diligent official assessment procedure. Value can be realized as the ratio of the summation of the positive and negative features of an entity. Equation (13.1) depicts this simplistic interpretation:

$$\text{Value} = \frac{\sum +}{\sum -} = \frac{\text{Worth}}{\text{Cost}} = \frac{\text{Function}}{\text{Cost}} = \frac{\text{Satisfaction of Needs}}{\text{Use of Resources}} \tag{13.1}$$

In reality, this equation is more complex, since we are dealing with many variables of different magnitude. A more descriptive representation is given in Equation (13.2).

$$\text{Value} = \frac{mb_1 + mb_2 + \ldots + mb_n}{mc_1 + mc_2 + \ldots + mc_n} \tag{13.2}$$

where b are factors associated with functions and c are factors associated with cost.

Equations (13.1) and (13.2) are both general in nature and can be used to measure the value of items and alternatives. The field of the psychophysics has provided us the means to measure the parameters of the equations (13.1) and (13.2). The quantification of these parameters allows value, as expressed in these equations, to be measured.

13.3.1 The VE process

The VE process is a rational and structured process using an interdisciplinary team which follows the following steps:

(1) Select the proper project or product for analysis in terms of time invested in the study.

(2) Display and measure the current value of a product or its components in terms of functions that fulfill a user's needs, goals or objectives.

(3) Develop and evaluate new alternatives to eliminate or improve component areas of low value.

(4) Match the new alternatives with the best way of accomplishing them.

The process begins with the origination phase, where a VE study group is created and a project is chosen and described. The product and all its components are inspected thoroughly with an objective to gain a comprehensive understanding of their functionality.

The familiarization leads to the information phase, where the function(s) of the product and its parts are documented by function analysis techniques. Constraints that dictate the original design, material, components or procedures are challenged for validity. The importance and cost of functions are quantified by various value measurement techniques. The output of the information phase is a well-arranged list of functions or items, arranged from the highest to lowest relative values as they exist. The low-value items become candidates for value improvement.

These candidates are for the innovation phase, where various creative techniques are used to generate new alternatives for their replacement or improvement. The task then becomes one of reducing the large list of alternatives for development and recommendation.

The objective of the evaluation phase is to pre-screen the large list using various information reduction techniques. The highest-scoring alternatives that emerge from the pre-screening are subjected to further evaluation and discriminating value measurement techniques as those used in the State 1 analysis of the information phase. Value measurement in this phase referred to as State 2, or further state, measurement. The highest valued alternatives that emerge, now numbering only two or three, are then further examined for economic and technical feasibility, for their ability to perform the desired function satisfactorily and for their ability to meet other standards, such as accuracy, quality, reliability, safety, ease of repair and environmental regulations.

In the implementation phase, a report is generated to describe the study, document inferences and specify the proposal for the decision makers. Program and action plans are established to generate and apply the choices that are obtained from analysis phase. Techniques used here are in the realm of the production and project management. The implementation of VE change-proposals is observed and followed up in order to offer support, clear any misunderstandings and confirm that the recommended actions are achieved.

13.4 ORIGINATION PHASE

This is the first step of the VE process.

13.4.1 Organization

Management support is important for the initiation and success of a VE study. The following questions should be addressed at the outset: What is the scope of the study? Who funds the study and how much funding is required? Who approves the study? Who is the requestor? What are the study's start and completion dates? Who should be on the study team? Who are the intended users of the study results? What is the expected format? Oral? Written? What level of the organization should be involved? What geographic area? What are the expected time, personnel requirements, and cost? Once the study parameters are reviewed, endorsement must be taken from the management and the participants should have the appropriate time available to participate.

13.4.2 Project Selection

Many VE studies arise from a necessity in a specific, well-defined area. Consequently, a pre-defined project may obviate the need for formal project selection. In general, the resources that can be allocated to a VE study are often limited. The criteria that a VE study should look into are to:

(1) Solve a problem
(2) Have a good probability of success and implementation
(3) Have objectives that are credible
(4) Be important to the people in the area being studied

(5) Have the commitment of the requestor and the team members

(6) Have receptivity

13.4.3 The VE Team

A VE team normally comprises seven members. Odd numbers of members in the team are preferred to avoid a split decision. The following characteristics are very important for constituting a team:

(1) It should be interdisciplinary.

(2) Members must be at equivalent levels in the organization hierarchy.

(3) They should appoint a decision maker among themselves.

(4) It is necessary that one or more members are well-versed in the VE process or they should have an external consultant to supply the VE methodology.

(5) At least one member should be an "expert" on the product or subject being studied.

13.5 PROJECT OR STUDY MISSION

Once the VE study has been selected and defined and the team formed, it is very helpful for the team to formulate a mission statement. The statement must be short and provide a broad definition of what is to be accomplished by the project or study and why. Although the statement may appear to be unreasonable or impossible, it does help to provide a challenge and generate a positive, creative attitude among the members.

13.5.1 Product Definition and Documentation

The VE team begins by collecting information on the subject being studied. It defines the product and its components through a review of facts. All possible sources of information should be pursued. It is necessary to have sufficient factual data in order to minimize personal opinions and bias, which adversely affect the final results.

13.6 INFORMATION PHASE

In this phase all the requisite information is gathered by the VE team.

13.6.1 Qualitative Analysis of Value: Function Analysis

The product and all its components are studied to determine their functions. Function analysis consists of definitional and structural techniques that employ a semantic clarification of functions which provides the basis for deriving a quantitative analysis of value.

A verb and a noun are needed in this process to define the functions. Functions are described according to the following guidelines:

(1) Each user's requirement for a service or product should be listed down.

(2) To illustrate a function, only one noun and one verb should be used.

(3) Secondary verbs should be avoided in the descriptions.

(4) Ignore sentences or expressions which direct only toward objectives or aim.

(5) Write many two-word pairs and then choose the most appropriate one.

Function description should be created for the product and all of its parts. Functions are frequently classified as either basic or secondary. A basic function is the prime reason for the existence of the product. Secondary functions, on the other hand, support the basic function(s) and allow them to occur.

13.6.2 Function Analysis Systems Technique (FAST)

The usage of Function Analysis distinguishes VE from other problem solving methods. A well-known and powerful structural technique of function analysis used by VE practitioners is Functional Analysis Systems Technique (FAST). It was created by Charles Bytheway in 1964. It is basically based on "Hows" and "Why". Three main questions with respect to the function analysis are: (1) What should it do? (primary requirements, basic needs); (2) What else will do that? (develop alternatives); and (3) What's the best alternative? (select the best, optimum option).

The technique allows two-word function definitions to be arranged in an order of importance on the basis of action and reaction. It expands upon the reason for the existence of the product, or the basic function. A FAST diagram is prepared as follows:

(1) All the functions accomplished by the product and its elements are described by using a pair of words which define the function. Each function is recorded on a separate small card to facilitate the construction of the diagram.

(2) The card that best describes the basic function is selected among the many cards.

(3) A branching tree structure should be created from the basic function by asking the question "How do I (verb) (noun)?" A more de-personalized branching question would be, "How does (the product) do this?" The "how" question will result in branching and is repeated until branching has stopped and the function order is in a logical sequence.

(4) The logic structure is verified in the reverse direction by asking question "Why do I (verb) (noun)?" for each function in the logic sequence. The how-why questions are used to test the logic of the entire diagram. Answers to the how and why logic sequence questions must make sense in both directions, that is, answers to how questions must logically flow from left-hand to right-hand and answers to why questions must read logically from right-side to left-side.

(5) A "vital function path" is an outcome from the logic sequence of the elementary and subsidiary functions. Any function not on this path is a prime target for redesign, elimination, and cost reduction.

(6) "Scope lines" are used in the FAST diagram. These lines bounds both ends of the diagram and outlines the boundaries

In simple words it is a "Why" and "How" logic diagram, as shown in Fig. 13.4.

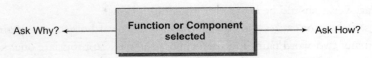

Fig 13.4   "Why" and "How" logic

A relationship between the functions at upper stage and lower stage is determined after answering all the rational queries of "How?" and "Why?" in the list. The FAST diagram depicts the hierarchy of these functions. Then a vital function path is established. The functions that are the outcome of other functions to be accomplished are classified by the critical or vital path. Let us take an example of a classroom transparency projector. The chosen characteristic was to "show diagram". The rational questions framed was "Why do we show a diagram?" and the response was, "To teach students." How is the "show diagram" achieved? "By reflecting image." How do we attain the task "reflect image?" "By projecting image." This is depicted in Fig. 13.5.

Figure 13.5   "Why" and "How" diagram of a classroom projector

Simultaneously, the query "Why?" should be replied. Why do we "project image"? "To reflect image". Why "reflect image"? To "show diagram" and so on. Then, it will be asked, "How is the project image function performed?" The answer is, "Illuminating transparency". The subsequent query, "How do we illuminate transparency?" results in two replies: (1) Through a light source and (2) When the transparency is placed in an appropriate manner. Thus, "provide light" and "place transparency", both are required to reply the query "How?" Then, what about the "Why?" query. The response of both the queries, "Why do we place transparency?" and "Why do we provide light?" is "To illuminate transparency". The reasoning queries are fulfilled, and the FAST diagram's blocks are arranged as shown in Fig. 13.6.

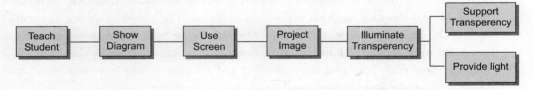

Figure 13.6   Basic FAST diagram

With the purpose of making boundaries of the FAST diagram for a particularly decided issue, the domain is determined and bounded on the FAST diagram by the "scope lines" (vertical straight lines depicted as dashed lines). The left-hand side line is positioned amid the elementary function which is under focus and the top-level elementary function. Alternatively, it can be said that the function on the left-hand side of this line is the subsequent top-level function that will not be entirely fulfilled by the solution of this product. The function positioned at the right-hand side is the elementary function that should be fulfilled by this solution. Here, in the example under consideration, the overhead projector should adequately "show diagram", however it will accomplish the "teach student" function only to some extent.

The right-hand side scope line is positioned to the left-hand side of a function which provides suitable interface to the solution under study. This means an appropriate input to the arrangement. In the considered example, the "provide power" interface function is at the bottom-most stage. The maker of these projectors believes that the user will arrange for an alternative electrical connection and the transparency slides. The decision to select the scope is subjective, and it is very important to rethink the scope toward the finishing of the FAST diagram.

The function to the right-hand side of a chosen function informs us about how it is executed. The function to the left-hand side specifies why the function is executed. The functions which are performed at the same period of time are positioned above one another. Lastly, the FAST shows which functions are executed by which component, if the components are presented, as shown in Fig. 13.7.

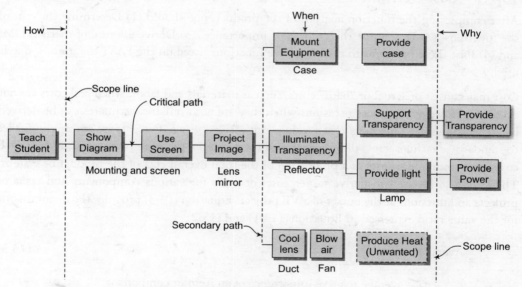

Figure 13.7   FAST diagram for classroom transparency projector

A FAST diagram by itself is not very useful and appears to be confusing and formidable to those not involved in its construction. The main value of the diagram lies in the intensive questioning and penetrating analysis required for its development. The FAST diagram becomes even more valuable and useful in the value measurement phase if costs and importance are allocated and posted for the functions on the diagram. The function analysis categorizes the components based on its functions; either there is a primary or a secondary function of components. FAST is an efficient technique to find the relation between functions. Investigating a complete system then provides an enhanced knowledge about the relation between costs and functions.

13.6.3 Constraints Analysis

Constraint analysis is helpful to challenge the constraints or reasons that dictate a particular component, material, design or procedure that is currently being used. Too often products become over-designed because the original constraints that dictated the design are no longer valid or are misinterpreted. The questions to answer are, "Why do we use what we use?" and "Is this reason still valid?" Once identified, costs due to invalid constraints should be derived and combined with function cost.

13.7 QUANTITATIVE ANALYSIS OF VALUE – STATE 1 VALUE MEASUREMENT

The next step in VE study is to determine the costs and their merits or importance and value.

13.7.1 Cost Derivation

After completing the function analysis of the product, one should (1) Determine the cost of the functions, (2) Determine their worth or importance, (3) Derive a figure of merit (FOM) and (4) Post these measures to the respective functions listed on the FAST diagram, if one is used.

Cost may consist of actual or "hard" costs, such as materials and labor, when real costs are not available. However, for those occasions when they are not available, estimates can be derived by the cost appraisal methods that produce subjective figures. Percentages are used in order to standardize ratings across different raters if there is more than one participant. Normalized costs show individual costs as a proportion of the total overall cost of the product or system. They can also be used to derive target costs or goals for various components and tasks of projects and missions at the outset of a VE project. Equation (13.3) provides the denominator for the value ratio expressed in Equations (13.1) and (13.2).

$$\text{Average} = \frac{C_1 + C_2 + \ldots + C_n}{n} \qquad (13.3)$$

C = average relative importance of an item or component

$C_{(1, 2, \ldots, n)}$ = individual participants' estimates of importance

n = total number of individuals rating importance

13.7.2 Worth or Importance Derivation

The functions are also studied for worth or importance. The worth or importance of functions, like cost, is determined indirectly by deriving the worth and importance of the items or components that collectively provide those functions. Worth is established by a technique of external cost comparison and is defined as the lowest cost that will reliably achieve the required function. It is determined by creatively comparing the cost of the function of the item or part to the cost of the function of analogous external items that can also reliably perform the same function. Importance, on the other hand, is established by an internal comparison of the items or components by comparing and rating each of them relative to the others. Equation (13.4) represents the derivation of importance as a percentage and provides the numerator for the value ration expressed

$$\text{Average} = \frac{I_1 + I_2 + \ldots + I_n}{n} \tag{13.4}$$

I = average relative importance of an item or component

$I_{(1, 2, \ldots, n)}$ = individual participants' estimates of importance

n = total number of individual rating importance

Both worth and importance may also be posted on their respective functions on the FAST diagram. It is important to emphasize that function comparison, whether it is for worth, importance or cost, should be performed at the same hierarchical level.

13.7.3 The Value Index

Once cost and importance are derived and posted for each function, they are used to compute in the numerator; whichever is used is situational. Similarly, the denominator may consist of actual hard costs, relative hard costs, or subjective estimated soft costs, although the former are preferred.

$$\text{Value Index} = \frac{\text{absolute worth}}{\text{absolute cost}} = \frac{\text{relative } I}{\text{relative } C}$$

The value index is a dimensionless number that allows one to array a system of functions in order of perceived value. Generally value index greater than 1.0 represents good value; an index less than 1.0 can indicate a function or component that needs attention and improvement.

13.7.4 Value Measurement Techniques

Value measurement techniques of paired comparisons and direct magnitude estimation (DME) are the most useful and descriptive process. The usual evaluation process employs a comparison of items where an individual chooses from a pair of functions or items the one that has a higher level of some specific characteristic. This procedure is iterated over all possible pairs and subsequently the numbers associated with each function or item are summed up. This process can be cumbersome and impractical if there are a large number items to evaluate and in this case DME (direct magnitude estimation) is recommended.

DME is a method in which people assign numbers to items in direct proportion to the magnitude of characteristic that the items possess. DME also allows one to express the magnitude of differences between items which is effectively used for less than 30 items.

The value measurement phase culminates in the quantification of functions and components according to value, importance and cost. These parameters are used to select likely candidates for redesign, improvement or elimination. Value for all the functions is individually computed from the importance and cost summation at any point in the diagram.

13.8   INNOVATION PHASE

This is the inventive phase where imaginative ideas are generated.

13.8.1   Improvement of Value

The innovation phase is the creative part of the VE process and a vital step in the redesign process. Activities in this phase are geared toward creating alternative ways of accomplishing functions. There are many ideas for collecting methods and one must decide what type to use. All criticism and evaluation are eliminated from the idea producing stage to allow maximum output by allowing everybody to participate. One may decide to use the traditional brainstorming or the nominal group technique (NGT). Nominal groups involve individuals who work in the presence of others, but who do not interact verbally with one another. Various combinations of function innovations produced several alternatives design are evaluated under evaluation phase to produce new design.

13.9   EVALUATION PHASE

This phase involves the evaluation of the ideas generated.

13.9.1   Pre-screening: Qualitative Analysis of Value

The evaluation phase entails a selection process in which ideas produced in the innovation phase are examined and a small number of ideas are selected. Pre-screening is required to screen the ideas from the innovation phase. Two effective, simple and fast, with high degree of discrimination screening methods are Pareto voting and Q-sort. Candidates for further VE study can be selected from the highest-valued items.

13.9.2   Quantitative Analysis of Value

The select group of items that emerges from the pre-screening is subjected to a more rigorous and discriminating evaluation to arrive at two or three candidates for possible development. The most suitable discriminating technique often used here is criteria analysis. Criteria analysis is a matrix scoring technique designed for evaluating alternatives by judging their individual merit against a set of criteria important for their end-use. The most important advantage of criteria analysis is that the process utilizes a common reference set of criteria. With paired comparison or DME, no common set of evaluation criteria is established. Criteria analysis also displays how each alternative satisfies each criterion. Decision rationale is documented and can be repeated in the light of new information or challenge.

13.10 IMPLEMENTATION PHASE

This phase concerns the preparation of the VE team's recommendations to the management. A report is prepared that describes the proposal(s) and lists suggested action plan for implementation. To minimize rejection, careful consideration should be given to report preparation. When a written document given to the decision makers, a brief executive summary should be included in the report to make it more effective. An oral presentation is often an excellent supplement to a written document.

Actual implementation of the recommendation is sometimes carried out by the VE team, but is often carried out by others. The study team provides the necessary inputs for the decision makers and the team members, along with the detailed report, which are available to assist the department or individuals responsible for implementation. Having the decision maker as a team member will ease the acceptance of the study. Planning and scheduling methods such as PERT (Program Evaluation Review Technique) are very useful in implementing VE recommendations.

PROBLEMS

1. Define value and explain how it is measured.
2. What is Value Engineering (VE)? When is it used and how is it applied?
3. Explain the steps involved in VE. Discuss the advantages and application areas of VE.
4. Differentiate between conventional and VE approaches.
5. Discuss the difference between VE and cost reduction techniques.
6. Differentiate between Value Engineering (VE) and Value Analysis (VA).
7. Explain how a team approach helps in value analysis in an organization.
8. Explain the Function Analysis Technique (FAST) with the help of a suitable example.
9. Draw FAST diagram of brewing tea.
10. Explain the various phases of VE Analysis.

❑❑❑

14

Forecasting

14.1 INTRODUCTION

In a competitive business environment, the success of an organization, be it manufacturing or service sector, depends to a great extent on its ability to correctly estimate the future demand of its products or services. The organizations can accomplish the task of estimation of future demand by using forecasting methods. Forecasting helps organizations in proper planning and control of operations and other resources which in turn ensures fulfillment of customers' future demand. Several methods of forecasting have been developed to deal with forecasting problems of different nature. In this chapter, those methods of forecasting that specifically suit to production or operations situations are discussed and described.

There is seldom a single forecasting method that can produce satisfactory results. One organization may find one method effective for its product or service, another may use some other method and a third may use a combination of methods. Forecasts of different time spans are required to serve as the basis for operating plans developed for different planning horizons. These include (1) plans for current operations and for the immediate future, (2) intermediate-range plans to provide for the capacities of personnel, materials and equipment required for the next 1 to 12 months, and (3) long-range plans for capacity, locations, changing product and service mix, and the development of new products and services.

The forecast required for the planning of daily or weekly operations is called short-term forecasting on the other hand. The forecast required for the planning horizon of 1 to 12 months and 1 to 10 years or more is termed as intermediate-term and long-term forecasting respectively. While selecting a forecasting method it is necessary to consider a method that produces reasonably accurate results for the planning horizon under consideration. For example, if the decision will deal with activities over the next three months, a one month forecast would be valueless. On the other hand, it is unwise to select forecasting model for daily or weekly decisions.

14.2 BASIC CATEGORIES OF FORECASTING METHODS

Forecasting methods can be divided into following three categories:
- Extrapolative or time series method

- Causal or explanatory method
- Qualitative or judgmental methods

In some situations, a combination of methods may be more appropriate than a single method.

Extrapolative or time series methods require past history of demand to forecast for the future. These methods, indeed, identify the pattern in historic data and extrapolate this pattern for the future. These methods perform well when the planning horizon is short. In other words, these methods are quite effective for short-term forecasting.

Causal methods of forecasting assume that the demand for an item depends on one or more independent factor, for example, price, advertising, competitor's price etc. These methods seek to establish a relationship between the variable to be forecasted and independent variables. Once the relationship is established, future values can be forecasted by simply plugging in the appropriate values for independent variables.

Judgmental methods depend on experts' opinion in making a prediction for the future. These methods perform well for medium to long-range forecasting tasks. These methods appear to be unscientific but in the absence of past data these are perhaps only methods that can be used for forecasting.

14.3   EXTRAPOLATIVE METHODS

The objective of extrapolative methods is to identify patterns in the past data so as to extrapolate it to obtain future estimates. Most of the patterns present in the past data depend on four components of demand: horizontal, trend, seasonal, and cyclical. These components are discussed below.

14.3.1   Components of Demand

The past history of data exhibits the horizontal component of demand when the demand fluctuates around an average demand. The average demand remains constant and does not consistently increase or decrease.

The trend component of demand exists when there is a sustained increase or decrease in the demand from one period to the next. For example, if the average monthly demand for a product has increased 5 to 10% in each of the past 15 years, then an upward trend in demand exists.

The seasonal component of demand refers to the impact of seasonal factors that affect demand positively or negatively. For example, the sales of air conditioners will be higher in summer months and lower in winter months every year, indicating a seasonal component of the demand for air conditioners.

The cyclical component of demand is similar to the seasonal component except that seasonality occurs at regular intervals and is of constant length, whereas the cyclic component varies in both time and duration of occurrence. For example, the influence of economic recession on the demand for a product will be reflected by the cyclic component. Recessions occur at irregular intervals and the length of time a recession lasts varies. This component is exhibited by most economic data, such as GDP, personal income, and industry sales of such consumer durables as automobiles and major appliances.

14.3.2 Moving Average Method

This is the simplest extrapolative method of forecasting. In this method, the following steps are needed to make forecast for the next period from the past data:

Step 1   Select the number of periods for which moving averages will be computed. This number N is called the order of moving average.

Step 2   Compute the average demand for the most recent N periods. This average demand then becomes the forecast for the next period.

Example 14.1 illustrates this method.

Example 14.1

The demand for a product in the months of January, February, March, April and May was 100, 90, 130, 110 and 90 units respectively. Forecast the demand for the month of June.

Solution

Step 1   Let us select number of periods $N = 3$. Since $N = 3$, the method is termed as three-period moving average. Other values of N can also be chosen. Larger N values will have a greater smoothing effect on random fluctuations in demand. Smaller N values will emphasize the more recent demand history. It may be noted that $N = 1$ will result in the present period demand as the forecast for the next period.

Step 2   Compute the average demand for the most recent 3 periods, $N = 3$

$$\text{Moving average} = \frac{\text{Demand for March, April, May}}{3}$$

$$= \frac{130 + 110 + 90}{3}$$

$$= 110 \text{ units}$$

Thus, forecast for June is 110 units

Now, suppose the actual demand for June turns out to be 115 units. The forecast for July will be computed by taking the average demand for April, May, and June. This forecast will be:

$$\frac{110 + 90 + 115}{3} = 105 \text{ units}$$

Once N is selected, the new moving average for each future period is computed by taking the average demand for the most recent N periods. This method is very simple in operation but the major disadvantage of this method is that it requires the storage of demand data for N periods for each item. Thus, in a situation where forecasts for a large number of items are to be made, the storage requirement is quite significant. Further, this method will not provide good forecasts if demand data reflect trend or seasonal components. For example, if there is an upward trend in the data, then the forecast made by this method will underestimate the actual demand.

The moving average method gives equal weight to the demand in each of the most recent N periods. However, this method can be modified by assigning a different weight to the demand of each previous period. Weighted moving average and exponential smoothing methods, which are discussed next, are suitable for accomplishing the differential weighting of demand in previous periods. In addition, the exponential smoothing methods can also incorporate trend and seasonality components of demand in forecasting.

14.3.3   Weighted Moving Average Method

This method of forecasting is similar to the moving average method except that the demand for each of the most recent N periods is given different weights according to the importance of the data. In this method, the following steps are needed to make forecast for the next period from past data:

Step 1 Select the number of periods N for which moving averages will be computed.

Step 2 Choose appropriate weights to be given to the demand for each number of periods N. The weights should be chosen in such a way that the sum of all weights is equal to 1.

Step 3 Use the following relation to compute the average demand for the most recent N periods. This average demand then becomes the forecast for the next period.

$$\text{Forecast for the next period} = \sum_{t=1}^{t=N} C_t D_t \qquad (14.1)$$

where C_t is the weight assigned to the demand and D_t is the demand. It should be noted that $\sum C_t = 1$.

Example 14.2 illustrates this method.

Example 14.2

Consider the demand data of Example 14.1. What is the forecast for June, if the weights assigned to data are 0.5, 0.3, and 0.2 respectively?

Solution

Forecast for June $= (0.5 \times 90 + 0.3 \times 110 + 0.2 \times 130)$

Forecast for June $= 104$ units.

14.3.4 Exponential Smoothing Methods

This method is known as the exponential smoothing method because the weight assigned to a previous period's demand decreases exponentially as that data gets older. Thus, recent demand data gets a higher weight than the older demand data. The demand may have a horizontal component with random fluctuations around an average demand called "base". The demand may have an increasing or decreasing trend or it may have a seasonality component. The demand may have both the trend and a seasonality component. Exponential smoothing models are discussed for the following conditions:

- Steady demand with no trend or seasonality component
- Demand with a linear trend
- Demand with a seasonality component
- Demand with both the trend and seasonality component

14.3.4.1 When to Use Exponential Smoothing Methods

These methods are suited for production and operations applications that involve forecasting for a large number of items. These methods perform well under the following conditions:

(1) The forecasting horizon is relatively short, for example, a daily, weekly, or monthly demand needs to be forecasted.

(2) When the information about cause and effect relationships between the demand of an item and independent factors that affect it are not available.

(3) Small effort in forecasting is desired. Effort is measured both by a method's ease of application and by the computational requirements (such as time, storage) needed to implement it.

(4) Updating of the forecast as new data becomes early available and can be accomplished by simply inputting the new data.

14.3.4.2 Simple Exponential Smoothing Model

This method is used when the past history of the demand data does not show the presence of trend or seasonality component. Thus, only the horizontal component of demand is present and, due to randomness, the demand fluctuates around an *average demand*, which is known as the *base*. If the base is constant from period to period, then all fluctuations in demand must be due to randomness. In reality, fluctuations in demand are caused by both changes in the base and random noise. The exponential smoothing method estimates the base and uses this for forecasting future demand.

In the simple exponential smoothing model, the base for any period is estimated by the following relation:

New base = Previous base + α (New demand – Previous base)

The above relation can be stated in symbols as:

$$S_t = S_{t-1} + \alpha(D_t - S_{t-1}) \qquad (14.2)$$

where,

S_t is the new base

D_t is the new demand

S_{t-1} is the previous base

α is a constant called smoothing coefficient $(0 \leq \alpha \leq 1.0)$

Equation (14.2) can be expressed as

$$S_t = \alpha D_t + (1 - \alpha)S_{t-1} \qquad (14.3)$$

It can be seen from Equation (14.3) that the new base is estimated by simply taking a weighted average of the new demand D_t and the previous base S_{t-1}

A high value of α places more importance to the new demand whereas a low value of α gives more importance to the previous base and ignores the new demand to a large extent. The appropriate value of α is a compromise between the two.

The new base S_t calculated by Equation (14.3) is taken as forecast for the next period or future. Thus,

$$F_{t+1} = S_t \qquad (14.4)$$

where F_t is the forecast for period t and F_{t+1} is the forecast for the next period.

From Equations (14.3) and (14.4), we can write:

$$F_{t+1} = \alpha D_t + (1 - \alpha)F_t \qquad (14.5)$$

$$F_t = \alpha D_{t-1} + (1 - \alpha)F_{t-1} \qquad (14.6)$$

From Equations (14.5) and (14.6), we have:

$$F_{t+1} = \alpha D_t + (1 - \alpha)[\alpha D_{t-1} + (1 - \alpha)F_{t-1}]$$

$$= \alpha D_t + \alpha(1 - \alpha)D_{t-1} + (1 - \alpha)^2 F_{t-1}$$

$$= \alpha D_t + \alpha(1 - \alpha)D_{t-1} + (1 - \alpha)^2[\alpha D_{t-2} + (1 - \alpha)F_{t-2}]$$

$$F_{t+1} = \alpha D_t + \alpha(1 - \alpha)D_{t-1} + \alpha(1 - \alpha)^2 D_{t-2} + (1 - \alpha)^3 F_{t-2} \qquad (14.7)$$

Thus, it can be seen from Equation (14.7) that the demand D_k i.e. the demand of k periods back, is given a weight of $\alpha(1 - \alpha)^k$. Since $(1 - \alpha)$ is a proper fraction, the weights go on decreasing as k increases. Thus, in this method the forecast is obtained by calculating weighted average of all the available data, giving more weights to the recent data and reducing the weight continuously as the data gets older. Therefore, it can be observed that the weights given to the available data decreases in exponential manner and due to this reason this method is known as exponential smoothing model.

It is useful to split the process of forecasting using simple exponential smoothing model in the following two steps:

Step 1 Estimate the new base, S_t using Equation (14.3).

Step 2   Forecast T, periods ahead by extrapolating the base computed in step 1. The forecast made in period t for T periods ahead, F_{t+T} is simply S_t, i.e.

$$F_{t+T} = S_t$$

Usually $T = 1$; that is, the forecast is made for only one period ahead.

Example 14.3 illustrates this method of forecasting.

Example 14.3

The monthly demand for a product is given below. Forecast the demand for each month using simple exponential smoothing model with smoothing coefficient $\alpha = 0.3$. What is the forecast for August?

Month	Actual Demand, D_t
Jan	200
Feb	260
Mar	210
Apr	220
May	230
Jun	270
Jul	290
Aug	–

Solution

In order to forecast the demand for Jan i.e. F_{Jan}, we need to have the base, S_t, for one period before (Initial) i.e. $S_{Initial}$. Let us assume that the base for the initial period is equal to 220. The computation of forecast for other months is shown below in Table 14.1.

Table 14.1   Calculations for simple exponential smoothing model with $\alpha = 0.3$

Month	Actual Demand, D_t	Base, S_t	Forecast, F_t
Initial		220 (Assumed)	–
Jan	200	$S_{Jan} = \alpha D_{Jan} + (1 - \alpha)S_{Initial}$ $= 0.3 \times 200 + 0.7 \times 220$ $= 214$	220
Feb	260	$S_{Feb} = \alpha D_{Feb} + (1 - \alpha)S_{Jan}$ $= 0.3 \times 260 + 0.7 \times 214$ $= 227.80$	$F_{Feb} = S_{Jan} = 214$
Mar	210	$S_{Mar} = \alpha D_{Mar} + (1 - \alpha)S_{Feb}$ $= 0.3 \times 210 + 0.7 \times 227.80$ $= 222.46$	$F_{Mar} = S_{Feb} = 227.80$

(Contd.)

Month	Actual Demand, D_t	Base, S_t	Forecast, F_t
Apr	220	$S_{Apr} = \alpha D_{Apr} + (1 - \alpha)S_{Mar}$ $= 0.3 \times 220 + 0.7 \times 222.46$ $= 221.72$	$F_{Apr} = S_{Mar} = 222.46$
May	230	$S_{May} = \alpha D_{May} + (1 - \alpha)S_{Apr}$ $= 0.3 \times 230 + 0.7 \times 221.72$ $= 224.21$	$F_{May} = S_{Apr} = 221.72$
Jun	270	$S_{Jun} = \alpha D_{Jun} + (1 - \alpha)S_{May}$ $= 0.3 \times 270 + 0.7 \times 224.21$ $= 237.94$	$F_{Jun} = S_{May} = 224.21$
Jul	290	$S_{Jul} = \alpha D_{Jul} + (1 - \alpha)S_{Jun}$ $= 0.3 \times 290 + 0.7 \times 237.94$ $= 253.56$	$F_{Jul} = S_{Jun} = 237.94$
Aug	–	–	$F_{Aug} = S_{Jul} = 253.56$

14.3.4.3 Linear or Additive Trend Model

The difference between new base and previous base i.e. $S_t - S_{t-1}$ gives a raw measure of linear trend in exponentially smoothed averages of the demand data. If an attempt is made to compensate for the trend using this raw measure, then an unstable correction may be observed due to the presence of random effects. Sometimes a negative trend may also be recorded when, indeed, the general trend is positive. The raw trend measure needs to be stabilized in the same way as the actual demand data are stabilized by applying exponential smoothing model so as to minimize these irregular effects.

In linear or additive trend model, the $S_t - S_{t-1}$ series is smoothed with the smoothing coefficient β. The value of β need not be same as α used in smoothing D_t.

The computation of new base S_t must now reflect the trend, so Equation (14.2) is modified by simply adding the old smoothed trend to the previous base as follows:

$$S_t = \alpha D_t + (1 - \alpha)(S_{t-1} + T_{t-1}) \tag{14.8}$$

The updated value of T_t, the smoothed trend, is:

$$T_t = \beta(S_t - S_{t-1}) + (1 - \beta)T_{t-1} \tag{14.9}$$

The forecast for the next period is obtained by adding T_t to the current smoothed average S_t as follows:

$$\text{Forecast for the next period} = F_{t+1} = S_t + T_t \tag{14.10}$$

This model is quite useful when trend and random variations are present in the data. In addition, the concept is useful in some of the more complex models that combine estimates of average demand, trend and seasonal components.

Example 14.4 illustrates the computation of forecasts using this model.

Example 14.4

Compute the forecast for each month for the data given in Example 14.3 using the linear or additive trend model with $\alpha = 0.3$ and $\beta = 0.2$.

Solution

In order to forecast the demand for Jan i.e. F_{Jan}, we need to know the values of the base S_t and trend T_t for one period before (Initial) i.e. $S_{Initial}$ and $T_{Initial}$. Let us take $S_{Initial} = 220$ and $T_{Initial} = 0$. The computation of forecast for other months is shown below:

$$S_{Jan} = \alpha D_{Jan} + (1 - \alpha)(S_{Initial} + T_{Intial})$$

$$= 0.3 \times 200 + 0.7 \times (220 + 0)$$

$$S_{Jan} = 214$$

$$T_{Jan} = \beta(S_{Jan} - S_{Initial}) + (1 - \beta)T_{Initial}$$

$$= 0.2 \times (214 - 220) + 0.8 \times 0$$

$$T_{Jan} = -1.2$$

$$F_{Feb} = S_{Jan} + T_{Jan}$$

$$= 214 - 1.2$$

$$F_{Feb} = 212.8$$

$$S_{Feb} = \alpha D_{Feb} + (1 - \alpha)(S_{Jan} + T_{Jan})$$

$$= 0.3 \times 260 + 0.7 \times (214 - 1.2)$$

$$S_{Feb} = 226.96$$

$$T_{Feb} = \beta(S_{Feb} - S_{Jan}) + (1 - \beta)T_{Jan}$$

$$= 0.2 \times (226.96 - 214) + 0.8 \times (-1.2)$$

$$T_{Feb} = 1.63$$

$$F_{Mar} = S_{Feb} + T_{Feb}$$

$$= 226.96 + 1.63$$

$$F_{Mar} = 228.59$$

$$S_{Mar} = \alpha D_{Mar} + (1 - \alpha)(S_{Feb} + T_{Feb})$$

$$= 0.3 \times 210 + 0.7 \times (226.96 + 1.63)$$

$$S_{Mar} = 223.01$$

$$T_{Mar} = \beta(S_{Mar} - S_{Feb}) + (1 - \beta)T_{Feb}$$

$$= 0.2 \times (223.01 - 226.96) + 0.8 \times 1.63$$

$$T_{\text{Mar}} = 0.51$$

$$F_{\text{Apr}} = S_{\text{Mar}} + T_{\text{Mar}}$$

$$= 223.01 + 0.51$$

$$F_{\text{Apr}} = 223.52$$

$$S_{\text{Apr}} = \alpha D_{\text{Apr}} + (1 - \alpha)(S_{\text{Mar}} + T_{\text{Mar}})$$

$$= 0.3 \times 220 + 0.7 \times (223.01 + 0.51)$$

$$S_{\text{Apr}} = 222.46$$

$$T_{\text{Apr}} = \beta(S_{\text{Apr}} - S_{\text{Mar}}) + (1 - \beta)T_{\text{Mar}}$$

$$= 0.2 \times (222.46 - 223.01) + 0.8 \times 0.51$$

$$T_{\text{Apr}} = 0.30$$

$$F_{\text{May}} = S_{\text{Apr}} + T_{\text{Apr}}$$

$$= 222.46 + 0.30$$

$$F_{\text{May}} = 222.76$$

$$S_{\text{May}} = \alpha D_{\text{May}} + (1 - \alpha)(S_{\text{Apr}} + T_{\text{Apr}})$$

$$= 0.3 \times 230 + 0.7 \times (222.46 + 0.30)$$

$$S_{\text{May}} = 224.93$$

$$T_{\text{May}} = \beta(S_{\text{May}} - S_{\text{Apr}}) + (1 - \beta)T_{\text{Apr}}$$

$$= 0.2 \times (224.93 - 222.46) + 0.8 \times 0.30$$

$$T_{\text{May}} = 0.73$$

$$F_{\text{Jun}} = S_{\text{May}} + T_{\text{May}}$$

$$= 224.93 + 0.73$$

$$F_{\text{Jun}} = 225.76$$

$$S_{\text{Jun}} = \alpha D_{\text{Jun}} + (1 - \alpha)(S_{\text{May}} + T_{\text{May}})$$

$$= 0.3 \times 270 + 0.7 \times (224.93 + 0.730)$$

$$S_{\text{Jun}} = 238.96$$

$$T_{\text{Jun}} = \beta(S_{\text{Jun}} - S_{\text{May}}) + (1 - \beta)T_{\text{May}}$$

$$= 0.2 \times (238.96 - 224.93) + 0.8 \times 0.73$$

$$T_{\text{Jun}} = 3.39$$

$$F_{Jul} = S_{Jun} + T_{Jun}$$

$$= 238.96 + 3.39$$

$$F_{Jul} = 242.35$$

$$S_{Jul} = \alpha D_{Jul} + (1 - \alpha)(S_{Jun} + T_{Jun})$$

$$= 0.3 \times 290 + 0.7 \times (238.96 + 3.39)$$

$$S_{Jul} = 256.65$$

$$T_{Jul} = \beta(S_{Jul} - S_{Jun}) + (1 - \beta)T_{Jun}$$

$$= 0.2 \times (256.65 - 238.96) + 0.8 \times 3.39$$

$$T_{Jul} = 6.25$$

$$F_{Aug} = S_{Jul} + T_{Jul}$$

$$= 256.65 + 6.25$$

$$F_{Aug} = 262.90$$

The results of the computations are listed in Table 14.2.

Table 14.2 Results for linear or additive trend model with $\alpha = 0.3$ and $\beta = 0.2$

Month	Actual Demand, D_t	Base, S_t	Trend, T_t	Forecast, F_t
Initial		220 (Assumed)	0	–
Jan	200	214	– 1.2	220
Feb	260	226.96	1.63	212.8
Mar	210	223.01	0.51	228.59
Apr	220	222.46	0.30	223.52
May	230	224.93	0.73	222.76
Jun	270	238.96	3.39	225.76
Jul	290	256.65		242.35
Aug	–	–		262.90

14.3.4.4 Ratio Seasonality Model

There are products for which the demand is affected by seasons. To forecast demand for such products, we need to find the seasonal index from the demand history. For example, the quarterly demand for a product during last three years is shown in Table 14.3.

Table 14.3 Quarterly demand for a product

Year	Quarter	Demand
2015	I	540
	II	840
	III	960
	IV	730
2016	I	510
	II	760
	III	1030
	IV	670
2017	I	560
	II	820
	III	940
	IV	770

It appears from this table that the demand for this product is maximum during the third quarter. In order to forecast the demand for such a product, we need to define the seasonality index for each quarter. The demand is de-seasonalized by dividing the total annual demand by 4 which is the number of quarters in a year. For the year 2015, the total demand is 3070. Had the demand been uniformly distributed over the four quarters, the demand in each quarter would have been 767.50 (3070)/4. The seasonality index for a quarter is defined as the ratio of the actual demand to the de-seasonalized demand. The seasonality index for each of the four quarters of the year 2015 is shown in Table 14.4.

Table 14.4 Seasonality indices

Year	Quarter	Demand	Seasonality index
2015	I	540	540/767.50 = 0.70
	II	840	840/767.50 = 1.09
	III	960	960/767.50 = 1.25
	IV	730	730/767.50 = 0.95

The seasonality indices shown in Table 14.4 are only used to initialize the process. A process for updating the indices will be used in the succeeding years so that they will reflect changes that may occur.

The actual demand for the years 2016 and 2017 can be normalized by dividing them by the previous year's seasonal index for that period, I_{t-L}. (L is the number of periods in one cycle, 4 if data are by quarter, 12 if data are by month). Thus, if actual demand for second quarter of the year 2016 is $D_t = 760$, we divide it by the index for the second quarter of the previous year, $I_{t-4} = 1.09$, to obtain $760/1.09 = 697.25$. The result of this process is to de-seasonalize by decreasing adjusted demand during high-demand periods and increasing it during low-demand periods. The de-seasonalized smoothed average, S_t, is then given by Equation (14.11).

$$S_t = \alpha \left(\frac{D_t}{I_{t-L}} \right) + (1-\alpha)S_{t-1} \qquad (14.11)$$

It should be noted that the seasonal indices for the year 2015 reflect only that year's seasonal experience. If the seasonal cycle repeats itself consistently each year, using the indices of the year 2015 each year would be appropriate. However, there may be a component of noise in the seasonality index and it needs to be smoothed for forecasting the future demand. Therefore, Equation (14.12) is used to update the seasonal indices.

$$I_t = \gamma \left(\frac{D_t}{S_t} \right) + (1-\gamma) I_{t-L} \qquad (14.12)$$

where, I_t is the seasonality index for the period t

γ is the smoothing coefficient

It can be seen from Equation (14.12) that the actual demand D_t is divided by the new smoothed average S_t, computed by Equation (14.11), to reflect the amount by which D_t exceeds or falls short of the de-seasonalized average. This variation from the de-seasonalized average is weighted by the smoothing coefficient. The old seasonal index is last year's index and is weighted by $1 - \gamma$.

The forecast for the upcoming period $(t + 1)$ is obtained by modifying the most current smoothed average S_t by the seasonal index for the upcoming period I_{t-L+1} as given in Equation (14.13).

Forecast for the next period $= F_{t+1} = S_t \times I_{t-L+1}$ \qquad (14.13)

The forecast for m periods ahead can be computed by Equation (14.14).

$$F_{t+m} = S_t \times I_{t-L+m} \qquad (14.14)$$

Example 14.5 illustrates the computation of forecast by this method.

Example 14.5

Compute the forecast for each quarter of the years 2016 and 2017 for the data shown in Table 14.3 using ratio seasonality model with $\alpha = 0.3$ and $\gamma = 0.2$. What is the forecast for the first quarter of the year 2018?

Solution

Table 14.5 shows the computation of the forecasts.

Table 14.5 Sample computations for ratio seasonality model with $\alpha = 0.3$ and $\gamma = 0.2$

Year	Quarter	Actual Demand D_t	S_t Equation (14.11)	I_t Equation (14.12)	$F_t + 1$ Equation (14.13)
2015	I	540		0.70	
	II	840		1.09	
	III	960		1.25	
	IV	730	720 (Assumed)	0.95	—
2016	I	510	722.57	0.70	—
	II	760	714.97	1.08	787.60
	III	1030	747.68	1.27	893.71
	IV	670	734.95	0.94	710.29
2017	I	560	754.46	0.71	514.46
	II	820	755.90	1.08	814.82
	III	940	751.18	1.27	959.99
	IV	770	771.57	0.95	706.11
2018	I	—	—	—	547.81

14.3.4.5   *Additive Trend and Ratio Seasonality Model*

This model is used to forecast demand for products that show seasonal impact and also either increasing or decreasing trend. The two models i.e. linear or additive and ratio seasonality models, discussed above can be combined together to take into account the trend and seasonality component in demand.

The basic equations which define the trend and seasonal indices i.e. Equations (14.9) and (14.12) remain the same and they are reproduced below:

$$T_t = \beta(S_t - S_{t-1}) + (1 - \beta)T_{t-1} \tag{14.15}$$

$$I_t = \gamma \left(\frac{D_t}{S_t} \right) + (1 - \gamma) I_{t-L} \tag{14.16}$$

The equation which defines the base for a period t needs to be modified to incorporate the trend and seasonality corrections. The equation for calculating the base for a period t may be written as Equation (14.17).

$$S_t = \alpha \left(\frac{D_t}{I_{t-L}} \right) + (1 - \alpha)(S_{t-1} + T_{t-1}) \tag{14.17}$$

The forecast for the upcoming period ($t + 1$) can be expressed as Equation (14.18).

$$F_{t+1} = (S_t + T_t)(I_{t-L+1})$$ (14.18)

The forecast for m periods ahead can be written as Equation (14.19).

$$F_{t+m} = (S_t + mT_t)(I_{t-L+m})$$ (14.19)

Example 14.6 illustrates the computation of forecast using this model.

Example 14.6

Compute the forecast for each quarter of the years 2016 and 2017 for the data shown in Table 14.3 using additive trend and ratio seasonality model with $\alpha = 0.3$, $\beta = 0.2$ and $\gamma = 0.1$. What is the forecast for the first quarter of the year 2018?

Solution

Table 14.6 shows the computation of the forecasts.

Table 14.6 Sample computations for additive trend and ratio seasonality model with $\alpha = 0.3$, $\beta = 0.2$ and $\gamma = 0.1$

Year	Quarter	Actual Demand D_t	S_t Equation (14.17)	T_t Equation (14.15)	I_t Equation (14.16)	F_{t+1} Equation (14.18)
2015	I	540			0.70	
	II	840			1.09	
	III	960			1.25	
	IV	730	720 (Assumed)	1.00 (Assumed)	0.95	–
2016	I	510	723.27	1.45	0.69	–
	II	760	716.48	–0.198	1.08	504.71
	III	1030	748.59	6.265	1.25	789.94
	IV	670	739.98	3.29	0.94	943.58
2017	I	560	760.29	6.69	0.69	706.11
	II	820	764.66	6.23	1.08	536.89
	III	940	763.43	4.74	1.25	832.56
	IV	770	780.88	7.28	0.94	967.89
2018	I	–	–	–		729.76

14.3.5 Adaptive Methods

It is a common practice to use reasonably small values of α in exponential smoothing models in order to reduce random variations in demand. These forecasting models can effectively track the changes in situations where demand rates either increase or decrease gradually. However, if demand changes suddenly, such forecasting models that use a small value for α will lag substantially behind the actual change. To overcome this problem, researchers have proposed adaptive response systems.

The adaptive smoothing systems basically monitor the forecast error and based on pre-set rules they react to large errors by increasing the value of α. Trigg and Leach (1967) proposed an adaptive smoothing model in which the value of α is set equal to the absolute value of a tracking signal. They have defined the tracking signal as:

$$\text{Tracking signal} = \frac{\text{Smoothed forecast error}}{\text{Smoothed absolute forecast error}}$$

If the error is small, α will be small. If, due to a sudden shift in demand, the error is large, then the tracking signal will be large. The large error signals that α should increase so as to give more weight to current demand. The forecast would then reflect the change in actual demand. When actual demand stabilizes at the new level, the adaptive system would re-set the value of α to a lower level that filters out random variations effectively. A number of adaptive systems have been proposed by researchers, for example, Chow (1965), Trigg and Leach (1967), and Roberts and Reed (1969).

Adaptive models perform well in situations where demand data exhibits rapid shifts and little randomness. This method may be quite useful for forecasting demand for new products for which little historical demand data are available.

14.4 CAUSAL OR EXPLANATORY METHODS

It has been observed that there are certain factors or variables that cause the demand for a particular product or service to vary from one period to another. Such factors are called causal or independent factors. For example, the factors such as price, income, price of the related goods, taste, number of customers etc. may significantly affect the level of demand for a consumer product. The basic idea behind causal methods is to identify such causal factor(s) and subsequently to develop an empirical relationship between this causal or independent factor (s) and the demand which is a dependent factor. The relationship, so developed, is used to forecast future values of dependent variable simply by plugging in the appropriate values for the independent variable(s).

14.4.1 Regression Analysis

Regression analysis is a technique used to develop a forecasting function called regression equation. The regression equation expresses the relationship between the factor/variable (called dependent variable) to be forecasted and the factor(s) (called independent variables) that presumably cause the dependent variable to increase or decrease. For example, for a consumer good, we might postulate that sales are related to disposable personal income, if disposable personal income increases, sales will increase, and if people have less money to spend then sales will decrease. In addition, we might also postulate that the good's sales are controlled by its own price, price of the related goods, taste and number of customers. Regression analysis is used to develop the relationship between dependent and independent variables.

Once we know the relationship between the dependent variable (company sales) and the independent variables (good's own price, disposable personal income, price of related goods, taste and number of customers), we can use this relationship to forecast company sales.

To sum up, regression analysis seeks to identify independent variables that affect the dependent variable to be forecasted and establish the relationship between them so that the known values of independent variables can be used to estimate the forecasted value of the dependent variable.

If only one independent variable is used to estimate the dependent variable, the relationship between the two is established using *simple regression analysis*. We will illustrate this method with a simple example. However, if more than one independent variable influence the dependent variable, the relationship between them is established using *multiple regression analysis*.

14.4.2 Simple Regression Analysis

Suppose the annual profits of a company depends on the research and development (R & D) expenditure. The annual profit data for the past 8 years is given in Table 14.7. These data show that in year 2015, the company's R&D expenditure is ₹ 4 million and its annual profit is ₹ 33 million.

Table 14.7 Data and Intermediate Computations for Regression Analysis

Year	R & D Expenditure, X (Millions of ₹)	Annual Profit, Y (Millions of ₹)	XY	X^2	Y^2
2012	6	32	192	36	1024
2013	10	39	390	100	1521
2014	5	31	155	25	961
2015	4	33	132	16	1089
2016	2	24	48	4	576
2017	3	21	63	9	441
Sum	30	180	980	190	5612

The objective is to develop a relationship between Annual Profit Y and R & D Expenditure X. This relationship is then used to forecast the value of dependent variable Y from the independent variable X.

The first step in regression analysis is to plot the data in Table 14.7 as a scatter diagram. The scatter diagram with the line of best fit is shown in Figure 14.1.

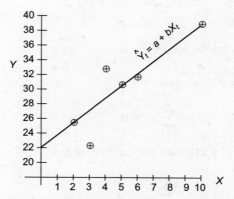

Figure 14.1 Scatter diagram with the line of best fit for the data in Table 14.7

The method of least squares is used to fit a line to the data in Figure 14.1. This method minimizes the sum of the squares of the vertical deviations separating the observed values of the dependent variable Y from the fitted line. The least squares line is represented by the equation of a straight line as in Equation (14.20).

$$\hat{Y}_t = a + bX_t \tag{14.20}$$

where X is the observed value of the independent variable, $\hat{Y}$ is the estimated value of the dependent variable, a and b, are the intercept and slope of the fitted line respectively. The values of a and b are such that $\sum (Y_t - \hat{Y}_t)^2$ is minimum. The values of a and b are determined as in Equation (14.21).

$$\sum_{t=1}^{n} e_t^2 = \sum_{t=1}^{n} (Y_t - a - bX_t)^2 \tag{14.21}$$

where e_t is the error i.e. difference between the observed value of Y and the estimated value of Y.

Differentiating Equation (14.21) with respect to a, we get:

$$\frac{\partial \sum_{t=1}^{n} e_t^2}{\partial a} = \sum_{t=1}^{n} -2(Y_t - a - bX_t)$$

where, n is the number of observations in the sample data.

For minimum error: $\dfrac{\partial \sum_{t=1}^{n} e_t^2}{\partial a} = 0$

Thus,

$$\sum_{t=1}^{n} -2(Y_t - a - bX_t) = 0 \tag{14.22}$$

Simplifying Equation (14.22), we get:

$$\sum_{t=1}^{n} Y_t - na - b\sum_{t=1}^{n} X_t = 0 \tag{14.23}$$

Differentiating Equation (14.21) with respect to b we get:

$$\frac{\partial \sum_{t=1}^{n} e_t^2}{\partial b} = \sum_{t=1}^{n} -2X_t(Y_t - a - bX_t) \tag{14.24}$$

For minimum error: $\dfrac{\partial \sum_{t=1}^{n} e_t^2}{\partial a} = 0$

Thus,

$$\sum_{t=1}^{n} -2X_t(Y_t - a - bX_t) = 0 \tag{14.25}$$

Simplifying Equation (14.25), we get:

$$\sum_{t=1}^{n} X_t Y_t - a\sum_{t=1}^{n} X_t - b\sum_{t=1}^{n} X_t^2 = 0 \tag{14.26}$$

To determine the values of a and b, Equations (14.23) and (14.26) are solved as:

Multiplying Equation (14.23) by $\sum_{t=1}^{n} X_t$, we get:

$$\sum_{t=1}^{n} X_t \sum_{t=1}^{n} Y_t - na\sum_{t=1}^{n} X_t - b\left(\sum_{t=1}^{n} X_t\right)^2 = 0 \tag{14.27}$$

Multiplying Equation (14.26) by n, we get:

$$n\sum_{t=1}^{n} X_t Y_t - na\sum_{t=1}^{n} X_t - nb\sum_{t=1}^{n} X_t^2 = 0 \tag{14.28}$$

Subtracting Equation (14.28) from Equation (14.27), we get:

$$\sum_{t=1}^{n} X_t \sum_{t=1}^{n} Y_t - na\sum_{t=1}^{n} X_t - b\left(\sum_{t=1}^{n} X_t\right)^2 - n\sum_{t=1}^{n} X_t Y_t + na\sum_{t=1}^{n} X_t + nb\sum_{t=1}^{n} X_t^2 = 0 \tag{14.29}$$

Simplifying Equation (14.29), we get:

$$b = \frac{\sum_{t=1}^{n} X_t Y_t - n\overline{X}\,\overline{Y}}{\sum_{t=1}^{n} X_t^2 - \overline{X}^2 \times n} \qquad (14.30)$$

where $\overline{X}$ and $\overline{Y}$ are the arithmetic mean of the variables X and Y. $\overline{X}$ and $\overline{Y}$ are calculated as:

$$\overline{X} = \frac{\sum_{t=1}^{n} X_t}{n} \quad \text{and} \quad \overline{Y} = \frac{\sum_{t=1}^{n} Y_t}{n}$$

From Equation (14.23), we get:

$$a = \frac{\sum_{t=1}^{n} Y_t}{n} - \frac{b\sum_{t=1}^{n} X_t}{n} \qquad (14.31)$$

or

$$a = \overline{Y} - b\overline{X} \qquad (14.32)$$

In Table 14.7, the intermediate computations for estimating a and b are shown. From Table 14.7, we compute:

$$\overline{X} = \frac{\sum_{t=1}^{n} X_t}{n} = \frac{30}{6} = 5$$

$$\overline{Y} = \frac{\sum_{t=1}^{n} Y_t}{n} = \frac{180}{6} = 30$$

Now we substitute $n = 6$, $\sum_{t=1}^{n} X_t Y_t = 980$, $\sum_{t=1}^{n} X_t^2 = 190$, $\overline{X} = 5$, and $\overline{Y} = 30$ in Equations (14.29) and (14.32) to obtain:

$$b = \frac{980 - (6)(5)(30)}{190 - (5)^2 \times 6} = 2$$

$$a = 30 - (2)(5) = 20$$

The regression line is:

$$\hat{Y}_t = 20 + 2X_t \tag{14.33}$$

To use this regression line to forecast the value of the dependent variable, we simply substitute the value of X_t and compute $\hat{Y}_t$. For example, if the R & D expenditure in year 2018 is estimated to be ₹7.0 million, then the forecast for annual profit in year 2018 will be:

$$\hat{Y}_t = 20 + (2)(7)$$

$$= ₹34 \text{ million}$$

$$= ₹3,40,00,000$$

14.4.2.1 Reliability of Forecasts

Since regression line is used to forecast the value of the dependent variable, it is quite possible that the actual value, once it becomes known, will be different from the forecasted value. It is expected that, on an average, the deviation between the forecasted value and the actual value will be larger if the observed data points are widely scattered around the regression line and smaller if the data points are close to the line. A measure of the extent to which data points are scattered around a regression line can be obtained using the *standard error of estimate* S_e as in Equation (14.34).

$$S_e = \sqrt{\frac{\sum\limits_{t=1}^{n}(Y_t - \hat{Y}_t)^2}{n-2}} \tag{14.34}$$

where Y_t and $\hat{Y}_t$ are the observed and the estimated values of the dependent variable in period t. For the ease of calculation, a short-cut method for finding the *standard error of estimate* S_e is given in Equation (14.35).

$$S_e = \sqrt{\frac{\sum\limits_{t=1}^{n}Y_t^2 - a\sum\limits_{t=1}^{n}Y_t - b\sum\limits_{t=1}^{n}X_t Y_t}{n-2}} \tag{14.35}$$

In the example in Table 14.7, the *standard error of estimate* S_e is $3.61 \times ₹10,00,000 = ₹36,10,000$. If the observed points are assumed to be normally distributed around the regression line, we can expect to find 68% of the points within $\pm 1\ S_e$, 95.5% of the points within $\pm 2\ S_e$ and 99.7% of the points within $\pm 3\ S_e$. Now, applying this concept to the example given in Table 14.7, we are roughly 68% confident that that the actual annual profit of the company will be within $\pm 1\ S_e$ from $\hat{Y}_t$. We can calculate the upper and lower limits of this prediction interval for the annual profit of year 2018 in which the company spends ₹7 million on R & D as follows:

$$\hat{Y}_t + 1\,S_e = ₹\,3,40,00,000 + ₹\,36,10,000$$

$$= ₹\,3,76,10,000 \leftarrow \text{Upper limit of prediction interval}$$

and

$$\hat{Y}_t - 1\,S_e = ₹\,3,40,00,000 - ₹\,36,10,000$$

$$= ₹\,3,03,90,000 \leftarrow \text{Lower limit of prediction interval}$$

Hence, for $X = ₹7$ million in the year 2018, we can say with 68% confidence that the forecasted annual income could be between ₹3,03,90,000 and ₹3,76,10,000.

For 95.5% confidence, the prediction interval would be as follows:

$$\hat{Y}_t + 2\,S_e = ₹\,3,40,00,000 + 2 \times ₹\,36,10,000$$

$$= ₹\,4,12,20,000 \leftarrow \text{Upper limit of prediction interval}$$

and

$$\hat{Y}_t - 2\,S_e = ₹\,3,40,60,000 - 2 \times ₹\,36,10,000$$

$$= ₹\,2,67,80,000 \leftarrow \text{Lower limit of prediction interval}$$

For 99.7% confidence, the prediction interval would be as follows:

$$\hat{Y}_t + 3\,S_e = ₹\,3,40,00,000 + 3 \times ₹\,36,10,000$$

$$= ₹\,4,48,30,000 \leftarrow \text{Upper limit of prediction interval}$$

and

$$\hat{Y}_t - 3\,S_e = ₹\,3,40,00,000 - 3 \times ₹\,36,10,000$$

$$= ₹\,2,31,70,000 \leftarrow \text{Lower limit of prediction interval}$$

14.4.2.2 Tests of Significance of Parameter Estimates

In the above example in Table 14.7, we estimated the slope of the regression line b from one sample of $R\,\&\,D$ expenditure-annual profits data of the company. If we had used a different sample (say, data for a different 6-year period), we would have obtained a somewhat different estimate of b. The greater is the dispersion of (i.e., the more spread out are) the estimated values of b (that we would obtain if we are to actually run many regression for different data samples), the smaller is the confidence that we have in our single estimated value of the b.

To test the hypothesis that b is statistically significant (i.e., the $R\,\&\,D$ expenditure positively affects annual profit), we first of all need to calculate the *standard error* of b. The *standard error* of b is given by Equation (14.36).

$$S_b = \sqrt{\frac{\sum\limits_{t=1}^{n}(Y_t - \hat{Y}_t)^2}{(n-k)\sum\limits_{t=1}^{n}(X_t - \overline{X})^2}} \qquad (14.36)$$

where Y_t and X_t are the actual sample observations of the dependent and independent variable in period t, $\hat{Y}_t$ is the value of dependent variable in period t estimated from the regression line, $\overline{X}$ is the mean value of the independent variable, n is the number of observations or data points used in the estimation of the regression line, and k is the number of estimated coefficients in the regression. The value of $(n-k)$ is called the *degrees of freedom* (df). Since in simple regression analysis, we estimate two parameters, a and b, the value of k is 2, and the degrees of freedom is $n-2$.

The value of S_b for the example in Table 14.7 is calculated by substituting the values from Table 14.8 into Equation (14.36). In Table 14.8, the values of $\hat{Y}_t$ in column 4 are obtained by substituting the various R & D expenditures of column 2 into Equation (14.33). Column 5 is obtained by subtracting the values in column 4 from the corresponding values in column 3, column 6 is obtained by squaring the values in column 5, column 7 is obtained by subtracting average value of X from the values in column 2 and column 8 is obtained by squaring the values in column 7.

Table 14.8   Calculations to estimate the Standard Error of b

1 Year	2 X_t	3 Y_t	4 $\hat{Y}_t$	5 $Y_t - \hat{Y}_t$	6 $(Y_t - \hat{Y}_t)^2$	7 $X_t - \overline{X}$	8 $(X_t - \overline{X})^2$
2012	6	32	32.00	0.00	0.0	1.00	1
2013	10	39	40.00	-1.00	1.00	5.00	25
2014	5	31	30.00	1.00	1.00	0.00	0
2015	4	33	28.00	5.00	25.00	-1.00	1
2016	2	24	24.00	0.00	0.00	-3.00	9
2017	3	21	26.00	-5.00	25.00	-2.00	4
$n=6$	$\sum X_t = 30$ $\overline{X} = 5$	$\sum Y_t = 180$ $\overline{Y} = 30$			$\sum(Y_t - \hat{Y}_t)^2$ $=52.00$		$\sum(X_t - \overline{X})^2$ $= 40$

Thus, the value of S_b is equal to:

$$S_b = \sqrt{\frac{\sum\limits_{t=1}^{n}(Y_t - \hat{Y}_t)^2}{(n-k)\sum\limits_{t=1}^{n}(X_t - \overline{X})^2}} = \sqrt{\frac{52.0}{(6-2)(40)}} = 0.57$$

After obtaining the value of S_b, we next calculate the ratio b/S_b. This is called the t *statistic* or t *ratio*. The higher this calculated t ratio is, the more confident we are that the true but unknown value of b that we are seeking is not equal to zero (i.e., that there is a significant relationship between R & D expenditure and annual sales). For our example, we have:

$$t = \frac{2.0}{0.57} = 3.51$$

In order to conduct a significance test for b, we compare the calculated t ratio to the critical value of t distribution with $n - k = 6 - 2 = 4$ degrees of freedom given by t-table (Appendix C). This so called t test of the statistical significance of the estimated coefficient is usually performed at the 5% level of significance. Thus, we go down the column headed 0.05 in t-table (Appendix C) we reach 4 degrees of freedom. This gives the critical value of $t = 2.776$. Since our calculated value of $t = 3.51$ is greater than the critical value of $t = 2.776$ for the 5% level of significance with 4 degrees of freedom, we reject the null hypothesis that there is no relationship between X (R & D expenditure) and Y (annual sales). To say that there is a statistically significant relationship between X and Y at the 5% level means that we are 95% confident that such a relationship does exist.

The above concepts can also be used to determine confidence intervals for the b coefficient. Thus, using the tabular value of $t = 2.776$ for the 5% level of significance (2.5 percent in each tail) and 4 degrees of freedom , we can say that we are 95% confident that the true value of b will be between:

$$b \pm 2.776 S_b$$

$$2.0 \pm 2.776(0.57)$$

$$2.0 \pm 1.58$$

That is, we are 95% confident that the true value of b lies between 0.42 and 3.58. This interval of the value of b is called *confidence interval*.

14.4.2.3   Coefficient of Determination

In addition to testing for the statistical significance of the slope parameter b of the regression line, we can also test for the overall explanatory power of the entire regression. This is achieved by calculating the *coefficient of determination*, which is usually denoted by R^2. The coefficient of determination (R^2) is defined as the proportion of the total variation or dispersion in the dependent variable (about its mean) that is explained by the variation in the independent variable(s) in the regression. In terms of our R & D expenditure-annual profit example, R^2 measures how much of the variation in the company's annual profit is explained by the variation in its R & D expenditure. The closer the observed data points fall to the regression line, the greater is the proportion of the variation in the company's annual profit explained by the variation in its R & D expenditure, and the larger is the value of the coefficient of determination (R^2). The coefficient of determination R^2 measures the strength of the linear relationship between the dependent and independent variables. We can calculate R^2 from the following formula:

$$R^2 = 1 - \frac{\sum_{t=1}^{n}(Y_t - \hat{Y}_t)^2}{\sum_{t=1}^{n}(Y_t - \overline{Y})^2} = 1 - [(\text{Unexplained variation})/(\text{Total variation})]$$

For calculation simplicity, an alternative expression for R^2 is:

$$R^2 = \frac{a\sum_{t=1}^{n}Y_t + b\sum_{t=1}^{n}X_t\,Y_t - n\overline{Y}^2}{\sum_{t=1}^{n}Y_t^2 - n\overline{Y}^2} \tag{14.37}$$

For the data in Table 14.7, R^2 is calculated as:

Year	X_t	Y_t	X_tY_t	Y_t^2
2012	6	32	192	1024
2013	10	39	390	1521
2014	5	31	155	961
2015	4	33	132	1089
2016	2	24	48	576
2017	3	21	63	441
Sum	30	180	980	5612

$a = 20.0$; $b = 2.0$; $\sum_{t=1}^{n}Y_t = 180$ $\sum_{t=1}^{n}X_t\,Y_t = 980$; $n = 6$; $\sum_{t=1}^{n}Y_t^2 = 5612$; $\overline{Y} = 30$

Thus, $R^2 = \dfrac{(20)(180) + (2)(980) - (6)(30)^2}{5612 - (6)(30)^2} = 0.7547$.

This means that only 75.47% of the variation in the annual profit is explained by the variation in R & D expenditure and 24.53% has not been explained. It should be noted that if R^2 is low, we should either look for another independent variable or include other variables in the analysis.

The square root of the coefficient of determination (R^2) is the *coefficient of correlation*, which is denoted by R, That is,

$$R = \sqrt{R^2} \tag{14.38}$$

This is simply a measure of the degree of association or covariation that exists between variables X and Y. For our R & D expenditure and annual profit example:

$$R = \sqrt{R^2} = \sqrt{0.7547} = 0.8687$$

This means that variables X and Y vary together 86.87% of the time. The coefficient of correlation ranges in value between -1 (if all the sample observation points fall on a negatively sloped straight line) and 1 (for perfect positive linear correlation).

14.4.3 Multiple Regression Analysis

The concept of simple regression analysis can be extended to multiple regression analysis. Multiple regression analysis is used to develop relationship between dependent variable and more than one independent variable. For example, the company's annual profit may be assumed to depend not only on the company's $R \& D$ expenditure (as given in the above example) but also on its advertising expenditure. The regression model for this situation can be written as

$$\hat{Y}_t = a + b_1 X_{1t} + b_2 X_{2t} \tag{14.39}$$

where $\hat{Y}_t$ is the forecasted value of the dependent variable referring to the company's annual profit, X_{1t} refers to the company's $R \& D$ expenditure, and X_{2t} refers to its advertising expenditure. T coefficients a, b_1, and b_2 are the parameters to be estimated.

The model can also be generalized to any number of independent or explanatory variables (k), as indicated in Equation (14.40)

$$\hat{Y}_t = a + b_1 X_{1t} + b_2 X_{2t} + \cdots + b_k X_{kt} \tag{14.40}$$

The process of estimating the parameters or coefficients i.e. a, b_1, b_2, . . . , b_k of a multiple regression equation is in principle the same as in simple regression analysis, but since the calculations are much more complex and time-consuming, they are done with computers. The computer also routinely provides the standard error of the estimates, the t statistics, the coefficient of multiple determination, etc. All that is required is to be able to set up the regression analysis, feed the data into the computer, and interpret the results.

14.5 QUALITATIVE OR JUDGMENTAL METHODS

The qualitative methods are of considerable significance in situations where we do not have a past history of demand data of a product or a service or even if the data exists, they may not be representative of future conditions. These methods incorporate intuitive judgments, opinions and subjective probability estimates. There are many different qualitative methods for making forecasts. We discuss briefly the build-up method, surveys methods, test markets, and the panel of experts.

14.5.1 Build-up Method

As the name suggests, the build-up method provides an overall estimate by obtaining estimate from the people who are at the bottom of an organization and then adding together estimates from people at higher level in the hierarchy. For example, if this method is used for predicting revenue, the first step is to have each sales representative estimate his or her revenue. These estimates are then passed on to the next higher level in the organization for review and evaluation. Estimates that are too high or too low are discussed with the representative so that management can understand the logic that supports the prediction. If the representative cannot convince the supervisor, a new prediction based on this discussion is made. The prediction is then passed on to the next in the organization.

As these subjective judgments are passed up the organization, they are reviewed and refined until they become, in total, the revenue forecast for the organization. Subsequently, it is the responsibility of the top management to make a final judgment on the forecast's validity.

14.5.2 Survey Method

Surveys are often used to prepare forecasts when historical data is not available, or when historical data is judged not to be indicative of the future. Surveys can also be used to verify the results of another forecasting technique. Surveys are used by organizations to gather information externally. A survey is a systematic effort to extract information from specific groups and is usually conducted via a written questionnaire, a phone interview or the internet. The target of the survey could be consumers, purchasing agents, economists and the like. A survey may attempt to determine how many consumers would buy a new brand of shampoo or consider a maintenance service that comes to your home or work place to do repairs on your car. Surveys of purchasing agents are conducted to assess the health of the economy.

14.5.3 Test Markets

In a test market, the forecaster places a new or redesigned product in a city or cities believed to be representative of the organization's overall market. For example, an organization that want to test the oil change service either at home or at workplace of the consumer may offer the services in one or two cities to determine how customers might react. The forecaster examines the sales behavior in the test markets and uses it to predict sales in other markets. Test marketing is expensive if it is extended for a long period. The results of test market are relatively more accurate than those of a survey because the consumers in a test market actually use the product.

14.5.4 Panel of Experts

A panel of experts involves a group of people who are knowledgeable about the subject being considered to make a forecast. This group attempts to make a forecast by building consensus. In an organization, this process may involve executives who are trying to predict the level of information technology applied to banking operations, or store managers who are trying to estimate labor cost in retail operations. The panel can be used for a wide variety of forecasts, and with this method, forecasts can often be made very quickly.

The Delphi technique uses a panel of experts in a way that eliminates the potential dominance of the most prestigious, the most verbal, and the best salespeople. The objective is to obtain expert opinion in the form of a consensus instead of a compromise. The Delphi technique was first developed by the RAND Corporation as a means of achieving a unanimous decision through consensus. The panel of experts often includes individuals from both inside and outside the organization. Each member is an expert on some aspect of the problem, but no one is an expert on the entire problem. The procedure for the Delphi technique involves the following steps:

(1) Each expert is provided with a sequence of questionaire which is directed at the same item or a set of items.

(2) Each expert makes independent predictions in the form of brief statements.

(3) The coordinator edits and clarifies these statements.

(4) The coordinator provides a new sequence of questionnaire to the experts that includes feedback supplied by the other experts.

(5) Steps 2 to 4 are repeated several times until a consensus forecast is obtained.

14.6 FORECAST ERRORS

A forecast is never perfectly accurate. Forecasts will always deviate from the actual demand. This difference between the actual demand and the forecasted demand is the forecast error. Although forecast error is inevitable, the objective of forecasting is that it should be as low as possible. Mathematically the forecast error is defined as:

$$e_t = D_t - F_t$$

where, e_t is the forecast error for period t

D_t is the actual demand for period t

F_t is the forecast for period t

Forecast errors are used to measure accuracy of forecasting models. In addition, they also provide a basis for comparing the performance of alternate models. Following are the commonly used forecast errors measures:

- Average error $(AE) = \dfrac{1}{n} \sum_{t=1}^{n} e_t$

- Mean absolute deviation $(MAD) = \dfrac{1}{n} \sum_{t=1}^{n} |e_t|$

- Mean squared error $(MSE) = \dfrac{1}{n} \sum_{t=1}^{n} e_t^2$

- Mean absolute percentage error $(MAPE) = \dfrac{1}{n} \sum_{t=1}^{n} \left| \dfrac{e_t}{D_t} \times 100 \right|$

For a large sample, the average error (AE) should be zero, otherwise the model exhibits bias. The forecasting model that shows a systematic tendency to forecast high or low values of demand is said to be biased.

The mean absolute deviation (MAD) provides additional information that is used in selecting a forecasting model and its parameters such as α in the exponential smoothing model.

The mean squared error (MSE) penalizes larger errors and it also provides information similar to MAD.

Mean absolute percentage error (MAPE) gives the decision maker an idea of how much the forecast is off as a percentage of demand. Table 14.9 shows the computation of four measures of forecast error.

One or more of these four measures could be used to compare the performance of alternate forecasting models.

Table 14.9 Computation of four measures of Forecast Error

Period	Demand (D_t)	Forecast (F_t)	Error (e_t)	(Error)²	% Error
1	25	23	+ 2	4	8.00
2	35	30	+ 5	25	14.29
3	15	20	− 5	25	− 33.33
4	45	35	+ 10	100	22.22
5	35	40	− 5	25	− 14.29
6	15	10	+ 5	25	33.33
7	45	35	+ 10	100	22.22
8	25	30	− 5	25	− 20.00
9	35	30	+ 5	25	14.29
10	15	20	− 5	25	− 33.33

$$\text{Average error } (AE) = \frac{2+5-5+10-5+5+10-5+5-5}{10} = \frac{17}{10} = 1.7$$

$$\text{Mean absolute deviation } (MAD) = \frac{2+5+5+10+5+5+10+5+5+5}{10} = \frac{57}{10} = 5.7$$

$$\text{Mean squared error } (MSE) = \frac{4+25+25+100+25+25+100+25+25+25}{10} = \frac{379}{10} = 37.9$$

Mean absolute percentage error (MAPE)

$$= \frac{8.00+14.29+33.33+22.22+14.29+33.33+22.22+20.00+14.29+33.33}{10}$$

$$= \frac{379}{10} = 21.53\%$$

PROBLEMS

1. A Bajaj motorcycle dealer in Delhi wants to forecast the demand for Bajaj Pulsar during the next month. From sales records, the dealer has accumulated the following data for the past year.

    (a) Compute a three-month moving average forecast of demand for April through January (of the next year).

    (b) Compute a five-month moving average forecast of demand for June through January (of the next year).

Month	Motorcycle Sales
January	14
February	12
March	15
April	13
May	10
June	17
July	15
August	16
September	17
October	15
November	19
December	21

2. The demand for a product during the last 10 weeks has been 40, 45, 43, 55, 47, 49, 61, 57, 53 and 55. Make the forecast for the 11th week by each of the following methods:

    (a) 3-weekly moving average

    (b) 3-weekly moving average with weights attached to the past demands such that the most recent demand is weighted three times as heavily as each of the preceding demands.

3. The manager of a gasoline station wants to forecast the demand for unleaded gasoline next month so that proper number of liters can be ordered from the distributors. The manager has accumulated the following data on demand for unleaded gasoline from sales team during the past ten months:

Month	Unleaded gasoline demand (liters)
January	810
February	735
March	640
April	510
May	655
June	700
July	740
August	820
September	1210
October	990

(a) Compute the three-month moving average forecast for April through October.

(b) Compute the weighted three-month moving average forecast for April through October. Assign weights of 0.5, 0.3, and 0.2 to the months in sequence, starting with the most recent month.

(c) Compute an exponentially smoothed forecast using α equal to 0.2.

(d) Compare the forecasts obtained from (a), (b), and (c) using MAD. Which forecast appears to be more accurate?

4. A company requires a quarterly forecast of its product for the upcoming year. The following data are available from the past demand history:

Year	Quarter	Demand
2015	1	20
	2	40
	3	50
	4	20
2016	1	20
	2	60
	3	60
	4	20
2017	1	20
	2	60
	3	70
	4	30
2018	1	
	2	
	3	
	4	

(a) Using the linear trend exponential smoothing model, compute the 2018 quarterly forecast. Assume $\alpha = \beta = 0.5$, $S_{Q3,\ 2017} = 60$, and $T_{Q3,\ 2017} = 20$.

(b) What will be quarterly forecast for 2018 if the linear trend and ratio seasonality model is employed? Assume $\alpha = \beta = \gamma = 0.5$, $S_{Q3,\ 2017} = 60$, $T_{Q3,\ 2017} = 20$, and the seasonality index for the four quarters are $I_{Q1} = 0.5$, $I_{Q2} = 1.3$, $I_{Q3} = 2.0$, and $I_{Q4} = 0.5$.

(c) If 2018 demands actually turn out to be $Q1 = 40$, $Q2 = 60$, $Q3 = 65$, and $Q4 = 40$. What are the forecast errors of the two models in part (a) and (b)?

5. Students' enrollment in Engineering Economy course is higher in the first semester than in the second semester. Using the following past data, forecast students enrollment in semester 1 and semester 2 of the upcoming year.

Year	Semester	Enrollment
2016	1	150
	2	100
2017	1	140
	2	80
2018	1	
	2	

Use the ratio seasonality model with $\alpha = \gamma = 0.2$, $S_{\text{semester 1, 2017}} = 125$, $I_{\text{semester 1, 2017}} = 1.5$, and $I_{\text{semester 2, 2017}} = 0.7$.

6. The following data give the ages and blood pressure of 10 women:

Age	: 56	42	36	47	49	42	60	72	63	55
Blood pressure	: 147	125	118	128	145	140	155	160	149	150

    (a) Determine the least squares regression equation of blood pressure on age.

    (b) Estimate the blood pressure of a woman whose age is 45 years.

7. A company wants to assess the impact of R & D expenditure (₹ in 1,000s) on its annual profit (₹ in 1,000s). The following table presents the information for the last eight years:

Year	R & D expenditure	Annual profit
2010	9	45
2011	7	42
2012	5	41
2013	10	60
2014	4	30
2015	5	34
2016	3	25
2017	2	20

    Estimate the regression equation and predict the annual profit for the year 2021 for an allocated sum of ₹ 1,00,000 as R & D expenditure.

8. The personnel manager of an electronics manufacturing company devises a manual test for job applicants to predict their production rating in the assembly department. In order to do this, the manager selects a random sample of 10 applicants. They are given the test and later assigned a production rating. The results are as follows:

Worker	A	B	C	D	E	F	G	H	I	J
Test score	53	36	88	84	86	64	45	48	39	69
Production rating	45	43	89	79	84	66	49	48	43	76

    Fit a linear least square regression equation of production rating on test score.

9. The following data relates to the scores obtained by salespersons of a company in an intelligence test and their weekly sales (in ₹ 1,000s):

Salesperson	A	B	C	D	E	F	G	H	I
Test score	50	60	50	60	80	50	80	40	70
Weekly sales	30	60	40	50	60	30	70	50	60

    (a) Obtain the regression equation of sales on intelligence test scores of the salesperson.

    (b) If the intelligence test score of a salesperson is 65, what would be his or her expected weekly sales?

10. The following data relating to the number of weeks of experience in a job involving the wiring of an electric motor and the number of motors rejected during the past week for 12 randomly selected workers:

Worker	Experience (weeks)	No. of Rejects
1	2	26
2	9	20
3	6	28
4	14	16
5	8	23
6	12	18
7	10	24
8	4	26
9	2	38
10	11	22
11	1	32
12	8	25

(a) Determine the linear regression equation for estimating the number of components rejected given the number of weeks of experience. Comment on the relationship between the two variables as indicated by the regression equation.

(b) Use the regression equation to estimate the number of motors rejected for an employee with 3 weeks of experience in the job.

(c) Determine the 95% approximate prediction interval for estimating the number of motors rejected for an employee with 3 weeks of experience in the job, using only the standard error of estimate.

11. A financial analyst has gathered the following data about the relationship between income and investment in securities in respect of 8 randomly selected families:

Income (in ₹ 1,000s)	8	12	9	24	43	37	19	16
Percent invested in securities	36	25	35	15	28	19	20	22

(a) Develop an estimating equation that best describes this data.

(b) Find the coefficient of determination and interpret it. Also find coefficient of correlation.

(c) Calculate the standard error of estimate for this relationship.

(d) Find the approximate 90% confidence interval for the percentage of income invested in securities by a family earning ₹ 25,000 monthly.

12. A financial analyst obtained the following information relating to return on security A and that of market M for the past 8 years:

Year	1	2	3	4	5	6	7	8
Return A	10	15	18	14	16	16	18	4
Market M	12	14	13	10	9	13	14	7

(a) Develop an estimating equation that best describes this data.

(b) Find the coefficient of determination and interpret it. Also find coefficient of correlation.

13. The total cost of production for a product is related to the number of units produced. The data for the past ten months is as follows:

Month	Production cost (in ₹ 1,000s)	Number of units produced (in 1,000 units)
1	35	10
2	56	15
3	51	13
4	27	9
5	42	11
6	74	17
7	26	9
8	50	12
9	70	18
10	60	15

(a) Using regression analysis, determine the relationship between production cost and units produced.

(b) If the company is planning a production level of 16,000 units for the upcoming month, what is your forecast for the total production cost that the company will incur at this production level?

(c) Compute the standard error of estimate for the regression line. What is its interpretation in the context of this problem?

(d) What percentage of variation in production cost is explained by the units produced?

14. The forecasted and the actual demand for a product during the last five periods are shown in the following table:

Period	Forecasted demand	Actual demand
1	400	430
2	435	390
3	420	425
4	428	440
5	435	438

Calculate the following forecast errors:

(a) Average error (AE)

(b) Mean absolute deviation (MAD)

(c) Mean squared error (MSE)

(d) Mean absolute percentage error ($MAPE$)

❑❑❑

15

Cost Estimation

15.1 INTRODUCTION

In today's competitive environment, success of an organization, be it a manufacturing or a service organization, depends to a great extent on its ability to estimate costs as accurately as possible. For timely and economical completion of an engineering or business project, it is essential to accurately estimate costs associated with different stages of the project. In engineering projects, the estimation of costs is considered to be more important than revenue estimation. Therefore, different aspects of cost estimation are discussed in this chapter.

Out of two types of costs, i.e., direct costs and indirect costs, it may be noted that indirect costs are not associated with a specific function, department or process. Hence, some logical basis should be used for allocating indirect costs to the different functions. For this purpose, both traditional as well as Activity-Based Costing (ABC) are also discussed in this chapter.

15.2 HOW DOES AN ORGANIZATION ESTIMATE COST?

Organizations are involved in either producing tangible products or providing services for customers for which serious efforts are made by them. In order to perform the required activities to produce goods or provide services, capital is required. In other words, for every activity there is an associated cost and, therefore, the organizations need to make cost estimates of the activities. The estimates should be as accurate as possible, so that the organization can arrange the capital required well in advance. Generally, cost estimations are required to be made in the initial stages of implementation of a project or a system. Maintenance and upgradation costs are estimated as a percentage of the first cost and they are subsequently added to make the trial cost estimates. Typical examples of a project include physical items such as building, bridge, manufacturing plant etc.

On the other hand, a system refers to an operational design involving non-physical items such as processes, software, etc. Typical examples of a system include ERP software, GPS system etc.

Cost comprises direct costs and indirect costs. Direct costs refer to costs that are incurred due to the involvement of several resources, such as people, material and machines, in the activities. Indirect costs are mainly incurred due to the factors that support the activities. Cost estimation is a complex activity and to successfully perform this activity the organizations must address the following issues: (i) Cost components to be estimated, (ii) Approach to cost estimation, (iii) Accuracy of cost estimates and (iv) Techniques for cost estimation.

15.2.1 Cost Estimates

Depending upon the nature of a project, that is whether it is simple or complex, several cost components are estimated. For example, for a project which involves a single piece of equipment, the cost components are the first cost P and annual operating cost (*AOC*). *AOC* is also called maintenance and operating (*M & O*) cost. First cost *P* has several cost elements, some of them are directly estimated whereas others are estimated based on the examination of records of simple projects. Typical cost elements of the first cost are equipment cost, delivery charges, installation cost, taxes and insurance and cost of training of personnel for the use of the equipment.

Similarly, the cost elements of *AOC* are direct labor cost, direct materials, maintenance cost and rework and rebuild cost. Out of the numerous cost elements, some can be estimated with a fair amount of accuracy whereas others may be difficult to estimate.

15.2.2 Cost Estimation Approach

Two approaches are generally followed for cost estimation, viz., "bottom up" and "top down".

As the name suggests, the "bottom up" approach builds the total cost by first considering the equipment and capital recovery and then adding various estimated costs to it. These include direct material, direct labor, *M & O* and indirect costs. Subsequently, the profit desired by the organization is added to the total cost to estimate the required price of the product or services. The schematic of the "bottom up" approach for cost estimation is shown in Fig. 15.1. This approach of cost estimation is generally employed by industry, business and the public sector. In this approach, required price for the product is treated as an output variable and the various cost estimates as input variables.

In the "top down" approach, the price of the product set by the organization determines the total cost required to produce it. The schematic of the "top down" approach is shown in Fig. 15.2. This is an effective approach, particularly in the early stages of a product design. This approach is useful in encouraging innovation, new design, process improvement, etc. In this approach, the competitive price is treated as an input variable and the cost estimates as output variables.

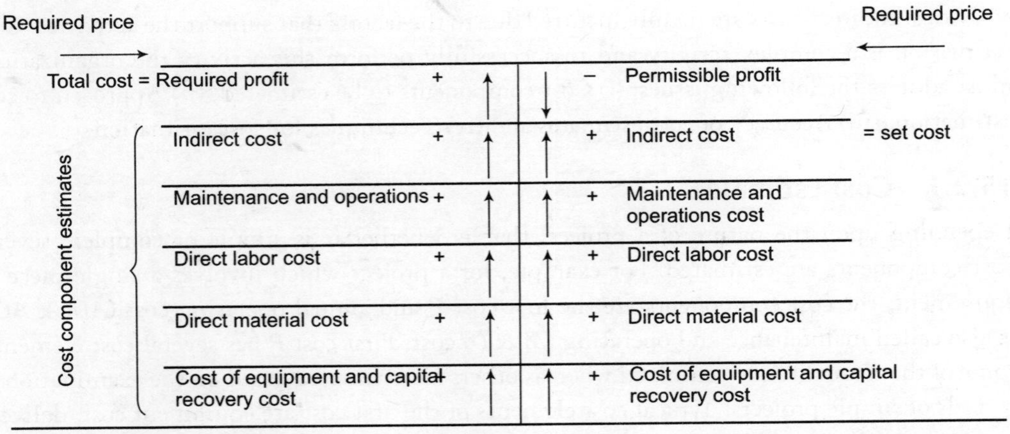

Figure 15.1 Schematic of bottom up approach   **Figure 15.2** Schematic of top down approach

15.2.3   Accuracy of Estimates

Cost estimates are seldom exact but reasonably accurate estimates are required for effective economic evaluation. As the project progresses, accurate cost estimates are required for budget allocation. In the initial stages of any project, cost estimation accuracy of the order of ±20% of actual cost may be sufficient. However, in the final stage of the project, a narrow range of accuracy, which is ±5% of the actual cost, is required for effective and efficient execution of the project.

15.2.4   Cost Estimation Methods

Techniques that lead to excellent cost estimates are expert opinion and comparison with the existing projects. Methods such as unit method and cost index provide present cost estimates based on the past cost experiences, with due consideration to inflation. Other relatively simple mathematical techniques of cost estimation are cost-capacity equations, the factor method and the learning curve. They are primarily employed at the preliminary design stage. Industries and corporations use dedicated software systems for cost estimations pertaining to their activities. Various cost estimation techniques are described in subsequent sections.

15.3   UNIT METHOD

This is a very simple but highly popular cost estimation technique among professionals of almost all organizations. The total estimated cost is obtained by using Equation (15.1).

$$C_T = U \times N \qquad\qquad\qquad (15.1)$$

where, C_T = the total estimated cost

U = per unit cost factor

N = number of units

The value of U must be regularly updated to incorporate changes in costs and inflation. Some typical cost factors and their values are given below:

Cost to construct a building for offices (₹ 20,000 per square meter)

Cost to construct residential buildings (₹ 12,000 per square meter)

Total average cost of operating an automobile (₹ 16 per kilometer)

Cost of furnishing an office with wood-work (₹ 6,000 per square meter)

Using the typical values of unit cost factor given above, we can obtain the total estimated cost. For example, if a residential house is constructed in 1,000 square meters and the unit construction cost is ₹ 12,000 per square meter, then the total estimated construction cost (C_T) is obtained as:

$$C_T = ₹\, 12,000 \times 1,000 = ₹\, 1,20,00,000$$

If a project or a system comprises several components, then to obtain total cost, the following steps are followed:

Step 1   Estimate the unit cost factors for each component.

Step 2   Identify the number of units of each factor.

Step 3   Multiply the unit cost factor with the respective number of unit of the factor and subsequently sum all products to get the total cost.

Example 15.1 illustrates the total cost estimation by unit method.

Example 15.1

A Ludhiana based casting industry has received an order for 2,000 castings of an automobile part. Estimate the cost of entire order based on unit method. Various components expenses involved are as follows:
Cost of raw material ₹ 450 per kg with the amount of bulk material being 3 tons, Cost of machining and tooling = ₹ 2,000 per hour with total estimated hours to be equal to 1,000. The labor costs associated with the order are as follows: molding, casting and cleaning at the rate of ₹ 500 per hour with total estimated time equal to 4,000 hours. Cost associated with finishing, packaging and shipping = ₹ 400 per hour with total estimated equivalent number of hours to be 1,000. There is also involvement of indirect labor worth 500 hours costing ₹ 750 per hour.

Solution

Apply Equation (15.1) to each of the five areas and sum the results to obtain total cost C_T. The details of calculation are shown in Table 15.1.

Table 15.1   Total cost estimate

Resource	Amount, N	Unit Cost Factor, U	Cost estimate = $U \times N$
Material	3 tons = 3,000 kg	₹450 per kg	₹13,50,000
Machinery and Tooling	1,000 hours	₹2,000 per hour	₹20,00,000
Labor, Molding, Casting and Cleaning	4,000 hours	₹500 per hour	₹20,00,000
Labor, Finishing, Packaging and Shipping	1,000 hours	₹400 per hour	₹4,00,000
Labor, Indirect	500 hours	₹750 per hour	₹3,75,000
Total Cost Estimate			₹61,25,000

15.4   COST INDEXES

The ratio of cost of something today to its cost sometime in the past is referred as the cost index. Being the ratio of the two costs, it represents a dimensionless number which shows a relative change in the cost of something with time. Consumer Price Index (CPI) is one such popular index with which many people are familiar. This index typically demonstrates the relationship between present and past costs of several consumer goods and services. Engineering discipline is concerned with other indexes such as Engineering New Record (ENR) construction cost Index and Marshell and swift (M & S) cost index which is also known as installed equipment cost index. In a country, there are agencies which compile the current and past values of several kinds of indexes and, therefore, the values of these indexes may be obtained from them. Cost of something in present time "t", i.e., C_t may be estimated by using Equation (15.2).

$$C_t = C_o (I_t / I_o) \tag{15.2}$$

where, C_o = Cost at previous time to

      I = value of the cost index at present time "t".

      I_o = value of cost index at previous (base) time "o".

Example 15.2 illustrates the use of cost index method to upgrade the present cost.

Example 15.2

A civil engineer is interested in estimating the cost of skilled labor for the job done in a construction project. He finds that a project of a similar complexity and magnitude was completed 5 years ago at a skilled labor cost of ₹4,58,640. The ENR Skilled labor index was 510 then and now is 818. What is the estimated labor cost?

Solution

Given data:

$$C_o = ₹4,58,640; \; I_5 = 818; \; I_o = 510$$

Using Equation (15.2), the present cost estimate is obtained as,

$$C_5 = ₹\, 4,58,640 \,(818/510) = ₹\, 7,35,623.$$

15.5 COST ESTIMATION RELATIONSHIPS

Various cost relationships to estimate costs are described in the subsequent sections.

15.5.1 Cost-Capacity Equation

Plants, equipments and construction costs can be estimated by using Cost Estimating Relationship (CER). CER predicts costs based on the variables associated with a project in its early design stages.

Cost capacity equation is one of the most widely used CER models. It is basically an equation that relates the cost of a component, system or plant to its capacity. This is also known as the *power law and sizing model*. The cost capacity equation is given by Equation (15.3).

$$C_2 = C_1 \,(S_2/S_1)^n \tag{15.3}$$

where, C_1 = cost at capacity at S_1

C_2 = cost at capacity S_2

n = correlating exponent

The value of the exponent for various components, systems or entire plant can be obtained from different sources available on the internet or from the database of a relevant agency. In case the value of the exponent for a particular unit is not known, it is a common practice to use the average value of 0.6.

The value of "n" lies between 0 and 1, i.e., $0 < n \le 1$. When $n < 1$, the economies of scale prevail; when $n = 1$, a linear relationship is present; and when $n > 1$, dis-economies of scale exist where a larger size is expected to be more costly than that of a purely linear relation. When Equation (15.2) is combined with Equation (15.3), the cost at time "t" and capacity level 2 can be given by Equation (15.4).

$$C_2 = C_1(S_2/S_1)^n(I_t/I_o) \tag{15.4}$$

Example 15.3 illustrates the use of cost capacity equation to estimate cost.

Example 15.3

Total drilling cost for crude oil at a tapping rate of 0.8 million gallons per day was ₹ 12 lakh in 2002. Estimate the cost of drilling at a tapping rate of 2 million gallons per day as on today. Take the value of exponent for tapping rate ranging from 0.1 to 20 million gallons per day as 0.24. The cost index of 15 in 2002 has been updated to 92 in the current year.

Solution

Equation (15.3) is used to estimate the cost of drilling crude oil in 2002, but it must be updated by the cost index to today's rupee value. Equation (15.4) is used to estimate the updated cost by considering both operations. The estimated cost in the current year in rupee value is obtained as:

$$C_2 = 12,00,000 \, (2.0/0.8)^{0.24} \, (92/15) = ₹ \, 91,70,297.32$$

15.5.2   Factor Method

Factor method is another very widely used model of Cost Estimating Relationship (CER). This method was developed for estimating total cost of process plant. This method provides a reliable total plant cost by multiplying the cost of the major equipment by a certain factor. This factor is commonly referred to as Lang factors after Hans J. Lang, who first proposed this method. The relation given by Equation (15.5) is used to estimate total plant cost by factor method.

$$C_T = h \times C_E \tag{15.5}$$

where, C_T = Total plant cost

h = Overall cost factor or sum of individual cost factors

C_E = Total cost of major equipment.

Generally, the value of "h" is obtained by summing the cost factors of individuals cost components, such as construction, maintenance, direct labor, materials and indirect cost elements.

In literature, the overall factor for some types of plants are available which are as follows: solid process plants, 3.10; solid fluid process plants, 3.63; and fluid process plants, 4.74. A value of $h = 4$ is used by the consulting engineering firms to obtain highly accurate overall cost estimates.

Example 15.4 illustrates the use of factor method to estimate total cost.

Example 15.4

A sugar industry intends to undergo a major vertical and horizontal plant expansion. Estimated cost of equipment and material needed for the expansion is ₹4.25 crores. The overall cost of expansion, however, is not known at the moment but if the overall cost factor $h = 4$, then estimate the overall cost of expansion.

Solution

For estimating the total plant expansion cost, we can use Equation (15.5).

$$C_T = h \times C_E = 4 \times ₹4.25 \times 10^7 = ₹ \, 17 \text{ crores}$$

Direct costs are directly related to a particular product, function or process whereas indirect costs are not directly to single function but are shared by several. Examples of indirect costs include administration, computer services, security and several other support functions. As we know that the total cost consists of both direct and indirect costs. Therefore, separate factors for direct and indirect cost components are used to determine total cost factor "h" which is given by Equation (15.6).

$$h = 1 + \sum_{i=1}^{n} f_i \qquad (15.6)$$

where, f_i = factor for each cost component of indirect cost

n = number of components.

If the indirect cost factor is applied to the total direct cost, only the direct cost factors are added to obtain "h". Thus, Equation (15.6) is modified and is given by Equation (15.7).

$$C_T = \left[C_E \left(1 + \sum_{i=1}^{n} f_i \right) \right] (1 + f_I) \qquad (15.7)$$

where, f_i = factors for direct cost components only

f_I = Indirect cost factors

Example 15.5 illustrates the application of Equation (15.7) to estimate the total cost.

Example 15.5

A rice mill is planning to install a captive power plant. The estimated cost of the equipment required for the plant is ₹6.0 crores. The exact details of other cost components are not known, but the cost factor for installation of accessories is 0.68. The construction cost factor is 0.55 and indirect cost factor is 0.24. Determine the total plant cost if the indirect cost is applied to (a) the cost of equipment and (b) the total direct cost.

Solution

(a) Total cost of equipment is ₹6.0 crores. Since both the direct and indirect cost factors are applied to only the equipment, the overall cost factor from Equation (15.6) is $h = 1 + 0.68 + 0.55 + 0.24 = 2.47$. Thus, the total captive power plant cost estimate $C_T = 2.47 \times ₹6.0 = ₹14.82$ crores.

(b) Now the total direct cost is calculated first, and Equation (15.7) is used to estimate the total plant cost.

$$h = 1 + \sum_{i=1}^{n} f_i = 1 + 0.68 + 0.55 = 2.23$$

$$C_T = [₹6.0 \times 2.23] \times (1.24) = ₹16.59 \text{ crores.}$$

15.5.3 Learning Curve

It has been observed that completion time of repetitive operations decreases as the number of times operations are performed increases. This happens due to the learning acquired for the operations. It leads to the conclusion that efficiency and improved performance do occur with more units. Learning curve is another important CER model which is mainly used to predict the time to complete a specific repeated unit. According to this model, there is a constant decrease in completion time every time the production is doubled. To illustrate this concept, suppose the time to prepare and test the hardness of a sample is 60 minutes and 32 test samples are to be prepared and their hardness is to be measured. Suppose, the learning involved in this operation leads to a reduction of 10% in completion time and thus the learning rate is 90%. Each time double the numbers of samples are prepared and tested, the completion times are 90% of the previous level. Thus the preparation and testing of the second sample takes $60(0.90) = 54$ minutes, the fourth takes 48.6 minutes, and so on.

The relationship given in Equation (15.8) is used to estimate the constant reduction of estimated time and cost for doubled production.

$$T_n = T_1(n)^m \tag{15.8}$$

where, n = unit number

T_1 = time or cost for the first unit

T_n = time or cost for the nth unit

m = learning curve slope parameters

Equation (15.8) can be used to estimate the time to compute a specific unit (1st or 2nd, 3rd or 4th . . . or nth). The relation given in Equation (15.8) plots as an exponentially decreasing curve on xy coordinates. The slope "m" can be defined as:

$$m = \frac{\log (\text{learning rate})}{\log 2} \tag{15.9}$$

Take the logarithm and the plot is a straight line on log-log paper.

$$\log T_n = \log T_1 + m \log T_n \tag{15.10}$$

Table 15.2 shows the per unit completion time estimates for 32 samples.

Table 15.2   Per unit completion time estimates

Unit, n	Time (minutes)
1	60.0
2	54.0
4	48.6
8	43.7
16	39.4
32	35.4

Figure 15.3 shows the plot of the result shown in the above Table.

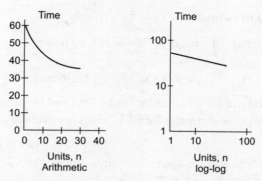

Figure 15.3 Plots of time estimates for a learning rate of 90% for 32 units with $T_1 = 60$ minutes

The slope of learning curve for this illustrative example can be obtained from Equation (15.9) as:

$$m = \frac{\log(0.90)}{\log 2} = -0.152$$

Thus, the slope is a negative number.

The cost C_n for the nth unit or the total cost C_T for all n units can be obtained by multiplying the cost per unit C by the appropriate time estimate.

$$C_n = (\text{cost per unit})(\text{time for nth unit}) = C \times T_n \tag{15.11}$$

$C_T = (\text{cost per unit})(\text{total time for all } n \text{ units})$

$$C_T = C(T_1 + T_2 + \ldots + T_n) \tag{15.12}$$

The learning curve provides good time estimates for larger scale projects where the batch size is relatively small, say 5 to 100 or 200. When a particular production process is repeated for a very large number of times then it has been observed that learning does not continue to increase and in this situation learning curve model may underestimate costs. Learning curve is also known by other names such as manufacturing progress function, experience curve, and improvement curve.

Example 15.6

Central Pollution Control Board (CPCB), New Delhi, has ordered 20 sophisticated equipments for gathering and analyzing the emission data at 12 critical locations in Delhi. The supplier of the equipment took 360 hours to build the equipment. If the direct and indirect costs average around ₹2,000 per hour and about 75% learning rate is assumed, then estimate (a) the time needed to complete unit 3 and 12 and (b) the total labor cost of unit 12.

Solution

(a) Equation (15.9) is used to find the learning curve slope "m" for 75% learning as:

$$m = \frac{\log(0.75)}{\log 2} = -0.415$$

Equation (15.8) is used to estimate the time for a specific unit.

$$T_3 = 60(3)^{-0.415} = 228.19 \text{ hours}$$

$$T_{12} = 360(12)^{-0.415} = 128.36 \text{ hours}$$

(b) The total cost estimate for all 12 units can be obtained by using Equation (15.12). But for this, individual time estimates for all 12 units are made first and then multiplied by ₹ 2,000.

$$T_1 = 360(1)^{-0.415} = 360 \text{ hours} ; \qquad T_2 = 360(2)^{-0.415} = 270 \text{ hours} ;$$

$$T_3 = 360(3)^{-0.415} = 228.19 \text{ hours} \qquad T_4 = 360(4)^{-0.415} = 202.51 \text{ hours} ;$$

$$T_5 = 360(5)^{-0.415} = 184.59 \text{ hours} ; \qquad T_6 = 360(6)^{-0.415} = 171.15 \text{ hours} ;$$

$$T_7 = 360(7)^{-0.415} = 160.54 \text{ hours} ; \qquad T_8 = 360(8)^{-0.415} = 151.88 \text{ hours} ;$$

$$T_9 = 360(9)^{-0.415} = 144.64 \text{ hours} ; \qquad T_{10} = 360(10)^{-0.415} = 138.45 \text{ hours} ;$$

$$T_{11} = 360(11)^{-0.415} = 133.08 \text{ hours} ; \qquad T_{12} = 360(12)^{-0.415} = 128.36 \text{ hours}$$

Using Equation (15.12):

$$C_T = C \,(T_1 + T_2 + \ldots + T_n)$$
$$C_T = ₹\,2,000 \,(T_1 + T_2 + T_3 + T_4 + T_5 + T_6 + T_7 + T_8 + T_9 + T_{10} + T_{11} + T_{12})$$
$$C_T = ₹\,2,000 \times 2,273.39 = ₹\,45,46,780$$

15.6   ESTIMATION AND ALLOCATION OF INDIRECT COST

Before the economic evaluation of a project, it is extremely important to estimate both direct and indirect costs associated with the project. In the preceding sections, direct costs have been discussed. This section deals with indirect costs, i.e., the costs that are not directly connected with a project or a system. Indirect costs are also referred to as overhead. These costs are required to meet the indirect expenses. Examples of indirect costs include operation and maintenance (O&M) costs, costs related to the safety and welfare of employees, taxes, utilities and a number of several other costs. It is very difficult to track in detail the indirect costs and, therefore, for allocation of indirect costs the following methods are used:

- Inclusion of indirect cost as an inherent element in the cost estimation. The factor method discussed in the previous section implements this by including an indirect cost factor.
- Use of some reasonable basis to calculate the applicable indirect cost rates.

- Utilization of Activity-Based-Costing (ABC). This method is especially applicable to high indirect cost industries.

Indirect costs are traditionally estimated by an indirect cost rate which is allocated on some basis. For example, direct labor hours may be used as a basis for estimating indirect cost rate of a machine shop for accounting and supervision related expenses. Similarly, indirect cost rates for purchasing, receiving, inspection, etc., can be calculated on the basis of material cost. In general, indirect or overhead cost rate can be calculated from the relation given in Equation (15.13).

$$\text{indirect/overhead cost rate} = \frac{\text{estimated total indirect cost}}{\text{estimated basis level}} \qquad (15.13)$$

Example 15.7 illustrates the use of Equation (15.13) in calculating the indirect cost rate.

Example 15.7

A consumer durables company intends to launch a luxury air conditioner. The company is evaluating a make-or-buy decision for some of the parts. The procurement division has provided the planning engineer with the standard rate so that the planning department can estimate the indirect cost.

Machine	Indirect cost rate
Hydraulic press for bending	₹ 300 per labor rupees
Vertical milling machine	₹ 1,000 per labor hour
Brazing machine	₹ 100 per material rupees

The planning division intends to apply a single basket rate per direct labor hour until the procurement division provided it with the wide ranging rates given above and the bases and the usage data given in the following table. Elaborate how the indirect cost rates were estimated.

Table 15.3 Indirect cost bases and activity

Machine	Indirect cost basis	Expected annual activity	Annual indirect budgeted costing
Hydraulic press for bending	Direct labor cost	₹ 4,50,000	₹ 1,500
Vertical milling machine	Direct labor hour	250 hours	₹ 2,50,000
Brazing machine	Material cost	₹ 15,000	₹ 1,50,000

Solution

The rates would be determined at the beginning of the year using the total indirect costs budgeted for each machine. Equation (15.13) is used to determine the rates provided to the company.

Hydraulic press for bending: Rate = ₹ 4,50,000/₹ 1,500 = ₹ 300 per direct labor rupees

Vertical milling machine: ₹ 2,50,000/250 = ₹ 1,000 per direct labor hour

Brazing machine: Rate = ₹ 1,50,000/₹ 15,000 = ₹ 100 per material rupees

It is a general practice to determine a blanket rate of indirect cost using a common base for allocating the indirect costs to multiple projects. For example, a blanket rate of indirect cost for multiple projects based on the material cost can be computed as given in Equation (15.14).

$$\frac{\text{Blanket indirect}}{\text{Overhead cost rate}} = \frac{\text{Expected indirect costs of all projects}}{\text{Estimated material cost of all projects}} \quad (15.14)$$

It can be realized that it is easy to calculate and apply blanket indirect cost rates. However, they do not account for the different accomplishments and functions of assets or people in the same department. For example, consider a production system which comprises of few automated and few manually operated machines. If a blanket rate of indirect costs is determined for this system, it will unnecessarily over-accumulate indirect costs for manually operated machines. Therefore, more than one basis should be normally used to allocate overhead. This has led to the development of Activity Based Costing (ABC) method. For economic evaluation of projects, it is essential to correctly estimate their indirect costs along with their direct costs and both should be considered in the evaluation process. Example 15.8 illustrates this concept for a make or buy evaluation.

Example 15.8

A Delhi based famous chain of restaurants wishes to undergo upgradation of its kitchen facility for which it has purchased a universal cooking line at an annual cost of ₹50,00,000. It also considers an opinion to make the necessary equipment in-house. For its departments involved in decision making, the annual indirect cost rates and hours plus estimated direct material and direct labor costs are given in Table 15.4. The allocated hours in the table represent the time required to produce the cooking equipment for a year. Parts must be purchased with the following estimates: First cost of ₹75,00,000; salvage value of ₹1,80,000 and life of 10 years. Evaluate between make or buy decision at market M.A.R.R. of 15% per year.

Table 15.4 In house making cost estimates

| Department | Indirect Costs | | | Direct Material cost | Direct Labor cost |
	Basis, Hours	Rate per hour	Allocated hours		
P	Labor	₹1,200	20,000	₹45,00,000	₹45,00,000
Q	Machine	₹600	20,000	₹2,50,000	₹10,00,000
R	Labor	₹2,800	8,000	₹3,00,000	₹5,00,000
			Total	₹50,50,000	₹60,00,000

Solution

For making the components in-house, the AOC consists of direct labor, direct material, and indirect costs. We use the data given in Table 15.4 to distribute indirect costs.

Department P: 20,000 (₹1,200) = ₹2,40,00,000

Department Q: 20,000 (₹600) = ₹1,20,000

Department R: 8,000 (₹ 2,800) = ₹ 2,24,00,000

Total of these = ₹ 5,84,00,000

AOC = direct labor + direct material + indirect costs

AOC = ₹ 60,00,000 + ₹ 50,50,000 + ₹ 5,84,00,000 = ₹ 6,94,50,000

The make alternative annual worth is the total of capital recovery and AOC

$AW_{make} = -P(A/P, i\%, n) + S(A/F, i\%, n) - AOC$

$AW_{make} = -₹ 75,00,000(A/P, 15\%, 10) + ₹ 1,80,000(A/F, 15\%, 10) - ₹ 6,94,50,000$

$AW_{make} = -₹ 75,00,000(0.19925) + ₹ 1,80,000(0.04925) - ₹ 6,94,50,000$

$AW_{make} = -₹ 7,09,35,510$

Currently, the universal cooking line is purchased at $AW_{buy} = -₹ 50,00,000$. It is cheaper to purchase from outside because the AW of costs is less.

For industries expected to incur high indirect costs, ABC is the best method for the allocation of indirect costs. ABC approach requires identification of activities and cost drivers. Following steps are involved in the implementation of ABC method:

Step 1   Identification of each activity and its associated total cost

Step 2   Identification of the cost drivers and their usage volumes

Step 3   Computation of the overhead rate for each activity using the following relation:

$$\text{ABC overhead rate} = \frac{\text{total cost of activity}}{\text{total volume of cost drivers}} \qquad (15.15)$$

It may be noted that activity refers to support departments or functions, e.g., purchase, quality, personnel, etc. Cost drivers are generally expressed in quantity or volumes, e.g., number of purchase orders, number of construction approvals, number of machine set-ups, etc. To understand this concept, suppose a company makes two models of car (cost centers) and has three primary support departments (activities), that is, purchasing, quality and personnel. Purchasing is an overhead activity with a cost driver of number of purchase orders issued. The ABC allocation rate, in rupees per purchase order, is used to distribute budgeted indirect costs to the two other cost centers. Example 15.9 illustrates the use of ABC method.

Example 15.9

A major equipment manufacturer has four plants located at different locations across India. It plans and executes the business travel for its workforce in each plant. Last year ₹ 32,00,000 were disbursed in travel expenses at a rate of ₹ 32,00,000/2,800 = ₹ 1,142.86 per employee.

Location	Employees	Allocation
A	900	₹14,00,000
B	750	₹8,50,000
C	600	₹5,50,000
D	550	₹4,00,000

The ERP system allocates the amount ₹32,00,000 using *ABC* method on the basis of number of travel vouchers, entitlement of travelers etc. In *ABC* terminology, the travel is an activity and the travel voucher is the cost driver. The following table indicates the details of distribution of 500 vouchers to each employee category, i.e., their entitlement. Use the *ABC* method to allocate travel expenses to each category of employee across all locations. Compare allocations for different locations based on workforce size and number of travel vouchers.

Table 15.5 Distribution of travel vouchers among employees of different categories

Location	Employee Category					Total
	1	2	3	4	5	
A	25	50				75
B	70		35		35	140
C	80	35		30		145
D					140	140
Total	175	85	35	30	175	500

Solution

The *ABC* allocation is to the employee category, not to the location of plants. The *ABC* allocation rate per travel voucher can be calculated using Equation (15.15) as:

ABC allocation rate = ₹32,00,000/500 = ₹6,400 per voucher.

Using this rate the allocation of travel expenses to the employees of different categories can be determined as:

Allocation to the employees of category 1 = ₹6,400 × 175 = ₹11,20,000

Allocation to the employees of category 2 = ₹6,400 × 85 = ₹5,44,000

Allocation to the employees of category 3 = ₹6,400 × 35 = ₹2,24,000

Allocation to the employees of category 4 = ₹6,400 × 30 = ₹1,92,000

Allocation to the employees of category 5 = ₹6,400 × 175 = ₹11,20,000

Similarly, allocation to location *A* = ₹6,400 × 75 = ₹4,80,000

Allocation to location *B* = ₹6,400 × 140 = ₹8,96,000

Allocation to location *C* = ₹6,400 × 145 = ₹9,28,000

Allocation to location $D = ₹6,400 \times 140 = ₹8,96,000$

The comparison between ABC allocation and traditional allocation is given in the following table:

Location	ABC Allocation	Traditional Allocation
A	₹4,80,000	₹14,00,000
B	₹8,96,000	₹8,50,000
C	₹9,28,000	₹5,50,000
D	₹8,96,000	₹4,00,000

From the above calculations, it is evident that employees of category 1 and 5 dominate the ABC allocation. Comparison of location-wise allocation for ABC with the respective traditional method indicates a significant difference in the amounts for all locations, except location B. This comparison supports the assumption that employees' category, not the locations, drive travel expenses and that travel vouchers are a good cost driver for ABC allocation.

PROBLEMS

1. The cost of an office communication management system in 2017 was ₹7,00,000. If the equipment cost index for a similar system is 1200, estimate its cost, considering the cost index in 2017 as 1027.5.

2. Labor cost index for institutional building construction was 79 in the year 2008 and 210 in 2017. If the labor cost for constructing a multistoried institutional building was ₹5,00,00,000 in 2017, what would have been its cost in 2018?

3. Estimate the equipment cost index in 2017 if it was 860 in 2002 and it increases by 2% per year.

4. The commissioning and execution cost for a 50 ton per day capacity petrochemical plant was ₹5,00,00,000. If the cost for a similar plant with 2,000 ton per day capacity was ₹9,00,00,000, estimate the value of cost-capacity exponent.

5. The cost of the fermentation system of a sugar plant having 200 ton per day capacity was estimated as ₹5,00,00,000. If a smaller similar unit costed ₹1,80,00,000 and cost-capacity equation index is 0.46, estimate the size of smaller plant based on which the cost projection was done.

6. The cost of a 1.5 MW gas turbine in the year 2002 was ₹80,00,000. Estimate the cost of 7.5 MW gas turbine in the year 2017 if the cost-capacity equation exponent is 0.42. The equipment cost index, however, increased by 42% in this duration.

7. Engine assembly, transmission system, painting and sheet metal lines of an automobile industry all are considered as separate cost units for allocating indirect cost. If ₹20,00,000 are to be allocated in the automobile company next year, determine the indirect cost rate for each line. Consider the operating hours as the basis of allocation. Indirect cost allocation is given in the following table:

Cost Center	Indirect Cost Allocated, ₹	Estimated Operating Hours
Engine Assembly	7,50,000	1,000
Transmission System	5,50,000	750
Painting	3,00,000	150
Sheet Metal	4,00,000	400

8. Prevailing indirect cost and allocation basis for various divisions of a turnkey project enterprise are given in the following table:

Division	Allocation Basis	Allocation Rate, ₹	Direct Labor Hours	Direct Labor Cost, ₹	Machine Hours
1	MH	1.25	5,000	40,000	3,200
2	DLC	410	10,000	45,000	25,000
3	DLC	1,250	12,000	55,000	20,000
4	DLH	0.95	16,000	35,000	25,000
5	DLC	800	8,000	56,000	12,000
6	DLH	1.10	7,500	65,000	50,000

*MH – Machine Hour, DLC – Daily Labor Cost, DLH – Daily Labor Hour

In order to restructure the allocation process, cost elements and their allocations were reviewed and it was decided that for all future purposes, a simple average of the DLC rate in division 2, 3 and 5 must be used to calculate indirect actual cost. Determine the expenditure for all six divisions and the total variance relative to indirect cost allocation budget of ₹1,20,00,000.

9. An onboard caterer distributes the cost of food based on the type of airlines. For the current year, the total budget and allocation of ₹75,00,000 worth of indirect cost of food are distributed at the rate of 10% of the entire budget.

	Airline			
	A	B	C	D
Budget, ₹	1,50,00,000	4,50,00,000	5,00,00,000	1,00,00,000
Allocation, ₹	15,00,000	25,00,000	30,00,000	5,00,000

(a) Use ABC allocation method with a cost pool of ₹75,00,000 in food cost. The activity is the number of total flyers using a particular airline.

Airline	A	B	C	D
Number of Flyers	3,500	4,000	8,000	1,000

(b) Using ABC method of allocation but considering the activities as VIP lounge users per day, the average number of lounge users by the flyers as follows:

Airline	A	B	C	D
VIP Lounge Users	3.0	2.5	1.25	4.75

(c) Comment on the cost distribution for the two types of activities. Identify, if any other activity may also involve towards cost that might be considered for ABC method which may reflect in realistic allocation of indirect costs.

16

Decision Making

16.1 INTRODUCTION

People make several decisions every day which have an impact on their lives. Decisions such as which academic institution to choose for education, where to invest capital, which doctor to consult for treatment, whether to make or buy a product, whether to rent or use an owned resource and so on are some important decisions for which people would desire to make the correct choice. A decision, in general, may be defined as the selection by the decision maker of an act, considered to be the best according to some pre-designated standard, from among the several available options. Decision making is the process of making a choice for a particular action when several alternative actions are available in a given decision make environment. The purpose of decision making is to select the best alternative action, i.e., the one with the highest expected value.

16.2 TERMINOLOGIES USED IN DECISION MAKING

Course of action: The decision taken by the decision maker is called a course of action.

State of nature: Future events that are beyond the control of the decision maker are called state of nature.

Pay-off: A pay-off is an outcome, expressed in numerical values, that results from each possible combination of alternatives and states of nature.

16.3 STEPS IN DECISION MAKING

The following steps are involved in decision making:

Step 1   Determine the various alternative courses of action from which the best one is to be selected.

Step 2   Identify the possible outcomes, called the states of nature or events for the decision problem.

Step 3 Construct a pay-off table, i.e., a table representing profit, benefits and so on for each combination of alternative course of action and state of nature.

If there are m alternative courses of action $A_1, A_2, \ldots, A_m$ and n states of nature $S_1, S_2, \ldots, S_n$, then the pay-off matrix will be represented as presented in Table 16.1.

Table 16.1 Pay-off table

States of nature	Courses of action (Decision alternative)			
	A_1	A_2	. . .	A_m
S_1	x_{11}	x_{12}	. . .	x_{1m}
S_2	x_{21}	x_{22}	. . .	x_{2m}
. . .	. . .	. . .	. . .	. . .
S_n	x_{n1}	x_{n2}	. . .	x_{nm}

where x_{ij} is the pay-off resulting from the combination of ith state of nature or event and j^{th} action.

Step 4 Choose the criterion which results in largest pay-off. The criterion may be economic, quantitative or qualitative (for example, market share, profit, taste of a food and so on.)

Example 16.1

A manufacturing firm produces three types of products. The fixed and variable costs are given below:

Type of Product	Fixed Cost (₹)	Variable Cost per unit (₹)
A	35,000	20
B	45,000	15
C	63,000	12

The likely demand (units) of the products is:
Low demand: 5,000
Moderate demand 10,000
High demand 15,000
If the unit selling price of each type of product is ₹ 40, then prepare the pay-off matrix.

Solution

Let P_1, P_2 and P_3 denote profit for low (D_1), moderate (D_2) and high (D_3) demands respectively. Then the pay-off is given by:

Pay-off = Sales revenue – Total cost

Pay-off = Sales revenue – (Fixed cost + Variable cost)

The calculations for pay-off for each pair of alternate demand (course of action) and the types of product (states of nature) are given below:

P_1 for product $A = 5,000 \times 40 - (35,000 + 20 \times 5,000)$

$$= 2,00,000 - (35,000 + 1,00,000)$$

$$= 2,00,000 - 1,35,000 = ₹ 65,000$$

P_1 for product $B = 5,000 \times 40 - (45,000 + 15 \times 5,000)$

$$= 2,00,000 - (45,000 + 75,000)$$

$$= 2,00,000 - 1,20,000 = ₹ 80,000$$

P_1 for product $C = 5,000 \times 40 - (63,000 + 12 \times 5,000)$

$$= 2,00,000 - (63,000 + 60,000)$$

$$= 2,00,000 - 1,23,000 = ₹ 77,000$$

P_2 for product $A = 10,000 \times 40 - (35,000 + 20 \times 10,000)$

$$= 4,00,000 - 2,35,000 = ₹ 1,65,000$$

P_2 for product $B = 10,000 \times 40 - (45,000 + 15 \times 10,000)$

$$= 4,00,000 - 1,95,000 = ₹ 2,05,000$$

P_2 for product $C = 10,000 \times 40 - (63,000 + 12 \times 10,000)$

$$= 4,00,000 - 1,83,000 = ₹ 2,17,000$$

P_3 for product $A = 15,000 \times 40 - (35,000 + 20 \times 15,000)$

$$= 6,00,000 - 3,35,000 = ₹ 2,65,000$$

P_3 for product $B = 15,000 \times 40 - (45,000 + 15 \times 15,000)$

$$= 6,00,000 - 2,70,000 = ₹ 3,30,000$$

P_3 for product $C = 15,000 \times 40 - (63,000 + 12 \times 15,000)$

$$= 6,00,000 - 2,43,000 = ₹ 3,57,000$$

The pay-off matrix is shown in Table 16.2.

Table 16.2   Pay-off Matrix

Type of Product (Event)	Alternative Demand		
	D_1	D_2	D_3
A	₹ 65,000	₹ 1,65,000	₹ 2,65,000
B	₹ 80,000	₹ 2,05,000	₹ 3,30,000
C	₹ 77,000	₹ 2,17,000	₹ 3,57,000

16.4 DECISION MAKING ENVIRONMENT

There are three different types of environment in which decision makers operate.

Decision Making under Certainty   In this condition, there is complete certainty about the future and the decision maker has perfect information of the consequence of every decision choice with certainty. In this situation, it is obvious that the decision maker would select the alternative that has the largest pay-off for the known state of the nature. For example, consider a person who wishes to invest his capital. Consider the various alternatives such as post office deposit, fixed deposit, RBI bonds, etc., for investment. For all these alternatives, the investor has complete information about maturity of the invested capital.

Decision Making under uncertainty   This refers to the situation in which more than one state of nature exists and the decision maker does not have sufficient knowledge about them and therefore, he or she is unable to assign any probability to their occurrence.

Decision Making under Risk   This refers to the situation in which more than one state of nature exists and the decision maker has sufficient knowledge about and therefore, he or she is able to assign probabilities to the occurrence of each state.

16.5 DECISION MAKING UNDER UNCERTAINTY

The following five criteria are commonly used for decision making under uncertainty:
1. Optimistic criterion or the rule of optimism Maximax
2. Pessimistic criterion or the rule of pessimism Maximin
3. Regret criterion Minimax
4. Realism criterion (Hurwicz's rule)
5. Criterion of Insufficient Reason or Equally likely criterion (Laplace's rule)

16.5.1   The Maximax Criterion

This criterion reflects the optimistic view of the decision maker and he or she will select the best option for each strategy or choice available to him or her. Example 16.2 illustrates the application of this criterion in decision making.

Example 16.2

Consider the pay-off matrix of Example 16.1. Select the type of product using optimistic criterion.

Solution

The pay-off matrix of Example 16.1 is reproduced in Table 16.3.

Table 16.3 Pay-off matrix

Type of Product (Alternative Strategies)	Alternative Demand (States of nature)		
	D_1	D_2	D_3
A	₹65,000	₹1,65,000	₹2,65,000
B	₹80,000	₹2,05,000	₹3,30,000
C	₹77,000	₹2,17,000	₹3,57,000

Select the maximum pay-off for each alternative, i.e., type of product.

For product A, maximum pay-off is ₹2,65,000

For product B, maximum pay-off is ₹3,30,000

For product C, maximum pay-off is ₹3,57,000

Now, choose the alternative with the maximum pay-off in this group. So according to this criterion, the manufacturing firm would select to produce product C.

16.5.2 The Maximin Criterion

According to this criterion, the best of the worst is chosen.

For the pay-off matrix of Example 16.2, the manufacturing firm would attempt to maximize its minimum possible pay-off. It would first list the minimum pay-off possible for each alternative and then select the alternative within this group that gives the maximum pay-off. In Table 16.4, the minimum pay-off of product A, B and C are ₹65,000, ₹80,000 and ₹77,000 respectively. Now, within this group, the maximum pay-off is that of product B, which is ₹80,000. Thus, according to this criterion, the manufacturing firm would decide to produce product B.

Table 16.4 Pay-off matrix

Type of Product (Alternative Strategies)	Alternative Demand (States of nature)		
	D_1	D_2	D_3
A	₹65,000	₹1,65,000	₹2,65,000
B	₹80,000	₹2,05,000	₹3,30,000
C	₹77,000	₹2,17,000	₹3,57,000

16.5.3 The Minimax Regret Criterion

In this approach, the decision maker calculates the opportunity loss or regret for not taking the best decision under each state of nature. For example, if the demand turned out to be low, i.e. D_1 then, the manufacturing firm would suffer a loss of ₹15,000 (₹80,000 − ₹65,000) for not having chosen to produce product B which would have given a maximum profit of

₹80,000. The decision maker would select the maximum regret for each alternative and would attempt to minimize the opportunity loss or regret by choosing the alternative which gives the least regret. For each state of nature, the highest pay-off value is taken and all other pay-off values for that state of nature are subtracted from it. Table 16.5 shows the regret values.

Table 16.5   Regret values

Type of Product (Alternative Strategies)	Alternative Demand (States of nature)		
	D_1	D_2	D_3
A	₹15,000	₹52,000	₹92,000
B	₹0	₹12,000	₹27,000
C	₹3,000	₹0	₹0

The maximum regret for product A, B and C are ₹92,000, ₹27,000 and ₹3,000 respectively and the minimum of these regrets is ₹3,000. Thus, the manufacturer should produce product C.

16.5.4   The Realism Criterion (Hurwicz's Rule)

According to this criterion, a measure of realism is calculated for each alternative by using the following formula:

$$\text{Measure of realism} = \alpha\,(\text{Maximum pay-off}) + (1 - \alpha)\,(\text{Minimum pay-off}).$$

Here α represents co-efficient or index of optimism and its value lies between 0 and 1. If α is equal to 0, it implies that the decision maker is pessimist about the state of nature. On the other hand, a value of α equal to 1 indicates optimism about the state of nature. After calculating the measure of realism for each alternative, the alternative with the highest measure of realism value is chosen.

Let the manufacturing firm feel highly optimistic and take $\alpha = 0.9$. The measure of realism values for the three alternatives are:

Product A: 0.9 (2,65,000) + 0.1 (65,000) = ₹2,32,000

Product B: 0.9 (3,30,000) + 0.1 (80,000) = ₹2,89,000

Product C: 0.9 (3,57,000) + 0.1 (77,000) = ₹3,13,600

The manufacturing firm would now choose the alternative of product C as it gives maximum value of the measure of realism.

16.5.5   Criterion of Insufficient Reason (Laplace's Rule)

It is also called the Laplace's rule, which states that in the absence of any evidence to the contrary, all states of nature are fairly assumed to have an equal probability of occurrence.

In this approach, the expected value for each alternative is calculated. The calculation of expected value for each alternative of the manufacturing firm is shown below. Since there are three states of nature, each is assumed to occur with a probability of $\frac{1}{3}$.

Product A: $\frac{1}{3}$ (65,000) + $\frac{1}{3}$ (1,65,000) + $\frac{1}{3}$ (2,65,000) = ₹ 1,65,000

Product B: $\frac{1}{3}$ (80,000) + $\frac{1}{3}$ (2,05,000) + $\frac{1}{3}$ (3,30,000) = ₹ 2,05,000

Product C: $\frac{1}{3}$ (77,000) + $\frac{1}{3}$ (2,17,000) + $\frac{1}{3}$ (3,57,000) = ₹ 2,17,000

The manufacturing firm should choose the alternative of product C as it has the highest expected value.

16.6 DECISION MAKING UNDER RISK

In this case, it is assumed that the decision maker has sufficient information about the states of nature. Therefore, he can easily list the various states of nature and can also assign probabilities to their occurrence. For assigning probabilities, the decision maker may use the past experience or may use subjective judgement. For making decision under the condition of risk, the most commonly used criterion is the criterion of expected value or the Baye's criterion.

16.6.1 Expected Monetary Value (EMV)

In order to arrive at the best possible decision, expected pay-off for each alternative is calculated as the sum of the weighted pay-offs for the alternatives. The weights are the probability values which the decision-maker assigns to the different states of nature. Consider the manufacturing firm's problem given in Example 16.1. Suppose the management of the firm assigns probability value of 0.1, 0.3 and 0.6 to the demands, D_1, D_2 and D_3 respectively. Using this information, we can calculate expected value for each alternative, i.e., product A, B and C as shown in Table 16.6.

Table 16.6   Pay-off and expected values

Type of Product (Alternative Strategies)	Alternative Demand (States of nature)			Expected Value
	D_1	D_2	D_3	
Probability	0.1	0.3	0.6	
Product A	₹ 65,000	₹ 1,65,000	₹ 2,65,000	₹ 2,15,000
Product B	₹ 80,000	₹ 2,05,000	₹ 3,30,000	₹ 2,67,000
Product C	₹ 77,000	₹ 2,17,000	₹ 3,57,000	₹ 2,87,000

Expected Value for Product $A = 0.1 \times 65,000 + 0.3 \times 1,65,000 + 0.6 \times 2,65,000$
$$= ₹2,15,000$$

Expected Value for Product $B = 0.1 \times 80,000 + 0.3 \times 2,05,000 + 0.6 \times 3,30,000$
$$= ₹2,67,500$$

Expected Value for Product $C = 0.1 \times 77,000 + 0.3 \times 2,17,000 + 0.6 \times 3,57,000$
$$= ₹2,87,000$$

The manufacturing firm should opt for product C as this option will give it the maximum expected value.

It may be noted that the decision maker may not necessarily assign the same probability to different states of nature for all the alternatives. The decision maker may feel that the probability of occurrence of different states of nature may vary with the alternative and, therefore, he or she may assign different probabilities to different states of nature for different alternatives. Let us reconsider the problem of the manufacturing firm and assign different probabilities for different alternatives as shown in Table 16.7. The expected value for the alternatives can be calculated in the same way as explained earlier and the values are also shown in Table 16.7.

Table 16.7 Alternatives with different probabilities assigned to different states of nature

Type of Product (Alternative Strategies)	States of nature			Expected Value
	D_1	D_2	D_3	
Probability	0.1	0.3	0.6	
Product A	₹65,000	₹1,65,000	₹2,65,000	₹2,15,000
Probability	0.3	0.2	0.5	
Product B	₹80,000	₹2,05,000	₹3,30,000	₹2,30,000
Probability	0.3	0.3	0.4	
Product C	₹77,000	₹2,17,000	₹3,57,000	₹2,31,000

The manufacturing firm should opt for product C as this option gives the highest expected value.

Example 16.3

A retailer purchases a particular product at the rate of ₹200 per kg daily and sells it at the rate of ₹250 per kg. The unsold product at the end of the day can be disposed off the next day at the salvage value of ₹170 per kg. Past sales have ranged from 25 to 28 kg per day. Table 16.8 shows the sales record for the past 100 days.

Table 16.8 Sales record

Amount sold (kg)	25	26	27	28
Number of days	30	40	20	10

Determine the quantity of the product in kg which the retailer should purchase to optimize the profit.

Solution

Let A_i ($i = 1, 2, 3, 4$) be the possible alternative strategies, i.e., the quantity of the product purchased in kg and S_j ($j = 1, 2, 3, 4$) be the possible states of nature, i.e., daily demand.

Profit per kg can be defined as:

Profit per kg = selling price per kg – cost per kg

$$= ₹250 - ₹200 = ₹50 \text{ per kg}$$

In case of unsold product, there would be loss which is:

Loss per kg = ₹200 – ₹170 = ₹30 per kg

Conditional Profit = Profit × amount sold (kg) – Loss × amount unsold (kg)

$$= \begin{cases} 50A & \text{if } S \geq A \\ (250 - 200)S - 30(A - S) = 80S - 30A & \text{if } A > S \end{cases}$$

The pay-off table is shown in Table 16.9.

Table 16.9   Pay-off for Example 16.3

	State of nature (Demand per day, S_j)				Expected pay-off of alternative strategy = $\sum_{j=1}^{4}$ Prob. P_j × pay-off of alternative strategy i (P_{ij})
Probability	0.3	0.4	0.20	0.1	
Alternative strategy, i.e., purchase per day, A_i	25	26	27	28	
25	1,250	1,250	1,250	1,250	1,250
26	1,220	1,300	1,300	1,300	1,276
27	1,190	1,270	1,350	1,350	1,270
28	1,160	1,240	1,320	1,400	1,248

Expected monetary value (EMV) is 1,276 because this is the maximum profit among the profits from various alternative strategies. Thus, the daily purchase should be 26kg.

16.6.2   Expected Opportunity Loss (EOL)

In this approach, the decision maker, first of all, calculates the opportunity loss for each state of nature and then calculates the expected value of each alternative by the sum of the product of opportunity loss and probability. The alternative with the minimum expected value is selected as the best alternative. Mathematically, expected opportunity loss (EOL) is defined as:

$$EOL \text{ (state of nature, } S_j) = \sum_{j=1}^{S} l_{ij} P_j$$

Where l_{ij} is the opportunity loss due to state of nature (S_j) and alternative strategy (A_i) and P_j is the probability of occurrence of state of nature, S_j

Example 16.4

Using the information given in Example 16.3, find expected opportunity loss for each alternative strategy and select the best alternative.

Solution

The pay-off table 16.9 is shown below:

	State of nature (Demand per day, S_j)				Expected pay-off of alternative strategy = $\sum_{j=1}^{4}$ Prob. $P_j \times$ pay-off of alternative strategy i (P_{ij})
Probability	0.3	0.4	0.2	0.1	
Alternative strategy, i.e., purchase per day, A_i	25	26	27	28	
25	1,250	1,250	1,250	1,250	1,250
26	1,220	1,300	1,300	1,300	1,276
27	1,190	1,270	1,350	1,350	1,270
28	1,160	1,240	1,320	1,400	1,248

The opportunity loss can be calculated as the difference between maximum pay-off and individual pay-off for a particular state of nature as shown in Table 16.10.

Table 16.10　Opportunity loss

	State of nature (Demand per day, S_j)				EOL $= \sum_{j=1}^{4} l_{ij} P_j$
Probability	0.3	0.4	0.2	0.1	
Alternative strategy, i.e., purchase per day, A_i	25	26	27	28	
25	1,250 − 1,250 = 0	1,300 − 1,250 = 50	1,350 − 1,250 = 100	1,400 − 1,250 = 150	55
26	1,250 − 1,220 = 30	1,300 − 1,300 = 0	1,350 − 1,300 = 50	1,400 − 1,300 = 100	29 (optimal decision)
27	1,250 − 1,190 = 60	1,300 − 1,270 = 30	1,350 − 1,350 = 0	1,400 − 1,350 = 50	35
28	1,250 − 1,160 = 90	1,300 − 1,240 = 60	1,350 − 1,320 = 30	1,400 − 1,400 = 0	57

The minimum opportunity loss is 29. Thus, the daily purchase of 26 kg of the product is recommended

16.6.3 Expected Value of Perfect Information (EVPI)

In the example 16.3, if the retailer can correctly estimate the daily demand of the product then he or she can remove all uncertainty and would have no losses. But for this the retailer has to obtain perfect information regarding occurrence of various states of nature. This does not mean that the daily demand of the product should not vary from 25 to 28 kg. Demand would be 25 kg per day 30% of the time, 26 kg per day 40% of the time, 27 kg per day 20% of the time and 28 kg per day 10% of the time. But with perfect information, the retailer will know in advance what the demand would be for the following day and can stock accordingly. When the demand is 25 kg, the retailer will stock 25 kg and will get a profit of ₹ 1,250 (refer Table 16.9). Table 16.11 shows the profit (pay-off) when the retailer has perfect information.

Table 16.11 Pay-off (profits) with perfect information

Stock Policy	Demand				Expected Value
	25	26	27	28	
Probability	0.3	0.4	0.2	0.1	
25	1,250				375
26		1,300			520
27			1,350		270
28				1,400	140
				Total	1,305

The maximum profit that the retailer can expect with perfect information is ₹ 1,305. In the absence of perfect information, the retailer's expected value was ₹ 1,276 as shown in Example 16.3. The value of perfect information is 1,305 − 1,276 = ₹ 29. The retailer should, therefore, not spend more than ₹ 29 every day in gathering this perfect information or else he or she will not attain the maximum expected profit of ₹ 1,305.

16.7 MARGINAL ANALYSIS

In the retailer's problem given in Example 16.3, he or she has 4 stocking alternatives and 4 demand levels, giving a total of 16 combinations. If the number of options increases, the computational effort would increase tremendously and constructions of conditional profit or loss tables becomes extremely difficult. However, the marginal approach overcomes this difficulty. In case of addition of a unit in the stock, only two things are possible: it will either get sold or will not be sold. The sum of the probabilities of these two events must be 1. If the probability of selling the additional unit is 0.7, then the probability that it will not be sold is 0.3.

If P represents the probability of selling an additional unit then $(1 - P)$ is the probability that the unit will not be sold. If the additional unit is sold, it results in some profit and if it is not sold, then some loss occurs. This profit or loss is called marginal profit (MP) or marginal loss (ML). If the retailer stocked 25 kg and had a demand of 25 kg he or she would have no loss. Suppose the retailer stocks an additional 1 kg of the product, i.e., 26 kg and is unable to sell more than 25kg. Then he or she would suffer a loss of ₹ 200 on the additional weight of the product. This is marginal loss on the 26th kg.

The expected profit from an additional unit would be $P(MP)$ and the expected loss would be $(1 - P)(ML)$. It would be worthwhile stocking an additional unit as long as the expected marginal profit is greater than the expected marginal loss. The stock would be maintained up to the point where:

$$P(MP) = (1 - P)(ML)$$

$$\text{or} \quad P = \frac{ML}{MP + ML}$$

Let us apply this to the retailer's problem in Example 16.3. We know that the probability of selling at least 25 kg of the product is 1.0 as the daily demand varies between 25 and 28 kg. The probability that the retailer will be able to sell an additional unit, i.e., the probability of selling at least 26 kg of product is:

Probability of selling at least 26 kg = 1 – Probability of selling 25 kg

$$= 1 - 0.3 = 0.7$$

It may be seen that as the demand increases, the probability of selling the additional kg decreases. This is shown in Table 16.12.

Table 16.12   Probability values with increasing demand

Demand	Probability that demand will be	Probability that the demand will be equal to or greater than
25	0.3	1.0
26	0.4	0.7
27	0.2	0.3
28	0.1	0.1

P represents the minimum required probability of selling at least an additional unit to justify the stocking of that additional unit. Let us calculate P for the retailer. The marginal profit per unit is ₹ 50 and the marginal loss per unit is ₹ 200. Hence:

$$P = \frac{ML}{MP + ML} = \frac{200}{50 + 200} = \frac{200}{250} = 0.8$$

We can now compare this value with the probability values shown in Table 16.12. The retailer should not stock 26 kg as the probability of selling of it is only 0.7, whereas the minimum required probability of selling at least an additional kg in 0.8

16.8 DECISION TREES

The decision problems we have discussed so far are single stage problems where the decision maker has to select the best alternative or course of action based on information available at a point of time. There are many complex situations in which it is required to make decisions at multiple stages. The problems of this nature are called sequential decision problems and they are solved by taking a sequence of decisions in such a way that each decision gives rise to the occurrence of an event which, in turn, influences the subsequent decision.

Multiple stage decision situations can be easily analyzed using a diagram called a decision tree. A decision tree consists of nodes, branches, probability estimates and pay-offs. Nodes are of two types:
(1) Decision nodes denoted by a square (□)
(2) Chance nodes represented by a circle (○)

Decision node refers to a point where the decision maker must select one action from the available courses of action. The branches emerging from a decision node represent the various alternative courses of action. Each course of action leads to a chance node which shows the response of the decision taken by the decision maker.

Various nodes of the decision tree are connected by the branches which are of two types:
(1) Decision branches
(2) Chance branches

A branch emerging from a decision node represents a course of action to be chosen at a decision point. A branch which emerges from a chance node represents the state of nature of a set of chance events. The probabilities of the occurrence of the chance events are also shown in the diagram and they are written alongside the particular branch.

A branch which is not followed by either a decision or a chance node is called a *terminal branch* and it represents either a course of action or a chance outcome. The terminal points are mutually exclusive points, i.e., exactly one course of action would be chosen for each terminal point. The pay-off can be either positive (i.e., sales or profit) or negative (i.e., expenditure or cost) and can be associated either with decision branches or chance branches.

The expected return for a chance node is calculated as the sum of the products of the probabilities of the branches emanating from it and their respective expected returns. Similarly, for a decision node, the expected return is calculated for each of its branches and the highest return is selected. Example 16.5 illustrates construction of a decision tree.

Example 16.5

An air conditioner manufacturing company has to decide whether to set up a large plant or a small plant for its new range of air conditioners. A large plant will cost the company ₹500 lakhs while a small plant will cost ₹240 lakhs. A survey conducted by the marketing team of the company reveals the following estimates for sales over the next 10 years:

High demand   Probability = 0.5
Moderate demand   Probability = 0.3
Low demand   Probability = 0.2

(a) A large pant with a high demand will yield an annual profit of ₹200 lakhs.
(b) A large plant with a moderate demand will yield an annual profit of ₹120 lakhs.
(c) A large plant with a low demand will lose ₹400 lakhs annually because of production inefficiencies.
(d) A small plant with a high demand will yield a profit of ₹500 lakhs annually, taking into account the cost of lost sales due to inability to meet demand.
(e) A small plant with a moderate demand will yield a profit of ₹70 lakhs, as the losses due to lost sales will be lower.
(f) A small plant with a low demand will yield a profit of ₹90 lakhs annually as the plant capacity and demand will match.

Solution

Using the information, let us construct the decision tree. The decision tree will start with a decision node and from this node two branches will emanate. One branch will represent "set up a large plant" and the other branch will represent "set up a small plant" as shown in Fig. 16.1.

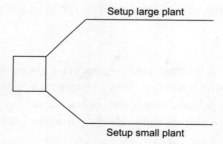

Setup large plant

Setup small plant

Figure 16.1   Branches of Decision Tree

If the company builds either a large plant or a small plant, it can result in three outcomes, i.e., the market demand can be high, moderate or low. These outcomes are now added to the tree at the end of chance nodes as shown in Fig. 16.2.

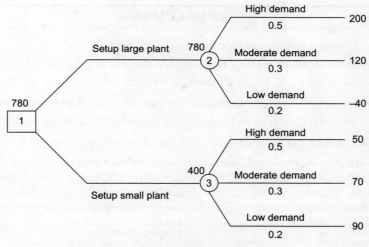

Figure 16.2 Decision Tree with chance nodes

The decision tree shown in Fig. 16.2 also shows the probability of each state of nature (0.5, 0.3 and 0.2) and the pay-off for each outcome (200, 120, −40, 50, 70 and 90). In order to calculate the value of the decisions, the tree is rolled back. The expected value at chance node 2 is computed as:

Expected value = 0.5 × 200 + 0.3 × 120 + 0.2 × (−40) = 128

The expected value in 10 years is ₹ 128 × 10 = ₹ 1,280 lakhs. The cost for setting large plant is ₹ 500 lakhs.

Net expected gains at the end of 10 years is ₹ 1,280 − ₹ 500 = ₹ 780 lakhs. The expected value is written on the decision node 2 of the tree.

Similarly, expected value of chance node 3 can be found as:

Expected value = 0.5 × 50 + 0.3 × 70 + 0.2 × 90 = 64

The expected value in 10 years is ₹ 10 × 64 = ₹ 640 lakhs. The cost of building a small plant is ₹ 240 lakhs.

Net expected gains at the end of 10 years is₹ ₹640 − ₹ 240 = ₹ 400 lakhs. This value is also written on decision node 3. The chance node with the higher value is carried backward to the decision box and the decision is taken accordingly. The decision in this case is that the company should setup a large plant as the expected value (₹ 780 lakhs) is higher than the expected value (₹ 400 lakhs) for a small plant.

PROBLEMS

1. Briefly explain the different decision making-environments.

2. In the context of decision making, briefly explain the following terms:

   (i) Course of action, (ii) States of nature (iii) Pay-off

3. Explain the different decision rules used in the context of decision making under uncertainty.

4. Briefly explain the various methods used in the context of decision making under risk.

5. What is a decision tree? Discuss its use in the decision making process.

6. A person is planning to invest ₹2 lakh in the shares of three companies X, Y and Z. The growth in capital and return on capital, which is a measure of pay-off, depends upon the existing economic conditions. The person's expectations of net earnings (in ₹000s) on his investment of ₹2 lakhs under three economic conditions, i.e., recession, stable situation and growth are given below:

Company	Economic Condition		
	Recession	Stable	Growth
X	−30	12	20
Y	8	15	16
Z	13	12	30

   Select the optimal strategy for investment under the following conditions:

   (a) Maximin criterion

   (b) Maximax criterion

   (c) Minimax regret

   (d) Laplace's rule

   (e) Hurwicz's rule, $\alpha = 0.7$

7. A mango merchant buys mangoes for ₹200 per box of 5 kg and sells them for ₹300 per box. Any unsold quality is thrown away at the end of the day. The merchant wishes to find the number of boxes he should stock at the beginning of the day. The past 100 days sales are given below:

Daily sales (no. of boxes)	20	22	24	26
No. of days	10	15	45	30

8. In a FMCG store, the cost of an item in stock for a week is ₹100 per unit and the cost of unit shortage is ₹250. The probability distribution of the weekly sales of the item as follows:

Weekly Sales	0	1	2	3	4	5	6
Probability	0.15	0.10	0.20	0.25	0.10	0.15	0.05

   How many units per week should the dealer order? Also, find the expected value.

9. The management of a manufacturing firm is required to choose between product X and product Y for manufacturing. The probability matrix for the two products is as follows:

Courses of action	States of nature		
	Good	Fair	Poor
Product X	70%	20%	10%
Product Y	75%	15%	10%

The expected profit by selling these products is given below:

Courses of action	States of nature		
	Good	Fair	Poor
Product X	70,000	30,000	10,000
Product Y	1,00,000	40,000	−6,000

Calculate the expected value of the alternatives and suggest which product the firm should produce.

10. Public health engineering department (PHE) is consideringing to drill a bore-well in an arid area. Past experience of the engineer shows that only 60% of the bore-wells were successful when drilled for 250 feet in the area. The survey data in the neighborhood shows that if the depth of drilling is 300 feet, then the probability of success of getting water is 15%. The cost of drilling is ₹1,000 per foot. An estimate of the PHE department shows that if the bore is not drilled there successfully, then the cost of water supply is ₹25,00,000 over the next 10 years. Draw the decision tree and advise the best strategy to be adopted by the PHE department.

❑❑❑

APPENDIX A

Table 1 Discrete Cash Flow: Compound Interest Factors

	Single Payments		Uniform Series Payments				Arithmetic Gradients	
n	Compound Amount F/P	Present Worth P/F	Sinking Fund A/F	Compound Amount F/A	Capital Recovery A/P	Present Worth P/A	Gradient Present Worth P/G	Gradient Uniform Series A/G
1	1.00250	0.99750	1.00000	1.00000	1.00250	0.99750		
2	1.00500	0.99500	0.49938	2.00250	0.50188	1.99250	0.99500	0.49940
3	1.00750	0.99250	0.33250	3.00750	0.33500	2.98510	2.98010	0.99830
4	1.01000	0.99010	0.24906	4.01500	0.25156	3.97510	5.95030	1.49690
5	1.01260	0.98760	0.19900	5.02510	0.20150	4.96270	9.90070	1.99500
6	1.01510	0.98510	0.16563	6.03760	0.16813	5.94780	14.82630	2.49270
7	1.01760	0.98270	0.14179	7.05270	0.14429	6.93050	20.72230	2.99000
8	1.02020	0.98020	0.12391	8.07040	0.12674	7.91070	27.58390	3.48690
9	1.02270	0.97780	0.11000	9.09050	0.11250	8.88850	35.40610	3.98340
10	1.02530	0.97530	0.09888	10.11330	0.10138	9.86390	44.18420	4.47940
11	1.02780	0.97290	0.08978	11.13850	0.09228	10.83680	53.91330	4.97500
12	1.03040	0.97050	0.08219	12.16640	0 08469	11.80730	64.58860	5.47020
13	1.03300	0.96810	0.07578	13.19680	0.07828	12.77530	76.20530	5.96500
14	1.03560	0.96560	0.07028	14.22980	0.07278	13.74100	88.75870	6.45940
15	1.03820	0.96320	0.06551	15.26540	0.06801	14.70420	102.24410	6.95340
16	1.04080	0.96080	0.06134	16.30350	0.06384	15.66500	116.65670	7.44690
17	1.04340	0.95840	0.05766	17.34430	0.06016	16.62350	131.99170	7.94010
18	1.04600	0.95610	0.05438	18.38760	0.05688	17.57950	148.24460	8.43280
19	1.04860	0.95370	0.05146	19.43360	0.05396	18.53320	165.41060	8.92510
20	1.05120	0.95130	0.04882	20.48220	0.05132	19.48450	183.48510	9.41700
21	1.05380	0.94890	0.04644	21.53340	0.04894	20.43340	202.46340	9.90850
22	1.05650	0.94660	0.04427	22.58720	0.04677	21.38000	222.34100	10.39950
23	1.05910	0.94420	0.04229	23.64370	0.04379	22.32410	243.11300	10.89010

	Single Payments		Uniform Series Payments				Arithmetic Gradients	
n	Compound Amount F/P	Present Worth P/F	Sinking Fund A/F	Compound Amount F/A	Capital Recovery A/P	Present Worth P/A	Gradient Present Worth P/G	Gradient Uniform Series A/G
24	1.06180	0.94180	0.04048	24.70280	0.04298	23.26600	264.77530	11.38040
25	1.06440	0.93950	0.03881	25.76460	0.41310	24.20550	287.32300	11.87020
26	1.06710	0.93710	0.03727	26.82900	0.03977	25.14260	310.75160	12.35960
27	1.06970	0.93480	0.03585	27.89610	0.03835	26.07740	335.05660	12.84850
28	1.07240	0.93250	0.03452	28.96580	0.03702	27.00990	360.23340	13.33710
29	1.07510	0.93010	0.03329	30.03820	0.03579	27.94000	386.27760	13.82520
30	1.07780	0.92780	0.03214	31.11330	0.03464	28.86790	413.18470	14.31300
36	1.09410	0.91400	0.02658	37.62060	0.02908	34.38650	592.49880	17.23060
40	1.10500	0.90500	0.02380	42.01320	0.02630	38.01990	728.73990	19.16730
48	1.12730	0.88710	0.01963	50.93120	0.02213	45.17870	1040.06000	23.02090
50	1.13300	0.88260	0.01881	53.18870	0.02130	46.94620	1125.78000	23.98020
52	1.13860	0.87820	0.01803	55.45750	0.02053	48.70480	1214.59000	24.93770
55	1.14720	0.87170	0.01698	58.88190	0.01948	51.32640	1353.53000	26.37100
60	1.16160	0.86090	0.01547	64.64670	0.01797	55.65240	1600.08000	28.75140
72	1.19690	0.83550	0.01269	78.77940	0.01519	65.81690	2265.56000	34.42210
75	1.20590	0.82920	0.01214	82.37920	0.01464	68.31080	2447.61000	35.83050
84	1.23340	0.81080	0.01071	93.34190	0.01321	75.68130	3029.76000	40.03310
90	1.25200	0.79870	0.00992	100.78850	0.01242	80.50380	3446.87000	42.81620
96	1.27090	0.78690	0.00923	108.34740	0.01173	85.25460	3886.28000	45.58440
100	1.28360	0.77900	0.00881	113.45000	0.01131	88.38250	4191.24000	47.42160
108	1.30950	0.76360	0.00808	123.80930	0.01058	94.54530	4829.01000	51.07620
120	1.34940	0.74110	0.00716	139.74140	0.00966	103.56180	5852.11000	56.50840
132	1.39040	0.71920	0.00640	156.15820	0.00890	112.31210	6950.01000	61.88130
144	1.43270	0.69800	0.00578	173.07430	0.00828	120.80410	8117.41000	67.19490
240	1.82080	0.54920	0.00305	328.30200	0.00555	180.31090	19399.00000	107.58630
360	2.45680	0.40700	0.00172	582.73690	0.00422	237.18940	36264.00000	152.89020
480	3.31510	0.30160	0.00108	926.05950	0.00358	279.34180	53821.00000	192.66990

0.5% **Table 2** Discrete Cash Flow: Compound Interest Factors **0.5%**

	Single Payments		Uniform Series Payments				Arithmetic Gradients	
n	Compound Amount F/P	Present Worth P/F	Sinking Fund A/F	Compound Amount F/A	Capital Recovery A/P	Present Worth P/A	Gradient Present Worth P/G	Gradient Uniform Series A/G
1	1.00500	0.99500	1.00000	1.00000	1.00500	0.99500		
2	1.01000	0.99010	0.49875	2.00500	0.50375	1.98510	0.99010	0.49880
3	1.01510	0.98510	0.33167	3.01500	0.33667	2.97020	2.96040	0.99670
4	1.02020	0.98020	0.24813	4.03010	0.25313	3.95050	5.90110	1.49380
5	1.02530	0.97540	0.19801	5.05030	0.20301	4.92590	9.80260	1.99000
6	1.03040	0.97050	0.16460	6.07550	0.16960	5.89640	14.65520	2.48550
7	1.03550	0.96570	0.14073	7.10590	0.14573	6.86210	20.44930	2.98010
8	1.04070	0.96090	0.12283	8.14140	0.12783	7.82300	27.17550	3.47380
9	1.0459	0.95610	0.10891	9.18210	0.11391	8.77910	34.82440	3.96680
10	1.05110	0.95130	0.09777	10.22800	0.10277	9.73040	43.38650	4.45890
11	1.05640	0.94660	0.08866	11.27920	0.09366	10.67700	52.85260	4.95010
12	1.06170	0.94190	0.08107	12.33560	0.08607	11.61890	63.21360	5.44060
13	1.06700	0.93720	0.07464	13.39720	0.07964	12.55620	74.46020	5.93020
14	1.07230	0.93260	0.06914	14.46420	0.07414	13.48870	86.58350	6.41900
15	1.07770	0.92790	.0.06436	15.53650	0.06936	14.41660	99.57430	6.90690
16	1.08310	0.92330	0.06019	16.61420	0.06519	15.33990	113.42380	7.39400
17	1.08850	0.91870	0.05651	17.69730	0.06151	16.25860	128.12310	7.88030
18	1.09390	0.91410	0.05323	18.78580	0.05823	17.17280	143.66340	8.36580
19	1.09940	0.90960	0.05030	19.87970	0.05530	18.08240	160.03600	8.85040
20	1.10490	0.90510	0.04767	20.97910	0.05267	18.98740	177.23220	9.33420
21	1.11040	0.90060	0.04528	22.08400	0.05028	19.88800	195.24340	9.81720
22	1.11600	0.89610	0.04311	23.19440	0.04811	20.78410	214.06110	10.29930
23	1.12160	0.89160	0.04113	24.31040	0.04613	21.67570	233.67680	10.78060
24	1.12720	0.88720	0.03932	25.43200	0.04432	22.56290	254.08200	11.26110
25	1.13280	0.88280	0.03765	26.55910	0.04265	23.44560	275.26860	11.74070

n	Single Payments		Uniform Series Payments				Arithmetic Gradients	
	Compound Amount F/P	Present Worth P/F	Sinking Fund A/F	Compound Amount F/A	Capital Recovery A/P	Present Worth P/A	Gradient Present Worth P/G	Gradient Uniform Series A/G
26	1.13850	0.87840	0.03611	27.69190	0.04111	24.32400	297.22810	12.21950
27	1.14420	0.87400	0.03469	28.83040	0.03969	25.19800	319.95230	12.69750
28	1.14990	0.86970	0.03336	29.97450	0.03836	26.06770	343.43320	13.17470
29	1.15560	0.86530	0.03213	31.12440	0.03713	26.93300	367.66250	13.65100
30	1.16140	0.86100	0.03098	32.28000	0.03598	27.79410	392.63240	14.12650
36	1.19670	0.83560	0.02542	39.33610	0.03042	32.87100	557.55980	16.96210
40	1.22080	0.81910	0.02265	44.15880	0.02765	36.17220	681.33470	18.83590
48	1.27050	0.78710	0.01849	54.09780	0.02349	42.58030	959.91880	22.54370
50	1.28320	0.77930	0.01765	56.64520	0.02265	44.14280	1035.70000	23.46240
52	1.29610	0.77160	0.01689	59.21800	0.02189	45.68970	1113.82000	24.37780
55	1.31560	0.76010	0.01584	63.12580	0.02084	47.98140	1235.27000	25.74470
60	1.34890	0.74140	0.01433	69.77000	0.01933	51.72560	1448.65000	28.00640
72	1.43200	0.69830	0.01157	86.40890	0.01657	60.33950	2012.35000	33.35040
75	1.45360	0.68790	0.01102	90.72650	0.01602	62.41360	2163.75000	34.66790
84	1.52040	0.65770	0.00961	104.07390	0.01461	68.45300	2640.66000	38.57630
90	1.56660	0.63830	0.00883	113.31090	0.01383	72.33130	2976.08000	41.14510
96	1.61410	0.61950	0.00814	122.82850	0.01314	76.09520	3324.18000	43.68450
100	1.64670	0.60730	0.00773	129.33370	0.01273	78.54260	3562.79000	45.36130
108	1.71370	0.58350	0.00701	142.73990	0.01201	83.29340	4054.37000	48.67580
120	1.81940	0.54960	0.00610	163.87930	0.01110	90.07350	4823.51000	53.55080
132	1.93160	0.51770	0.00537	186.32260	0.01037	96.45960	5624.59000	58.31030
144	2.05080	0.48760	0.00476	210.15020	0.00976	102.47470	6451.31000	62.95510
240	3.31020	0.30210	0.00216	462.04090	0.00716	139.58080	13416.00000	96.11310
360	6.02260	0.16600	0.00100	1004.52000	0.00600	166.79160	21403.00000	128.32360
480	10.95750	0.09130	0.00050	1991.49000	0.00550	181.74760	27588.00000	151.79490

0.75%   **Table 3**   Discrete Cash Flow: Compound Interest Factors   **0.75%**

	Single Payments		Uniform Series Payments				Arithmetic Gradients	
n	Compound Amount F/P	Present Worth P/F	Sinking Fund A/F	Compound Amount F/A	Capital Recovery A/P	Present Worth P/A	Gradient Present Worth P/G	Gradient Uniform Series A/G
1	1.00750	0.99260	1.00000	1.00000	1.00750	0.99260		
2	1.01510	0.98520	0.49813	2.00750	0.50563	1.97770	0.98520	0.49810
3	1.02270	0.97780	0.33085	3.02260	0.33835	2.95560	2.94080	0.99500
4	1.03030	0.97060	0.24721	4.04520	0.25471	3.92610	5.85250	1.49070
5	1.03810	0.96330	0.19702	5.07560	0.20452	4.88940	9.70580	1.98510
6	1.04590	0.95620	0.16357	6.11360	0.17107	5.84560	14.48660	2.47820
7	1.05370	0.94900	0.13967	7.15950	0.14717	6.79460	20.18080	2.97010
8	1.06160	0.94200	0.12176	8.21320	0.12926	7.73660	26.77470	3.46080
9	1.06960	0.93500	0.10782	9.27480	0.11532	8.67160	34.25440	3.95020
10	1.07760	0.92800	0.09667	10.34430	0.10417	9.59960	42.60640	4.43840
11	1.08570	0.92110	0.08755	11.42190	0.09505	10.52070	51.81740	4.92530
12	1.09380	0.91420	0.07995	12.50760	0.08745	11.43490	61.87400	5.41100
13	1.10200	0.90740	0.07352	13.60140	0.08102	12.34230	72.76320	5.89540
14	1.11030	0.90070	0.06801	14.70340	0.07551	13.24300	84.47200	6.37860
15	1.11860	0.89400	0.06324	15.81370	0.07074	14.13700	96.98760	6.86060
16	1.12700	0.88730	0.05906	16.93230	0.06656	15.02430	110.29730	7.34130
17	1.13540	0.88070	0.05537	18.05930	0.06287	15.90500	124.38870	7.82070
18	1.14400	0.87420	0.05210	19.19470	0.05960	16.77920	139.24940	8.29890
19	1.15250	0.86760	0.04917	20.33870	0.05667	17.64680	171.22970	8.77590
20	1.16120	0.86120	0.04653	21.49120	0.05403	18.50800	171.22970	9.25160
21	1.16990	0.85480	0.04415	22.65240	0.05165	19.36280	188.32530	9.72610
22	1.17870	0.84840	0.04198	23.82230	0.04948	20.21120	206.14200	10.19940
23	1.18750	0.84210	0.04000	25.00100	0.04750	21.05330	224.89230	10.67140
24	1.19640	0.83580	0.03818	26.18850	0.04568	21.88910	243.89230	11.14220
25	1.20540	0.82960	0.03652	27.38490	0.04402	22.71880	263.80290	11.61170

	Single Payments		Uniform Series Payments				Arithmetic Gradients	
n	Compound Amount F/P	Present Worth P/F	Sinking Fund A/F	Compound Amount F/A	Capital Recovery A/P	Present Worth P/A	Gradient Present Worth P/G	Gradient Uniform Series A/G
26	1.21440	0.82340	0.03498	28.59030	0.04248	23.54220	284.38880	12.08000
27	1.22350	0.81730	0.03355	29.80470	0.04105	24.35950	305.63870	12.54700
28	1.23270	0.81120	0.03223	31.02820	0.03973	25.17070	327.54160	13.01280
29	1.24200	0.80520	0.03100	32.26090	0.03850	25.97590	350.08670	13.47740
30	1.25130	0.79920	0.02985	33.50290	0.03735	26.77510	373.26310	13.94070
36	1.30860	0.76410	0.02430	41.15270	0.03180	31.44680	524.99240	16.69460
40	1.34830	0.74160	0.02153	46.44650	0.02903	34.44690	637.46930	18.50580
48	1.43140	0.69860	0.01739	57.52070	0.02903	40.18480	886.84040	22.06910
50	1.45300	0.68830	0.01656	60.39430	0.02406	41.56640	953.84860	22.94760
52	1.47480	0.67800	0.01580	63.31110	0.02330	42.92760	1022.59000	23.82110
55	1.50830	0.66300	0.01476	67.76880	0.02226	44.93160	1128.79000	25.12230
60	1.56570	0.63870	0.01326	75.42410	0.02076	48.17340	1313.52000	27.26650
72	1.71260	0.58390	0.01053	95.00700	0.01803	55.47680	1791.25000	32.28820
75	1.75140	0.57100	0.00998	100.18330	0.01748	57.20270	1917.22000	33.51630
84	1.87320	0.53380	0.00859	116.42690	0.01609	62.15400	2308.13000	37.13570
90	1.95910	0.51040	0.00782	127.87900	0.01532	65.27460	2578.00000	39.49460
96	2.04890	0.48810	0.00715	139.85620	0.01417	68.25840	2853.94000	41.81070
100	2.11110	0.47370	0.00675	148.14450	0.01425	70.17460	3040.75000	43.33110
108	2.24110	0.44620	0.00604	165.48320	0.01354	73.83940	3419.90000	46.31540
120	2.45140	0.40790	0.00517	193.51430	0.01267	78.94170	3998.56000	50.65210
132	2.68130	0.37300	0.00446	224.17480	0.01196	83.60640	4583.57000	54.82320
144	2.93280	0.34100	0.00388	257.71160	0.01138	87.87110	5169.58000	58.83140
240	6.00920	0.16640	0.00150	667.88690	0.00900	111.14500	9494.12000	85.42100
360	14.73060	0.06790	0.00055	1830.74000	0.00805	124.28190	13312.00000	107.11450
480	36.10990	0.02770	0.00021	4681.32000	0.00771	129.64090	15513.00000	119.66200

1%   **Table 4**   Discrete Cash Flow: Compound Interest Factors   **1%**

	Single Payments		Uniform Series Payments				Arithmetic Gradients	
n	Compound Amount F/P	Present Worth P/F	Sinking Fund A/F	Compound Amount F/A	Capital Recovery A/P	Present Worth P/A	Gradient Present Worth P/G	Gradient Uniform Series A/G
1	1.01000	0.99010	1.00000	1.00000	1.01000	0.99010		
2	1.02010	0.98030	0.49751	2.01000	0.50751	1.97040	0.98030	0.49750
3	1.03030	0.97060	0.33002	3.03010	0.34002	2.94100	2.92150	0.99340
4	1.04060	0.96100	0.24628	4.06040	0.25628	3.90200	5.80440	1.48760
5	1.05100	0.95150	0.19604	5.10100	0.20604	4.85340	9.61030	1.98010
6	1.06150	0.94200	0.16255	6.15200	0.17255	5.79550	14.32050	2.47100
7	1.07210	0.93270	0.13863	7.21350	0.14863	6.72820	19.91680	2.96020
8	1.0.829	0.92350	0.12069	8.28570	0.13069	7.65170	26.38120	3.44780
9	1.09370	0.91430	0.10674	9.36850	0.11674	8.56600	33.69590	3.93370
10	1.10460	0.90530	0.09558	10.46220	0.10558	9.47130	41.84350	4.41790
11	1.11570	0.89630	0.08645	11.56680	0.09645	10.36760	50.80670	4.90050
12	1.12680	0.88740	0.07885	12.68250	0.08885	11.25510	60.56870	5.38150
13	1.13810	0.87870	0.07241	13.80930	0.08241	12.13370	71.11260	5.86070
14	1.14950	0.87000	0.06690	14.94740	0.07690	13.00370	82.42210	6.33840
15	1.16100	0.86130	0.06212	16.09690	0.07212	13.86510	94.48100	6.81430
16	1.17260	0.85280	0.05794	17.25790	0.06794	14.71790	107.27340	7.28860
17	1.18430	0.84440	0.05426	18.43040	0.06426	15.56230	120.78340	7.76130
18	1.19610	0.83600	0.05098	19.61470	0.06098	16.39830	134.99570	8.23230
19	1.20810	0.82770	0.04805	20.81090	0.05805	17.22600	149.89500	8.70170
20	1.22020	0.81950	0.04542	22.01900	0.05542	18.04560	165.46640	9.16940
21	1.23240	0.81140	0.04303	23.23920	0.05303	18.85700	181.69500	9.63540
22	1.24470	0.80340	0.04086	24.47160	0.05086	19.66040	198.56630	10.09980
23	1.25720	0.79540	0.03889	25.71630	0.04889	20.45580	216.06600	10.56260
24	1.26970	0.78760	0.03707	26.97350	0.04707	21.24340	234.18000	11.02370
25	1.28240	0.77980	0.03541	28.24320	0.04541	22.02320	252.89450	11.48310

	Single Payments		Uniform Series Payments				Arithmetic Gradients	
n	Compound Amount F/P	Present Worth P/F	Sinking Fund A/F	Compound Amount F/A	Capital Recovery A/P	Present Worth P/A	Gradient Present Worth P/G	Gradient Uniform Series A/G
26	1.29530	0.77200	0.03387	29.52560	0.04387	22.79520	272.19570	11.94090
27	1.30820	0.76440	0.03245	30.82090	0.04245	23.55960	292.07020	12.39710
28	1.32130	0.75680	0.03112	32.12910	0.04112	24.31640	312.50470	12.85160
29	1.33450	0.74930	0.02990	33.45040	0.03990	25.06580	333.48630	13.30440
30	1.34780	0.74190	0.02875	34.78490	0.03875	25.80770	355.00210	13.75570
36	1.43080	0.69890	0.02321	43.07690	0.03321	30.10750	494.62070	16.42850
40	1.48890	0.67170	0.02046	48.88640	0.03046	32.83470	596.85610	18.17760
48	1.61220	0.62030	0.01633	61.22260	0.02633	37.97400	820.14600	21.59760
50	1.64460	0.60800	0.01551	64.46320	0.02551	39.19610	879.41760	22.43630
52	1.67770	0.59610	0.01476	67.76890	0.02476	40.39420	939.91750	23.26860
55	1.72850	0.57850	0.01373	72.85250	0.02373	42.14720	1032.81000	24.50490
60	1.81670	0.55040	0.01224	81.66970	0.02224	44.95500	1192.81000	26.53330
72	2.04710	0.48850	0.00955	104.70990	0.01955	51.15040	1597.87000	31.23860
75	2.10910	0.47410	0.00902	110.91280	0.01902	52.58710	1702.73000	32.37930
84	2.30670	0.43350	0.00765	130.67230	0.01765	56.64850	2023.32000	35.71700
90	2.44860	0.40840	0.00690	144.86330	0.01690	59.16090	2240.57000	37.87240
96	2.59930	0.38470	0.00625	159.92730	0.01625	61.52770	2459.43000	39.97270
100	2.70480	0.36970	0.00587	170.48140	0.01587	63.02890	2605.78000	41.34260
108	2.92890	0.34140	0.00518	192.89260	0.01518	65.85780	2898.42000	44.01030
120	3.30040	0.30300	0.00435	230.03870	0.01435	69.70050	3334.11000	47.83490
132	3.71900	0.26890	0.00368	271.89590	0.01368	73.11080	3761.69000	51.45200
144	4.19060	0.23860	0.00101	319.06160	0.01313	76.13720	4177.47000	54.86760
240	10.89260	0.09180	0.00101	989.25540	0.01101	90.81940	6878.60000	75.73930
360	35.94960	0.02780	0.00029	3494.96000	0.01029	97.21830	8720.43000	89.69950
480	118.64770	0.00840	0.00008	11765.00000	0.01008	99.15720	9511.16000	95.92000

1.25%   **Table 5**   Discrete Cash Flow: Compound Interest Factors   **1.25%**

	Single Payments		Uniform Series Payments				Arithmetic Gradients	
n	Compound Amount *F/P*	Present Worth *P/F*	Sinking Fund *A/F*	Compound Amount *F/A*	Capital Recovery *A/P*	Present Worth *P/A*	Gradient Present Worth *P/G*	Gradient Uniform Series *A/G*
1	1.01250	0.98770	1.00000	1.00000	1.01250	0.98770		
2	1.02520	0.97550	0.49680	2.01250	0.50939	1.96310	0.97550	0.49690
3	1.03800	0.96340	0.32920	3.03770	0.34170	2.92650	2.90230	0.99170
4	1.05090	0.95150	0.24536	4.07560	0.25786	3.87810	5.75690	1.48450
5	1.06410	0.93980	0.19506	5.12660	0.20756	4.81780	9.51600	1.97520
6	1.07740	0.92820	0.16153	6.19070	0.17403	5.74600	14.15690	2.46380
7	1.09090	0.91670	0.13759	7.26800	0.15009	6.66270	19.65710	2.95030
8	1.10450	0.90540	0.11963	8.35890	0.13213	7.56810	25.99490	3.43480
9	1.11830	0.89420	0.10567	9.46340	0.11817	8.46230	33.14870	3.91720
10	1.13230	0.88320	0.09450	10.58170	0.10700	9.34550	41.09730	4.39750
11	1.14640	0.87230	0.08537	11.71390	0.09787	10.21780	49.82010	4.87580
12	1.16080	0.86150	0.07726	12.86040	0.09026	11.07930	59.29670	5.35200
13	1.17530	0.85090	0.07132	14.02110	0.08382	11.93020	69.50720	5.82620
14	1.19000	0.84040	0.06581	15.19640	0.07831	12.77060	80.43200	6.29820
15	1.20480	0.83000	0.06103	16.38630	0.07353	13.60050	92.05190	6.76820
16	1.21990	0.81970	0.05685	17.59120	0.06935	14.42030	104.34810	7.23620
17	1.23510	0.80960	0.05316	18.81110	0.06566	15.22990	117.30210	7.70210
18	1.25060	0.79960	0.04988	20.04620	0.06238	16.02950	130.89580	8.16590
19	1.26620	0.78980	0.04696	21.29680	0.05946	16.81930	145.11150	8.62770
20	1.28200	0.78000	0.04432	22.56300	0.05682	17.59930	159.93160	9.08740
21	1.29810	0.77040	0.04194	23.84500	0.05444	18.36970	175.33920	9.54500
22	1.31430	0.76090	0.03977	25.14310	0.05227	19.13060	191.31740	10.00060
23	1.33070	0.75150	0.03780	26.45740	0.05030	19.88200	207.84990	10.45420
24	1.34740	0.74220	0.03599	27.78810	0.04849	20.62420	224.92040	10.90560
25	1.36420	0.73300	0.03432	29.13540	0.04682	21.35730	242.51320	11.35510

n	Single Payments		Uniform Series Payments				Arithmetic Gradients	
	Compound Amount F/P	Present Worth P/F	Sinking Fund A/F	Compound Amount F/A	Capital Recovery A/P	Present Worth P/A	Gradient Present Worth P/G	Gradient Uniform Series A/G
26	1.38120	0.72400	0.03279	30.49960	0.04529	22.08130	260.61280	11.80240
27	1.39850	0.71500	0.03137	31.88090	0.04387	22.79630	279.20400	12.24780
28	1.41600	0.70620	0.03005	33.27940	0.04255	23.50250	298.27190	12.69110
29	1.43370	0.69750	0.02882	34.69540	0.04132	24.20000	317.80190	13.13230
30	1.45160	0.68890	0.02768	36.12910	0.04018	24.88890	337.77970	13.57150
36	1.56390	0.63940	0.02217	45.11550	0.03467	28.84730	466.28300	16.16390
40	1.64360	0.60840	0.01942	51.48960	0.03192	31.32690	559.23200	17.85150
48	1.81540	0.55090	0.01533	65.22840	0.02783	35.93150	759.22960	21.12990
50	1.86100	0.53730	0.01452	68.88180	0.02702	37.01290	811.67380	21.92950
52	1.90780	0.52420	0.01377	72.62710	0.02627	38.06770	864.94090	22.72110
55	1.98030	0.50500	0.01275	78.42250	0.02525	39.60170	946.22770	23.89360
60	2.10720	0.47460	0.01129	88.57450	0.02379	42.03460	1084.84000	25.80830
72	2.44590	0.40880	0.00865	115.67360	0.02115	47.29250	1428.46000	30.20470
75	2.53880	0.39390	0.00812	123.10350	0.02062	48.48900	1515.79000	31.26050
84	2.83910	0.35220	0.00680	147.12900	0.01930	51.82220	1778.84000	34.32580
90	3.05880	0.32690	0.00607	164.70500	0.01857	53.84610	1953.83000	36.28550
96	3.29550	0.30340	0.00545	183.64110	0.01795	55.72460	2127.52000	38.17930
100	3.46340	0.28870	0.00507	197.07230	0.01757	56.90130	2242.24000	39.40580
108	3.82530	0.26140	0.00442	226.02260	0.01692	59.08650	2468.26000	41.77370
120	4.44020	0.22520	0.00363	275.21710	0.01613	61.98280	2796.57000	45.11840
132	5.15400	0.19400	0.00301	332.31980	0.01551	64.47810	3109.35000	48.22340
144	5.98250	0.16720	0.00251	398.60210	0.01551	66.62770	3404.61000	51.09900
240	19.71550	0.05070	0.00067	1497.24000	0.01317	75.94230	5101.53000	67.17640
360	87.54100	0.01140	0.00014	6923.28000	0.01264	79.08610	5997.90000	75.84010
480	388.70070	0.00260	0.00003	31016.00000	0.01253	79.79420	6284.74000	78.76190

1.5% **Table 6** Discrete Cash Flow: Compound Interest Factors **1.5%**

	Single Payments		Uniform Series Payments				Arithmetic Gradients	
n	Compound Amount F/P	Present Worth P/F	Sinking Fund A/F	Compound Amount F/A	Capital Recovery A/P	Present Worth P/A	Gradient Present Worth P/G	Gradient Uniform Series A/G
1	1.0150	0.9852	1.00000	1.0000	1.01500	0.9852		
2	1.0302	0.9707	0.49628	2.0150	0.51128	1.9559	0.9707	0.4963
3	1.0457	0.9563	0.32838	3.0452	0.34338	2.9122	2.8833	0.9901
4	1.0614	0.9422	0.24444	4.0909	0.25944	3.8544	5.7098	1.4814
5	1.0773	0.9283	0.19409	5.1523	0.20909	4.7826	9.4229	1.9702
6	1.0934	0.9145	0.16053	6.2296	0.17553	5.6972	13.9956	2.4566
7	1.1098	0.9010	0.13656	7.3230	0.15156	6.5982	19.4018	2.9405
8	1.1265	0.8877	0.11858	8.4328	0.13358	7.4859	25.6157	3.4219
9	1.1434	0.8746	0.10461	9.5593	0.11961	8.3605	32.6125	3.9008
10	1.1605	0.8617	0.09343	10.7027	0.10843	9.2222	40.3675	4.3772
11	1.1779	0.8489	0.08429	11.8633	0.09929	10.0711	48.8568	4.8512
12	1.1956	0.8364	0.07668	13.0412	0.09168	10.9075	58.0571	5.3227
13	1.2136	0.8240	0.07024	14.2368	0.08524	11.7315	67.9454	5.7917
14	1.2318	0.8118	0.06472	15.4504	0.07972	12.5434	78.4994	6.2582
15	1.2502	0.7999	0.05994	16.6821	0.07494	13.3432	89.6974	6.7223
16	1.2690	0.7880	0.05577	17.9324	0.07077	14.1313	101.5178	7.1839
17	1.2880	0.7764	0.05208	19.2014	0.06708	14.9076	113.9400	7.6431
18	1.3073	0.7649	0.04881	20.4894	0.06381	15.6726	126.9435	8.0997
19	1.3270	0.7536	0.04588	21.7967	0.06088	16.4262	140.5084	8.5539
20	1.3469	0.7425	0.04325	23.1237	0.05825	17.1686	154.6154	9.0057
21	1.3671	0.7315	0.04087	24.4705	0.05587	17.9001	169.2453	9.4550
22	1.3876	0.7207	0.03870	25.8376	0.05370	18.6208	184.3798	9.9018
23	1.4084	0.7100	0.03673	27.2251	0.05173	19.3309	200.0006	10.3462
24	1.4295	0.6995	0.03492	28.6335	0.04992	20.0304	216.0901	10.7881
25	1.4509	0.6892	0.03326	30.0630	0.04826	20.7196	232.6310	11.2276

n	Single Payments		Uniform Series Payments				Arithmetic Gradients	
	Compound Amount F/P	Present Worth P/F	Sinking Fund A/F	Compound Amount F/A	Capital Recovery A/P	Present Worth P/A	Gradient Present Worth P/G	Gradient Uniform Series A/G
26	1.4727	0.6790	0.03173	31.5140	0.04673	21.3986	249.6065	11.6646
27	1.4948	0.6690	0.03032	32.9867	0.04532	22.0676	267.0002	12.0992
28	1.5172	0.6591	0.02900	34.4815	0.04400	22.7267	284.7958	12.5313
29	1.5400	0.6494	0.02778	35.9987	0.04278	23.3761	302.9779	12.9610
30	1.5631	0.6398	0.02664	37.5387	0.04164	24.0158	321.5310	13.3883
36	1.7091	0.5851	0.02115	47.2760	0.03615	27.6607	439.8303	15.9009
40	1.8140	0.5513	0.01843	54.2679	0.03343	29.9158	524.3568	17.5277
48	2.0435	0.4894	0.01437	69.5652	0.02937	34.0426	703.5462	20.6667
50	2.1052	0.4750	0.01357	73.6828	0.02857	34.9997	749.9636	21.4277
52	2.1689	0.4611	0.01283	77.9249	0.02783	35.9287	796.8774	22.1794
55	2.2679	0.4409	0.01183	84.5296	0.02683	37.2715	868.0285	23.2894
60	2.4432	0.4093	0.01039	96.2147	0.02539	39.3803	988.1674	25.0930
72	2.9212	0.3423	0.00781	128.0772	0.02281	43.8447	1279.79	29.1893
75	3.0546	0.3274	0.00730	136.9728	0.02230	44.8416	1352.56	30.1631
84	3.4926	0.2863	0.00602	166.1726	0.02102	47.5786	1568.51	32.9668
90	3.8189	0.2619	0.00532	187.9299	0.02032	49.2099	1709.54	34.7399
96	4.1758	0.2395	0.00472	211.7202	0.01972	50.7017	1847.47	36.4381
100	4.4320	0.2256	0.00437	228.8030	0.01937	51.6247	1937.45	37.5295
108	4.9927	0.2003	0.00376	266.1778	0.01876	53.3137	2112.13	39.6171
120	5.9693	0.1675	0.00302	331.2882	0.01802	55.4985	2359.71	42.5185
132	7.1370	0.1401	0.00244	409.1354	0.01744	57.3257	2588.71	45.1579
144	8.5332	0.1172	0.00199	502.2109	0.01699	58.8540	2798.58	47.5512
240	35.6328	0.0281	0.00043	2308.85	0.01543	64.7957	3870.69	59.7368
360	212.7038	0.0047	0.00007	14114	0.01507	66.3532	4310.72	64.9662
480	1269.70	0.0008	0.00001	84580	0.01501	66.6142	4415.74	66.2883

2% **Table 7** Discrete Cash Flow: Compound Interest Factors 2%

	Single Payments		Uniform Series Payments				Arithmetic Gradients	
n	Compound Amount F/P	Present Worth P/F	Sinking Fund A/F	Compound Amount F/A	Capital Recovery A/P	Present Worth P/A	Gradient Present Worth P/G	Gradient Uniform Series A/G
1	1.0200	0.9804	1.00000	1.0000	1.02000	1.9804		
2	1.0404	0.9612	0.49505	2.0200	0.51505	1.9416	0.9612	0.4950
3	1.0612	0.9423	0.32675	3.0604	0.34675	2.8839	2.8458	0.9868
4	1.0824	0.9238	0.24262	4.1216	0.26262	3.8077	5.6173	1.4752
5	1.1041	0.9057	0.19216	5.2040	0.21216	4.7135	9.2403	1.9604
6	1.1262	0.8880	0.15853	6.3081	0.17853	5.6014	13.6801	2.4423
7	1.1487	0.8706	0.13451	7.4343	0.15451	6.4720	18.9035	2.9208
8	1.1717	0.8535	0.11651	8.5830	0.13651	7.3255	24.8779	3.3961
9	1.1951	0.8368	0.10252	9.7546	0.12252	8.1622	31.5720	3.8681
10	1.2190	0.8203	0.09133	10.9497	0.11133	8.9826	38.9551	4.3367
11	1.2434	0.8043	0.08218	12.1687	0.10218	9.7868	46.9977	4.8021
12	1.2682	0.7885	0.07456	13.4121	0.09456	10.5753	55.6712	5.2642
13	1.2936	0.7730	0.06812	14.6803	0.08812	11.3484	64.9475	5.7231
14	1.3195	0.7579	0.06260	15.9739	0.08260	12.1062	74.7999	6.1786
15	1.3459	0.7430	0.05783	17.2934	0.07783	12.8493	85.2021	6.6309
16	1.3728	0.7284	0.05365	18.6393	0.07365	13.5777	96.1288	7.0799
17	1.4002	0.7142	0.04997	20.0121	0.06997	14.2919	107.5554	7.5256
18	1.4282	0.7002	0.04670	21.4123	0.06670	14.9920	119.4581	7.9681
19	1.4568	0.6864	0.04378	22.8406	0.06378	15.6785	131.8139	8.4073
20	1.4859	0.6730	0.04116	24.2974	0.06116	16.3514	144.6003	8.8433
21	1.5157	0.6598	0.03878	25.7833	0.05878	17.0112	157.7959	9.2760
22	1.5460	0.6468	0.03663	27.2990	0.05663	17.6580	171.3795	9.7055
23	1.5769	0.6342	0.03467	28.8450	0.05467	18.2922	185.3309	10.1317
24	1.6084	0.6217	0.03287	30.4219	0.05287	18.9139	199.6305	10.5547
25	1.6406	0.6095	0.03122	32.0303	0.05122	19.5235	214.2592	10.9745

	Single Payments		Uniform Series Payments				Arithmetic Gradients	
n	Compound Amount F/P	Present Worth P/F	Sinking Fund A/F	Compound Amount F/A	Capital Recovery A/P	Present Worth P/A	Gradient Present Worth P/G	Gradient Uniform Series A/G
26	1.6734	0.5976	0.02970	33.6709	0.04970	20.1210	229.1987	11.3910
27	1.7069	0.5859	0.02829	35.3443	0.04829	20.7069	244.4311	11.8043
28	1.7410	0.5744	0.02699	37.0512	0.04699	21.2813	259.9392	12.2145
29	1.7758	0.5631	0.02578	38.7922	0.04578	21.8444	275.7064	12.6214
30	1.8114	0.5521	0.02465	40.5681	0.04465	22.3965	291.7164	13.0251
36	2.0399	0.4902	0.01923	51.9944	0.03923	25.4888	392.0405	15.3809
40	0.2080	0.4529	0.01656	60.4020	0.03656	27.3555	461.9931	16.8885
48	2.5871	0.3865	0.01260	79.3535	0.03260	30.6731	605.9657	19.7556
50	2.6916	0.3715	0.01182	84.5794	0.03182	31.4236	642.3606	20.4420
52	2.8003	0.3571	0.01111	90.0164	0.03111	32.1449	678.7849	21.1164
55	2.9717	0.3365	0.01014	98.5865	0.03014	33.1748	733.3527	22.1057
60	3.2810	0.3048	0.00877	114.0512	0.02877	34.7609	823.6975	23.6961
72	4.1611	0.2403	0.00633	158.0570	0.02633	37.9841	1034.06	27.2234
75	4.4158	0.2265	0.00586	170.7918	0.02586	38.6771	1084.64	28.0434
84	5.2773	0.1895	0.00468	213.8666	0.02468	40.5255	1230.42	30.3616
90	5.9431	0.1683	0.00405	247.1567	0.02405	41.5869	1322.17	31.7929
96	6.6929	0.1494	0.00351	284.6467	0.02351	42.5294	1409.30	33.1370
100	7.2446	0.1380	0.00320	312.2323	0.02320	43.0984	1464.75	33.9863
108	8.4883	0.1178	0.00267	374.4129	0.02267	44.1095	1569.30	35.5774
120	10.7652	0.0929	0.00205	488.2582	0.02205	45.3554	1710.42	37.7114
132	13.6528	0.0732	0.00158	632.6415	0.02158	46.3378	1833.47	39.5676
144	17.3151	0.0578	0.00123	815.7545	0.02123	47.1123	1939.79	41.1738
240	115.8887	0.0086	0.00017	5744.44	0.02017	49.5686	2374.88	47.9110
360	1247.56	0.0008	0.00002	62328	0.02002	49.9599	2482.57	49.7112
480	13430	0.0001			0.02000	49.9963	2498.03	49.9643

3%   **Table 8**   Discrete Cash Flow: Compound Interest Factors   **3%**

n	Single Payments		Uniform Series Payments				Arithmetic Gradients	
	Compound Amount F/P	Present Worth P/F	Sinking Fund A/F	Compound Amount F/A	Capital Recovery A/P	Present Worth P/A	Gradient Present Worth P/G	Gradient Uniform Series A/G
1	1.0300	0.9709	1.00000	1.0000	1.03000	0.9709		
2	1.0609	0.9426	0.49261	2.0300	0.52261	1.9135	0.9426	0.4926
3	1.0927	0.9151	0.32353	3.0909	0.35353	2.8286	2.7729	0.9803
4	1.1255	0.8885	0.23903	4.1836	0.26903	3.7171	5.4383	1.463
5	1.1593	0.8626	0.18835	5.3091	0.21835	4.5797	8.8888	1.9409
6	1.941	0.8375	0.15460	6.4684	0.18460	5.4172	13.0762	2.4138
7	1.2299	0.8131	0.13051	7.6625	0.16051	6.2303	17.9547	2.8819
8	1.2668	0.7894	0.11246	8.8923	0.14246	7.0197	23.4806	3.3450
9	1.3048	0.7664	0.09843	10.1591	0.12843	7.7861	29.6119	3.8032
10	1.3439	0.7441	0.08723	11.4639	0.11723	8.5302	36.3088	4.2565
11	1.3842	0.7224	0.07808	12.8078	0.10808	9.2526	43.5330	4.7049
12	1.4258	0.7014	0.07046	14.1920	0.10046	9.9540	51.2482	5.1485
13	1.4685	0.6810	0.06403	15.6178	0.09403	10.6350	59.4196	5.5872
14	1.5126	0.6611	0.05853	17.0863	0.08853	11.2961	68.0141	6.0210
15	1.5580	0.6419	0.05377	18.5989	0.08377	11.9379	77.0002	6.4500
16	1.6047	0.6232	0.04961	20.1569	0.07961	12.5611	86.3477	6.8742
17	1.6528	0.6050	0.04595	21.7616	0.07595	13.1661	96.0280	7.2936
18	1.7024	0.5874	0.04271	23.4144	0.07271	13.7535	106.0137	7.7081
19	1.7535	0.5703	0.03981	25.1169	0.06981	14.3238	116.2788	8.1179
20	1.8061	0.5537	0.03722	26.8704	0.06722	14.8775	126.7987	8.5229
21	1.8603	0.5375	0.03487	28.6765	0.06487	15.4150	137.5496	8.9231
22	1.9161	0.5219	0.03275	30.5368	0.06275	15.9369	148.5094	9.3186
23	1.9736	0.5067	0.03081	32.4529	0.06081	16.4436	159.6566	9.7093
24	2.0328	0.4919	0.02905	34.4265	0.05905	16.9355	170.9711	10.0954
25	2.0938	0.4776	0.02743	36.4593	0.05743	17.4131	182.4336	10.4768

n	Single Payments		Uniform Series Payments				Arithmetic Gradients	
	Compound Amount F/P	Present Worth P/F	Sinking Fund A/F	Compound Amount F/A	Capital Recovery A/P	Present Worth P/A	Gradient Present Worth P/G	Gradient Uniform Series A/G
26	2.1566	0.4637	0.02594	38.5530	0.05594	17.8768	194.0260	10.8535
27	2.2213	0.4502	0.02456	40.7096	0.05456	18.3270	205.7309	11.2255
28	2.2879	0.4371	0.02329	42.9309	0.05329	18.7641	217.5320	11.5930
29	2.3566	0.4243	0.02211	45.2189	0.05211	19.1885	229.4137	11.9558
30	2.4273	0.4120	0.02102	47.5754	0.05102	19.6004	241.3613	12.3141
31	2.5001	0.4000	0.02000	50.0027	0.05000	20.0004	253.3609	12.6678
32	2.5751	0.3883	0.01905	52.5028	0.04905	20.3888	265.3993	13.0169
33	2.6523	0.3770	0.01816	55.0778	0.04816	20.7658	277.4642	13.3616
34	2.7319	0.3660	0.01732	57.7302	0.04732	21.1318	289.5437	13.7018
35	2.8139	0.3554	0.01654	60.4621	0.04654	21.4872	301.6267	14.0375
40	3.2620	0.3066	0.01326	75.4013	0.04326	23.1148	361.7499	15.6502
45	3.7816	0.2644	0.01079	92.7199	0.04079	24.5187	420.6325	17.1556
50	4.3839	0.2281	0.00887	112.7969	0.03887	25.7298	477.4803	18.5575
55	5.0821	0.1968	0.00735	136.0716	0.03735	26.7744	531.7411	19.8600
60	5.8916	0.1697	0.00613	163.0534	0.03613	27.6756	583.0526	21.0674
65	6.8300	0.1464	0.00515	194.3328	0.03515	28.4529	631.2010	22.1841
70	7.9178	0.1263	0.00434	230.5941	0.03434	29.1234	676.0869	23.2145
75	9.1789	0.1089	0.00367	272.6309	0.03367	29.7018	717.6978	24.1634
80	10.6409	0.0940	0.00311 S	321.3630	0.03311	30.2008	756.0865	25.0353
84	11.9764	0.0835	0.00273	365.8805	0.03273	30.5501	784.5434	25.6806
85	12.3357	0.0811	0.00265	377.8570	0.03265	30.6312	791.3529	25.8349
90	14.3005	0.0699	0.00226	443.3489	0.03226	31.0024	823.6302	26.5667
96	17.0755	0.0586	0.00187	535.8502	0.03187	31.3812	858.6377	27.3615
108	24.3456	0.0411	0.00129	778.1863	0.03129	31.9642	917.6013	28.7072
120	34.7110	0.0288	0.00089	1123.70	0.03089	32.3730	963.8635	29.7737

4%　　　　　　　　　　　**Table 9**　　Discrete Cash Flow: Compound Interest Factors　　　　　　**4%**

	Single Payments		Uniform Series Payments				Arithmetic Gradients	
n	Compound Amount F/P	Present Worth P/F	Sinking Fund A/F	Compound Amount F/A	Capital Recovery A/P	Present Worth P/A	Gradient Present Worth P/G	Gradient Uniform Series A/G
1	1.0400	0.9615	1.00000	1.0000	1.04000	0.9615		
2	1.0816	0.9246	0.49020	2.0400	0.53020	1.8861	0.9246	0.4902
	1.1249	0.8890	0.32035	3.1216	0.36035	2.7751	2.7025	0.9739
4	1.1699	0.8548	0.23549	4.2465	0.27549	3.6299	5.2670	1.4510
5	1.2167	0.8219	0.18463	5.4163	0.22463	4.4518	8.5547	1.9216
6	1.2653	0.7903	0.15076	6.6330	0.19076	5.2421	12.5062	2.3857
7	1.3159	0.7599	0.12661	7.8983	0.16661	6.0021	17.0657	2.8433
8	1.3686	0.7307	0.10853	9.2142	0.14853	6.7327	22.1806	3.2944
9	1.4233	0.7026	0.09449	10.5828	0.13449	7.4353	27.8013	3.7391
10	1.4802	0.6756	0.08329	12.0061	0.12329	8.1109	33.8814	4.1773
11	1.5395	0.6496	0.07415	13.4864	0.11415	8.7605	40.3772	4.6090
12	1.6010	0.6246	0.06655	15.0258	0.10655	9.3851	47.2477	5.0343
13	1.6651	0.6006	0.06014	16.6268	0.10014	9.9856	54.4546	5.4533
14	1.7317	0.5775	0.05467	18.2919	0.09467	10.5631	61.9618	5.8659
15	1.8009	0.5553	0.04994	20.0236	0.08994	11.1184	69.7355	6.2721
16	1.8730	0.5339	0.04582	21.8245	0.08582	11.6523	77.7441	6.6720
17	1.9479	0.5134	0.04220	23.6975	0.08220	12.1657	85.9581	7.0656
18	2.0258	0.4936	0.03899	25.6454	0.07899	12.6593	94.3498	7.4530
19	2.1068	0.4746	0.03614	27.6712	0.07614	13.1339	102.8933	7.8342
20	2.1911	0.4564	0.03358	29.7781	0.07358	13.5903	111.5647	8.2091
21	2.2788	0.4388	0.03128	31.9692	0.07128	14.0292	120.3414	8.5779
22	2.3699	0.4220	0.02920	34.2480	0.06920	14.4511	129.2024	8.9407
23	2.4647	0.4057	0.02731	36.6179	0.06731	14.8568	138.1284	9.2973
24	2.5633	0.3901	0.02559	39.0826	0.06559	15.2470	147.1012	9.6479
25	2.6658	0.3751	0.02401	41.6459	0.06401	15.6221	156.1040	9.9925

n	Single Payments		Uniform Series Payments				Arithmetic Gradients	
	Compound Amount F/P	Present Worth P/F	Sinking Fund A/F	Compound Amount F/A	Capital Recovery A/P	Present Worth P/A	Gradient Present Worth P/G	Gradient Uniform Series A/G
26	2.7725	0.3607	0.02257	44.3117	0.06257	15.9828	165.1212	10.3312
27	2.8834	0.3468	0.02124	47.0842	0.06124	16.3296	174.1385	10.6640
28	2.9987	0.3335	0.02001	49.9676	0.06001	16.6631	183.1424	10.9909
29	3.1187	0.3207	0.01888	52.9663	0.05888	16.9837	192.1206	11.3120
30	3.2434	0.3083	0.01783	56.0849	0.05783	17.2920	201.0618	11.6274
31	3.3731	0.2965	0.01686	59.3283	0.05686	17.5885	209.9556	11.9371
32	3.5081	0.2851	0.01595	62.7015	0.05595	17.8736	218.7924	12.2411
33	3.6484	0.2741	0.01510	66.2095	0.05510	18.1476	227.5634	12.5396
34	3.7943	0.2636	0.01431	69.8579	0.05431	18.4112	236.2607	12.8324
35	3.9461	0.2534	0.01358	73.6522	0.05358	18.6646	244.8768	13.1198
40	4.8010	0.2083	0.01052	95.0255	0.05052	19.7928	286.5303	14.4765
45	5.8412	0.1712	0.00826	121.0294	0.04826	20.7200	325.4028	15.7047
50	7.1067	0.1407	0.00655	152.6671	0.04655	21.4822	361.1638	16.8122
55	8.6464	0.1157	0.00523	191.1592	0.04523	22.1086	393.6890	17.8070
60	10.5196	0.0951	0.00420	237.9907	0.04420	22.6235	422.9966	18.6972
65	12.7987	0.0781	0.00339	294.9684	0.04339	23.0467	449.2014	19.4909
70	15.5716	0.0642	0.00275	364.2905	0.04275	23.3945	472.4789	20.1961
75	18.9453	0.0528	0.00223	448.6314	0.04223	23.6804	493.0408	20.8206
80	23.0498	0.0434	0.00181	551.2450	0.04181	23.9154	511.1161	21.3718
85	28.0436	0.0357	0.00148	676.0901	0.04148	24.1085	526.9384	21.8569
90	34.1193	0.0293	0.00121	827.9833	0.04121	24.2673	540.7369	22.2826
96	43.1718	0.0232	0.00095	1054.30	0.04095	24.4209	554.9312	22.7236
108	69.1195	0.0145	0.00059	1702.99	0.04059	24.6383	576.8949	23.4146
120	110.6626	0.0090	0.00036	2741.56	0.04036	24.7741	592.2428	23.9057
144	283.6618	0.0035	0.00014	7066.55	0.04014	24.9119	610.1055	24.4906

5% **Table 10** Discrete Cash Flow: Compound Interest Factors **5%**

	Single Payments		Uniform Series Payments				Arithmetic Gradients	
n	Compound Amount F/P	Present Worth P/F	Sinking Fund A/F	Compound Amount F/A	Capital Recovery A/P	Present Worth P/A	Gradient Present Worth P/G	Gradient Uniform Series A/G
1	1.0500	0.9524	1.00000	1.0000	1.05000	0.9524		
2	1.1025	0.9070	0.48780	2.0500	0.53780	1.8594	0.9070	0.4878
3	1.1576	0.8638	0.31721	3.1525	0.36721	2.7232	2.6347	0.9675
4	1.2155	0.8227	0.23201	4.3101	0.28201	3.5460	5.1028	1.4391
5	1.2763	0.7835	0.18097	5.5256	0.23097	4.3295	8.2369	1.9025
6	1.3401	0.7462	0.14702	6.8019	0.19702	5.0757	11.9680	2.3579
7	1.4071	0.7107	0.12282	8.1420	0.17282	5.7864	16.2321	2.8052
8	1.4775	0.6768	0.10472	9.5491	0.15472	6.4632	20.9700	3.2445
9	1.5513	0.6446	0.09069	11.0266	0.14069	7.1078	26.1268	3.6758
10	1.6289	0.6139	0.07950	12.5779	0.12950	7.7217	31.6520	4.0991
11	1.7103	0.5847	0.07039	14.2068	0.12039	8.3064	37.4988	4.5144
12	1.7959	0.5568	0.06283	15.9171	0.11283	8.8633	43.6241	4.9219
13	1.8856	0.5303	0.05646	17.7130	0.10646	9.3936	49.9879	5.3215
14	1.9799	0.5051	0.05102	19.5986	0.10102	9.8986	56.5538	5.7133
15	2.0789	0.4810	0.04634	21.5786	0.09634	10.3797	63.2880	6.0973
16	2.1829	0.4581	0.04227	23.6575	0.09227	10.8378	70.1597	6.4736
17	2.2920	0.4363	0.03870	25.8404	0.08870	11.274	77.1405	6.8423
18	2.4066	0.4155	0.03555	28.1324	0.08555	11.6896	84.2043	7.2034
19	2.5270	0.3957	0.03275	30.5390	0.08275	12.0853	91.3275	7.5569
20	2.6533	0.3769	0.03024	33.0660	0.08024	12.4622	98.4884	7.9030
21	2.7860	0.3589	0.02800	35.7193	0.07800	12.8212	105.6673	8.2416
22	2.9253	0.3418	0.02597	38.5052	0.07597	13.1630	112.8461	8.5730
23	3.0715	0.3256	0.02414	41.4305	0.07414	13.4886	120.0087	8.8971
24	3.2251	0.3101	0.02247	44.5020	0.07247	13.7986	127.1402	9.2140
25	3.3864	0.2953	0.02095	47.7271	0.07095	14.0939	134.2275	9.5238

n	Single Payments		Uniform Series Payments				Arithmetic Gradients	
	Compound Amount F/P	Present Worth P/F	Sinking Fund A/F	Compound Amount F/A	Capital Recovery A/P	Present Worth P/A	Gradient Present Worth P/G	Gradient Uniform Series A/G
26	3.5557	0.2812	0.01956	51.1135	0.06956	14.3752	141.2585	9.8266
27	3.7335	0.2678	0.01829	54.6691	0.06829	14.6430	148.2226	10.1224
28	3.9201	0.2551	0.01712	58.4026	0.06712	14.8981	155.1101	10.4114
29	4.1161	0.2429	0.01605	62.3227	0.06605	15.1411	161.9126	10.6936
30	4.3219	0.2314	0.01505	66.4388	0.06505	15.3725	168.6226	10.9691
31	4.5380	0.2204	0.01413	70.7608	0.06413	15.5928	175.2333	11.2381
32	4.7649	0.2099	0.01328	75.2988	0.06328	15.8027	181.7392	11.5005
33	5.0032	0.1999	0.01249	80.0638	0.06249	16.0025	188.1351	11.7566
34	5.2533	0.1904	0.01176	85.0670	0.06176	1 16.1929	194.4168	12.0063
35	5.5160	0.1813	0.01107	90.3203	0.06107	16.3742	200.5807	12.2498
40	7.0400	0.1420	0.00828	120.7998	0.05828	17.1591	229.5452	13.3775
45	8.9850	0.1113	0.00626	159.7002	0.05626	17.7741	255.3145	14.3644
50	11.4674	0.0872	0.00478	209.3480	0.05478	18.2559	277.9148	15.2233
55	14.6356	0.0683	0.00367	272.7126	0.05367	18.6335	297.5104	15.9664
60	18.6792	0.0535	0.00283	353.5837	0.05283	18.9293	314.3432	16.6062
65	23.8399	0.0419	0.00219	456.7980	0.05219	19.1611	328.6910	17.1541
70	30.4264	0.0329	0.00170	588.5285	0.05170	19.3427	340.8409	17.6212
75	38.8327	0.0258	0.00132	756.6537	0.05132	19.4850	351.0721	18.0176
80	49.5614	0.0202	0.00103	971.2288	0.05103	19.5965	359.6460	18.3526
85	63.2544	0.0158	0.00080	1245.09	0.05080	19.6838	366.8007	18.6346
90	80.7304	0.0124	0.00063	1594.61	0.05063	19.7523	372.7488	18.8712
95	103.0347	0.0097	0.00049	2040.69	0.05049	19.8059	377.6774	19.0689
96	108.1864	0.0092	0.00047	2143.73	0.05047	19.8151	378.5555	19.1044
98	119.2755	0.0084	0.00042	2365.51	0.05042	19.8323	380.2139	19.1714
100	131.5013	0.0076	0.00038	2610.03	0.05038	19.8479	381.7492	19.2337

6%   **Table 11**   Discrete Cash Flow: Compound Interest Factors   6%

	Single Payments		Uniform Series Payments				Arithmetic Gradients	
n	Compound Amount F/P	Present Worth P/F	Sinking Fund A/F	Compound Amount F/A	Capital Recovery A/P	Present Worth P/A	Gradient Present Worth P/G	Gradient Uniform Series A/G
1	1.0600	0.9434	1.00000	1.0000	1.06000	0.9434		
2	1.1236	0.8900	0.48544	2.0600	0.54544	1.8334	0.8900	0.4854
3	1.1910	0.8396	0.31411	3.1836	0.37411	2.6730	2.5692	0.9612
4	1.2625	0.7921	0.22859	4.3746	0.28859	3.4651	4.9455	1.4272
5	1.3382	0.7473	0.17740	5.6371	0.23740	4.2124	7.9345	1.8836
6	1.4185	0.7050	0.14336	6.9753	0.20336	4.9173	11.4594	2.3304
7	1.5036	0.6651	0.11914	8.3938	0.17914	5.5824	15.4497	2.7676
8	1.5938	0.6274	0.10104	9.8975	0.16104	6.2098	19.8416	3.1952
9	1.6895	0.5919	0.08702	11.4913	0.14702	6.8017	24.5768	3.6133
10	1.7908	0.5584	0.07587	13.1808	0.13587	7.3601	29.6023	4.0220
11	1.8983	0.5268	0.06679	14.9716	0.12679	7.8869	34.8702	4.4213
12	2.0122	0.4970	0.05928	16.8699	0.11928	8.3838	40.3369	4.8113
13	2.1329	0.4688	0.05296	18.8821	0.11296	8.8527	45.9629	5.1920
14	2.2609	0.4423	0.04758	21.0151	0.10758	9.2950	51.7128	5.5635
15	2.3966	0.4173	0.04296	23.2760	0.10296	9.7122	57.5546	5.9260
16	2.5404	0.3936	0.03895	25.6725	0.09895	10.1059	63.4592	6.2794
17	2.6928	0.3714	0.03544	28.2129	0.09544	10.4773	69.4011	6.6240
18	2.8543	0.3503	0.03236	30.9057	0.09236	10.8276	75.3569	6.9597
19	3.0256	0.3305	0.02962	33.7600	0.08962	11.1581	81.3062	7.2867
20	3.2071	0.3118	0.02718	36.7856	0.08718	11.4699	87.2304	7.6051
21	3.3996	0.2942	0.02500	39.9927	0.08500	11.7641	93.1136	7.9151
22	3.6035	0.2775	0.02305	43.3923	0.08305	12.0416	98.9412	8.2166
23	3.8197	0.2618	0.02128	46.9958	0.08128	12.3034	104.7007	8.5099
24	4.0489	0.2470	0.01968	50.8156	0.07968	12.5504	110.3812	8.7951
25	4.2919	0.2330	0.01823	54.8645	0.07823	12.7834	115.9732	9.0722

n	Single Payments		Uniform Series Payments				Arithmetic Gradients	
	Compound Amount *F/P*	Present Worth *P/F*	Sinking Fund *A/F*	Compound Amount *F/A*	Capital Recovery *A/P*	Present Worth *P/A*	Gradient Present Worth *P/G*	Gradient Uniform Series *A/G*
26	4.5494	0.2198	0.01690	59.1564	0.07690	13.0032	121.4684	9.3414
27	4.8223	0.2074	0.01570	63.7058	0.07570	13.2105	126.8600	9.6029
28	5.1117	0.1956	0.01459	68.5281	0.07459	13.4062	132.1420	9.8568
29	5.4184	0.1846	0.01358	73.6398	0.07358	13.5907	137.3096	10.1032
30	5.7435	0.1741	0.01265	79.0582	0.07265	13.7648	142.3588	10.3422
31	6.0881	0.1643	0.01179	84.8017	0.07179	13.9291	147.2864	10.5740
32	6.4534	0.1550	0.01100	90.8898	0.07100	14.0840	152.0901	10.7988
33	6.8406	0.1462	0.01027	97.3432	0.07027	14.2302	156.7681	11.0166
34	7.2510	0.1379	0.00960	104.1838	0.06960	14.3681	161.3192	11.2276
35	7.6861	0.1301	0.00897	111.4348	0.06897	14.4982	165.7427	11.4319
40	10.2857	0.0972	0.00646	154.7620	0.06646	15.0463	185.9568	12.3590
45	13.7646	0.0727	0.00470	212.7435	0.06470	15.4558	203.1096	13.1413
50	18.4202	0.0543	0.00344	290.3359	0.06344	15.7619	217.4574	13.7964
55	24.6503	0.0406	0.00254	394.1720	0.06254	15.9905	229.3222	14.3411
60	32.9877	0.0303	0.00188	533.1282	0.06188	16.1614	239.0428	14.7909
65	44.1450	0.0227	0.00139	719.0829	0.06139	16.2891	246.9450	15.1601
70	59.0759	0.0169	0.00103	967.9322	0.06103	16.3845	253.3271	15.4613
75	79.0569	0.0126	0.00077	1300.95	0.06077	16.4558	258.4527	15.7058
80	105.7960	0.0095	0.00057	1746.60	0.06057	16.5091	262.5493	15.9033
85	141.5789	0.0071	0.00043	2342.98	0.06043	16.5489	265.8096	16.0620
90	189.4645	0.0053	0.00032	3141.08	0.06032	16.5787	268.3946	16.1891
95	253.5463	0.0039	0.00024	4209.10	0.06024	16.6009	270.4375	16.2905
96	268.7590	0.0037	0.00022	4462.65	0.06022	16.6047	270.7909	16.3081
98	301.9776	0.0033	0.00020	5016.29	0.06020	16.6115	271.4491	16.3411
100	339.3021	0.0029	0.00018	5638.37	0.06018	16.6175	272.0471	16.3711

7%   **Table 12**   Discrete Cash Flow: Compound Interest Factors   **7%**

	Single Payments		Uniform Series Payments				Arithmetic Gradients	
n	Compound Amount F/P	Present Worth P/F	Sinking Fund A/F	Compound Amount F/A	Capital Recovery A/P	Present Worth P/A	Gradient Present Worth P/G	Gradient Uniform Series A/G
1	1.0700	0.9346	1.00000	1.0000	1.07000	0.9346		
2	1.1449	0.8734	0.48309	2.0700	0.55309	1.8080	0.8734	0.4831
3	1.2250	0.8163	0.31105	3.2149	0.38105	2.6243	2.5060	0.9549
4	1.3108	0.7629	0.22523	4.4399	0.29523	3.3872	4.7947	1.4155
5	1.4026	0.7130	0.17389	5.7507	0.24389	4.1002	7.6467	1.8650
6	1.5007	0.6663	0.13980	7.1533	0.20980	4.7665	10.9784	2.3032
7	1.6058	0.6227	0.11555	8.6540	0.18555	5.3893	14.7149	2.7304
8	1.7182	0.5820	0.09747	10.2598	0.16747	5.9713	18.7889	3.1465
9	1.8385	0.5439	0.08349	11.9780	0.15349	6.5152	23.1404	3.5517
10	1.9672	0.5083	0.07238	13.8164	0.14238	7.0236	27.7156	3.9461
11	2.1049	0.4751	0.06336	15.7836	0.13336	7.4987	32.4665	4.3296
12	2.2522	0.4440	0.05590	17.8885	0.12590	7.9427	37.3506	4.7025
13	2.4098	0.4150	0.04965	20.1406	0.11965	8.3577	42.3302	5.0648
14	2.5785	0.3878	0.04434	22.5505	0.11434	8.7455	47.3718	5.4167
15	2.7590	0.3624	0.03979	25.1290	0.10979	9.1079	52.4461	5.7583
16	2.9522	0.3387	0.03586	27.8881	0.10586	9.4466	57.5271	6.0897
17	3.1588	0.3166	0.03243	30.8402	0.10243	9.7632	62.5923	6.4110
18	3.3799	0.2959	0.02941	33.9990	0.09941	10.0591	67.6219	6.7225
19	3.6165	0.2765	0.02675	37.3790	0.09675	10.3356	72.5991	7.0242
20	3.8697	0.2584	0.02439	40.9955	0.09439	10.5940	77.5091	7.3163
21	4.1406	0.2415	0.02229	44.8652	0.09229	10.8355	82.3393	7.5990
22	4.4304	0.2257	0.02041	49.0057	0.09041	11.0612	87.0793	7.8725
23	4.7405	0.2109	0.01871	53.4361	0.08871	11.2722	91.7201	8.1369
24	5.0724	0.1971	0.01719	58.1767	0.08719	11.4693	96.2545	8.3923

n	Single Payments		Uniform Series Payments				Arithmetic Gradients	
	Compound Amount F/P	Present Worth P/F	Sinking Fund A/F	Compound Amount F/A	Capital Recovery A/P	Present Worth P/A	Gradient Present Worth P/G	Gradient Uniform Series A/G
25	5.4274	0.1842	0.01581	63.2490	0.08581	11.6536	100.6765	8.6391
26	5.8074	0.1722	0.01456	68.6765	0.08456	11.8258	104.9814	8.8773
27	6.2139	0.1609	0.01343	74.4838	0.08343	11.9867	109.1656	9.1072
28	6.6488	0.1504	0.01239	80.6977	0.08239	12.1371	113.2264	9.3289
29	7.1143	0.1406	0.01145	87.3465	0.08145	12.2777	117.1622	9.5427
30	7.6123	0.1314	0.01059	94.4608	0.08059	12.4090	120.9718	9.7487
31	8.1451	0.1228	0.00980	102.0730	0.07980	12.5318	124.6550	9.9471
32	8.7153	0.1147	0.00907	110.2182	0.07907	12.6466	128.2120	10.1381
33	9.3253	0.1072	0.00841	118.9334	0.07841	12.7538	131.6435	10.3219
34	9.9781	0.1002	0.00780	128.2588	0.07780	12.8540	134.9507	10.4987
35	10.6766	0.0937	0.00723	138.2369	0.07723	12.9477	138.1353	10.6687
40	14.9745	0.0668	0.00501	199.6351	0.07501	13.3317	152.2928	11.4233
45	21.0025	0.0476	0.00350	285.7493	0.07350	13.6055	163.7559	12.0360
50	29.4570	0.0339	0.00246	406.5289	0.07246	13.8007	172.9051	12.5287
55	41.3150	0.0242	0.00174	575.9286	0.07174	13.9399	180.1243	12.9215
60	57.9464	0.0173	0.00123	813.5204	0.07123	14.0392	185.7677	13.2321
65	81.2729	0.0123	0.00087	1146.76	0.07087	14.1099	190.1452	13.4760
70	113.9894	0.0088	0.00062	1614.13	0.07062	14.1604	193.5185	13.6662
75	159.8760	0.0063	0.00044	2269.66	0.07044	14.1964	196.1035	13.8136
80	224.2344	0.0045	0.00031	3189.06	0.07031	14.2220	198.0748	13.9273
85	314.5003	0.0032	0.00022	4478.58	0.07022	14.2403	199.5717	14.0146
90	41.1030	0.0023	0.00016	6287.19	0.07016	14.2533	200.7042	14.0812
95	18.6697	0.0016	0.00011	8823.85	0.07011	14.2626	201.5581	
96	51.9766	0.0015	0.00011	9442.52	0.07011	14.564		
98	7.8970	0.0013	0.00009	10813				
100	7.7163	0.0012						

8%   **Table 13**   Discrete Cash Flow: Compound Interest Factors   8%

	Single Payments		Uniform Series Payments				Arithmetic Gradients	
n	Compound Amount F/P	Present Worth P/F	Sinking Fund A/F	Compound Amount F/A	Capital Recovery A/P	Present Worth P/A	Gradient Present Worth P/G	Gradient Uniform Series A/G
1	1.0800	0.9259	1.00000	1.0000	1.08000	0.9259		
2	1.1664	0.8573	0.48077	2.0800	0.56077	1.7833	0.8573	0.4808
3	1.2597	0.7938	0.30803	3.2464	0.38803	2.5771	2.4450	0.9487
4	1.3605	0.7350	0.22192	4.5061	0.30192	3.3121	4.6501	1.4040
5	1.4693	0.6806	0.17046	5.8666	0.25046	3.9927	7.3724	1.8464
6	1.5869	0.6302	0.13632	7.3359	0.21632	4.6229	10.5233	2.2763
7	1.7138	0.5835	0.11207	8.9228	0.19207	5.2064	14.0242	2.6937
8	1.8509	0.5403	0.09401	10.6366	0.17401	5.7466	17.8061	3.0985
9	1.9990	0.5002	0.08008	12.4876	0.16008	6.2469	21.8081	3.4910
10	2.1589	0.4632	0.06903	14.4866	0.14903	6.7101	25.9768	3.8713
11	2.3316	0.4289	0.06008	16.6455	0.14008	7.1390	30.2657	4.2395
12	2.5182	0.3971	0.05270	18.9771	0.13270	7.5361	34.6339	4.5957
13	2.7196	0.3677	0.04652	21.4953	0.12652	7.9038	39.0463	4.9402
14	2.9372	0.3405	0.04130	24.2149	0.12130	8.2442	43.4723	5.2731
15	3.1722	0.3152	0.03683	27.1521	0.11683	8.5595	47.8857	5.5945
16	3.4259	0.2919	0.03298	30.3243	0.11298	8.8514	52.2640	5.9046
17	3.7000	0.2703	0.02963	33.7502	0.10963	9.1216	56.5883	6.2037
18	3.9960	0.2502	0.02670	37.4502	0.10670	9.3719	60.8426	6.4920
19	4.3157	0.2317	0.02413	41.4463	0.10413	9.6036	65.0134	6.7697
20	4.6610	0.2145	0.02185	45.7620	0.10185	9.8181	69.0898	7.0369
21	5.0338	0.1987	0.01983	50.4229	0.09983	10.0168	73.0629	7.2940
22	5.4365	0.1839	0.01803	55.4568	0.09803	10.2007	76.9257	7.5412
23	5.8715	0.1703	0.01642	60.8933	0.09642	10.3711	80.6726	7.7786
24	6.3412	0.1577	0.01498	66.7648	0.09498	10.5288	84.2997	8.0066
25	6.8485	0.1460	0.01368	73.10159	0.09368	10.6748	87.8041	8.2254

	Single Payments		Uniform Series Payments				Arithmetic Gradients	
n	Compound Amount F/P	Present Worth P/F	Sinking Fund A/F	Compound Amount F/A	Capital Recovery A/P	Present Worth P/A	Gradient Present Worth P/G	Gradient Uniform Series A/G
26	7.3964	0.1352	0.01251	79.9544	0.09251	10.8100	91.1842	8.4352
27	7.9881	0.1252	0.01145	87.3508	0.09145	10.9352	94.4390	8.6363
28	8.6271	0.1159	0.01049	95.3388	0.09049	11.0511	97.5687	8.8289
29	9.3173	0.1073	0.00962	103.9659	0.08962	11.1584	100.5738	9.0133
30	10.0627	0.0994	0.00883	113.2832	0.08883	11.2578	103.4558	9.1897
31	10.8677	0.0920	0.00811	123.3459	0.08811	11.3498	106.2163	9.3584
32	11.7371	0.0852	0.00745	134.2135	0.08745	11.4350	108.8575	9.5197
33	12.6760	0.0789	0.00685	145.9506	0.08685	11.5139	111.3819	9.6737
34	13.6901	0.0730	0.00630	158.6267	0.08630	11.5869	113.7924	9.8208
35	14.7853	0.0676	0.00580	172.3168	0.08580	11.6546	116.0920	9.9611
40	21.7245	0.0460	0.00386	259.0565	0.08386	11.9246	126.0422	10.5699
45	31.9204	0.0313	0.00259	386.5056	0.08259	12.1084	133.7331	11.0447
50	46.9016	0.0213	0.00174	573.7702	0.08174	12.2335	139.5928	11.4107
55	68.9139	0.0145	0.00118	848.9232	0.08118	12.3186	144.0065	11.6902
60	101.2571	0.0099	0.00080	1253.21	0.08080	12.3766	147.3000	11.9015
65	148.7798	0.0067	0.00054	1847.25	0.08054	12.4160	149.7387	12.0602
70	218.6064	0.0046	0.00037	2720.08	0.08037	12.4428	151.5326	12.1783
75	321.2045	0.0031	0.00025	4002.56	0.08025	12.4611	152.8448	12.2658
80	471.9548	0.0021	0.00017	5886.94	0.08017	12.4735	153.8001	12.3301
85	693.4565	0.0014	0.00012	8655.71	0.08012	12.4820	154.4925	12.3772
90	1018.92	0.0010	0.00008	12724	0.08008	12.4877	154.9925	12.4116
95	1497.12	0.0007	0.00005	18702	0.08005	12.4917	155.3524	12.4365
96	1616.89	0.0006	0.00005	20199	0.08005	12.4923	155.4112	12.4406
98	1885.94	0.0005	0.00004	23562	0.08004	12.4934	155.5176	12.4480
100	2199.76	0.0005	0.00004	27485	0.08004	12.4943	155.6107	12.4545

9%   **Table 14**   Discrete Cash Flow: Compound Interest Factors   **9%**

	Single Payments		Uniform Series Payments				Arithmetic Gradients	
n	Compound Amount *F/P*	Present Worth *P/F*	Sinking Fund *A/F*	Compound Amount *F/A*	Capital Recovery *A/P*	Present Worth *P/A*	Gradient Present Worth *P/G*	Gradient Uniform Series *A/G*
1	1.0900	0.9174	1.00000	1.0000	1.09000	0.9174		
2	1.1881	0.8417	0.47847	2.0900	0.56847	1.7591	0.8417	0.4785
3	1.2950	0.7722	0.30505	3.2781	0.39505	2.5313	2.3860	0.9426
4	1.4116	0.7084	0.21867	4.5731	0.30867	3.2397	4.5113	1.3925
5	1.5386	0.6499	0.16709	5.9847	0.25709	3.8897	7.1110	1.8282
6	1.6771	0.5963	0.13292	7.5233	0.22292	4.4859	10.0924	2.2498
7	1.8280	0.5470	0.10869	9.2004	0.19869	5.0330	13.3746	2.6571
8	1.9926	0.5019	0.09067	11.0285	0.18067	5.5348	16.8877	3.0512
9	2.1719	0.4604	0.07680	13.0210	0.16680	5.9952	20.5711	3.4312
10	2.3674	0.4224	0.06582	15.1929	0.15582	6.4177	24.3728	3.7978
11	2.5804	0.3875	0.05695	17.5603	0.14695	6.8052	28.2481	4.1510
12	2.8127	0.3555	0.04965	20.1407	0.13965	7.1607	32.1590	4.4910
13	3.0658	0.3262	0.04357	22.9534	0.13357	7.4869	36.0731	4.8182
14	3.3417	0.2992	0.03843	26.0192	0.12843	7.7862	39.9633	5.1326
15	3.6425	0.2745	0.03406	29.3609	0.12406	8.0607	43.8069	5.4346
16	3.9703	0.2519	0.03030	33.0034	0.12030	8.3126	47.5849	5.7245
17	4.3276	0.2311	0.02705	36.9737	0.11705	8.5436	51.2821	6.0024
18	4.7171	0.2120	0.02421	41.3013	0.11421	8.7556	54.8860	6.2687
19	5.1417	0.1945	0.02173	46.0185	0.11173	8.9501	58.3868	6.5236
20	5.6044	0.1784	0.01955	51.1601	0.10955	9.1285	61.7770	6.7674
21	6.1088	0.1637	0.01762	56.7645	0.10762	9.2922	65.0509	7.0006
22	6.6586	0.1502	0.01590	62.8733	0.10590	9.4424	68.2048	7.2232
23	7.2579	0.1378	0.01438	69.5319	0.10438	9.5802	71.2359	7.4357
24	7.9111	0.1264	0.01302	76.7898	0.10302	9.7066	74.1433	7.6384
25	8.6231	0.1160	0.01181	84.7009	0.10181	9.8226	76.9265	7.8316

n	Single Payments		Uniform Series Payments				Arithmetic Gradients	
	Compound Amount F/P	Present Worth P/F	Sinking Fund A/F	Compound Amount F/A	Capital Recovery A/P	Present Worth P/A	Gradient Present Worth P/G	Gradient Uniform Series A/G
26	9.3992	0.1064	0.01072	93.3240	0.10072	9.9290	79.5863	8.0156
27	10.2451	0.0976	0.00973	102.7231	0.09973	10.0266	82.1241	8.1906
28	11.1671	0.0895	0.00885	112.9682	0.09885	10.1161	84.5419	8.3571
29	12.1722	0.0822	0.00806	124.1354	0.09806	10.1983	86.8422	8.5154
30	13.2677	0.0754	0.00734	136.3075	0.09734	10.2737	89.0280	8.6657
31	14.4618	0.0691	0.00669	149.5752	0.09669	10.3428	91.1024	8.8083
32	15.7633	0.0634	0.00610	164.0370	0.09610	10.4062	93.0690	8.9436
33	17.1820	0.0582	0.00556	179.8003	0.09556	10.4644	94.9314	9.0718
34	18.7284	0.0534	0.00508	196.9823	0.09508	10.5178	96.6935	9.1933
35	20.4140	0.0490	0.00464	215.7108	0.09464	10.5668	98.3590	9.3083
40	31.4094	0.0318	0.00296	337.8824	0.09296	10.7574	105.3762	9.7957
45	48.3273	0.0207	0.00190	525.8587	0.09190	10.8812	110.5561	10.1603
50	74.3575	0.0134	0.00123	815.0836	0.09123	10.9617	114.3251	10.4295
55	114.4083	0.0087	0.00079	1260.09	0.09079	11.0140	117.0362	10.6261
60	176.0313	0.0057	0.00051	1944.79	0.09051	11.0480	118.9683	10.7683
65	270.8460	0.0037	0.00033	2998.29	0.09033	11.0701	120.3344	10.8702
70	416.7301	0.0024	0.00022	4619.22	0.09022	11.0844	121.2942	10.9427
75	641.1909	0.0016	0.00014	7113.23	0.09014	11.0938	121.9646	10.9940
80	986.5517	0.0010	0.00009	10951	0.09009	11.0998	122.4306	11.0299
85	1517.93	0.0007	0.00006	16855	0.09006	11.1038	122.7533	11.0551
90	2335.53	0.0004	0.00004	25939	0.09004	11.1064	122.9758	11.0726
95	3593.50	0.0003	0.00003	39917	0.09003	11.1080	123.1287	11.0847
96	3916.91	0.0003	0.00002	43510	0.09002	11.1083	123.1529	11.0866
98	4653.68	0.0002	0.00002	51696	0.09002	11.1087	123.1963	11.0900
100	5529.04	0.0002	0.00002	61423	0.09002	11.1091	123.2335	11.0930

10%   **Table 15**   Discrete Cash Flow: Compound Interest Factors   **10%**

n	Single Payments		Uniform Series Payments				Arithmetic Gradients	
	Compound Amount F/P	Present Worth P/F	Sinking Fund A/F	Compound Amount F/A	Capital Recovery A/P	Present Worth P/A	Gradient Present Worth P/G	Gradient Uniform Series A/G
1	1.1000	0.9091	1.00000	1.0000	1.10000	0.9091		
2	1.2100	0.8264	0.47619	2.1000	0.57619	1.7355	0.8264	0.4762
3	1.3310	0.7513	0.30211	3.3100	0.40211	2.4869	2.3291	0.9366
4	1.4641	0.6830	0.21547	4.6410	0.31547	3.1699	4.3781	1.3812
5	1.6105	0.6209	0.16380	6.1051	0.26380	3.7908	6.8618	1.8101
6	1.7716	0.5645	0.12961	7.7156	0.22961	4.3553	9.6842	2.2236
7	1.9487	0.5132	0.10541	9.4872	0.20541	4.8684	12.7631	2.6216
8	2.1436	0.4665	0.08744	11.4359	0.18744	5.3349	16.0287	3.0045
9	2.3579	0.4241	0.07364	13.5795	0.17364	5.7590	19.4215	3.3724
10	2.5937	0.3855	0.06275	15.9374	0.16275	6.1446	22.8913	3.7255
11	2.8531	0.3505	0.05396	18.5312	0.15396	6.4951	26.3963	4.0641
12	3.1384	0.3186	0.04676	21.3843	0.14676	6.8137	29.9012	4.3884
13	3.4523	0.2897	0.04078	24.5227	0.14078	7.1034	33.3772	4.6988
14	3.7975	0.2633	0.03575	27.9750	0.13575	7.3667	36.8005	4.9955
15	4.1772	0.2394	0.03147	31.7725	0.13147	7.6061	40.1520	5.2789
16	4.5950	0.2176	0.02782	35.9497	0.12782	7.8237	43.4164	5.5493
17	5.0545	0.1978	0.02466	40.5447	0.12466	8.0216	46.5819	5.8071
18	5.5599	0.1799	0.02193	45.5992	0.12193	8.2014	49.6395	6.0526
19	6.1159	0.1635	0.01955	51.1591	0.11955	8.3649	52.5827	6.2861
20	6.7275	0.1486	0.01746	57.2750	0.11746	8.5136	55.4069	6.5081
21	7.4002	0.1351	0.01562	64.0025	0.11562	8.6487	58.1095	6.7189
22	8.1403	0.1228	0.01401	71.4027	0.11401	8.7715	60.6893	6.9189
23	8.9543	0.1117	0.01257	79.5430	0.11257	8.8832	63.1462	7.1085
24	9.8497	0.1015	0.01130	88.4973	0.11130	8.9847	65.4813	7.2881
25	10.8347	0.0923	0.01017	98.3471	0.11017	9.0770	67.6964	7.4580
26	11.9182	0.0839	0.00916	109.1818	0.10916	9.1609	69.7940	7.6186

n	Single Payments		Uniform Series Payments				Arithmetic Gradients	
	Compound Amount F/P	Present Worth P/F	Sinking Fund A/F	Compound Amount F/A	Capital Recovery A/P	Present Worth P/A	Gradient Present Worth P/G	Gradient Uniform Series A/G
27	13.1100	0.0763	0.00826	121.0999	0.10826	9.2372	71.7773	7.7704
28	14.4210	0.0693	0.00745	134.2099	0.10745	9.3066	73.6495	7.9137
29	15.8631	0.0630	0.00673	148.6309	0.10673	9.3696	75.4146	8.0489
30	17.4494	0.0573	0.00608	164.4940	0.10608	9.4269	77.0766	8.1762
31	19.1943	0.0521	0.00550	181.9434	0.10550	9.4790	78.6395	8.2962
32	21.1138	0.0474	0.00497	201.1378	0.10497	9.5264	80.1078	8.4091
33	23.2252	0.0431	0.00450	222.2515	0.10450	9.5694	81.4856	8.5152
34	25.5477	0.0391	0.00407	245.4767	0.10407	9.6086	82.7773	8.6149
35	28.1024	0.0356	0.00369	271.0244	0.10369	9.6442	83.9872	8.7086
40	45.2593	0.0221	0.00226	442.5926	0.10226	9.7791	88.9525	9.0962
45	72.8905	0.0137	0.00139	718.9048	0.10139	9.8628	92.4544	9.3740
50	117.3909	0.0085	0.00086	1163.91	0.10086	9.9148	94.8889	9.5704
55	189.0591	0.0053	0.00053	1880.59	0.10053	9.9471	96.5619	9.7075
60	304.4816	0.0033	0.00033	3034.82	0.10033	9.9672	97.7010	9.8023
65	490.3707	0.0020	0.00020	4893.71	0.10020	9.9796	98.4705	9.8672
70	789.7470	0.0013	0.00013	7887.47	0.10013	9.9873	98.9870	9.9113
75	1271.90	0.0008	0.00008	12709	0.10008	9.9921	99.3317	9.9410
80	2048.40	0.0005	0.00005	20474	0.10005	9.9951	99.5606	9.9609
85	3298.97	0.0003	0.00003	32980	0.10003	9.9970	99.7120	9.9742
90	5313.02	0.0002	0.00002	53120	0.10002	9.9981	99.8118	9.9831
95	8556.68	0.0001	0.00001	85557	0.10001	9.9988	99.8773	9.9889
96	9412.34	0.0001	0.00001	94113	0.10001	9.9989	99.8874	9.9898
98	11389	0.0001	0.00001		0.10001	9.9991	99.9052	9.9914
100	13781	0.0001	0.00001		0.10001	9.9993	99.9202	9.9927

11%   **Table 16**   Discrete Cash Flow: Compound Interest Factors   **11%**

	Single Payments		Uniform Series Payments				Arithmetic Gradients	
n	Compound Amount *F/P*	Present Worth *P/F*	Sinking Fund *A/F*	Compound Amount *F/A*	Capital Recovery *A/P*	Present Worth *P/A*	Gradient Present Worth *P/G*	Gradient Uniform Series *A/G*
1	1.1100	0.9009	1.00000	1.0000	1.11000	0.9009		
2	1.2321	0.8116	0.47393	2.1100	0.58393	1.7125	0.8116	0.4739
3	1.3676	0.7312	0.29921	3.3421	0.40921	2.4437	2.2740	0.9306
4	1.5181	0.6587	0.21233	4.7097	0.32233	3.1024	4.2502	1.3700
5	1.6851	0.5935	0.16057	6.2278	0.27057	3.6959	6.6240	1.7923
6	1.8704	0.5346	0.12638	7.9129	0.23638	4.2305	9.2972	2.1976
7	2.0762	0.4817	0.10222	9.7833	0.21222	4.7122	12.1872	2.5863
8	2.3045	0.4339	0.08432	11.8594	0.19432	5.1461	15.2246	2.9585
9	2.5580	0.3909	0.07060	14.1640	0.18060	5.5370	18.3520	3.3144
10	2.8394	0.3522	0.05980	16.7220	0.16980	5.8892	21.5217	3.6544
11	3.1518	0.3173	0.05112	19.5614	0.16112	6.2065	24.6945	3.9788
12	3.4985	0.2858	0.04403	22.7132	0.15403	6.4924	27.8388	4.2879
13	3.8833	0.2575	0.03815	26.2116	0.14815	6.7499	30.9290	4.5822
14	4.3104	0.2320	0.03323	30.0949	0.14323	6.9819	33.9449	4.8619
15	4.7846	0.2090	0.02907	34.4054	0.13907	7.1909	36.8709	5.1275
16	5.3109	0.1883	0.02552	39.1899	0.13552	7.3792	39.6953	5.3794
17	5.8951	0.1696	0.02247	44.5008	0.13247	7.5488	42.4095	5.6180
18	6.5436	0.1528	0.01984	50.3959	0.12984	7.7016	45.0074	5.8439
19	7.2633	0.1377	0.01756	56.9395	0.12756	7.8393	47.4856	6.0574
20	8.0623	0.1240	0.01558	64.2028	0.12558	7.9633	49.8423	6.2590
21	8.9492	0.1117	0.01384	72.2651	0.12384	8.0751	52.0771	6.4491
22	9.9336	0.1007	0.01231	81.2143	0.12231	8.1757	54.1912	6.6283
23	11.0263	0.0907	0.01097	91.1479	0.12097	8.2664	56.1864	6.7969
24	12.2392	0.0817	0.00979	102.1742	0.11979	8.3481	58.0656	6.9555
25	13.5855	0.0736	0.00874	114.4133	0.11874	8.4217	59.8322	7.1045

n	Single Payments		Uniform Series Payments				Arithmetic Gradients	
	Compound Amount *F/P*	Present Worth *P/F*	Sinking Fund *A/F*	Compound Amount *F/A*	Capital Recovery *A/P*	Present Worth *P/A*	Gradient Present Worth *P/G*	Gradient Uniform Series *A/G*
26	15.0799	0.0663	0.00781	127.9988	0.11781	8.4881	61.4900	7.2443
27	16.7386	0.0597	0.00699	143.0786	0.11699	8.5478	63.0433	7.3754
28	18.5799	0.0538	0.00626	159.8173	0.11626	8.6016	64.4965	7.4982
29	20.6237	0.0485	0.00561	178.3972	0.11561	8.6501	65.8542	7.6131
30	22.8923	0.0437	0.00502	199.0209	0.11502	8.6938	67.1210	7.7206
31	25.4104	0.0394	0.00451	221.9132	0.11451	8.7331	68.3016	7.8210
32	28.2056	0.0355	0.00404	247.3236	0.11404	8.7686	69.4007	7.9147
33	31.3082	0.0319	0.00363	275.5292	0.11363	8.8005	70.4228	8.0021
34	34.7521	0.0288	0.00326	306.8374	0.11326	8.8293	71.3724	8.0836
35	38.5749	0.0259	0.00293	341.5896	0.11293	8.8552	72.2538	8.1594
40	65.0009	0.0154	0.00172	581.8261	0.11172	8.9511	75.7789	8.4659
45	109.5302	0.0091	0.00101	986.6386	0.11101	9.0079	78.1551	8.6763
50	184.5648	0.0054	0.00060	1668.77	0.11060	9.0417	79.7341	8.8185
55	311.0025	0.0032	0.00035	2818.20	0.11035	9.0617	80.7712	8.9135
60	524.0572	0.0019	0.00021	4755.07	0.11021	9.0736	81.4461	8.9762
65	883.0669	0.0011	0.00012	8018.79	0.11012	9.0806	81.8819	9.0172
70	1488.02	0.0007	0.00007	13518	0.11007	9.0848	82.1614	9.0438
75	2507.40	0.0004	0.00004	22785	0.11004	9.0873	82.3397	9.0610
80	4225.11	0.0002	0.00003	38401	0.11003	9.0888	82.4529	9.0720
85	7119.56	0.0001	0.00002	64714	0.11002	9.0896	82.5245	9.0790

12%   **Table 17**   Discrete Cash Flow: Compound Interest Factors   **12%**

	Single Payments		Uniform Series Payments				Arithmetic Gradients	
n	Compound Amount *F/P*	Present Worth *P/F*	Sinking Fund *A/F*	Compound Amount *F/A*	Capital Recovery *A/P*	Present Worth *P/A*	Gradient Present Worth *P/G*	Gradient Uniform Series *A/G*
1	1.1200	0.8929	1.00000	1.0000	1.12000	0.8929		
2	1.2544	0.7972	0.47170	2.1200	0.59170	1.6901	0.7972	0.4717
3	1.4049	0.7118	0.29635	3.3744	0.41635	2.4018	2.2208	0.9246
4	1.5735	0.6355	1.20923	4.7793	0.32923	3.0373	4.1273	1.3589
5	1.7623	0.5674	0.15741	6.3528	0.27741	3.6048	6.3970	1.7746
6	1.9738	0.5066	0.12323	8.1152	0.24323	4.1114	8.9302	2.1720
7	2.2107	0.4523	0.09912	10.0890	0.21912	4.5638	11.6443	2.5512
8	2.4760	0.4039	0.08130	12.2997	0.20130	4.9676	14.4714	2.9131
9	2.7731	0.3606	0.06768	14.7757	0.18768	5.3282	17.3563	3.2574
10	3.1058	0.3220	0.05698	17.5487	0.17698	5.6502	20.2541	3.5847
11	3.4785	0.2875	0.04842	20.6546	0.16842	5.9377	23.1288	3.8953
12	3.8960	0.2567	0.04144	24.1331	0.16144	6.1944	25.9523	4.1897
13	4.3635	0.2292	0.03568	28.0291	0.15568	6.4235	28.7024	4.4683
14	4.8871	0.2046	0.03087	32.3926	0.15087	6.6282	31.3624	4.7317
15	5.4736	0.1827	0.02682	37.2797	0.14682	6.8109	33.9202	4.9803
16	6.1304	0.1631	0.02339	42.7533	0.14339	6.9740	36.3670	5.2147
17	6.8660	0.1456	0.02046	48.8837	0.14046	7.1196	38.6973	5.4353
18	7.6900	0.1300	0.01794	55.7497	0.13794	7.2497	40.9080	5.6427
19	8.6128	0.1161	0.01576	63.4397	0.13576	7.3658	42.9979	5.8375
20	9.6463	0.1037	0.01388	72.0524	0.13388	7.4694	44.9676	6.0202
21	10.8038	0.0926	0.01224	81.6987	0.13224	7.5620	46.8188	6.1913
22	12.1003	0.0826	0.01081	92.5026	0.13081	7.6446	48.5543	6.3514

n	Single Payments		Uniform Series Payments				Arithmetic Gradients	
	Compound Amount F/P	Present Worth P/F	Sinking Fund A/F	Compound Amount F/A	Capital Recovery A/P	Present Worth P/A	Gradient Present Worth P/G	Gradient Uniform Series A/G
23	13.5523	0.0738	0.00956	104.6029	0.12956	7.7184	50.1776	6.5010
24	15.1786	0.0659	0.00846	118.1552	0.12846	7.7843	51.6929	6.6406
25	17.0001	0.0588	0.00750	133.3339	0.12750	7.8431	53.1046	6.7708
26	19.0401	0.0525	0.00665	150.3339	0.12665	7.8957	54.4177	6.8921
27	21.3249	0.0469	0.00590	169.3740	0.12590	7.9426	55.6369	7.0049
28	23.8839	0.0419	0.00524	190.6989	0.12524	7.9844	56.7674	7.1098
29	26.7499	0.0374	0.00466	214.5828	0.12466	8.0218	57.8141	7.2071
30	29.9599	0.0334 5	0.00414	241.3327	0.12414	8.0552	58.7821	7.2974
31	33.5551	0.0298	0.00369	271.2926	0.12369	8.0850	59.6761	7.3811
32	37.5817	0.0266	0.00328	304.8477	0.12328	8.1116	60.5010	7.4586
33	42.0915	0.0238	0.00292	342.4294	0.12292	8.1354	61.2612	7.5302
34	47.1425	0.0212	0.00260	384.5210	0.12260	8.1566	61.9612	7.5965
35	52.7996	0.0189	0.00232	431.6635	0.12232	8.1755	62.6052	7.6577
40	93.0510	0.0107	0.00130	767.0914	0.12130	8.2438	65.1159	7.8988
45	163.9876	0.0061	0.0074	1358.23	0.12074	8.2825	66.7342	8.0572
50	289.0022	0.0035	0.00042	2400.02	0.12042	8.3045	67.7624	8.1597
55	509.3206	0.0020	0.00024	4236.01	0.12024	8.3170	68.4082	8.2251
60	897.5969	0.0011	0.00013	7471.64	0.12013	8.3240	68.8100	8.2664
65	1581.87	0.0006	0.00008	13174	0.12008	8.3281	69.0581	8.2922
70	2787.80	0.0004	0.00004	23223	0.12004	8.3303	69.2103	8.3082
75	4913.06	0.0002	0.00002	40934	0.12002	8.3316	69.3031	8.3181
80	8658.48	0.0001	0.00001	72146	0.12001	8.3324	69.3594	8.3241
85	15259	0.0001	0.00001		0.12001	8.3328	69.3935	8.3278

14%   **Table 18**   Discrete Cash Flow: Compound Interest Factors   **14%**

	Single Payments		Uniform Series Payments				Arithmetic Gradients	
n	Compound Amount *F/P*	Present Worth *P/F*	Sinking Fund *A/F*	Compound Amount *F/A*	Capital Recovery *A/P*	Present Worth *P/A*	Gradient Present Worth *P/G*	Gradient Uniform Series *A/G*
1	1.1400	0.8772	1.00000	1.0000	1.14000	0.8772		
2	1.2996	0.7695	0.46729	2.1400	0.60729	1.6467	0.7695	0.4673
3	1.4815	0.6750	0.29073	3.4396	0.43073	2.3216	2.1194	0.9129
4	1.6890	0.5921	0.20320	4.9211	0.34320	2.9137	3.8957	1.3370
5	1.9254	0.5194	0.15128	6.6101	0.29128	3.4331	5.9731	1.7399
6	2.1950	0.4556	0.11716	8.5355	0.25716	3.8887	8.2511	2.1218
7	2.5023	0.3996	0.09319	10.7305	0.23319	4.2883	10.6489	2.4832
8	2.8526	0.3506	0.07557	13.2328	0.21557	4.6389	13.1028	2.8246
9	3.2519	0.3075	0.06217	16.0853	0.20217	4.9464	15.5629	3.1463
10	3.7072	0.2697	0.05171	19.3373	0.19171	5.2161	17.9906	3.4490
11	4.2262	0.2366	0.04339	23.0445	0.18339	5.4527	20.3567	3.7333
12	4.8179	0.2076	0.03667	27.2707	0.17667	5.6603	22.6399	3.9998
13	5.4924	0.1821	0.03116	32.0887	0.17116	5.8424	24.8247	4.2491
14	6.2613	0.1597	0.02661	37.5811	0.16661	6.0021	26.9009	4.4819
15	7.1379	0.1401	0.02281	43.8424	0.16281	6.1422	28.8623	4.6990
16	8.1372	0.1229	0.01962	50.9804	0.15962	6.2651	30.7057	4.9011
17	9.2765	0.1078	0.01692	59.1176	0.15692	6.3729	32.4305	5.0888
18	10.5752	0.0946	0.01462	68.3941	0.15462	6.4674	34.0380	5.2630
19	12.0557	0.0829	0.01266	78.9692	0.15266	6.5504	35.5311	5.4243
20	13.7435	0.0728	0.01099	91.0249	0.15099	6.6231	36.9135	5.5714
21	15.6676	0.0638	0.00954	104.7684	0.14954	6.6870	38.1901	5.7111
22	17.8610	0.0560	0.00830	120.4360	0.14830	6.7429	39.3658	5.8381
23	20.3616	0.0491	0.00723	138.2970	0.14723	6.7921	40.4463	5.9549
24	23.2122	0.0431	0.00630	158.6586	0.14630	6.8351	41.4371	6.0624
25	26.4619	0.0378	0.00550	181.8708	0.14550	6.8729	42.3441	6.1610

n	Single Payments		Uniform Series Payments				Arithmetic Gradients	
	Compound Amount F/P	Present Worth P/F	Sinking Fund A/F	Compound Amount F/A	Capital Recovery A/P	Present Worth P/A	Gradient Present Worth P/G	Gradient Uniform Series A/G
26	30.1666	0.0331	0.00480	208.3327	0.14480	6.9061	43.1728	6.2514
27	34.3899	0.0291	0.00419	238.4993	0.14419	6.9352	43.9289	6.3342
28	39.2045	0.0255	0.00366	272.8892	0.14366	6.9607	44.6176	6.4100
29	44.6931	0.0224	0.00320	312.0937	0.14320	6.9830	45.2441	6.4791
30	50.9502	0.0196	0.00280	356.7868	0.14280	7.0027	45.8132	6.5423
31	58.0832	0.0172	0.00245	407.7370	0.14245	7.0199	46.3297	6.5998
32	66.2148	0.0151	0.00215	465.8202	0.14215	7.0350	46.7979	6.6522
33	75.4849	0.0132	0.00188	532.0350	0.14188	7.0482	47.2218	6.6998
34	86.0528	0.0116	0.00165	607.5199	0.14165	7.0599	47.6053	6.7431
35	98.1002	0.0102	0.00144	693.5727	0.14144	7.0700	47.9519	6.7824
40	188.8835	0.0053	0.00075	1342.03	0.14075	7.1050	49.2376	6.9300
45	363.6791	0.0027	0.00039	2590.56	0.14039	7.1232	49.9963	7.0188
50	700.2330	0.0014	0.00020	4994.52	0.14020	7.1327	50.4375	7.0714
55	1348.24	0.0007	0.00010	9623.13	0.14010	7.1376	50.6912	7.1020
60	2595.92	0.0004	0.00005	18535	0.14005	7.1401	50.8357	7.1197
65	4998.22	0.0002	0.00003	35694	0.14003	7.1414	50.9173	7.1298
70	9623.64	0.0001	0.00001	68733	0.14001	7.1421	50.9632	7.1356
75	18530	0.0001	0.00001		0.14001	7.1425	50.9887	7.1388
80	35677				0.14000	7.1427	51.0030	7.1406
85	68693				0.14000	7.1428	51.0108	7.1416

15%   **Table 19**   Discrete Cash Flow: Compound Interest Factors   **15%**

	Single Payments		Uniform Series Payments				Arithmetic Gradients	
n	Compound Amount F/P	Present Worth P/F	Sinking Fund A/F	Compound Amount F/A	Capital Recovery A/P	Present Worth P/A	Gradient Present Worth P/G	Gradient Uniform Series A/G
1	1.1500	0.8696	1.00000	1.0000	1.15000	0.8696		
2	1.3225	0.7561	0.46512	2.1500	0.61512	1.6257	0.7561	0.4651
3	1.5209	0.6575	0.28798	3.4725	0.43798	2.2832	2.0712	0.9071
4	1.7490	0.5718	0.20027	4.9934	0.35027	2.8550	3.7864	1.3263
5	2.0114	0.4972	0.14832	6.7424	0.29832	3.3522	5.7751	1.7228
6	2.3131	0.4323	0.11424	8.7537	0.26424	3.7845	7.9368	2.0972
7	2.6600	0.3759	0.09036	11.0668	0.24036	4.1604	10.1924	2.4498
8	3.0590	0.3269	0.07285	13.7268	0.22285	4.4873	12.4807	2.7813
9	3.5179	0.2843	0.05957	16.7858	0.20957	4.7716	14.7548	3.0922
10	4.0456	0.2472	0.04925	20.3037	0.19925	5.0188	16.9795	3.3832
11	4.6524	0.2149	0.04107	24.3493	0.19107	5.2337	19.1289	3.6549
12	5.3503	0.1869	0.03448	29.0017	0.18448	5.4206	21.1849	3.9082
13	6.1528	0.1625	0.02911	34.3519	0.17911	5.5831	23.1352	4.1438
14	7.0757	0.1413	0.02469	40.5047	0.17469	5.7245	24.9725	4.3624
15	8.1371	0.1229	0.02102	47.5804	0.17102	5.8474	26.6930	4.5650
16	9.3576	0.1069	0.01795	55.7175	0.16795	5.9542	28.2960	4.7522
17	10.7613	0.0929	0.01537	65.0751	0.16537	6.0472	29.7828	4.9251
18	12.3755	0.0808	0.01319	75.8364	0.16319	6.1280	31.1565	5.0843
19	14.2318	0.0703	0.01134	88.2118	0.16134	6.1982	32.4213	5.2307
20	16.3665	0.0611	0.00976	102.4436	0.15976	6.2593	33.5822	5.3651
21	18.8215	0.0531	0.00842	118.8101	0.15842	6.3125	34.6448	5.4883
22	21.6447	0.0462	0.00727	137.6316	0.15727	6.3587	35.6150	5.6010
23	24.8915	0.0402	0.00628	159.2764	0.15628	6.3988	36.4988	5.7040
24	28.6252	0.0349	0.00543	184.1678	0.15543	6.4338	37.3023	5.7979
25	32.9190	0.0304	0.00470	212.7930	0.15470	6.4641	38.0314	5.8834

n	Single Payments		Uniform Series Payments				Arithmetic Gradients	
	Compound Amount *F/P*	Present Worth *P/F*	Sinking Fund *A/F*	Compound Amount *F/A*	Capital Recovery *A/P*	Present Worth *P/A*	Gradient Present Worth *P/G*	Gradient Uniform Series *A/G*
26	37.8568	0.0264	0.00407	245.7120	0.15407	6.4906	38.6918	5.9612
27	43.5353	0.0230	0.00353	283.5688	0.15353	6.5135	39.2890	6.0319
28	50.0656	0.0200	0.00306	327.1041	0.15306	6.5335	39.8283	6.0960
29	57.5755	0.0174	0.00265	377.1697	0.15265	6.5509	40.3146	6.1541
30	66.2118	0.0151	0.00230	434.7451	0.15230	6.5660	40.7526	6.2066
31	76.1435	0.0131	0.00200	500.9569	0.15200	6.5791	41.1466	6.2541
32	87.5651	0.0114	0.00173	577.1005	0.15173	6.5905	41.5006	6.2970
33	100.6998	0.0099	0.00150	664.6655	0.15150	6.6005	41.8184	6.3357
34	115.8048	0.0086	0.00131	765.3654	0.15131	6.6091	42.1033	6.3705
35	133.1755	0.0075	0.00113	881.1702	0.15113	6.6166	42.3586	6.4019
40	267.8635	0.0037	0.00056	1779.09	0.15056	6.6418	43.2830	6.5168
45	538.7693	0.0019	0.00028	3585.13	0.15028	6.6543	43.8051	6.5830
50	1083.66	0.0009	0.00014	7217.72	0.15014	6.6605	44.0958	6.6205
55	2179.62	0.0005	0.00007	14524	0.15007	6.6636	44.2558	6.6414
60	4384.00	0.0002	0.00003	29220	0.15003	6.6651	44.3431	6.6530
65	8817.79	0.0001	0.00002	58779	0.15002	6.6659	44.3903	6.6593
70	17736	0.0001	0.00001		0.15001	6.6663	44.4156	6.6627
75	35673				0.15000	6.6665	44.4292	6.6646
80	71751				0.15000	6.6666	44.4364	6.6656
85					0.15000	6.6666	44.4402	6.6661

Table 20 Discrete Cash Flow: Compound Interest Factors

	Single Payments		Uniform Series Payments				Arithmetic Gradients	
n	Compound Amount F/P	Present Worth P/F	Sinking Fund A/F	Compound Amount F/A	Capital Recovery A/P	Present Worth P/A	Gradient Present Worth P/G	Gradient Uniform Series A/G
1	1.1600	0.8621	1.00000	1.0000	1.16000	0.8621		
2	1.3456	0.7432	0.46296	2.1600	0.62296	1.6052	0.7432	0.4630
3	1.5609	0.6407	0.28526	3.5056	0.44526	2.2459	2.0245	0.9014
4	1.8106	0.5523	0.19738	5.0665	0.35738	2.7982	3.6814	1.3156
5	2.1003	0.4761	0.14541	6.8771	0.30541	3.2743	5.5858	1.7060
6	2.4364	0.4104	0.11139	8.9775	0.27139	3.6847	7.6380	2.0729
7	2.8262	0.3538	0.08761	11.4139	0.24761	4.0386	9.7610	2.4169
8	3.2784	0.3050	0.07022	14.2401	0.23022	4.3436	11.8962	2.7388
9	3.8030	0.2630	0.05708	17.5185	0.21708	4.6065	13.9998	3.0391
10	4.4114	0.2267	0.04690	21.3215	0.20690	4.8332	16.0399	3.3187
11	5.1173	0.1954	0.03886	25.7329	0.19886	5.0286	17.9941	3.5783
12	5.9360	0.1685	0.03241	30.8502	0.19241	5.1971	19.8472	3.8189
13	6.8858	0.1452	0.02718	36.7862	0.18718	5.3423	21.5899	4.0413
14	7.9875	0.1252	0.02290	43.6720	0.18290	5.4675	23.2175	4.2464
15	9.2655	0.1079	0.01936	51.6595	0.17936	5.5755	24.7284	4.4352
16	10.7480	0.0930	0.01641	60.9250	0.17641	5.6685	26.1241	4.6086
17	12.4677	0.0802	0.01395	71.6730	0.17395	5.7487	27.4074	4.7676
18	14.4625	0.0691	0.01188	84.1407	0.17188	5.8178	28.5828	4.9130
19	16.7765	0.0596	0.01014	98.6032	0.17014	5.8775	29.6557	5.0457
20	19.4608	0.0514	0.00867	115.3797	0.16867	5.9288	30.6321	5.1666
22	26.1864	0.0382	0.00635	157.4150	0.16635	6.0113	32.3200	5.3765
24	35.2364	0.0284	0.00467	213.9776	0.16467	6.0726	33.6970	5.5490
26	47.4141	0.0211	0.00345	290.0883	0.16345	6.1182	34.8114	5.6898
28	63.8004	0.0157	0.00255	392.5028	0.16255	6.1520	35.7073	5.8041
30	85.8499	0.0116	0.00189	530.3117	0.16189	6.1772	36.4234	5.8964
32	115.5196	0.0087	0.00140	715.7475	0.16140	6.1959	36.9930	5.9706
34	155.4432	0.0064	0.00104	965.2698	0.16104	6.2098	37.4441	6.0299
35	180.3141	0.0055	0.00089	1120.71	0.16089	6.2153	37.6327	6.0548
36	209.1643	0.0048	0.00077	1301.03	0.16077	6.2201	37.8000	6.0771
38	281.4515	0.0036	0.00057	1752.82	0.16057	6.2278	38.0799	6.1145
40	378.7212	0.0026	0.00042	2360.76	0.16042	6.2335	38.2992	6.1441
45	795.4438	0.0013	0.00020	4965.27	0.16020	6.2421	38.6598	6.1934
50	1670.70	0.0006	0.00010	10436	0.16010	6.2463	38.8521	6.2201
55	3509.05	0.0003	0.00005	21925	0.16005	6.2482	38.9534	6.2343
60	7370.20	0.0001	0.00002	46058	0.16002	6.2492	39.0063	6.2419

Table 21 Discrete Cash Flow: Compound Interest Factors

	Single Payments		Uniform Series Payments				Arithmetic Gradients	
n	Compound Amount F/P	Present Worth P/F	Sinking Fund A/F	Compound Amount F/A	Capital Recovery A/P	Present Worth P/A	Gradient Present Worth P/G	Gradient Uniform Series A/G
1	1.1800	0.8475	1.00000	1.0000	1.18000	0.8475		
2	1.3924	0.7182	0.45872	2.1800	0.63872	1.5656	0.7182	0.4587
3	1.6430	0.6086	0.27992	3.5724	0.45992	2.1743	1.9354	0.8902
4	1.9388	0.5158	0.19174	5.2154	0.37174	2.6901	3.4828	1.2947
5	2.2878	0.4371	0.13978	7.1542	0.31978	3.1272	5.2312	1.6728
6	2.6996	0.3704	0.10591	9.4420	0.28591	3.4976	7.0834	2.0252
7	3.1855	0.3139	0.08236	12.1415	0.26236	3.8115	8.9670	2.3526
8	3.7589	0.2660	0.06524	15.3270	0.24524	4.0776	10.8292	2.6558
9	4.4355	0.2255	0.05239	19.0859	0.23239	4.3030	12.6329	2.9358
10	5.2338	0.1911	0.04251	23.5213	0.22251	4.4941	14.3525	3.1936
11	6.1759	0.1619	0.03478	28.7551	0.21478	4.6560	15.9716	3.4303
12	7.2876	0.1372	0.02863	34.9311	0.20863	4.7932	17.4811	3.6470
13	8.5994	0.1163	0.02369	42.2187	0.20369	4.9095	18.8765	3.8449
14	10.1472	0.0985	0.01968	50.8180	0.19968	5.0081	20.1576	4.0250
15	11.9737	0.0835	0.01640	60.9653	0.19640	5.0916	21.3269	4.1887
16	14.1290	0.0708	0.01371	72.9390	0.19371	5.1624	22.3885	4.3369
17	16.6722	0.0600	0.01149	87.0680	0.19149	5.2223	23.3482	4.4708
18	19.6733	0.0508	0.00964	103.7403	0.18964	5.2732	24.2123	4.5916
19	23.2144	0.0431	0.00810	123.4135	0.18810	5.3162	24.9877	4.7003
20	27.3930	0.0365	0.00682	146.6280	0.18682	5.3527	25.6813	4.7978
22	38.1421	0.0262	0.00485	206.3448	0.18485	5.4099	26.8506	4.9632
24	53.1090	0.0188	0.00345	289.4945	0.18345	5.4509	27.7725	5.0950
26	73.9490	0.0135	0.00247	405.2721	0.18247	5.4804	28.4935	5.1991
28	102.9666	0.0097	0.00177	566.4809	0.18177	5.5016	29.0537	5.2810
30	143.3706	0.0070	0.00126	790.9480	0.18126	5.5168	29.4864	5.3448
32	199.6293	0.0050	0.00091	1103.50	0.18091	5.5277	29.8191	5.3945
34	277.9638	0.0036	0.00065	1538.69	0.18065	5.5356	30.0736	5.4328
35	327.9973	0.0030	0.00055	1816.65	0.18055	5.5386	30.1773	5.4485
36	387.0368	0.0026	0.00047	2144.65	0.18047	5.5412	30.2677	5.4623
38	538.9100	0.0019	0.00033	2988.39	0.18033	5.5452	30.4152	5.4849
40	750.3783	0.0013	0.00024	4163.21	0.18024	5.5482	30.5269	5.5022
45	1716.68	0.0006	0.00010	9531.58	0.18010	5.5523	30.7006	5.5293
50	3927.36	0.0003	0.00005	21813	0.18005	5.5541	30.7856	5.5428
55	8984.84	0.0001	0.00002	49910	0.18002	5.5549	30.8268	5.5494
60	20555			114190	0.18001	5.5553	30.8465	5.5526

20%   **Table 22**   Discrete Cash Flow: Compound Interest Factors   **20%**

	Single Payments		Uniform Series Payments				Arithmetic Gradients	
n	Compound Amount F/P	Present Worth P/F	Sinking Fund A/F	Compound Amount F/A	Capital Recovery A/P	Present Worth P/A	Gradient Present Worth P/G	Gradient Uniform Series A/G
1	1.2000	0.8333	1.00000	1.0000	1.20000	0.8333		
2	1.4400	0.6944	0.45455	2.2000	0.65455	1.5278	0.6944	0.4545
3	1.7280	0.5787	0.27473	3.6400	0.47473	2.1065	1.8519	0.8791
4	2.0736	0.4823	0.18629	5.3680	0.38629	2.5887	3.2986	1.2742
5	2.4883	0.4019	0.13438	7.4416	0.33438	2.9906	4.9061	1.6405
6	2.9860	0.3349	0.10071	9.9299	0.30071	3.3255	6.5806	1.9788
7	3.5832	0.2791	0.07742	12.9159	0.27742	3.6046	8.2551	2.2902
8	4.2998	0.2326	0.06061	16.4991	0.26061	3.8372	9.8831	2.5756
9	5.1598	0.1938	0.04808	20.7989	0.24808	4.0310	11.4335	2.8364
10	6.1917	0.1615	0.03852	25.9587	0.23852	4.1925	12.8871	3.0739
11	7.4301	0.1346	0.03110	32.1504	0.23110	4.3271	14.2330	3.2893
12	8.9161	0.1122	0.02526	39.5805	0.22526	4.4392	15.4667	3.4841
13	10.6993	0.0935	0.02062	48.4966	0.22062	4.5327	16.5883	3.6597
14	12.8392	0.0779	0.01689	59.1959	0.21689	4.6106	17.6008	3.8175
15	15.4070	0.0649	0.01388	72.0351	0.21388	4.6755	18.5095	3.9588
16	18.4884	0.0541	0.01144	87.4421	0.21144	4.7296	19.3208	4.0851
17	22.1861	0.0451	0.00944	105.9306	0.20944	4.7746	20.0419	4.1976
18	26.6233	0.0376	0.00781	128.1167	0.20781	4.8122	20.6805	4.2975
19	31.9480	0.0313	0.00646	154.7400	0.20646	4.8435	21.2439	4.3861
20	38.3376	0.0261	0.00536	186.6880	0.20536	4.8696	21.7395	4.4643
22	55.2061	0.0181	0.00369	271.0307	0.20369	4.9094	22.5546	4.5941
24	79.4968	0.0126	0.00255	392.4842	0.20255	4.9371	23.1760	4.6943
26	114.4755	0.0087	0.00176	567.3773	0.20176	4.9563	23.6460	4.7709
28	164.8447	0.0061	0.00122	819.2233	0.20122	4.9697	23.9991	4.8291
30	237.3763	0.0042	0.00085	1181.88	0.20085	4.9789	24.2628	4.8731
32	341.8219	0.0029	0.00059	1704.11	0.20059	4.9854	24.4588	4.9061
34	492.2235	0.0020	0.00041	2456.12	0.20041	4.9898	24.6038	4.9308
35	590.6682	0.0017	0.00034	2948.34	0.20034	4.9915	24.6614	4.9406
36	708.8019	0.0014	0.00028	3539.01	0.20028	4.9929	24.7108	4.9491
38	1020.67	0.0010	0.00020	5098.37	0.20020	4.9951	24.7894	4.9627
40	1469.77	0.0007	0.00014	7343.86	0.20014	4.9966	24.8469	4.9728
45	3657.26	0.0003	0.00005	18281	0.20005	4.9986	24.9316	4.9877
50	9100.44	0.0001	0.00002	45497	0.20002	4.9995	24.9698	4.9945
55	22645		0.00001		0.20001	4.9998	24.9868	4.9976

Table 23   Discrete Cash Flow: Compound Interest Factors

	Single Payments		Uniform Series Payments				Arithmetic Gradients	
n	Compound Amount F/P	Present Worth P/F	Sinking Fund A/F	Compound Amount F/A	Capital Recovery A/P	Present Worth P/A	Gradient Present Worth P/G	Gradient Uniform Series A/G
1	1.2200	0.8197	1.00000	1.0000	1.22000	0.8197		
2	1.4884	0.6719	0.45045	2.2200	0.67045	1.4915	0.6719	0.4505
3	1.8158	0.5507	0.26966	3.7084	0.48966	2.0422	1.7733	0.8683
4	2.2153	0.4514	0.18102	5.5242	0.40102	2.4936	3.1275	1.2542
5	2.7027	0.3700	0.12921	7.7396	0.34921	2.8636	4.6075	1.6090
6	3.2973	0.3033	0.09576	10.4423	0.31576	3.1669	6.1239	1.9337
7	4.0227	0.2486	0.07278	13.7396	0.29278	3.4155	7.6154	2.2297
8	4.9077	0.2038	0.05630	17.7623	0.27630	3.6193	9.0417	2.4982
9	5.9874	0.1670	0.04471	22.6700	0.26411	3.7863	10.3779	2.7409
10	7.3046	0.1369	0.03489	28.6574	0.25489	3.9232	11.6100	2.9593
11	8.917	0.1122	0.02781	35.9620	0.24781	4.0354	12.7321	3.1551
12	10.8722	0.0920	0.02228	44.8737	0.24228	4.1274	13.7438	3.3299
13	13.2641	0.0754	0.01794	55.7459	0.23794	4.2028	14.6485	3.4855
14	16.1822	0.0618	0.01449	69.0100	0.23449	4.2646	15.4519	3.6233
15	19.7423	0.0507	0.01174	85.1922	0.23174	4.3152	16.1610	3.7451
16	24.0856	0.0415	0.00953	104.9345	0.22953	4.3567	16.7838	3.8524
17	29.3844	0.0340	0.00775	129.0201	0.22775	4.3908	17.3283	3.9465
18	35.8490	0.0279	0.00631	158.4045	0.22631	4.4187	17.8025	4.0289
19	43.7358	0.0229	0.00515	194.2535	0.22515	4.4415	18.2141	4.1009
20	53.3576	0.0187	0.00420	237.9893	0.22420	4.4603	18.5702	4.1635
22	79.4175	0.0126	0.00281	356.4432	0.22281	4.4882	19.1418	4.2649
24	118.2050	0.0085	0.00188	532.7501	0.22188	4.5070	19.5635	4.3407
26	175.9364	0.0057	0.00126	795.1653	0.22126	4.5196	19.8720	4.3968
28	261.8637	0.0038	0.00084	1185.74	0.22084	4.5281	20.0962	4.4381
30	389.7579	0.0026	0.00057	1767.08	0.22057	4.5338	20.2583	4.4683
32	580.1156	0.0017	0.00038	2632.34	0.22038	4.5376	20.3748	4.4902
34	863.4441	0.0012	0.00026	3920.20	0.22026	4.5402	20.4582	4.5060
35	1053.40	0.0009	0.00021	4783.64	0.22021	4.5411	20.4905	4.5122
36	1285.15	0.0008	0.00017	5837.05	0.22017	4.5419	20.5178	4.5174
38	1912.82	0.0005	0.00012	8690.08	0.22012	4.5431	20.5601	4.5256
40	2847.04	0.0004	0.00008	12937	0.22008	4.5439	20.5900	4.5314
45	7694.71	0.0001	0.00003	34971	0.22003	4.5449	20.6319	4.5396
50	20797		0.00001	94525	0.22001	4.5452	20.6492	4.5431
55	56207				0.22000	4.5454	20.6563	4.5445

24%   **Table 24**   Discrete Cash Flow: Compound Interest Factors   **24%**

	Single Payments		Uniform Series Payments				Arithmetic Gradients	
n	Compound Amount F/P	Present Worth P/F	Sinking Fund A/F	Compound Amount F/A	Capital Recovery A/P	Present Worth P/A	Gradient Present Worth P/G	Gradient Uniform Series A/G
1	1.2400	0.8065	1.00000	1.0000	1.24000	0.8065		
2	1.5376	0.6504	0.44643	2.2400	0.68643	1.4568	0.6504	0.4464
3	1.9066	0.5245	0.26472	3.7776	0.50472	1.9813	1.6993	0.8577
4	2.3642	0.4230	0.17593	5.6842	0.41593	2.4043	2.9683	1.2346
5	2.9316	0.3411	0.12425	8.0484	0.36425	2.7454	4.3327	1.5782
6	3.6352	0.2751	0.09107	10.9801	0.33107	3.0205	5.7081	1.8898
7	4.5077	0.2218	0.06842	14.6153	0.30842	3.2423	7.0392	2.1710
8	5.5895	0.1789	0.05229	19.1229	0.29229	3.4212	8.2915	2.4236
9	6.9310	0.1443	0.04047	24.7125	0.28047	3.5655	9.4458	2.6492
10	8.5944	0.1164	0.03160	31.6434	0.27160	3.6819	10.4930	2.8499
11	10.6571	0.0938	0.02485	40.2379	0.26485	3.7757	11.4313	3.0276
12	13.2148	0.0757	0.01965	50.8950	0.25965	3.8514	12.2637	3.1843
13	16.3863	0.0610	0.01560	64.1097	0.25560	3.9124	12.9960	3.3218
14	20.3191	0.0492	0.01242	80.4961	0.25242	3.9616	13.6358	3.4420
15	25.1956	0.0397	0.00992	100.8151	0.24992	4.0013	14.1915	3.5467
16	31.2426	0.0320	0.00794	126.0108	0.24794	4.0333	14.6716	3.6376
17	38.7408	0.0258	0.00636	157.2534	0.24636	4.0591	15.0846	3.7162
18	48.0386	0.0208	0.00510	195.9942	0.24510	4.0799	15.4385	3.7840
19	59.5679	0.0168	0.00410	244.0328	0.24410	4.0967	15.7406	3.8423
20	73.8641	0.0135	0.00329	303.6006	0.24329	4.1103	15.9979	3.8922
22	113.5735	0.0088	0.00213	469.0563	0.24213	4.1300	16.4011	3.9712
24	174.6306	0.0057	0.00138	723.4610	0.24138	4.1428	16.6891	4.0284
26	268.5121	0.0037	0.00090	1114.63	0.24090	4.1511	16.8930	4.0695
28	412.8642	0.0024	0.00058	1716.10	0.24058	4.1566	17.0365	4.0987
30	634.8199	0.0016	0.00038	2640.92	0.24038	4.1601	17.1369	4.1193
32	976.0991	0.0010	0.00025	4062.91	0.24025	4.1624	17.2067	4.1338
34	1500.85	0.0007	0.00016	6249.38	0.24016	4.1639	17.2552	4.1440
35	1861.05	0.0005	0.00013	7750.23	0.24013	4.1664	17.2734	4.1479
36	2307.71	0.0004	0.00010	9611.28	0.24010	4.1649	17.2886	4.1511
38	3548.33	0.0003	0.00007	14781	0.24007	4.1655	17.3116	4.1560
40	5455.91	0.0002	0.00004	22729	0.24004	4.1659	17.3274	4.1593
45	15995	0.0001	0.00002	66640	0.24002	4.1664	17.3483	4.1639
50	46890		0.00001		0.24001	4.1666	17.3563	4.1653
55					0.24000	4.1666	17.3593	4.1663

Appendix A

Table 25 Discrete Cash Flow: Compound Interest Factors

	Single Payments		Uniform Series Payments				Arithmetic Gradients	
n	Compound Amount F/P	Present Worth P/F	Sinking Fund A/F	Compound Amount F/A	Capital Recovery A/P	Present Worth P/A	Gradient Present Worth P/G	Gradient Uniform Series A/G
1	1.2500	0.8000	1.00000	1.0000	1.25000	0.8000		
2	1.5625	0.6400	0.44444	2.2500	0.69444	1.4400	0.6400	0.4444
3	1.9531	0.5120	0.26230	3.8125	0.51230	1.9520	1.6640	0.8525
4	2.4414	0.4096	0.17344	5.7656	0.42344	2.3616	2.8928	1.2249
5	3.0518	0.3277	0.12185	8.2070	0.37185	2.6893	4.2035	1.5631
6	3.8147	0.2621	0.08882	11.2588	0.33882	2.9514	5.5142	1.8683
7	4.7684	0.2097	0.06634	15.0735	0.31634	3.1611	6.7725	2.1424
8	5.9605	0.1678	0.05040	19.8419	0.30040	3.3289	7.9469	2.3872
9	7.4506	0.1342	0.03876	25.8023	0.28876	3.4631	9.0207	2.6048
10	9.3132	0.1074	0.03007	33.2529	0.28007	3.5705	9.9870	2.7971
11	11.6415	0.0859	0.02349	42.5661	0.27349	3.6564	10.8460	2.9663
12	14.5519	0.0687	0.01845	54.2077	0.26845	3.7251	11.6020	3.1145
13	18.1899	0.0550	0.01454	68.7596	0.26454	3.7801	12.2617	3.2437
14	22.7374	0.0440	0.01150	86.9495	0.26150	3.8241	12.8334	3.3559
15	28.4217	0.0352	0.00912	109.6868	0.25912	3.8593	13.3260	3.4530
16	35.5271	0.0281	0.00724	138.1085	0.25724	3.8874	13.7482	3.5366
17	44.4089	0.0225	0.00576	173.6357	0.25576	3.9099	14.1085	3.6084
18	55.5112	0.0180	0.00459	218.0446	0.25459	3.9279	14.4147	3.6698
19	69.3889	0.0144	0.00366	273.5558	0.25366	3.9424	14.6741	3.7222
20	86.7362	0.0115	0.00292	342.9447	0.25292	3.9539	14.8932	3.7667
22	135.5253	0.0074	0.00186	538.1011	0.25186	3.9705	15.2326	3.8365
24	211.7582	0.0047	0.00119	843.0329	0.25119	3.9811	15.4711	3.8861
26	330.8722	0.0030	0.00076	1319.49	0.25076	3.9879	15.6373	3.9212
28	516.9879	0.0019	0.00048	2063.95	0.25048	3.9923	15.7524	3.9457
30	807.7936	0.0012	0.00031	3227.17	0.25031	3.9950	15.8316	3.9628
32	1262.18	0.0008	0.00020	5044.71	0.25020	3.9968	15.8859	3.9746
34	1972.15	0.0005	0.00013	7884.61	0.25013	3.9980	15.9229	3.9828
35	2465.19	0.0004	0.00010	9856.76	.025010	3.9984	15.9367	3.9858
36	3081.49	0.0003	0.00008	12322	0.25008	3.9987	15.9481	3.9883
38	4814.82	0.0002	0.00005	19255	0.25005	3.9992	15.9651	3.9921
40	7523.16	0.0001	0.00003	30089	0.25003	3.9995	15.9766	3.9947
45	22959		0.00001	91831	0.25001	3.9998	15.9915	3.9980
50	70065				0.25000	3.9999	15.9969	3.9993
55					0.25000	4.0000	15.9989	3.9997

30%   **Table 26**   Discrete Cash Flow: Compound Interest Factors   **30%**

n	Single Payments		Uniform Series Payments				Arithmetic Gradients	
	Compound Amount F/P	Present Worth P/F	Sinking Fund A/F	Compound Amount F/A	Capital Recovery A/P	Present Worth P/A	Gradient Present Worth P/G	Gradient Uniform Series A/G
1	1.3000	0.7692	1.00000	1.0000	1.30000	0.7692		
2	1.6900	0.5917	0.43478	2.3000	0.73478	1.3609	0.5917	0.4348
3	2.1970	0.4552	0.25063	3.9900	0.55063	1.8161	1.5020	0.8271
4	2.8561	0.3501	0.16163	6.1870	0.46163	2.1662	2.5524	1.1783
5	3.7129	0.2693	0.11058	9.0431	0.41058	2.4356	3.6297	1.4903
6	4.8268	0.2072	0.07839	12.7560	0.37839	2.6427	4.6656	1.7654
7	6.2749	0.1594	0.05687	17.5828	0.35687	2.8021	5.6218	2.0063
8	8.1573	0.1226	0.04192	23.8577	0.34192	2.9247	6.4800	2.2156
9	10.6045	0.0943	0.03124	32.0150	0.33124	3.0190	7.2343	2.3963
10	13.7858	0.0725	0.02346	42.6195	0.32346	3.0915	7.8872	2.5512
11	17.9216	0.0558	0.01773	56.4053	0.31773	3.1473	8.4452	2.6833
12	23.2981	0.0429	0.01345	74.3270	0.31345	3.1903	8.9173	2.7952
13	30.2875	0.0330	0.01024	97.6250	0.31024	3.2233	9.3135	2.8895
14	39.3738	0.0254	0.00782	127.9125	0.30782	3.2487	9.6437	2.9685
15	51.1859	0.0195	0.00598	167.2863	0.30598	3.2682	9.9172	3.0344
16	66.5417	0.0150	0.00458	218.4722	0.30458	3.2832	10.1426	3.0892
17	86.5042	0.0116	0.00351	285.0139	0.30351	3.2948	10.3276	3.1345
18	112.4554	0.0089	0.00269	371.5180	0.30269	3.3037	10.4788	3.1718
19	146.1920	0.0068	0.00207	483.9734	0.30207	3.3105	10.6019	3.2025
20	190.0496	0.0053	0.00159	630.1655	0.30159	3.3158	10.7019	3.2275
22	321.1839	0.0031	0.00094	1067.28	0.30094	3.3230	10.8482	3.2646
24	542.8008	0.0018	0.00055	1806.00	0.30055	3.3272	10.9433	3.2890
25	705.6410	0.0014	0.00043	2348.80	0.30043	3.3286	10.9773	3.2979
26	917.3333	0.0011	0.00033	3054.44	0.30033	3.3297	11.0045	3.3050
28	1550.29	0.0006	0.00019	5164.31	0.30019	3.3312	11.0437	3.3153
30	2620.00	0.0004	0.00011	8729.99	0.30011	3.3321	11.0687	3.3219
32	4427.79	0.0002	0.00007	14756	0.30007	3.3326	11.0845	3.3261
34	7482.97	0.0001	0.00004	24940	0.30004	3.3329	11.0945	3.3288
35	9727.86	0.0001	0.00003	32423	0.30003	3.3330	11.0980	3.3297

Table 27 Discrete Cash Flow: Compound Interest Factors

	Single Payments		Uniform Series Payments				Arithmetic Gradients	
n	Compound Amount *F/P*	Present Worth *P/F*	Sinking Fund *A/F*	Compound Amount *F/A*	Capital Recovery *A/P*	Present Worth *P/A*	Gradient Present Worth *P/G*	Gradient Uniform Series *A/G*
1	1.3500	0.7407	1.00000	1.0000	1.35000	0.7407		
2	1.8225	0.5487	0.42553	2.3500	0.77553	1.2894	0.5487	0.4255
3	2.4604	0.4064	0.23966	4.1725	0.58966	1.6959	1.3616	0.8029
4	3.3215	0.3011	0.15076	6.6329	0.50076	1.9969	2.2648	1.1341
5	4.4840	0.2230	0.10046	9.9544	0.45046	2.2200	3.1568	1.4220
6	6.0534	0.1652	0.06926	14.4384	0.41926	2.3852	3.9828	1.6698
7	8.1722	0.1224	0.04880	20.4919	0.39880	2.5075	4.7170	1.8811
8	11.0324	0.0906	0.03489	28.6640	0.38489	2.5982	5.3515	2.0597
9	14.8937	0.0671	0.02519	39.6964	0.37519	2.6653	5.8886	2.2094
10	20.1066	0.0497	0.01832	54.5902	0.36832	2.7150	6.3363	2.3338
11	27.1439	0.0368	0.01339	74.6967	0.36339	2.7519	6.7047	2.4364
12	36.6442	0.0273	0.00982	101.8406	0.35982	2.7792	7.0049	2.5205
13	49.4697	0.0202	0.00722	138.4848	0.35722	2.7994	7.2474	2.5889
14	66.7841	0.0150	0.00532	187.9544	0.35532	2.8144	7.4421	2.6443
15	90.1585	0.0111	0.00393	254.7385	0.35393	2.8255	7.5974	2.6889
16	121.7139	0.0082	0.00290	344.8970	0.35290	2.8337	7.7206	2.7246
17	164.3138	0.0061	0.00214	466.6109	0.35214	2.8398	7.8180	2.7530
18	221.8236	0.0045	0.00158	630.9247	0.35158	2.8443	7.8946	2.7756
19	299.4619	0.0033	0.00117	852.7483	0.35117	2.8476	7.9547	2.7935
20	404.2736	0.0025	0.00087	1152.21	0.35087	2.8501	8.0017	2.8075
22	736.7886	0.0014	0.00048	2102.25	0.35048	2.8533	8.0669	2.8272
24	1342.80	0.0007	0.00026	3833.71	0.35026	2.8550	8.1061	2.8393
25	1812.78	0.0006	0.00019	5176.50	0.35019	2.8556	8.1194	2.8433
26	2447.25	0.0004	0.00014	6989.28	0.35014	2.8560	8.1296	2.8465
28	4460.11	0.0002	0.00014	12740	0.35008	2.8565	8.1435	2.8509
30	8128.55	0.0001	0.00004	23222	0.35004	2.8568	8.1517	2.8535
32	14814	0.0001	0.00002	42324	0.35002	2.8569	8.1565	2.8550
34	26999		0.00001	77137	0.35001	2.8570	8.1594	2.8559
35	36449		0.00001		0.35001	2.8571	8.1603	2.8562

40%   **Table 28**   Discrete Cash Flow: Compound Interest Factors   **40%**

	Single Payments		Uniform Series Payments				Arithmetic Gradients	
n	Compound Amount F/P	Present Worth P/F	Sinking Fund A/F	Compound Amount F/A	Capital Recovery A/P	Present Worth P/A	Gradient Present Worth P/G	Gradient Uniform Series A/G
1	1.4000	0.7143	1.00000	1.0000	1.40000	0.7143		
2	1.9600	0.5102	0.41667	2.4000	0.81667	1.2245	0.5102	0.4167
3	2.7440	0.3644	0.22936	4.3600	0.62936	1.5889	1.2391	0.7798
4	3.8416	0.2603	0.14077	7.1040	0.54077	1.8492	2.0200	1.0923
5	5.3782	0.1859	0.09136	10.9456	0.49136	2.0352	2.7637	1.3580
6	7.5295	0.1328	0.06126	16.3238	0.46126	2.1680	3.4278	1.5811
7	10.5414	0.0949	0.04192	23.8534	0.44192	2.2628	3.9970	1.7664
8	14.7579	0.0678	0.02907	34.3947	0.42907	2.3306	4.4713	1.9185
9	20.6610	0.0484	0.02034	49.1526	0.42034	2.3790	4.8585	2.0422
10	28.9255	0.0346	0.01432	69.8137	0.41432	2.4136	5.1696	2.1419
11	40.4957	0.0247	0.01013	98.7391	0.41013	2.4383	5.4166	2.2215
12	56.6939	0.0176	0.00718	139.2348	0.40718	2.4559	5.6106	2.2845
13	79.3715	0.0126	0.00510	195.9287	0.40510	2.4685	5.7618	2.3341
14	111.1201	0.0090	0.00363	275.3002	0.40363	2.4775	5.8788	2.3729
15	155.5681	0.0064	0.00259	386.4202	0.40259	2.4839	5.9688	2.4030
16	217.7953	0.0046	0.00185	541.9883	0.40185	2.4885	6.0376	2.4262
17	304.9135	0.0033	0.00132	759.7837	0.40132	2.4918	6.0901	2.4441
18	426.8789	0.0023	0.00094	1064.70	0.40094	2.4941	6.1299	2.4577
19	597.6304	0.0017	0.00067	1491.58	0.40067	2.4958	6.1601	2.4682
20	836.6826	0.0012	0.00048	2089.21	0.40048	2.4970	6.1828	2.4761
22	1639.90	0.0006	0.00024	4097.24	0.40024	2.4985	6.2127	2.4866
24	3214.20	0.0003	0.00012	8033.00	0.40012	2.4992	6.2294	2.4925
25	4499.88	0.0002	0.00009	11247	0.40009	2.4994	6.2347	2.4944
26	6299.83	0.0002	0.00006	15747	0.40006	2.4996	6.2387	2.4959
28	12348	0.0001	0.00003	30867	0.40003	2.4998	6.2438	2.4977
30	24201		0.00002	60501	0.40002	2.4999	6.2466	2.4988
32	47435		0.00001		0.40001	2.4999	6.2482	2.4993
34	92972				0.40000	2.5000	6.2490	2.4996
35					0.40000	2.5000	6.2493	2.4997

Table 29 Discrete Cash Flow: Compound Interest Factors

	Single Payments		Uniform Series Payments				Arithmetic Gradients	
n	Compound Amount F/P	Present Worth P/F	Sinking Fund A/F	Compound Amount F/A	Capital Recovery A/P	Present Worth P/A	Gradient Present Worth P/G	Gradient Uniform Series A/G
1	1.5000	0.6667	1.00000	1.0000	1.50000 1	0.6667		
2	2.2500	0.4444	0.40000	2.5000	0.90000	1.1111	0.4444	0.4000
3	3.3750	0.2963	0.21053	4.7500	0.71053	1.4074	1.0370	0.7368
4	5.0625	0.1975	0.12308	8.1250	0.62308	1.6049	1.6296	1.0154
5	7.5938	0.1317	0.07583	13.1875	0.57583	1.7366	2.1564	1.2417
6	11.3906	0.0878	0.04812	20.7813	0.54812	1.8244	2.5953	1.4226
7	17.0859	0.0585	0.03108	32.1719	0.53108	1.8829	2.9465	1.5648
8	25.6289	0.0390	0.02030	49.2578	0.52030	1.9220	3.2196	1.6752
9	38.4434	0.0260	0.01335	74.8867	0.51335	1.9480	3.4277	1.7596
10	57.6650	0.0173	0.00882	113.3301	0.50882	1.9653	3.5838	1.8235
11	86.4976	0.0116	0.00585	170.9951	0.50585	1.9769	3.6994	1.8713
12	129.7463	0.0077	0.00388	257.4927	0.50388	1.9846	3.7842	1.9068
13	194.6195	0.0051	0.00258	387.2390	0.50258	1.9897	3.8459	1.9329
14	291.9293	0.0034	0.00172	581.8585	0.50172	1.9931	3.8904	1.9519
15	437.8939	0.0023	0.00114	873.7878	0.50114	1.9954	3.9224	1.9657
16	656.8408	0.0015	0.00076	1311.68	0.50076	1.9970	3.9452	1.9756
17	985.2613	0.0010	0.00051	1968.52	0.50051	1.9980	3.9614	1.9827
18	1477.89	0.0007	0.00034	2953.78	0.50034	1.9986	3.9729	1.9878
19	2216.84	0.0005	0.00023	4431.68	0.50023	1.9991	3.9811	1.9914
20	3325.26	0.0003	0.00015	6648.51	0.50015	1.9994	3.9868	1.9940
22	7481.83	0.0001	0.00007	14962	0.50003	1.9997	3.9936	1.9971
24	16834	0.0001	0.00003	33666	0.50003	1.9999	3.9969	1.9986
25	25251		0.00002	50500	0.50002	1.9999	3.9979	1.9990
26	37877		0.00001	75752	0.50001	1.9999	3.9985	1.9993
28	85223		0.00001		0.50001	2.0000	3.9993	1.9997
30					0.50000	2.0000	3.9997	1.9998
32					0.50000	2.0000	3.9998	1.9999
34					0.50000	2.0000	3.9999	2.0000
35					0.50000	2.0000	3.9999	2.0000

APPENDIX B

TABLE 1 Standard Normal Curve Areas

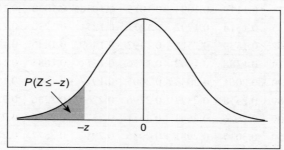

$P(Z \leq -z)$

$-z$ 0

Entries in this table provide cumulative probabilities, that is, the area under the curve to the left of $-z$. For example, $P(Z \leq -1.52) = 0.0643$.

z	0.00	0.01	0.02	0.03	0.04	0.05	0.06	0.07	0.08	0.09
−3.9	0.0000	0.0000	0.0000	0.0000	0.0000	0.0000	0.0000	0.0000	0.0000	0.0000
−3.8	0.0001	0.0001	0.0001	0.0001	0.0001	0.0001	0.0001	0.0001	0.0001	0.0001
−3.7	0.0001	0.0001	0.0001	0.0001	0.0001	0.0001	0.0001	0.0001	0.0001	0.0001
−3.6	0.0002	0.0002	0.0001	0.0001	0.0001	0.0001	0.0001	0.0001	0.0001	0.0001
−3.5	0.0002	0.0002	0.0002	0.0002	0.0002	0.0002	0.0002	0.0002	0.0002	0.0002
−3.4	0.0003	0.0003	0.0003	0.0003	0.0003	0.0003	0.0003	0.0003	0.0003	0.0002
−3.3	0.0005	0.0005	0.0005	0.0004	0.0004	0.0004	0.0004	0.0004	0.0004	0.0003
−3.2	0.0007	0.0007	0.0006	0.0006	0.0006	0.0006	0.0006	0.0005	0.0005	0.0005
−3.1	0.0010	0.0009	0.0009	0.0009	0.0008	0.0008	0.0008	0.0008	0.0007	0.0007
−3.0	0.0013	0.0013	0.0013	0.0012	0.0012	0.0011	0.0011	0.0011	0.0010	0.0010
−2.9	0.0019	0.0018	0.0018	0.0017	0.0016	0.0016	0.0015	0.0015	0.0014	0.0014
2.8	0.0026	0.0025	0.0024	0.0023	0.0023	0.0022	0.0021	0.0021	0.0020	0.0019
−2.7	0.0035	0.0034	0.0033	0.0032	0.0031	0.0030	0.0029	0.0028	0.0027	0.0026
− 2.6	0.0047	0.0045	0.0044	0.0043	0.0041	0.0040	0.0039	0.0038	0.0037	0.0036
−2.5	0.0062	0.0060	0.0059	0.0057	0.0055	0.0054	0.0052	0.0051	0.0049	0.004a
2.4	0.0082	0.0080	0.0078	0.0075	0.0073	0.0071	0.0069	0.0068	0.0066	0.0064
2.3	0.0107	0.0104	0.0102	0.0099	0.0096	0.0094	0.0091	0.0089	0.0087	0.0084
2.2	0.0139	0.0136	0.0132	0.0129	0.0125	0.0122	0.0119	0.0116	0.0113	0.0110
−2.1	0.0179	0.0174	0.0170	0.0166	0.0162	0.0158	0.0154	0.0150	0.0146	0.0143
−2.0	0.0228	0.0222	0.0217	0.0212	0.0207	0.0202	0.0197	0.0192	0.0188	0.0183
−1.9	0.0287	0.0281	0.0274	0.0268	0.0262	0.0256	0.0250	0.0244	0.0239	0.0233

z	0.00	0.01	0.02	0.03	0.04	0.05	0.06	0.07	0.08	0.09
−1.8	0.0359	0.0351	0.0344	0.0336	0.0329	0.0322	0.0314	0.0307	0.0301	0.0294
−1.7	0.0446	0.0436	0.0427	0.0418	0.0409	0.0401	0.0392	0.0384	0.0375	0.0367
−1.6	0.0548	0.0537	0.0526	0.0516	0.0505	0.0495	0.0485	0.0475	0.0465	0.0455
−1.5	0.0668	0.0655	0.0643	0.0630	0.0618	0.0606	0.0594	0.0582	0.0571	0.0559
−1.4	0.0808	0.0793	0.0778	0.0764	0.0749	0.0735	0.0721	0.0708	0.0694	0.0681
−1.3	0.0968	0.0951	0.0934	0.C918	0.0901	0.0885	0.0869	0.0853	0.0838	0.0823
−1.2	0.1151	0.1131	0.1112	0.1093	0.1075	0.1056	0.1038	0.1020	0.1003	0.0985
−1.1	0.1357	0.1335	0.1314	0.1292	0.1271	0.1251	0.1230	0.1210	0.1190	0.1170
−1.0	0.1587	0.1562	0.1539	0.1515	0.1492	0.1469	0.1446	0.1423	0.1401	0.1379
−0.9	0.1841	0.1814	0.1788	0.1762	0.1736	0.1711	0.1685	0.1660	0.1635	0.1611
−0.8	0.2119	0.2090	0.2061	0.2033	0.2005	0.1977	0.1949	0.1922	0.1894	0.1867
−0.7	0.2420	0.2389	0.2358	0.2327	0.2296	0.2266	0.2236	0.2206	0.2177	0.2148
−0.6	0.2743	0.2709	0.2676	0.2643	0.2611	0.2578	0.2546	0.2514	0.2483	0.2451
−0.5	0.3085	0.3050	0.3015	0.2981	0.2946	0.2912	0.2877	0.2843	0.2810	0.2776
−0.4	0.3446	0.3409	0.3372	0.3336	0.3300	0.3264	0.3228	0.3192	0.3156	0.3121
−0.3	0.3821	0.3783	0.3745	0.3707	0.3669	0.3632	0.3594	0.3557	0.3520	0.3483
−0.2	0.4207	0.4168	0.4129	0.4090	0.4052	0.4013	0.3974	0.3936	0.3897	0.3859
−0.1	0.4602	0.4562	0.4522	0.4483	0.4443	0.4404	0.4364	0.4325	0.4286	0.4247
−0.0	0.5000	0.4960	0.4920	0.4880	0.4840	0.4801	0.4761	0.4721	0.4681	0.4641

TABLE 1   (Continued)

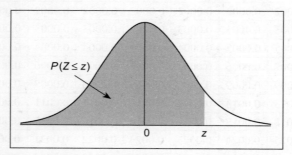

$P(Z \leq z)$

Entries in this table provide cumulative probabilities, that is, the area under the curve to the left of z. For example, $P(Z \leq 1.52) = 0.9357$.

z	0.00	0.01	0.02	0.03	0.04	0.05	0.06	0.07	0.08	0.09
0.0	0.5000	0.5040	0.5080	0.5120.	0.5160	0.5199	0.5239	0.5279	0.5319	0.5359
0.1	0.5398	0.5438	0.5478	0.5517	0.5557	0.5596	0.5636	0.5675	0.5714	0.5753
0.2	0.5793	0.5832	0.5871	0.5910	0.5948	0.5987	0.6026	0.6064	0.6103	0.6141
0.3	0.6179	0.6217	0.6255	0.6293	0.6331	0.6368	0.6406	0.6443	0.6480	0.6517
0.4	0.6554	0.6591	0.6628	0.6664	0.6700	0.6736	0.6772	0.6808	0.6844	0.6879

z	0.00	0.01	0.02	0.03	0.04	0.05	0.06	0.07	0.08	0.09
0.5	0.6915	0.6950	0.6985	0.7019	0.7054	0.7088	0.7123	0.7157	0.7190	0.7224
0.6	0.7257	0.7291	0.7324	0.7357	0.7389	0.7422	0.7454	0.7486	0.7517	0.7549
0.7	0.7580	0.7611	0.7642	0.7673	0.7704	0.7734	0.7764	0.7794	0.7823	0.7852
0.8	0.7881	0.7910	0.7939	0.7967	0.7995	0.8023	0.8051	0.8078	0.8106	0.8133
0.9	0.8159	0.8186	0.8212	0.8238	0.8264	0.8289	0.8315	0.8340	0.8365	0.8389
1.0	0.8413	0.8438	0.8461	0.8485	0.8508	0.8531	0.8554	0.8577	0.8599	0.8621
1.1	0.8643	0.8665	0.8686	0.8708	0.8729	0.8749	0.8770	0.8790	0.8810	0.8830
1.2	0.8849	0.8869	0.8888	0.8907	0.8925	0.8944	0.8962	0.8980	0.8997	0.9015
1.3	0.9032	0.9049	0.9066	0.9082	0.9099	0.9115	0.9131	0.9147	0.9162	0.9177
1.4	0.9192	0.9207	0.9222	0.9236	0.9251	0.9265	0.9279	0.9292	0.9306	0.9319
1.5	0.9332	0.9345	0.9357	0.9370	0.9382	0.9394	0.9406	0.9418	0.9429	0.9441
1.6	0.9452	0.9463	0.9474	0.9484	0.9495	0.9505	0.9515	0.9525	0.9535	0.9545
1.7	0.9554	0.9564	0.9573	0.9582	0.9591	0.9599	0.9608	0.9616	0.9625	0.9633
1.8	0.9641	0.9649	0.9656	0.9664	0.9671	0.9678	0.9686	0.9693	0.9699	0.9706
1.9	0.9713	0.9719	0.9726	0.9732	0.9738	0.9744	0.9750	0.9756	0.9761	0.9767
2.0	0.9772	0.9778	0.9783	0.9788	0.9793	0.9798	0.9803	0.9808	0.9812	0.9817
2.1	0.9821	0.9826	0.9830	0.9834	0.9838	0.9842	0.9846	0.9850	0.9854	0.9857
2.2	0.9861	0.9864	0.9868	0.9871	0.9875	0.9878	0.9881	0.9884	0.9887	0.9890
2.3	0.9893	0.9896	0.9898	0.9901	0.9904	0.9906	0.9909	0.9911	0.9913	0.9916
2.4	0.9918	0.9920	0.9922	0.9925	0.9927	0.9929	0.9931	0.9932	0.9934	0.9936
2.5	0.9938	0.9940	0.9941	0.9943	0.9945	0.9946	0.9948	0.9949	0.9951	0.9952
2.6	0.9953	0.9955	0.9956	0.9957	0.9959	0.9960	0.9961	0.9962	0.9963	0.9964
2.7	0.9965	0.9966	0.9967	0.9968	0.9969	0.9970	0.9971	0.9972	0.9973	0.9974
2.8	0.9974	0.9975	0.9976	0.9977	0.9977	0.9978	0.9979	0.9979	0.9980	0.9981
2:9	0.9981	0.9982	0.9982	0.9983	0.9984	0.9984	0.9985	0.9985	0.9986	0.9986
3.0	0.9987	0.9987	0.9987	0.9988	0.9988	0.9989	0.9989	0.9989	0.9990	0.9990
3.1	0.9990	0.9991	0.9991	0.9991	0.9992	0.9992	0.9992	0.9992	0.9993	0.9993
3.2	0.9993	0.9993	0.9994	0.9994	0.9994	0.9994	0.9994	0.9995	0.9995	0.9995
3.3	0.9995	0.9995	0.9995	0.9996	0.9996	0.9996	0.9996	0.9996	0.9996	0.9997
3.4	0.9997	0.9997	0.9997	0.9997	0.9997	0.9997	0.9997	0.9997	0.9997	0.9998
3.5	0.9998	0.9998	0.9998	0.9998	0.9998	0.9998	0.9998	0.9998	0.9998	0.9998
3.6	0.9998	0.9998	0.9999	0.9999	0.9999	0.9999	0.9999	0.9999	0.9999	0.9999
3.7	0.9999	0.9999	0.9999	0.9999	0.9999	0.9999	0.9999	0.9999	0.9999	0.9999
3.8	0.9999	0.9999	0.9999	0.9999	0.9999	0.9999	0.9999	0.9999	0.9999	0.9999
3.9	0.9999	0.9999	0.9999	0.9999	0.9999	0.9999	0.9999	0.9999	0.9999	0.9999

APPENDIX C

TABLE 1   Student's t distribution

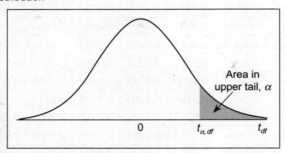

Entries in this table provide the values of $t_{\alpha,df}$ that correspond to a given upper-tail area a and a specified number of degrees of freedom df. For example for $\alpha = 0.05$ and $df = 10$, $P(T_{10} \geq 1.812) = 0.05$.

df	α					
	0.20	0.10	0.05	0.025	0.01	0.005
1	1.376	3.078	6.314	12.706	31.821	63.657
2	1.061	1.886	2.920	4.303	6.965	9.925
3	0.978	1.638	2.353	3.182	4.541	5.841
4	0.941	1.533	2.132	2.776	3.747	4.604
5	0.920	1.476	2.015	2.571	3.365	4.032
6	0.906	1.440	1.943	2.447	3.143	3.707
7	0.896	1.415	1.895	2.365	2.998	3.499
8	0.889	1.397	1.860	2.306	2.896	3.355
9	0.883	1.383	1.833	2.262	2.821	3.250
10	0.879	1.372	1.812	2.228	2.764	3.169
11	0.876	1.363	1.796	2.201	2.718	3.106
12	0.873	1.356	1.782	2.179	2.681	3.055
13	0.870	1.350	1.771	2.160	2.650	3.012
14	0.868	1.345	1.761	2.145	2.624	2.977
15	0.866	1.341	1.753	2.131	2.602	2.947
16	0.865	1.337	1.746	2.120	2.583	2.971
17	0.863	1.333	1.740	2.110	2.567	2.898
18	0.862	1.330	1.734	2.101	2.552	2.878
19	0.861	1.328	1.729	2.093	2.539	2.861
20	0.860	1.325	1.725	2.086	2.528	2.845

	α					
df	0.20	0.10	0.05	0.025	0.01	0.005
21	0.859	1.323	1.721	2.080	2.518	2.831
22	0.858	1.321	1.717	2.074	2.508	2.819
23	0.858	1.319	1.714	2.069	2.500	2.807
24	0.857	1.318	1.711	2.064	2.492	2.797
25	0.856	1.316	1.708	2.060	2.485	2.787
26	0.856	1.315	1.706	2.056	2.479	2.779
27	0.855	1.314	1.703	2.052	2.473	2.771
28	0.855	1.313	1.701	2.048	2.467	2.763
29	0.854	1.311	1.699	2.045	2.462	2.756
30	0.854	1.310	1.697	2.042	2.457	2.750
31	0.853	1.309	1.696	2.040	2.453	2.744
32	0.853	1.309	1.694	2.037	2.449	2.738
33	0.853	1.308	1.692	2.035	2.445	2.733
34	0.852	1.307	1.691	2.032	2.441	2.728
35	0.852	1.306	1.690	2.030	2.438	2.724
36	0.852	1.306	1.688	2.028	2.434	2.719
37	0.851	1.305	1.687	2.026	2.431	2.715
38	0.851	1.304	1.686	2.024	2.429	2.712
39	0.851	1.304	1.685	2.023	2.426	2.708
40	0.851	1.303	1.684	2.021	2.423	2.704,
41	0.850	1.303	1.683	2.020	2.421	2.701
42	0.850	1.302	1.682	2.018	2.418	2.698
43	0.850	1.302	1.681	2.017	2.416	2.695
44	0.850	1.301	1.680	2.015	2.414	2.692
45	0.850	1.301	1.679	2.014	2.412	2.690
46	0.850	1.300	1.679	2.013	2.410	2.687
47	0.849	1.300	1.678	2.012	2.408	2.685
48	0.849	1.299	1.677	2.011	2.407	2.682
49	0.849	1.299	1.677	2.010	2.405	2.680
50	0.849	1.299	1.676	2.009	2.403	2.678
51	0.849	1.298	1.675	2.008	2.402	2.676
52	0.849	1.298	1.675	2.007	2.400	2.674
53	0.848	1.298	1.674	2.006	2.399	2.672
54	0.848	1.297	1.674	2.005	2.397	2.670
55	0.848	1.297	1.673	2.004	2.396	2.668
56	0.848	1.297	1.673	2.003	2.395	2.667
57	0.848	1.297	1.672	2.002	2.394	2.665

	α					
df	0.20	0.10	0.05	0.025	0.01	0.005
58	0.848	1.296	1.672	2.002	2.392	2.663
59	0.848	1.296	1.671	2.001	2.391	2.662
60	0.848	1.296	1.671	2.000	2.390	2.660
80	0.846	1.292	1.664	1.990	2.374	2.639
100	0.845	1.290	1.660	1.984	2.364	2.626
150	0.844	1.287	1.655	1.976	2.351	2.609
200	0.843	1.286	1.653	1.972	2.345	2.601
500	0.842	1.283	1.648	1.965	2.334	2.586
1000	0.842	1.282	1.646	1.962	2.330	2.581
∞	0.842	1.282	1.645	1.960	2.326	2.576

Source: t values calculated with Excel.

Bibliography

Ariat, S. 1967. 'The Construction of a Utility Function from Expenditure Data.' *International Economic Review* 8 (1): 67-77.

Arrow, K., H. D. Block, and L. Hurwicz. 1959. 'On the Stability of the Competitve Equilibrium, II.' *Econometrica* 27: 82-109.

Arrow, K., and G. Debreu. 1954. 'Existence of Equilibrium for a Competitive Economy.' *Econometrica* 22 (3): 265-290.

Arrow, K., and F. H. Hahn. 1971. *General Competitive Analysis*. San Francisco: Holden Day.

Arrow, K., and L. Hurwicz. 1958. 'On the Stability of the Competitve Equilibrium, I.' *Econometrica* 26: 522-552.

Baye, M., J. Zhou, and Tian, G. 1993. 'Characterisations of the Existence of Equilibria in Games with Discontinuous and Non-Quasiconcave Payoffs.' *Review of Economic Studies* 60 (4): 935-948.

Bhatnagar, S. K. 1986. *Network Analysis Techniques*. New Delhi: Wiley Eastern Limited.

Blank, Leland, and Anthony Tarquin. 2005. *Engineering Economy*. New York: McGraw Hill Higher Education.

Blank, Leland, and Anthony Tarquin. 2013. *Basics of Engineering Economy*. New Delhi: McGraw Hill Education (India) Private Limited.

Border, K. C. 1985. *Fixed Point Theorems with Appplications to Economics and Game Theory*. Cambridge: Cambridge University Press.

Buffa, Elwood S., and Rakesh K. Sarin. 1984. *Modern Production/Operations Management*, Eighth Edition. New York: John Wiley & Sons, Inc.

Chow, W. M. 1965. 'Adaptive Control of the Exponential Smoothing Constant.' *The Journal of Industrial Engineering* 16 (5): 314-317.

Debreu, G. 1952. 'A Social Equilibrium Existence Theorem.' *Proceedings of the National Academy of Sciences of the U. S. A.* 38 (10): 386-393.

Debreu, G. 1959. *Theory of Value*. Wiley: New York.

Debreu, G. 1964. 'Continuity Properties of Paretian Utility.' *International Economic Review* 5 (3): 285-293.

Debreu, G. 1974. 'Excess Demand Functions.' *Journal of Mathematical Economics* 1 (1): 15-22.

DeGarmo, E. Paul, William G. Sullivan, and John R. Canada. 1984. *Engineering Economy*, Seventh Edition. New York: Macmillan; London: Collier Macmillan.

Diewert, E. 1974. 'Applications of Duality Theory.' In *Frontiers of Quantitative Economics*, edited by M. Intriligator and D. Kendrick. Amsterdam: North-Holland.

Fan, K. 1984. 'Some Properties of Convex Sets Related to Fixed Point Theorem.' *Mathematics Annuls* 266 (4): 519-537.

Gorman, T. 1953. 'Community Preference Fields.' *Econometrica* 21: 63-80.

Henderson, James M., and Quandt, Richard E, 2003. *Microeconomic theory: A Mathematical Approach*, 3rd Edition. New Delhi: Tata McGraw Hill Pub. Co.

Hicks, J. 1946. *Value and Capital*. Oxford: Clarendon Press.

Hotelling, H. 1932. 'Edgeworth's Taxation Paradox and the Nature of Demand and Supply Function.' *Political Economy* 40 (5): 577-616.

Hurwicz, L., and H. Uzawa. 1971. 'On the Integrability of Demand Functions.' In *Preferences, Utility, and Demand*, edited by J. S. Chipman, L. Hurwicz, M. Ritcher, and H. Sonnenshein. New York: Harcourt Brace Jovanovich.

Jehle, G. A., and P. Reny. 1998. *Advanced Microeconomic Theory*. London: Pearson Education Limited.

Koutsoyiannis, A. 2008. *Modern Microeconomics*, 2nd Edition, Revised Edition. Basingstoke: Macmillan Press Ltd.

Kumar, Pravin. 2015. *Industrial Engineering and Management*. New Delhi: Pearson India Education Services Pvt. Ltd..

Luenberger, D. 1995. *Microeconomic Theory*. New York: McGraw-Hill College.

Mas-Colell, A., M. D. Whinston, and J. Green. 1995. *Microeconomic Theory, Volume 1*. New York: Oxford University Press.

McFadden, D. 1978. 'Cost, Revenue, and Profit Functions.' In *Production Economics: A Dual Approach to Theory and Applications*, edited by M. Fuss and D. McFad-den. Amsterdam: North-Holland.

Michael, E. 1956. 'Continuous Selections. I.' *Annals of Mathematics* 63: 361-382.

Panneerselvam, R. 2002. *Engineering Economics*. New Delhi: Prentice Hall of India

Petersen, Craig H., and W. Cris Lewis, 1995. *Managerial Economics*, 3rd Edition. New Delhi: Prentice Hall of India Private Ltd.

Qian, Y. 2002. 'Understanding Modern Economics.' *Economic and Social System Comparison* 2: 1-12.

Quinzii, M. 1992. *Increasing Returns and Efficiency*. New York: Oxford University Press.

Rawls, J. 1971. *A Theory of Justice*. Cambridge: Harvard University Press.

Roberts, S. D., and Reed, R. 1969. 'The Development of a Self-Adaptive Forecasting Technique.' *AIEE Transactions* 1 (4): 314-322.

Roy, R. 1942. *De l'utilitt'e*. Paris: Hermann.

Roy, R. 1947. 'La distribution de revenu entre les divers biens.' *Econometrica* 15: 205-225.

Rubinstein, Ariel. 2006. *Lecture Notes in Microeconomics (The Economic Agent)*. Princeton: Princeton Univeristy Press.

Samuelson, P. 1947. *Foundations of Economic Analysis*. Cambridge: Harvard University Press.

Samuelson, P. 1948. 'Consumption Theory in Terms of Revealed Preference.' *Economica* 15 (60): 243-253.

Scarf, H. 1960. 'Some Examples of Global Instability of the Competitive Equilibrium.' *International Economic Review* 1 (3): 157-172.

Shafer, W., and H. Sonnenschein. 1975. 'Equilibrium in abstract economies without ordered preferences.' *Journal of Mathematical Economics* 2 (3): 345-348.

Shapiro, C. 1989. 'Theories of Oligopoly Behaviour.' In *Handbook of Industrial Organization*, Volume 1, edited by R. Schmalensee & R. Willig. Amsterdam: North-Holland.

Shephard, R. 1953. *Cost and Production Functions*. Princeton: Princeton University Press.

Singh, N., and X. Vives. 1984. 'Price and Quantity Competition in a Differentiated Duopoly.' *The RAND Journal of Economics* 15: 546-554.

Sonnenschein, H. 1968. 'The Dual of Duopoly is Complementary Monopoly: Or, Two of Cournot's Theories are One.' *Journal of Political Economy* 36: 316-318.

Sonnenschein, H. 1971. 'Demand Theory without Transitive Preferences, with Application to the Theory of Competitive Equilibrium.' In *Preferences, Utility, and Demand,* edited. by J. S. Chipman, L. Hurwicz, M. Richter, and H. Sonnenschein. New York: Harcourt Brace Jovanovich.

Srinath, L. S. 2001. *PERT and CPM: Principles and Applications,* 3rd Edition. New Delhi: Affiliated East-West Press Pvt. Ltd.

Stonier, A. W., and D. C. Hague. 1980. A Textbook of Economic Theory, 5th edition. London: Longman Group Ltd.

Takayama, A. 1985. *Mathematical Economics,* 2nd Edition. Cambridge: Cambridge University Press.

Trigg, D. W., and A. G. Leach. 1967. 'Exponential Smoothing with an Adaptive Response Rate.' *Operational Research Quarterly* 18 (1): 53-59.

Varian, H. R. 1992. *Microeconomic Analysis,* 3rd Edition. New York: W. W. Norton and Company.

Walras, L. 1874. *Elt'ements of d'Economie Politique Pure*. Lausanne: Gorbaz. [Translated as *Elements of Pure Economics*. Homewood, Ill: Irwin, 1954]

Wold, H. 1943. 'A Synthesis of Pure Demand Analysis, I-III.' *Skandinavisk Aktuarietid-skrift* 26 (1-2): 220–263.

Index

A.W. 190
absolute diminishing utility 64
actual value 408
adaptive models 403
adaptive smoothing model 403
annual cost 223
annual worth 229
annuity 136
arc elasticity 11
arithmetic gradient 142
arithmetic gradient series 142, 175
arrow diagrams 301, 305, 308
average demand 392
average error 415
average product 89
average product of labor 92
axiom of utility maximization 47

B/C ratio 267
backward integration 286
bar charts 298
base 392
base amount 142
benefit/cost (B/C) ratio 267
benefits 266
Bertrand game 120
Bertrand model 119, 120
Bertrand model of oligopoly 120
beta Distribution 352
book Depreciation 254
book value 255

Bordered-Hessian determinant 73
bottom up 423
breakeven analysis 34, 292
break-even chart 35, 292
break-even point 31, 35
budget constraint 72
budget line 72, 79
build-up method 414
burst events 304
business sector 5
buy decision 288

capital 3, 90
capital formation 4
capital recovery factor 137
capitalized cost 204, 234
cardinal marginal utility theory 49
cardinal utility analysis 65
cash flow diagrams 127
causal methods 389
ces function 94
ceteris paribus 52
challenger 242, 246
challenger First Cost 242
chance branches 451
change in demand 52, 54
change in quantity demanded 52
change in quantity supplied 85
change in supply 84, 85
change in the quantity supplied 84
circular flow 2

circular flow of income 4
Cobb-Douglas 104
Cobb-Douglas function 91, 92
Cobb-Douglas production function 91, 101
Cobb-Douglas technology 91
coefficient of correlation 412
coefficient of determination 411
collective consumption 4
collusive oligopoly mode 119
collusive oligopoly model 121
compensated demand curve 55
complementary goods 9, 19, 54
complements 54, 61
composite function 74, 102
compound interest 126
compounding frequency 169
compounding period 169
conditional input demand function 104
constant amount 142
constant elasticity of substitution 94
constant returns to scale 94, 104, 107
constant-value rupees 277
constrained optimization rules 102
constrained output maximization 94
constraint analysis 384
consumer price index 426
consumer surplus 80
consumer's equilibrium 72, 76
consumption 3
convex 93
cost 422
cost capacity equation 427
cost curve 96
cost estimating relationship 427, 428
cost estimation 423
cost function 104, 108
cost minimisation 105
cost minimization 83, 104
cost slope 359
cost volume profit 34
cost-demand relationship 30
costs 266

costs, Explicit 110
cost-volume-profit analysis 292
coterminated assumption 221
Cournot aggregation condition 62
Cournot equilibrium 119, 120
Cournot game 120
Cournot model 119
course of action 439
CPM 343
critical activities 312
critical path 300
critical path method 300, 301
cross elasticity 17
cross-price elasticity 61
cross-price elasticity of demand 60
cyclical component 390

debt capital 123
decision branches 451
decision making 439
decision node 451
decision tree 451
decisions 439
declining balance (db) method 257
decreasing returns 107
decreasing returns to scale 108
defender 242, 246
defender first cost 242
deferred annuity 156
definition of economics 45
deflation 278
Delphi technique 415
demand 7, 51
demand curve 9
demand equation 52
demand function 51, 105
demand schedule 8
demand shifters 52
dependency rule 305
depreciation 254
depreciation rate 255
detailed planning 298
determinants of demand 9

determinants of supply 20
diminishing marginal utility 50, 65
direct costs 358, 422
direct magnitude estimation 385
disbenefits 266
discrete cash flow 132
do-nothing 212
double declining balance 257
dummy activities 307
duopoly model 119

E.R.R. 199
E.R.R.R. 203
earliest event time 313
earliest finish time 325
earliest start time 325
economic activity 88
economic analysis 290
economic efficiency 24
economic equivalence 126
economic order quantity 290
economic profit 111
economic service life 241, 242
economic theory 46
economics 45
effective interest 168
effective interest rate 169
efficiency 24
elastic 12
elasticity 26, 56
elasticity of demand 10
elasticity of scale 108
elasticity of substitution 92, 100
elasticity of supply 86
Engel aggregation condition 62
engineering economics 22
engineering new record 426
entrepreneurship 3
equilibrium 85
equilibrium price 33, 115
equity capital 123
equivalent uniform annual cost 242

event 303
expected loss 450
expected opportunity loss 447
expected pay-off 445
expected profit 450
explicit costs 110
explicit reinvestment rate of return 219, 225,
    227, 231
exponential smoothing method 392
external rate of return 218, 224, 227, 230
extrapolative 389
extrapolative methods 389

factor method 428
factors of production 3
FAST diagram 381
feasible equilibrium points 86
firms 5
first cost 28, 254
first-order condition 73
fixed costs 28
float 330
forecast 415
forecast error 415
forecasted value 408
forward integration 286
forward pass 313
four sector model 6
free dispolsal 93
free float 331
function analysis 381
functional analysis systems 381
future 129
future worth 188, 229
future worth cost 215, 223

Gantt charts 299
Gossen's second law 65
government Sector 5
gradient series 175
gradient to present worth conversion factor 144
gradient to uniform series conversion factor 144

head event 304
Hooke's law 56
horizontal component 389
household sector 5
hypothesis-conclusion methods 377

I.R.R. 196
imperfect competition 117
imperfectly competitive markets 117
implicit cost 29, 110
income and leisure choice 76
income effect 56
income elasticity 15, 59
increasing returns 107
incremental 27
incremental b/c ratio 270
incremental cost 29
independent 212
independent float 332
indifference curve 56, 67
indifference curves approach 47
indifference map 69
indirect costs 358, 423
indirect or overhead cost rate 433
individual consumption 3
inelastic 12
inferior goods 16
inflated interest rate 277
inflation 276
inflation adjusted interest rate 281
inflation-adjusted rate 277
initial event 304
inner bound 79
innovation phase 386
input demand function 104
installed cost 254
insufficient reason (laplace's rule) 444
interest rate 124, 196
interfering float 332
internal rate of return 216, 224, 230
investment 4
investment cost 28

iso-cost line 101
isoquant 96, 97, 106

judgmental methods 389

labor 3
Lagrangian multiplier 74, 95, 102
land 3
Laplace's rule 444
latest event time 313
latest finish time 325
latest start time 326
law of demand 8
law of diminishing marginal returns 34
law of equilibrium utility 65
law of equi-marginal utility 65, 66
law of indifference 65, 66
law of maximum satisfaction 65, 66, 67
law of substitution 65, 66
law of supply 21, 83
learning curve 430
least squares line 405
Leontief technology 91, 106
Leontief utility function 48
lexicographic ordering 68
life-cycle cost 375
linear technology 106
linear utility function 48
long-run 90
long-run production possibilities set 91
lowest common multiple 221
luxuries 16, 25, 61
luxury goods 16, 60

M.A.R.R. 217
make decision 287
margin of safety 38
marginal 449
marginal loss 450
marginal product 89
marginal product of factors 91
marginal profit 450
marginal rate of substitution 70, 92

marginal rate of technical substitution 94, 97

marginal revenue 27, 63

marginal utilities 70

marginal utility 50, 64

market price 115

market value 255

Marshallian demand function 54

Matheson formula 257

maximax criterion 442, 443

maximization of utility 72

mean absolute deviation 415

mean absolute percentage error 415

mean squared error 415

merge events 304

minimax regret criterion 443

modified b/c ratio 267

monetary flow of income 5

money income 74

monopolist 115

monopolistic 118, 120

monopolistic competition 42, 117

monopoly 41, 115, 117

monotonic 48

monotonicity 49, 93

move on the function 52

movement on the supply function 84

moving average method 390

MRTS for a Cobb-Douglas Technology 100

multiple regression 413

multiplier 72

mutually exclusive 212

Nash equilibrium 119

necessities 16, 25

net cash flow 128, 138

net present value 129

net present worth 196

network 300

network analysis 300

network analysis technique 300

network diagrams 308

nodes 303

nominal 168

nominal group technique 386

nominal interest rate 168

non-collusive oligopoly model 119, 120

non-critical activities 312

no-return payback 207

normal duration 358

normal good 16, 60

normal profit 111

oligopoly 117, 118

oligopoly 41

opportunity cost 29, 110

optimal duration 358

ordinary demand function 54

outer bound 79

own elasticity of demand 61

P/V ratio 39

panel of experts 414

pareto sense 48

pareto voting 386

partial equilibrium 112

payback 207

payback period 207

pay-off 439

pay-off matrix 443

payoffs 120, 451

perfect competition 41, 117

perfect complements 71, 99

perfect substitutes 71, 98

perfectly competitive markets 113

perfectly price elastic 57

perfectly price inelastic 57

PERT 343

PERT computations 344

planning 297

point elasticity 11, 57

point of diminishing marginal utility 65

possibility of inaction 93

preference ordering 68

preferences 65

pre-planning 298

present value 129
present worth 129, 136, 186, 228
present worth cost 213, 222, 226
price elasticities 61
price elasticity 10, 56, 57
price out determination 116
price-taking behaviour 112
probability distribution curve 352, 353
producer's surplus 102
product differentiation 117
production 3, 88
production function 90, 95
production possibilities 89
production possibilities set 89
profit 30, 110
profit function 111
profit maximization 83, 114, 118
profit volume ratio 39
profitability index 196
profit-maximizing 112
profit-volume (P/V) graph 38
program evaluation and review technique 300
project crashing 358
project management 297
properties of indifference curves 70
properties of isoquant 97
proportionate rule 65
pure competition 113
pure monopoly 117

Q-sort 386
qualitative methods 413
quantity demanded 51
quasi-concavity 74

R2 411
random noise 392
rate of commodity substitution 70
ratio seasonality model 398
reaction function 119
real flow of income 5
real interest rate 277
realism criterion (Hurwicz's rule) 444

recovery rate 255
regression analysis 403
regression line 408
relatively price elastic 57
relatively price inelastic 57
repeatability assumptions 221
returns to scale 107
revealed preference 77
revealed preference hypothesis 47, 77
revealed preferred 79
revenue-demand relationship 30

salvage value 255, 262
satiation point 64
seasonal component 389
seasonal index 398
seasonality component 392
second-order condition 73
sequential decision problems 451
service output method 264
shift of the function 52
shift of the supply function 84
shifted uniform series 156
short-run 90
short-run average cost 109
short-run average fixed cost 109
short-run average variable cost 109
short-run marginal cost 109
short-run production possibilities set 90
short-run total cost 109
simple cost analysis 288
simple payback 207
single payment compound amount factor 133
single payment present worth factor 133
sinking fund factor 140
sinking fund method 262
Stackelberg equilibrium 120
Stackelberg model 120
standard deviation 352
standard error 409
standard error of estimate 408
straight Line (SL) method 255

strict preference 68

substitute goods 9, 54

substitutes 18, 61

substitution effect 56

succeeding event 304

sum-of-the-years'-digits (SYD) method 260

sunk cost 29

superior good. 60

supply 20

supply curve 22, 84

supply function 87

supply schedule 21, 83

survey method 414

surveys 414

SYD depreciation 260

tail event 304

tax depreciation 254

technical efficiency 24

technologically 88

technology 90

terminal branch 451

terminating event 304

test market 414

theory of pricing 63

three sector model 6

time series methods 389

time value of money 123

top down 423

total float 331

total product 89

total revenue 27, 57

total utility 50, 63

trend component 389

two sector model 6

unadjusted basis 254

uniform cash flow 178

uniform disbursement 135

uniform series 175

uniform series compound amount factor 140

uniform series present worth factor 137

UNIT method 424

unitary elastic 12

unitary price elastic 57

utility 47

utility function 47

utils 47

value engineering 375

value index 385

variable costs 28

variance 352

VE team 380

weighted moving average method 391

work-versus-time graphs 298